Lecture Notes in Biomathematics

Managing Editor: S. Levin

54

Mathematical Ecology

Proceedings of the Autumn Course (Research Seminars), held at the International Centre for Theoretical Physics, Miramare-Trieste, Italy, 29 November – 10 December 1982

Edited by S. A. Levin and T. G. Hallam

SPRINGER-VERLAG BERLIN
HEIDELBERG GMBH 1984

Editors

Simon A. Levin
Ecology & Systematics, E 347 Corson Hall
Cornell University, Ithaca, NY 14853-0239, USA

Thomas G. Hallam
Department of Mathematics, University of Tennessee
Knoxville, TN 37996-1300, USA

AMS Subject Classification (1980): 92-06, 92A10, 92A15, 92A17

ISBN 978-3-540-12919-6 ISBN 978-3-642-87422-2 (eBook)
DOI 10.1007/978-3-642-87422-2

Originally published by Springer-Verlag Berlin Heidelberg New York in 1984

2146/3140-543210

MATHEMATICAL ECOLOGY:

TRIESTE AUTUMN COURSE, 1982

Preface

These are the proceedings of the research part of the Autumn Course on Mathematical Ecology, held at the International Centre for Theoretical Physics, Miramare-Trieste, Italy, 29 November - 10 December 1982, under the sponsorship of the International Atomic Energy Agency (IAEA) and UNESCO. The research conference followed and overlapped somewhat a set of more basic lectures aimed at non-specialists, primarily from developing nations. The introductory lectures are currently being edited for publication as a textbook, and will be published as part of the Springer hardcover series in Biomathematics.

The active and dynamic nature of mathematical ecology is reflected in the wide range of papers found in this volume, covering a spectrum from autecology and evolutionary theory to ecosystems science, and ranging from studies closely tied to data to abstract formulations which develop the mathematical bases further.

Part I includes five papers on autecology, the direct interface between organism and environment. These include analyses of physiological, behavioral, and phenological responses to environment and to other organisms. Part II moves up one organizational level to the population of organisms, and treats population growth in stochastic environments, aspects of demographic structure, and coevolutionary interactions among species. In Part III, questions which arise at the community and ecosystem level are addressed, with attention to multi-species interactions and patterns; complexity, stability, and persistence; and the cycling of materials through systems. By this hierarchical transition through levels, one gets a broad picture of fundamental ecological research today.

Parts IV and V explore two of the major applied areas in mathematical ecology: fisheries ecology and the associated bioeconomic problems, and epidemiology. The latter represents one of the oldest areas in mathematical ecology, tracing back to the beginning of this century; whereas fisheries provided the bait which attracted the most famous mathematical ecologist of all time, Vito Volterra. Volterra was enticed by his son-in-law d'Ancona to provide mathematical explanations for the fluctuations of the Adriatic fisheries, and this sideline of Volterra's remains the core work in the mathematical literature in ecology.

Although applications of diffusion models to ecology and genetics also have had a long history, they received a great stimulus from the 1951 paper of Skellam; in the last decade, such models have formed one of the most active areas of current research in mathematics ecology. Chapters VI and VII explore various aspects of this research, focusing on both the dynamics and the patterns which result.

Involvement in this enterprise was an exciting activity for us and for all of the participants, and we hope that this volume can stimulate others as the lectures stimulated us at Trieste. We also look forward to a repeat of the course, and to a new set of research lectures, in the near future.

The editors gratefully acknowledge the support of the NSF and the EPA for their own contributions, and the sponsorship of UNESCO and the IAEA. Special thanks are due Mrs. Evelyn Cook for her exceptional job of assisting in the editing and preparation of this volume, and to Professors Abdus Salam and Giovanni Vidossich and the staff of the Centre for Theoretical Physics for their hospitality during the conference.

December 22, 1983

Simon A. Levin
Ithaca, New York

Thomas G. Hallam
Knoxville, Tennessee

LIST OF CONTRIBUTORS

Adu-Asamoah, Richard	Department of Agricultural & Resource Economics Oregon State University Corvallis, OR 97331 USA
Beauchat, Carol A.	Department of Physiology The University of Arizona Tucson, AZ 85721 USA
Beddington, John R.	International Institute for Environment and Development 10 Percy Street London, W1P ODR ENGLAND
Brauer, Fred	Mathematics Department University of Wisconsin VanVleck Hall - 480 Lincoln Drive Madison, WI 53706 USA
Brown, B. E.	National Marine Fisheries Service Northeast Fisheries Center Woods Hole, MA 02543 USA
Brown, Joel S.	Aerospace & Mechanical Engineering Department The University of Arizona Tucson, AZ 95721 USA
Busenberg, Stavros	Department of Mathematics Harvey Mudd College Claremont, CA 91711 USA
Capasso, Vincenzo	Department of Mathematics University of Bari Palazzo Ateneo 70121 Bari ITALY
Chesson, Peter L.	Department of Zoology The Ohio State University 1735 Neil Avenue Columbus, OH 43210 USA
Conrad, Jon M.	Department of Agricultural Economics 310 Warren Hall Cornell University Ithaca, NY 14853 USA
Cooke, J. G.	International Institute for Environment and Development 10 Percy Street London, W1P ODR ENGLAND

Cooke, Kenneth L.	Mathematics Department Pomona College Claremont, CA 91711 USA
Crowley, Philip H.	T. H. Morgan School of Biological Sciences University of Kentucky Lexington, KY 40506 USA
Cushing, James M.	Department of Mathematics and Program on Applied Mathematics The University of Arizona Building #89 Tucson, AZ 85721 USA
DeAngelis, Donald L.	Environmental Sciences Division Oak Ridge National Laboratory Oak Ridge, TN 37830 USA
DeMottoni, Piero	Instituto di Matematica Applicata Università dell 'Aquila Monteluco - Roio I-67040 Poggio di Roio L'Aquila ITALY
Diekmann, Odo	Mathematisch Centrum Kruislaan 413 1098 SJ Amsterdam THE NETHERLANDS
Dietz, Klaus	Institut für Medizinische Biometrie der Universität Tübingen Hallstattstr. 6 D 7400 Tübingen WEST GERMANY
Ellner, Stephen	Department of Mathematics and Program in Ecology University of Tennessee Knoxville, TN 37996-1300 USA
Gard, Thomas G.	Department of Mathematics University of Georgia Athens, GA 30602 USA
Gross, Louis J.	Department of Mathematics University of Tennessee Knoxville, TN 37996-1300 USA
Grosslein, M. D.	National Marine Fisheries Service Northeast Fisheries Center Woods Hole, MA 02543 USA
Gurney, William S. C.	Department of Applied Physics University of Strathclyde John Anderson Building - 107 Rottenrow Glasgow G4 ONG SCOTLAND

Hadeler, Karl P.	Lehrstuhl für Biomathematik Universität Tübingen Auf der Morgenstelle 28 D-7400 Tübingen WEST GERMANY
Haimovici, Adolf	Universitatea "Al. I. Cuza" Seminarul Matematic 6600 Iasi ROMANIA
Hastings, Alan	Department of Mathematics University of California Davis, CA 95616 USA
Hennemuth, R. C.	National Marine Fisheries Service Northeast Fisheries Center Woods Hole, MA 02543 USA
Iannelli, Mimmo	Departimento di Matematica Libera Universita Degli Studi di Trento 38050 Provo (Trento) ITALY
Jayakar, S. D.	Department of Genetics and Microbiology University of Pavia Pavia ITALY
Kareiva, Peter	Division of Biology & Medicine Brown University Providence, RI 02912 USA
(Present address)	Department of Zoology University of Washington Seattle, WA 98195 USA
Kawasaki, Kohkichi	Science and Engineering Research Institute Doshisha University Kyoto JAPAN
Kindlmann, Pavel	Czechoslovak Academy of Science Institute of Entomology Na sadkach 702 370 05 Ceske Budejovice CZECHOSLOVAKIA
Kitchell, J. A.	Department of Geology and Geophysics University of Wisconsin Madison, WI 53706 USA
Metz, J. A. J.	Institute of Theoretical Biology University of Leiden Groenhovenstraat 5 2311 BT Leiden THE NETHERLANDS

Mimura, Masayasu	Department of Mathematics Hiroshima University Hiroshima 730 JAPAN
Nisbet, Roger M.	Department of Applied Physics University of Strathclyde John Anderson Building - 107 Rottenrow Glasgow G4 ONG SCOTLAND
Okubo, Akira	Marine Sciences Research Center State University of New York Stony Brook, NY 11794 USA
Post, W. M.	Environmental Sciences Division Oak Ridge National Laboratory Oak Ridge, TN 37830 USA
Saleem, M.	Department of Mathematics and Program on Applied Mathematics The University of Arizona Building #89 Tucson, AZ 85721 USA
Schenzle, D.	Institut für Medizinische Biometrie der Universität Tübingen Hallstattstr. 6 D 7400 Tübingen WEST GERMANY
Segel, Lee A.	Department of Applied Mathematics The Weizmann Institute of Science Rehovot, 76100 ISRAEL
Shigesada, Nanako	Department of Biophysics Kyoto University Kyoto JAPAN
Silvert, William	Marine Ecology Laboratory Bedford Institute of Oceanography P.O. Box 1006 Dartmouth, Nova Scotia B2Y 4A2 CANADA
Sissenwine, Michael P.	National Marine Fisheries Service Northeast Fisheries Center Woods Hole, MA 02543 USA
Teramoto, Ei	Department of Biophysics Kyoto University Kyoto JAPAN
Travis, C. C.	Health and Safety Research Division Oak Ridge National Laboratory Oak Ridge, TN 37830 USA

Ulanowicz, Robert E.	University of Maryland Chesapeake Biological Laboratory Box 38 Solomons, MD 20688-0038 USA
vanBatenburg, F.H.D.	Institute of Theoretical Biology University of Leiden Groenhovenstraat 5 2311 BT Leiden THE NETHERLANDS
Vincent, Thomas L.	Aerospace & Mechanical Engineering Department The University of Arizona Tucson, AZ 95721 USA
White, George N. III	Fisheries Research Branch Marine Fish Division Bedford Institute of Oceanography P.O. Box 1006 Dartmouth, N.S. CANADA B2Y 4A2

TABLE OF CONTENTS

PART I

AUTECOLOGY

ON THE PHENOTYPIC PLASTICITY OF LEAF PHOTOSYNTHETIC CAPACITY

Louis J. Gross
Department of Mathematics
and
Graduate Programs in Ecology and Plant Physiology
University of Tennessee
Knoxville, Tennessee 37996-1300

Most of ecological and evolutionary theory considers environments which are static or changing on long-time scales relative to organism lifespans. Despite the recent flowering of interest in randomly varying environments as they affect population growth (Turelli, 1977) and population genetics in fine-grained environments (Templeton and Rothman, 1978), the theory remains far removed from an analysis of the direct response of organisms to environmental variation within their lifespan. My purpose here is to consider the effects of variation of a single environmental factor, light, on plant growth and attempt to draw some evolutionary conclusions. Here I specifically am interested in the potential for an organism to adapt to environmental variation within its lifespan, a phenomenon known as phenotypic plasticity. This refers to the capacity of a genotype to engender a range of potential phenotypes, depending upon environmental conditions during development. This capacity is generally ignored in genetics and population models, yet such plasticity is very important to understanding how plants function (Bradshaw, 1965; Harper, 1977). Most terrestrial plants are stuck in one location for the majority of their lifespan, and thus have to take the environment as it comes. Unlike animals, which have the capability through movement to spatially average the surrounding environment, plants can only respond to environmental variation within their lifespan through a flexible phenotype.

My goal here is to analyze the effects of variable light environments on photosynthetic capacity. All understory plants and those with an extensive canopy structure maintain leaves in environments with highly variable light levels (Gross and Chabot, 1979). It is through photosynthesis carried out in their leaves that the plants obtain all their energy for growth and reproduction. Individual plants produce morphologically and anatomically different leaves depending upon growth light levels (Boardman, 1977; Jurik et al., 1979). Sun leaves, which develop under high light conditions, tend to be thicker and have higher specific leaf weights, light-saturated photosynthetic rates, and dark respiration rates than shade leaves from the same plant, which develop under low light conditions (Boardman, 1977; Clough et al., 1980; Gross and Chabot, 1979; Jurik et al., 1979).

Similar differences in photosynthetic response to light have also been described for plants native to high light habitats relative to congeners from low light habitats (Björkman, 1968; Gauhl, 1976). This has led to the concept of sun and shade ecotypes, meaning plant populations which are genetically adapted to local habitat conditions. It has been demonstrated that these photosynthetic responses are heritable (Björkman and Holmgren, 1963; Holmgren, 1968), which provides evidence that natural selection acts not only on the types of leaves produced, but also on the range of leaf phenotypes a plant can display. Further studies however (Clough et al., 1979a,b, 1980) point out that the high and low light habitats used in studies of ecotypic differentiation differ from each other in many ways other than just light conditions and it is difficult to conclude that observed ecotypic differences are adaptive solely to differences in light conditions. These latter studies produced no evidence to indicate that sun and shade ecotypes existed in Solanum dulcamara, although since the plants were collected from a very restricted geographical area there is no assurance that the plants had time to evolve ecotypes. The available data clearly points out, however, that even if ecotypes do not exist, phenotypic plasticity of photosynthetic response does, is variable in magnitude from population to population, and is heritable. I here call the range of leaf phenotypes a plant can display the phenotypic plasticity range, and as Bradshaw (1965) noted for a number of plant characters, it is under specific genetic control.

The basic hypothesis of this paper is that there is a whole continuum of leaf types from sun to shade that a plant may produce. I proceed by determining which leaf types maximize photosynthetic gains in constant, deterministically varying, and randomly varying light environments. Using this, given any range of environments one can determine the optimal phenotypic plasticity range and at the end of this paper I briefly discuss what limits this range.

Acclimation in Constant Environments

The basic data used in this study are steady-state carbon exchange rates in response to light, illustrated in Figure 1, which approach a maximum photosynthetic capacity, K_2 , at high light levels. The curve fitted to this data is of Michaelis-Menton form (Thornley, 1976)

$$P = \frac{K_1 K_2 L}{K_1 L + K_2} - f(K_2) \qquad (1)$$

where P is photosynthetic rate per unit leaf area, L is the constant quantum flux in the photosynthetically active region of the spectrum, K_1 is the slope of the curve at $L = 0$, and $f(K_2)$ is the dark respiration rate measured during the light part of the day. Inherent in the formulation of (1) is the assumption that dark respiration rate is a function only of the maximum photosynthetic capacity of the leaf, which is reasonable since this is highly correlated with the density of mesophyll tissue (Chabot and Chabot, 1977). Leaves grown at high light levels tend to have higher values of both K_2 and $f(K_2)$ than those grown at lower levels. The data used here are for Fragaria virginiana, a wild strawberry (Gross and Chabot, 1979).

For leaves grown under constant daily light levels, the leaf type which maximizes total photosynthetic uptake in a day is the value of K_2 which maximizes

$$\overline{P} = \int_{day} P(t)\,dt - \int_{night} R(t)\,dt . \qquad (2)$$

Here $P(t)$ is the net photosynthetic rate, the night respiration rate $R(t)$ satisfies

$$\int_{night} R(t) = a \int_{day} P(t)\,dt + b(K_2)t_N , \qquad (3)$$

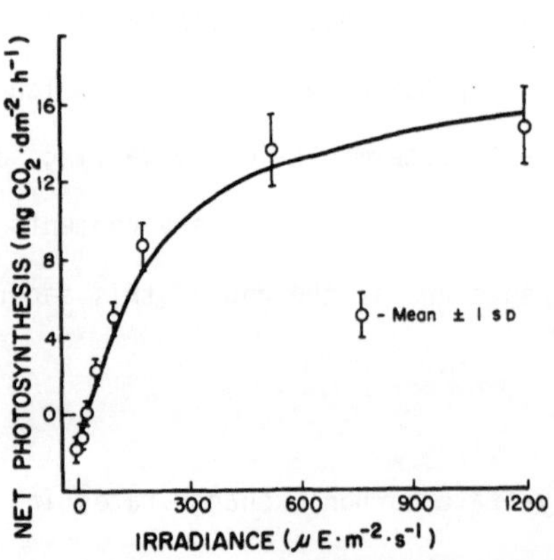

Figure 1: Steady-state photosynthetic response curve for Fragaria virginiana leaves grown at 290 μE m^{-2} sec^{-1} (from Gross, 1982).

$b(\cdot)$ is the basal metabolic rate, t_N is night length, and a is a constant. Some justification for this relationship between daily photosynthesis and total night respiration is available (McCree, 1970; Ludwig et al., 1975) although I have no data for F. virginiana. I further assume

$$f(K_2) = cK_2^{\alpha} , \; b(K_2) = dK_2^{\alpha} \tag{4}$$

where c, d, and α are positive constants. There is strong evidence that $f(\cdot)$ and $b(\cdot)$ are increasing (Bjorkman et al., 1972; Nobel et al., 1975; Chabot and Chabot, 1977; Patterson et al., 1978), though the specific form in (4) is chosen for convenience.

Then for a constant daily light level L, using (1)-(4), the problem is to choose K_2 to maximize

$$\overline{P} = (1 - a) \, t_D \left(\frac{K_1 K_2 L}{K_1 L + K_2} - cK_2^{\alpha}\right) - dK_2^{\alpha} \, t_N \tag{5}$$

where t_D is day length. This may be solved numerically using values for a, c, and d from Ludwig et al. (1975), t_D, t_N and L as in Gross and Chabot (1979), and a range of values for α. The choice $\alpha = 2.5$ provides the best fit to the available data, implying a strongly non-linear increase in respiration with photosynthetic capacity. There is conflict in the literature about the value of α, with some data implying $\alpha = 1$ (Björkman et al., 1972; Patterson, 1975), while others indicate $\alpha > 1$ (Patterson et al., 1978). The chief difficulty in using this data is that few studies measure total nightly respiration. Anatomical data do clearly point out non-linear increases in both specific leaf weight and mesophyll volume per unit leaf surface area, as a function of K_2 (Chabot and Chabot, 1977). Until further experiments are made on leaves grown at a number of different light levels, the results above will remain untested. If correct, they do indicate a strongly non-linear, concave-down increase in optimum K_2 with growth light level.

It should be kept in mind that the above approach has several limitations. Optimization models for evolutionary problems have been criticized for a number of reasons (Oster and Rocklin, 1979), including the difficulty in choosing any criteria which selection could be said to maximize. Although photosynthetic gain may seem the most reasonable fitness criteria for leaf production, there is relatively little evidence to point to a close relationship between photosynthesis and plant growth and reproduction. In general when comparisons are made over a wide range of species, leaf photosynthetic rates are not well correlated with crop yields (Elmore, 1980). Considering the wide range of environmental conditions involved in these comparisons and the fact that photosynthetic area per plant seems to be the dominant factor in controlling yield, the lack of correlation is hardly surprising. In environments with growth limitations imposed mostly by low light conditions, which are the focus of this paper, there is stronger evidence to support the use of photosynthetic gain as a fitness criteria. For example, Pearcy (1983) presents data on two understory tropical forest trees which show a strong correlation between relative growth rates and available light at the growing sites.

The above approach also ignores the different energy requirements needed to build leaves with different photosynthetic capacities. To adequately handle this, the time-scale of consideration should be that of a leaf's lifespan, rather than a single day. The development and senescence patterns for leaves with different photosynthetic capacities are not identical however (Jurik et al., 1979). My approach can be modified to include these complications, but the lack of adequate data to specify the model functional forms would not give much of a qualitative improvement over what can be done under the simpler assumptions. In what follows I therefore continue to ignore these difficulties, while realizing that the results may be only of qualitative rather than quantitative applicability.

Acclimation in Varying Environments

I next consider a more realistic environment in which light switches between two levels (see Figure 2), in a fixed manner throughout the day. To specify the photosynthetic gains in such an environment requires knowledge of the dynamics of photosynthetic response. The results of Gross (1982) indicate that a good approximation for carbon gain can be obtained for such an environment by assuming a first order response of photosynthesis to light increases, with time constant τ . Following a light decrease, there is an instantaneous reduction of uptake rate to the lower steady-state rate. Such an approximation is valid if the light alternations do not

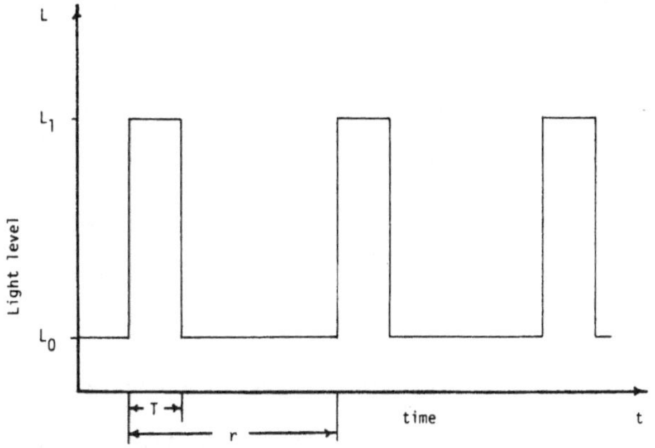

FIGURE 2. The simple varying light environment.

occur too rapidly. The type of environment illustrated in Figure 2 is typical of that below dense leaf canopies in which there are sudden increases in light, called sunflecks, above an ambient low light level.

The analog of (5) is

$$\overline{P} = \frac{t_D(1 - a)}{r} \{\frac{K_1K_2\ L_0}{K_1L_0 + K_2} [r - T + \tau(1 - e^{-T/\tau})] + \frac{K_1K_2\ L_1}{K_1L_1 + K_2} [T + \tau(e^{-T/\tau} - 1)]\} - [(1 - a)ct_D + t_N\ d]K_2^{\alpha} . \quad (6)$$

Here the first term represents uptake during low light periods, the second during high light periods, and the third term gives the respiratory losses throughout the day and night. In the case $T = 0$, (6) reduces to (5). Utilizing the same parameter values as in the previous section, with $\alpha = 2.5$, $\tau = 60$ sec., $L_0 = 50$ and $L_1 = 1200\ \mu E\ m^{-2}\ sec^{-1}$, the value of K_2 which maximizes $\overline{P}$, K_2 opt, can be calculated numerically. Some results for a variety of r and T values are given in Figure 3.

The main conclusions are that: (i) optimal photosynthetic capacity increases both as the sunfleck length T and the frequency of sunflecks, $\frac{1}{r}$, increases; (ii) the manner in which light energy is packaged affects the optimum photosynthetic capacity; and (iii) there is more efficient utilization of a long sunfleck

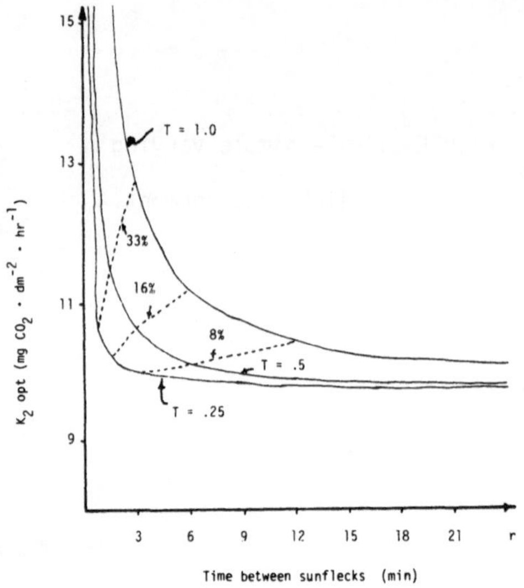

FIGURE 3: The solid lines give the optimal photosynthetic capacity calculated from (6) for the indicated values of T and r . Along the dashed lines there is a constant fraction of the day, as indicated, at the high light level of 1200 μE m^{-2} sec^{-1} .

as compared to several short ones with the same total quantum flux. Note that in Figure 3 for large r values K_2 opt approaches 9.7 mg CO_2 dm^{-2} hr^{-1} which is that obtained from the constant environment case, equation (5), with L = 50 . Similarly, as r approaches T , K_2 opt approaches 29 .

If one considers a range of potential environments in which a leaf could develop, then the optimal plasticity range, meaning the range of optimal values of K_2 for this range of environments, can be established from Figure 3. Figure 4 gives the optimal plasticity range, ΔK_2 opt , for a range of environments, each with a fixed

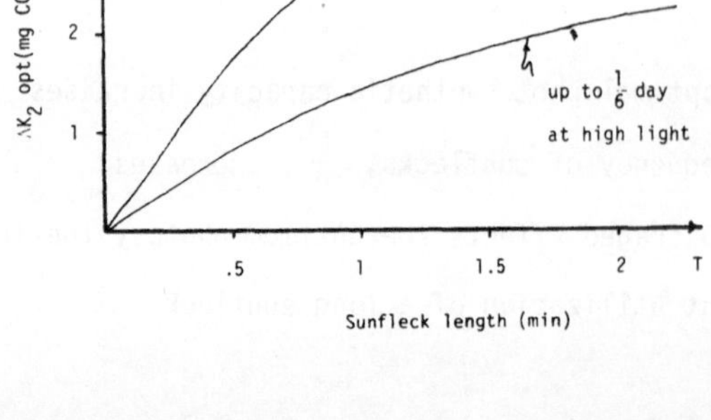

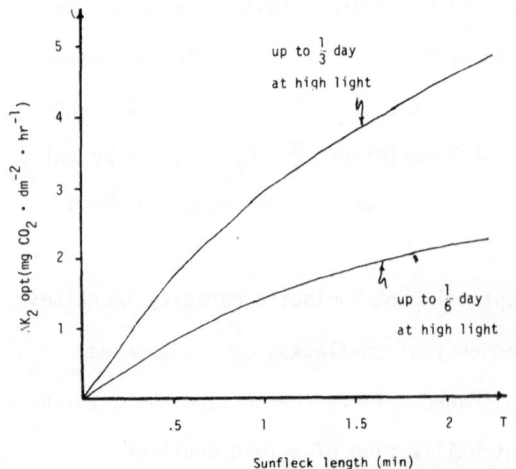

FIGURE 4: The optimal photosynthetic plasticity range for environments with fixed sunfleck lengths, but with the fraction of the day at high light variable up to the indicated level.

sunfleck length, but with varying times between sunflecks. This amounts to a range of environments with different fractions of the day spent at high light or, equivalently, with longer time periods of low light between sunflecks. As Figure 4 illustrates, the optimal plasticity range increases non-linearly with both sunfleck length and fraction of the day at high light. Note that the optimal plasticity range widens in environments with sunflecks of longer duration. From this, one might predict that a higher level of phenotypic plasticity might be observed in understory plants in environments such as temperate deciduous forests which tend to have longer sunflecks (Hutchison and Matt, 1977) than those in tropical forests (Pearcy, 1983). This prediction requires a similar range of proportions of the day at high light in the two habitats. My results provide a means to quantify the observations of differences in acclimation plasticity between sun and shade species (Björkman, 1968; Boardman, 1977; Bazzaz and Carlson, 1982; Zangerl and Bazzaz, 1983). The model results are consistent with the statement by Bazzaz and Carlson (1982) that the degree of population flexibility to acclimate to a variety of environments is related to the environmental variability of the habitat in which the population is normally found.

A Random Environment Analog

The above deterministically varying environment is rather unrealistic since periods of high and low light can be quite variable in the understory. I now relax the restriction that sunfleck length and frequency are fixed. Suppose that sunfleck length T is now a random variable Y_i and the time between sunflecks ($r - T$ in Figure 2) is a random variable X_i. Assume the environment is stationary throughout a day in the sense that $\{X_i\}$ and $\{Y_i\}$ are mutually independent processes and X_i and Y_i are independent, identically distributed with respective means, variances, and distributions δ, ρ^2, $F(x)$, and γ, ξ^2, $G(y)$. Then $Z_i = X_i + Y_i$ gives the sunfleck interarrival times and specifies a renewal process $\{N(t), t \geq 0\}$ which counts the number of sunflecks up to time t. Note that in this model I do not take the light levels L_0 and L_1 to be variable, although a Markov-renewal formulation could be used to accomplish this. The restriction to two light levels is reasonable for many understory environments with relatively constant diffuse light backgrounds, since the non-linearity of the photosynthetic response curve (Figure 1) acts as a filter in that all light levels above about 1/4 of full sunlight produce nearly identical photosynthetic rates.

The uptake from the ith sunfleck of length Y_i is

$$U_i = \frac{K_1 K_2 L_1}{K_1 L_1 + K_2}[Y_i + \tau(e^{-Y_i/\tau} - 1)] + \frac{K_1 K_2 L_0}{K_1 L_0 + K_2}\tau(1 - e^{-Y_i/\tau}) - cY_i K_2^{\alpha} \tag{7}$$

Letting $V(t) = \sum_{i=1}^{N(t)} U_i$ be the total carbon gain from sunflecks up to time t, and applying the renewal argument to $A(t) = E[V(t)]$, renewal theory (Karlin and Taylor, 1975) gives

$$A(t) = E[U_i](1 + E[N(t)]) . \tag{8}$$

The elementary renewal theorem then implies that

$$\lim_{t\to\infty} \frac{A(t)}{t} = \frac{E[U_i]}{\gamma + \delta}$$

so the mean uptake due to sunflecks in a day of length t_D is

$$A(t_D) \approx \frac{t_D}{\gamma + \delta} E[U_i] . \tag{9}$$

Utilizing the fact that the mean fraction of the day in sunflecks is $\frac{E[Y_i]}{E[Z_i]} = \frac{\gamma}{\gamma + \delta}$ and using a second order approximation for the $e^{-Y_i/\tau}$ term in (7), the mean total uptake through a day is

$$E[\overline{P}] \approx \frac{t_D(1 - a)}{\gamma + \delta} \left[\frac{K_1 K_2 L_1}{K_1 L_1 + K_2}\left(\frac{\gamma^2 + \xi^2}{2\tau}\right) - \frac{K_1 K_2 L_0}{K_1 L_0 + K_2}\left(\frac{\gamma^2 + \xi^2}{2\tau} - \gamma - \delta\right)\right] - [c(1 - a)t_D + dt_N]K_2^{\alpha} . \tag{10}$$

The mean optimal leaf type is that value of K_2 which maximizes (10). Note that (10) is the same as (6) with $r = \gamma + \delta$ and the exponential term replaced by a second order approximation involving the variances. Thus the mean maximization problem in the random environment essentially reduces to the deterministic problem already discussed.

On the "Cost" of Plasticity

The above development allows one to derive the optimal phenotypic range for any range of environments. However the approach provides no implications as to what limits plasticity, e.g. why the phenotypic plasticity range is restricted in some populations (Björkman, 1968; Gauhl, 1976) and why some species such as Douglas fir show genetic adaptation to local environments (Campbell, 1979), while others such as *Pinus monticola* adapt via phenotypic plasticity (Rehfeldt, 1979). The available information on the variability of photosynthetic response within a population (Clough et al., 1979a, b, 1980; Zangerl and Bazzaz, 1983) indicates that the majority of intra-populational variability is due to phenotypic plasticity rather than between-genotype variation. The essential observation is that individuals from shade-adapted populations tend to be less plastic than their congeners from more exposed habitats, although as remarked above Clough et al. (1979a) provide evidence that in some species there can be very little difference in the performance of individuals from sun and shade environments. A definite trend also exists for late successional species to be considerably less plastic than early successional ones (Bazzaz and Carlson, 1982; Zangerl and Bazzaz, 1983).

Bradshaw (1965) gives several potential explanations as to why a single genotype with infinite plasticity is not found. He discusses situations such as strong stabilizing selection, the possibility that acclimation responses are too slow relative to the time scale of environmental change, the potential for plasticity to involve too high a cost to the organism, and the lack of genetic variability controlling plasticity. For the case of photosynthetic plasticity, there is clear evidence for the existence of genetic variability (Björkman and Holmgren, 1963; Holmgren, 1968). Since the response time of plasticity is essentially the leaf development time, it does not occur on a slow time scale relative to major changes in understory light environments, excluding such sudden changes as light gaps formed by tree falls. Although the low light levels of shaded habitats select for leaves with shade photosynthetic response, it is unclear how this may be directly viewed as stabilizing selection on the plasticity range, since the high light phenotypes are never expressed in these habitats. Individuals with a wide plasticity range are not selected against unless there are attendant environmental cueing mechanisms which misread the environment.

Another possibility is to invoke some "cost" for the maintenance of plasticity, in that the ability to attain one phenotype reduces the fitness of other potential phenotypes within the plasticity range. However it is difficult to properly define such a cost unless it involves the loss of tissue when changing phenotype. In the case of photosynthetic capacity, I believe the most reasonable explanation for the evolution of genetic limitations on plasticity involve the irreversibility of leaf production. Although there is some potential for the photosynthetic capacity of leaves to be altered by environmental influences after the leaf has developed, this is limited (Jurik et al., 1979). In a non-stationary environment, as exists for example through the year in a forest understory, although a high light adapted leaf may be beneficial in the early spring, that same leaf will become maladapted as the canopy leafs out and the diffuse light level drops later in the season. One potential response to this, which does occur for example in _Fragaria_ species, is the senescence and subsequent replacement of early spring high light adapted leaves with low light adapted leaves after the canopy leafs out. If however all leaves are maintained throughout the growing season, then the initiation of a high light leaf in the spring may be maladaptive when viewed over the entire year. Note that the genetic limitations on plasticity of shade species may not involve the loss of plastic development sequences, but rather a change in the environmental cueing mechanisms which initiate high light leaves.

Plasticity allows adaptation to extremes, however the risks associated with having a phenotype initiated during one period of the year which is irreversible and maladaptive later in the year can select for genotypes with restricted plasticity ranges. Although this explanation needs further theoretical elaboration, it is consistent with the available transfer studies on sun and shade plants (Björkman, 1968; Gauhl, 1976). In all cases, shade adapted plants have a more restricted plasticity range than their congeners from exposed habitats. The shade plants do not have the capacity to produce high light leaves, which may have occured since the initiation of such leaves at any time in the season in the shade habitat is maladaptive when considered over the leaf lifespan.

Acknowledgements: Partial support for this work came from NSF grant MCS-8002963. The comments of Simon Levin and two anonymous reviewers are greatly appreciated.

Literature Cited

Bazzaz, F. and R.W. Carlson. (1982). Photosynthetic acclimation to variability in the light environment of early and late successional plants. Oecologia 54: 313-316.

Björkman, O. (1968). Further studies on differentiation of photosynthetic properties in sun and shade ecotypes of Solidago virgaurea. Physiol. Plant. 21: 84-99.

_________ and P. Holmgren. (1963). Adaptability of the photosynthetic apparatus to light intensity in ecotypes from exposed and shaded habitats. Physiol. Plant. 16: 889-914.

_________, N.K. Boardman, J.M. Anderson, S.W. Thorne, D.J. Goodchild and N.A. Pyliotus. (1972). Effect of light intensity during growth of Atriplex patula on the capacity of photosynthetic reactions, chloroplast compounds and structure. Carnegie Inst. Wash. Year Book 71: 115-135.

Boardman, N.K. (1977). Comparative photosynthesis of sun and shade plants. Ann. Rev. Plant Physiol. 28: 355-377.

Bradshaw, A.D. (1965). Evolutionary significance of phenotypic plasticity in plants. Adv. Genetics 13: 115-155.

Campbell, R.K. (1979). Genecology of Douglas-fir in a watershed in the Oregon Cascades. Ecol. 60: 1036-1050.

Chabot, B.F. and J.F. Chabot. (1977). Effects of light and temperature on leaf anatomy and photosynthesis in Fragaria vesca. Oecologia 26: 363-377.

Clough, J.M., J.A. Teeri, and R.S. Alberte. (1972a). Photosynthetic adaptation of Solanum dulcamara L. to sun and shade environments: I. A comparison of sun and shade populations. Oecologia 38: 13-22.

_________. (1979b). Photosynthetic adaptation of Solanum dulcamara L. to sun and shade environments: II. Physiological characterization of phenotypic response to environment. Plant. Physiol. 64: 25-30.

_________. (1980). Photosynthetic adaptation of Solanum dulcamara L. to sun and shade environments: III. Characterization of genotypes with differing photosynthetic performance. Oecologia 44: 221-225.

Elmore, C.D. (1980). The paradox of no correlation between leaf photosynthetic rates and crop yields. Pages 115-167 in J.D. Hesketh and J.W. Jones (editors), Predicting Photosynthesis for Ecosystem Models, Volume II. CRC Press, Boca Raton, Florida.

Gauhl, E. (1976). Photosynthetic response to varying light intensity in ecotypes of Solanum dulcamara L. from shaded and exposed habitats. Oecologia 22: 275-286.

Gross, L.J. (1982). Photosynthetic dynamics in varying light environments: a model and its application to whole leaf carbon gain. Ecology 63: 84-93.

_________ and B.F. Chabot. (1979). Time course of photosynthetic response to changes in incident light energy. Plant Physiol. 63: 1033-1038.

Harper, J.L. (1977). Population Biology of Plants. Academic Press, N.Y.

Holmgren, P. (1968). Leaf factors affecting light-saturated photosynthesis in ecotypes of Solidago virgaurea from exposed and shaded habitats. Physiol. Plant. 21: 676-698.

Hutchison, B.A. and D.R. Matt. (1977). The distribution of solar radiation within a deciduous forest. Ecol. Monog. 47: 185-207.

Jurik, T.W., J.F. Chabot, and B.F. Chabot. (1979). Ontogeny of photosynthetic performance in Fragaria virginiana under changing light regimes. Plant Physiol. 63; 542-547.

Karlin, S. and H.M. Taylor. (1975). A First Course in Stochastic Processes. 2nd edition, Academic Press, N.Y.

Ludwig, L.J., D.A. Charles-Edwards, and A.C. Withers. (1975). Tomato leaf photosynthesis and respiration in various light and carbon dioxide environments. P. 29-36 in R. Marcelle (editor), Environmental and Biological Control of Photosynthesis. Junk, The Hague.

McCree, K.J. (1970). An equation for the rate of respiration of white clover plants grown under controlled conditions. P. 221-229 in I. Setlik (editor), Prediction and Measurement of Photosynthetic Productivity. Center for Agric. Publ. and Docum., Wageningen.

Noble, P.S., L.J. Zaragoza, and W.K. Smith. (1975). Relation between mesophyll surface area, photosynthetic rate, and illumination level during development for leaves of Plectranthus parviflorus Henckel. Plant Physiol. 55: 1067-1070.

Oster, G.F. and S.M. Rocklin. (1979). Optimization models in evolutionary biology. P. 21-88 in S.A. Levin (editor), Some Mathematical Questions in Biology, X. Amer. Math. Soc., Providence, R.I.

Patterson, D.T. (1975). Photosynthetic acclimation to irradiance in Celastrus orbiculatus Thunb. Photosynthetica 9: 140-144.

Pearcy, R.W. (1983). The light environment and growth of C_3 and C_4 tree species in the understory of a Hawaiian forest. Oecologia 58: 19-25.

Rehfeldt, G.E. (1979). Ecotypic differentiation in populations of Pinus monticola in North Idaho - myth or reality? Amer. Natur. 114: 627-636.

Templeton, A.R. and E.D. Rothman. (1978). Evolution and fine-grained environmental runs. P. 131-183 in Hooker et al. (editors), Foundations and Applications of Decision Theory, Vol. II. Reidel, Dordrecht, Holland.

Thornley, J.H.M. (1976). Mathematical Models in Plant Physiology. Academic Press, London.

Turelli, M. (1977). Random environments and stochastic calculus. Theor. Popul. Biol. 13: 244-267.

Zangerl, A.R. and F.A. Bazzaz. (1983). Plasticity and genotypic variation in photosynthetic behavior of an early and a late successional species of Polygonum. Oecologia 57: 270-273.

A MODEL OF OPTIMAL THERMOREGULATION DURING GESTATION BY *Sceloporus jarrovi*, A LIVE-BEARING LIZARD

by

Stephen Ellner
Department of Mathematics and Program in Ecology
University of Tennessee, Knoxville

and

Carol A. Beuchat
Department of Physiology
University of Arizona, Tuscon

Reptiles are ectotherms; that is, they depend on external sources of heat to maintain their body temperature. Body temperature is regulated behaviorally rather than metabolically, primarily via habitat choice (sun vs. shade), orientation to thermal energy sources, and posture. Many of the processes involved in a reptile's ability to procure, process, and assimilate food, to escape from predation, and to reproduce vary in efficiency as a function of body temperature. These include the capacity for and ability to recover from activity, egestion and digestion rates, auditory sensitivity, renal function and gonadal growth (see Dawson 1975 for a review).

Behavioral maintenance of an appropriate body temperature (thermoregulation) is consequently of great importance to a reptile, and we may expect that selection pressure for thermoregulation around the fitness-maximizing optimal temperature is strong. If all physiological and ecological processes are optimized at a single body temperature T^* , then thermoregulation at or near T^* will be favored. There is evidence, however, that not all processes function best at a single body temperature. For example, many reptiles regulate higher body temperatures after feeding than at other times (see Huey 1982 for a review), and maximum activities of some reptile enzymes do not correspond well to preferred body temperatures (Licht 1964; F.H. Pough, personal communication). In most circumstances, no single process should be optimized at the expense of all others, so a compromise among the thermal optima for different processes would be favored (Huey 1982). Deviations from physiologically optimal body temperatures might also be favored by ecological factors, such as reduced energetic "costs" of thermoregulation (Huey and Slatkin 1976; Hainsworth and Wolf

1978; Huey 1982) or a reduced risk of predation.

The potential conflict between divergent thermal optima for physiological processes is perhaps clearest in reptiles that bear live young rather than laying eggs. The growth rate, survivorship, and frequency of morphological abnormalities of reptile embryos are influenced by temperature deviations of as little as 3°C from the optimum during development (e.g. Bustard 1969; Vinegar 1973; Sexton and Marion 1974; Muth 1980; reviewed in Beuchat 1982). A pregnant female therefore must thermoregulate through months of gestation for the mutual benefit of herself and the offspring developing within her.

This paper is a progress report on our attempts to model thermoregulation during gestation by *Sceloporus jarrovi* (Yarrow's spiny lizard), a live-bearing lizard native to the southwest United States and Mexico. Our efforts were prompted by the observation that the preferred body temperature is about 2°C lower in pregnant than in non-pregnant female *S. jarrovi* (± 32°C rather than ± 34°C, Beuchat 1982). A similar shift in preferred body temperature (from 30 to 28°C) has been noted in *S. cyanogenys* (Garrick 1974). Beuchat (1982) hypothesized that the shift in body temperatures in *S. jarrovi* reflected a compromise between the thermal requirements of the mother and those of her developing embryos. Other explanations for the shift are possible: for example, a body temperature of ± 32°C might be absolutely necessary for proper development of the embryos, or pregnancy might induce a 2°C shift in the mother's thermal optimum. The goal of our modelling effort is to test the compromise hypothesis by quantitatively predicting the body temperature that optimally compromises between conflicting thermal optima for mother and embryos.

Our goal is not yet achieved, because there are very few quantitative data available on the effects of body temperature during gestation for any live-bearing lizard. The physiological component of our model, which relates a female's survivorship and fecundity to her body temperature, is therefore somewhat speculative. Experiments aimed at quantifying the effects of body temperature on reproduction in laboratory populations of *S. jarrovi* are now in progress. For the present, we are nonetheless able to test the model qualitatively by comparing observed and predicted trends in preferred body temperature along an altitude gradient over which adult and juvenile vital rates vary drastically. The model is also of interest for the general approach, in which physiological and demographic sub-models involving experimentally measurable parameters are combined to yield predictions of optimal behavior.

The Model

The model is a version of the Leslie matrix adapted to the life history of *S. jarrovi* as described by Ballinger (1973, 1979, and personal communication). The model life-cycle is shown in Figure 1, and the notation is summarized in Table 1. In the model, the vital rates of a female are allowed to depend only on her age and her preferred body temperature during gestation; in particular, they are density-independent and are not subject to random fluctuations.

The population is censused in the spring immediately after birth takes place. Four types of individuals are then present: neonates, immature (non-reproductive) yearlings, mature yearlings, and adults (individuals 2 or more years old). Mature yearlings are yearlings that reproduce in their first year; in low-altitude populations about 70% of the yearlings are mature, but in high-altitude populations no yearlings are mature (Ballinger 1979). Birth is followed by growth over the summer: yearlings become adults, neonates become yearlings (of which a fraction g are mature), and adults increase in fecundity by a factor γ. This is followed by mating in autumn and overwinter gestation. We assume (with no loss of generality) that all mortality occurs during gestation. The survivorship of mature yearlings is reduced by a factor $\upsilon < 1$ relative to the survivorship π_0 of immature yearlings, representing a survival-cost of reproduction. The average survivorship of neonates to age 1 is thus $p_0 = \pi_0[1 - g + g\upsilon]$. Females give birth in the spring, beginning the next year's cycle. Estimates of the vital rates based on Ballinger's field-studies (1973, 1979, and personal communication) are summarized in Table 2.

The effects of body temperature are incorporated by allowing the growth, survivorship, and fecundity of a reproducing female to depend on her preferred body temperature (Ty in yearlings, Ta in adults). That is, υ, p_1, m_1, and m_2 are functions of the female's preferred body temperature (only the product of survivorship × growth appears in the model, so effects of temperature on growth can be absorbed into υ and p_1). We assume (as shown in Figure 2) that fecundity is

Figure 1. Model life-cycle of *S. jarrovi* (see text for explanation).

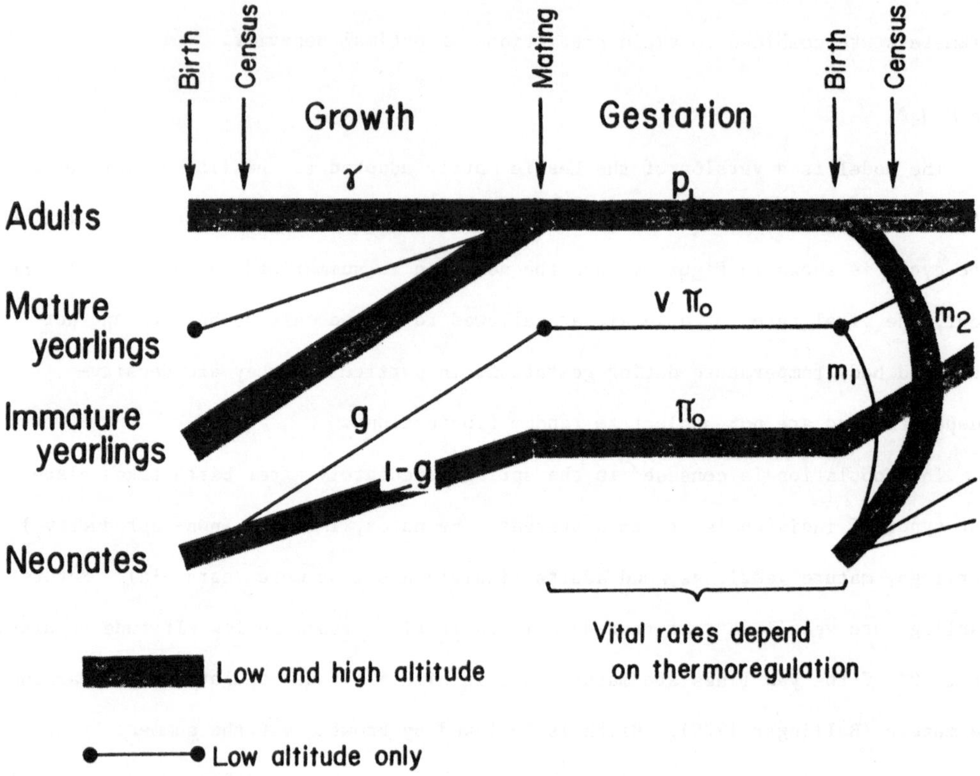

TABLE 1. Notation for vital rates in the model.

Vital Rate	*Definition*
π_0	Survivorship to age 1 of neonates that do not reproduce as yearlings.
v	Reproduction in survivorship of neonates due to reproduction as yearlings.
g	Fraction of neonates that reproduce as yearlings
p_0	$= \pi_0(1 - g + g\nu)$, average survivorship to age 1 of neonates.
p_1	Annual survivorship of yearlings and adults.
m_i	Annual fecundity (number of female offspring) of an age -i individual that reproduces.
γ	Annual growth rate of fecundity in adults; i.e., $m_k = \gamma^{k-2} m_2$ for $k > 2$.

FIGURE 2. Effects of body temperature on survivorship and fecundity (as a fraction of their maximum value) in the model.

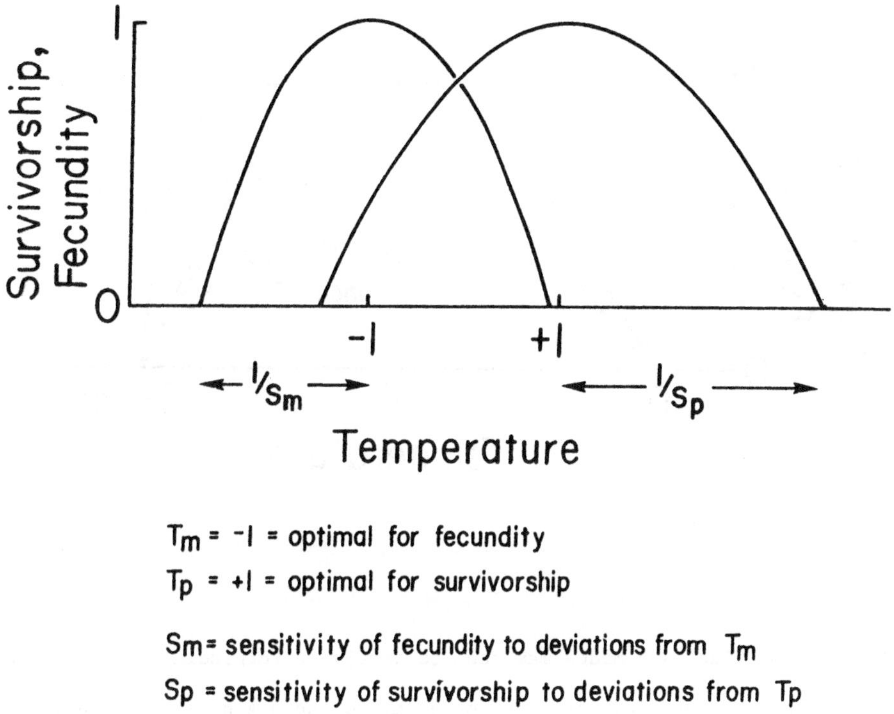

maximized at $T = T_m$, and survivorship is maximized at $T = T_p > T_m$; the temperature scale is then chosen so that $T_m = -1$, $T_p = +1$. The vital rates decrease as T deviates from T_m or T_p , reaching 0 at deviations $\pm 1/s_m$, $\pm 1/s_p$. s_m and s_p thus measure the sensitivity of fecundity and survivorship, respectively, to deviations from the temperature at which they are maximized.

The available data do not allow us to postulate a general functional form for the effects of body temperature on vital rates. Therefore, four simple forms were considered in numerical trials of the model: linear, quadratic, cosine, and octic (Figure 3). The octic form might be realistic for describing fecundity if small deviations merely slowed development, while deviations beyond some threshold induced fatal development abnormalities (this form of temperature-dependence was observed by Muth (1980) in eggs of *Dipsosaurus dorsalis*). The opposite extreme is the linear form, in which the loss per degree of deviation is constant. All these forms are symmetric about $T = T_m$ or T_p , but this entails no loss of generality: the optimal temperature must lie in $[T_m, T_p]$, so only the "inward" half of each curve enters into the

FIGURE 3. Functional forms for the dependence of vital rates on body temperature in the model ($T = T_m$ or T_p, $s = s_m$ or s_p) .

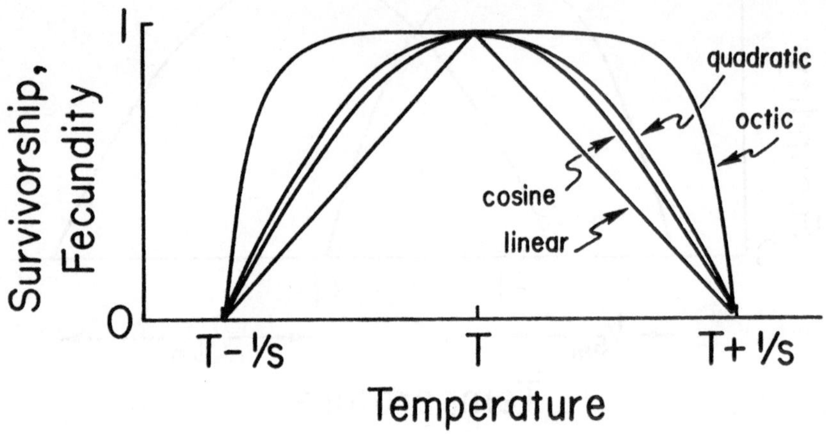

model.

The physiological sub-model may seem to involve a forbidding number of parameters, but this is only because lack of data forces us to use a general formulation. Once the effects of body temperature on vital rates have been measured, the physiological model will consist of just a few equations fit to the data.

Optimal Thermoregulation

The model describes the population dynamics of *S. jarrovi* as a matrix difference equation

$$n_{t+1} = A(Ty, Ta) \cdot n_t$$

where n_t is the vector of population densities at time t ; Ty, Ta are the preferred body temperatures of yearling and adult gestating females, and A is the population-projection matrix determined by the female's vital rates. The optimal body temperatures Ty, Ta for yearlings and adults are defined to be the temperatures that maximize the dominant eigenvalue of A . This optimality criterion is the basis for most theoretical studies of life-history evolution, and is generally appropriate for linear models of structured populations. For detailed discussions of the assumptions underlying the use of this criterion see Stearns (1977, 1980), Levin (1978), and Charlesworth (1980).

Analysis of the characteristic equation of A (summarized in the Appendix) shows that our assumptions imply the existence of a unique pair of temperatures $\hat{T}y$, $\hat{T}a$ at which the dominant eigenvalue is maximized. The model therefore predicts optimal body temperatures as a function of the following sets of data:

a) The values of the temperature-independent vital rates π_0, g, and γ;

b) The maximum values of the temperature-dependent vital rates ν, p_1, m_1 and m_2 (i.e., their values at $T = T_m$ or T_p);

c) The values of s_m and s_p for yearlings and adults; and

d) The functional forms of the temperature dependencies.

At present we have data on a), and are in the process of measuring c) and d) in laboratory populations. The problem is b). Field studies give us the actual values of the vital rates, not the potential maximum values, and protected, well-fed lab lizards may have higher survivorships and fecundities than their brethren in the field. So, the maximum values may be estimated from collectable data. We do this by requiring that the model lizards have the same vital rates as the field populations, when the model lizards behave optimally as defined above. *A priori* there could be several different sets of maximum values satisfying this requirement; whether or not this occurs depends on the functional form of the temperature-dependencies. For the functional forms in Figure 3 (and other qualitatively similar forms), there is only one set of maximum values for ν, p_1, m_1, and m_2 corresponding to a particular set of actual vital rates (Appendix). Unique optimal body temperatures are thus predicted as a function of c), d), and the observed vital rates of the populations in the field.

A Qualitative Test

The gist of the previous section is that quantitative predictions of optimal body temperatures for actual populations of *S. jarrovi* will be possible once we know how ν, p_1, m_1, and m_2 vary as functions of body temperature. At present these are all unknown. Fortunately, within biologically reasonable constraints on s_p and s_m, the effects of changes in the vital rates on the predicted optimal temperatures appear to be fairly insensitive to the details of the temperature dependencies. This allows us to make a preliminary test of the model by comparing observed and predicted shifts in adult body temperature between low and high-altitude populations of *S. jarrovi* in Arizona.

Burns (1970) and Beuchat (1982), using very different methods, found that body temperatures regulated by adults in the field at low and high altitudes were indistinguishable. To examine the model's predictions, we computed $\hat{T}a$ (optimal adult body temperature) for the low and high altitude vital rates in Table 2, using 12 different sets of s_p, s_m values in each of the 16 models defined by the four possible functional forms for survivorship and fecundity. The s_p, s_m values were $s_p = .1, .3, .5$ and $(s_p/s_m) = .5, 1, 2, 4$. Larger s_p values model reproduction by dead females, and the extreme (s_m, s_p) values usually resulted in Ta values near one of the theoretical limits T_p and T_m. We are therefore confident that these values are adequate to characterize the behavior of the model.

With few exceptions, the predicted shifts in adult body temperature were insignificantly small (Table 3; recall that $T_p - T_m = 2$ in the model's temperature scale, whereas data on the development of lizard embryos suggest $T_p - T_m$ is about 4 - 6°C (Beuchat 1982)). The larger predicted shifts occurred only in the extreme case of linearly temperature-dependent vital rates. On the whole then, the model conforms to the observed trend.

The question must still be asked, could the model have failed to conform to the trend? That is, could changes in vital rates mandate appreciable changes in Ta ? To find out, we constructed artificial sets of vital rates (Table 2). Artificial sets

TABLE 2. Values of vital rates. "Low" and "high" are estimated from the data of Ballinger (1973, 1979, and personal communication) for low and high altitude populations of *S. jarrovi* in Arizona. The other sets are hypothetical (see text for explanation).

Vital Rate	Low	High	A	B	A^+	B^-
p_0	.18	.31	.18	.31	.12	.31
p_1	.4	.4	.4	.4	.6	.2
g	.7	0	0	.7	0	.7
ν	.4 - 1	-	-	.7	0	.7
m_1	2.4	0	0	2.4	0	2.4
m_2	4.5	3.6	3.6	4.5	3	6
γ	1.1	1.2	1.2	1.1	1.3	1.05

TABLE 3. Difference in predicted optimal adult body temperature Ta between high and low altitude populations. Columns are defined by the functional form of survivorship, rows by the form of fecundity. Values reported are the average (maximum) of $\hat{Ta}$ (high) - $\hat{Ta}$ (low) over 12 combinations of δp, δm (see text).

	Linear	Quadratic	Cosine	Octic
Linear	.04(.24)	.04(.08)	.03(.08)	.003(.01)
Quadratic	.01(.05)	.02(.03)	.02(.03)	.003(.01)
Cosine	.02(.10)	.02(.04)	.02(.04)	.004(.02)
Octic	.005(.02)	.004(.01)	.004(.02)	.004(.01)

"A" and "B" are fictitious "intermediate altitude" populations in which compensating differences between low and high altitude vital rates are isolated. "A" differs from "low" only in ways favoring higher Ta (more rapid adult growth, fewer offspring) and "B" differs in ways favoring lower Ta (higher survivorship of neonates). In sets A^+ and B^- these differences are exaggerated within reasonable limits. A^+ was concocted from A by setting $p_1 = .6$, $\gamma = 1.3$ and reducing fecundity until a stable population (dominant eigenvalue = 1) was obtained. For B^- the adult fecundity in B was increased to 6, and γ, p_1 reduced to obtain a stable population. The values of p_1, γ, and m_2 used to define A^+ and B^- are all within the ranges reported for populations of *S. undulatus* (estimated by Stearns and Crandall (1981) from data of Tinkle and Ballinger (1972)).

The predicted optimal temperatures for A and B were more widely divergent than those for high and low altitude populations, but were still quite similar (average difference .10 for B vs. A, .02 for high vs. low, over all models and δ_p, δ_m values). Values for A^+ and B^- often differed by a significant portion of the theoretical maximum 2, and in the linear model achieved the maximum (Table 4).

In summary, reasonable changes in vital rates could have produced a serious mismatch of theory and data, and changes within the observed range could have produced more sizable shifts in optimal temperature than those obtained from the actual vital rates. Compensating differences in vital rates between low and high altitude populations lead to predicted optimal adult body temperatures that differ insignificantly or not at all, conforming to the observed trend.

TABLE 4. As in Table 3, but for the hypothetical sets of vital rates A^+ and B^-.

	Linear	Quadratic	Cosine	Octic
Linear	.47(2.0)	.48(1.3)	.51(1.1)	.08(.22)
Quadratic	.28(.75)	.32(.56)	.37(.60)	.09(.19)
Cosine	.28(.69)	.34(.56)	.33(.56)	.09(.18)
Octic	.12(.28)	.08(.19)	.08(.20)	.12(.60)

Discussion

We return to our original question: why do female *S. jarrovi* reduce their body temperature by 2°C during pregnancy? Our model shows that the available data are consistent with the hypothesis that females are compromising between their own optimal body temperature and the optimal temperature for embryo development. This supports the compromise hypothesis, but does not allow us to reject alternative hypotheses. In particular, the available data are equally consistent with the hypothesis that during gestation *S. jarrovi* maintain body temperatures optimal for embryo development.

The compromise hypothesis requires that the females' preferred body temperatures lie between divergent thermal optima for themselves and their embryos. To see if this is true, we are maintaining laboratory populations of *S. jarrovi* throughout gestation in a variety of controlled temperature regimes. The growth, survival, and fecundity of the females and the development rate of embryos will be measured at each temperature. These data should also allow us to model accurately the effects of body temperature on reproduction in *S. jarrovi*. This may well involve modifying the working hypotheses embodied in the present model. For example, body temperature might affect embryo development rates (rather than total fecundity) and thereby determine the offspring's chances of growing large enough to breed as yearlings. This could be modelled by letting the maturation fraction g depend on body temperature. However, once the necessary modifications are made, we should be able to test the revised hypotheses by quantitatively predicting the preferred body temperatures of gestating females.

The model presented here is also readily adaptable to more general life histories and to any physiological processes with divergent thermal optima. Such models could, in principle, make theoretical predictions about many aspects of thermoregulation in particular species and about interspecific differences in thermoregulation. Should animals of different ages or sizes thermoregulate differently? What are the effects of frequency of reproduction, clutch size, and age at first reproduction on thermoregulation? If precise thermoregulation is "costly" (Huey 1974), what range of deviations from the optimal temperature should be tolerated?

The main obstacle to generalizations and theoretical applications of the model is the dearth of information on temperature dependencies in lizards at the whole-animal level. The vast majority of studies demonstrating temperature dependence in reptiles have considered processes at the tissue, cellular, or molecular levels in isolation from other processes. These studies say little about ecologically relevant aspects of whole-animal function in natural populations (Licht 1967; Huey and Stevenson 1979). Optimal thermoregulation cannot be modelled without at least some idea of 1) which vital rates are most sensitive to variations in body temperature, and 2) the qualitative form of the temperature-dependence. We therefore feel that before developing a more general theory, it is important to validate the basic model framework by quantifying the "simple" model for *S. jarrovi* and testing as rigorously as possible its ability to predict actual thermoregulatory behaviors.

LITERATURE CITED

Ballinger, R.E. 1973. Comparative demography of two viviparous iguanid lizards (*Sceloporus jarrovi* and *Sceloporus poinsetti*). Ecology 54: 269-283.

Ballinger, R.E. 1979. Intraspecific variation in demography and life history of the lizard, *Sceloporus jarrovi*, along an altitudinal gradient in southeastern Arizona. Ecology 60: 901-909.

Beuchat, C.A. 1982. Physiological and ecological consequences of vaviparity in a lizard. Ph.D. thesis, Cornell University, Ithaca, New York.

Burns, T.A. 1970. Temperature of Yarrow's spiny lizard *Sceloporus jarrovi* at high altitudes. Herpetologica 26: 9-16.

Bustard, H.R. 1969. Tail abnormalites in reptiles resulting from high temperature egg incubation. British J. Herpetology 4: 121-123.

Charlesworth, B. 1980. Evolution in Age-Structured Populations. Cambridge Univ. Press, Cambridge, U.K.

Dawson, W.R. 1975. On the significance of the preferred body temperatures of reptiles. In: D.M. Gates and R.B. Schmerl (eds.), Perspectives of Biophysical Ecology. Ecological Studies v. 12, Springer-Verlag, New York, pp. 443-473.

Garrick, L.D. 1974. Reproductive influences on behavioral thermoregulation in the lizard, *Sceloporus cyanogenys*. Physiology and Behavior 12: 85-91.

Hainsworth, F.R. and L.L. Wolf. 1978. The economics of temperature regulation and torpor in nonmammalian organisms. In: C.H. Wang and J.W. Hudson (eds.),

Strategies in Cold: natural torpidity and thermogenesis. Academic Press, N.Y., pp. 147-184.

Huey, R.B. 1974. Behavioral thermoregulation in lizards: importance of associated costs. Science 184: 1001-1003.

Huey, R.B. 1982. Temperature, physiology, and the ecology of reptiles. In: C. Gans and F.H. Pough (eds.), Biology of the Reptilia, vol. 12. Academic Press, NY, pp. 25-91.

Huey, R.B. and M. Slatkin. 1976. Costs and benefits of lizard thermoregulation. Quart. Rev. Biol. 51: 363-384.

Huey, R.B. and R.D. Stevenson. 1979. Integrating thermal physiology and ecology of ectotherms: a discussion of approaches. Amer. Zool. 19: 357-366.

Levin, S.A. 1978. On the evolution of ecological parameters. In P.F. Brussard (ed.), Ecological Genetics: The Interface. Springer-Verlag, New York, pp. 3-26.

Licht, P. 1964. A comparative study of the thermal dependence of contractility in saurian skeletal muscle. Comp. Biochem. Physiol. 13: 27-34.

Licht, P. 1967. Thermal adaptation in the enzymes of lizards in relation to preferred body temperatures. In C.L. Prosser (ed.), Molecular Mechanisms of Temperature Regulation. AAAS Publ. 84, pp. 131-145.

Muth, A. 1980. Physiological ecology of desert iguana (*Dipsosaurus dorsalis*) eggs: temperature and water relations. Ecology 61: 1335-1343.

Sexton, O.J. and K.R. Marion. 1974. Duration of incubation of *Sceloporus undulatus* eggs at constant temperature. Physiol. Zool. 47: 91-98.

Stearns, S.C. 1977. The evolution of life history tactics: a critique of the theory and a review of the data. Ann. Rev. Ecol. Syst. 8 : 145-171.

Stearns, S.C. 1980. A new view of life history evolution. Oikos 35: 266-281.

Stearns, S.C. and R.E. Crandall. 1981. Quantitative predictions of delayed maturity. Evolution 35: 455-463.

Tinkle, D.W. and R.E. Ballinger. 1972. *Sceloporus undulatus*: a study of the intraspecific comparative demography of a lizard. Ecology 53: 570-584.

Vinegar, A. 1973. The effects of temperature on the growth and development of embryos of the Indian python, *Python molurus* (Reptilia: Serpentes: Boidae). Copeia 1973: 171-173.

Appendix. Summary analysis of optimal adult body temperature.

The characteristic equation of the population-projection matrix A is

$$\lambda^2 - \lambda(B_0 + p_1\gamma) + p_1(B_0\gamma - B_1) = 0 \qquad (1)$$

where $B_0 = gm_1p_0$, $B_1 = m_2p_0$ and the other symbols are defined in Table 1. For the analysis of (1) below, we assume that the temperature dependencies in the vital rates are C^2 on $(T - 1/s, T + 1/s)$ with non-positive second derivative, a maximum at T and no other critical points, as drawn in Figure 2. Temperature is scaled so that $T_m = -1$, $T_p = +1$. Values of $s_p \geq .5$ give $p_1 = 0$ but $m_1 > 0$ at $T_a = -1$, which is reproduction by dead females, so we assume $s_p < .5$. For now assume $s_m < .5$ also, which implies all vital rates are positive on $[-1, 1]$.

Suppose that all model parameters are specified. Equation (1) then defines the dominant eigenvalue of A as a function $\lambda(T_y, T_a)$. Some implicit differentiations of (1) show that for fixed T_a, the condition $\frac{\partial\lambda}{\partial T_a} = 0$ defines a unique $\hat{T}_y(T_a) \in (-1, 1)$ at which λ is maximized as a function of T_a. Similarly there is a unique $\hat{T}_a(T_y) \in (-1, 1)$ at which λ is maximized as a function of T_y alone. The curves $T = \hat{T}_a(t)$, $t = \hat{T}_y(T)$ must intersect at a point $(\hat{t}, \hat{T})$. At any such point, some implicit differentiation and manipulations of (1) show that

$$\frac{\partial^2}{\partial T_y \, \partial T_a} (p_1 B_1) = \frac{d\, p_1}{d\, T_a} \frac{d\, B_0}{d\, T_y}$$

from which it follows that $\frac{d\,\hat{T}_y}{d\,T_a} = \frac{d\,\hat{T}_a}{d\,T_y} = 0$. The intersection point $(\hat{t}, \hat{T})$ is therefore unique, and gives the optimal body temperatures. This definition of "optimal" body temperatures assumes that T_y and T_a can vary independently, which seems reasonable since body temperature is determined behaviorally rather than metabolically.

To obtain $\hat{T}_a$ values from observed vital rates, assume that we know the actual values of m_1, m_2, p_0, and p_1 rather than their theoretical maximum values. The criteria to be satisfied by $\hat{T}_a$ are 1) optimality $(\frac{\partial\lambda}{\partial T_a} = 0)$ and 2) matching the observed vital rates. Formally, these are

$$p_1' \, \tilde{\gamma}(\tilde{\lambda} - \tilde{B}_0) - \tilde{p}_0(p_1' \, \tilde{m}_2 + \tilde{p}_1 \, m_2') = 0 \qquad (2)$$

$$m_2(m_2^*, \hat{T}_a) = \tilde{m}_2 \qquad (3)$$

$$p_1(p_1^*, \hat{T}_a) = \tilde{p}_1 \qquad (4)$$

where $'$ denotes $\frac{d}{d\,T_a}$, $\sim$ denotes the observed value of a vital rate, and $*$ denotes the theoretical maximum value of a vital rate. Equations (2)-(4) are to be solved for m_2^*, p_1^* , and $\hat{T}_a$.

The left-hand side of (2) is of the form $p_1^* H(m_2^*, \hat{T}_a)$. Direct substitution shows that $H(m_2^*, -1) > 0$, $H(m_2^*, 1) < 0$ for any $m_2^* > 0$, giving a solution $\hat{T}_a(m_2^*)$ of (2). Equation (3) then reads

$$m_2(m_2^*, \hat{T}_a(m_2^*)) = \tilde{m}_2 \ . \tag{5}$$

The Implicit Function Theorem applied to $H(m_2^*, \hat{T}_a(m_2^*)) = 0$ shows that $\hat{T}_a$ is decreasing in m_2^* , and the left hand side of (5) is consequently increasing in m_2^* . Moreover for m_2^* sufficiently large $m_2(m_2^*, \hat{T}_a(m_2^*)) \geq m_2(m_2^*, \hat{T}_a(10)) \to \infty$ as $m_2^* \to +\infty$, and $m_2 \leq m_2^* \to 0$ as $m_2^* \to 0$. Equation (5) thus has a unique solution, which substituted into (2) and (4) determines $\hat{T}_a$ and p_1^* . There is no guarantee that $p_1^* \leq 1$; if (4) gives $p_1^* > 1$, the model cannot be fit to the data.

If $s_m > .5$, $\lambda(T_a)$ will be increasing on $[-1 + 1/s_m, 1]$, since a higher T_a increases survivorship without decreasing fecundity. If $\lambda(T_a)$ is maximized at $T_a = 1$, the optimal body temperature gives zero reproduction by adults. This situation arises in the limit $m_2^* \to 0$, with the result that equation (5) might fail to have a solution if $\tilde{m}_2$ is small. However, it is still the case that (2)-(4) have at most one solution.

A similar analysis is possible for T_y . However, we do not have a good estimate of υ for low-altitude populations, so numerical calculations of T_y are not possible. A full analysis for T_y and T_a will be presented elsewhere (Beuchat and Ellner, in preparation) for the model revised in light of the experimental data.

HOLLING'S 'HUNGRY MANTID' MODEL FOR THE INVERTEBRATE FUNCTIONAL RESPONSE CONSIDERED AS A MARKOV PROCESS. PART O: A SURVEY OF THE MAIN IDEAS AND RESULTS.

J.A.J. Metz and F.H.D. van Batenburg

Institute of Theoretical Biology, University of Leiden, the Netherlands

1. Introduction

In his marvelous 1966 paper 'On the functional response of invertebrate predators to prey density' Holling describes the results of a beautiful set of experiments on the predatory behaviour of the mantid *Hierodula crassa* together with a rather complicated simulation model in which all the experimental detail is combined into one overall picture. He also makes a case for simulation as opposed to analytical methods. This now is a bit of a challenge, and in this paper we shall try to convince you that analytical methods are somewhat more powerful than Holling seemed to be aware of.

There was also a less frivolous reason for us to embark upon the research summarized here. Holling's (1959) secretary or disk model for predation has instigated a spate of applications, experimental as well as theoretical, which is still in full flow today. This is much less the case for his hungry mantid model, even if for predation (as opposed to insect parasitism) this model probably is much closer to biological reality. The reason, no doubt, is its complexity. By the judicious use of analytical methods we can break down this complexity step by step to arrive at various simple end results. At what place we end depends on the relative orders of magnitude of the various parameters. For the original parameter values of Holling's mantid the end result turns out to be very simple indeed.

2. The general invertebrate predator

Figure 1 shows a representation of the prey catching process broken down into its main components according to Holling. The rectangular boxes correspond to the various directly observable activities of a generalized invertebrate predator, with between parentheses a reference to the particular form this activity takes in a preying mantid. The duration and/or success of each of these activities may be influenced by the predator's satiation (or, equivalently, hunger as used in Holling's original model formulation). Satiation itself increases during eating and decreases otherwise.

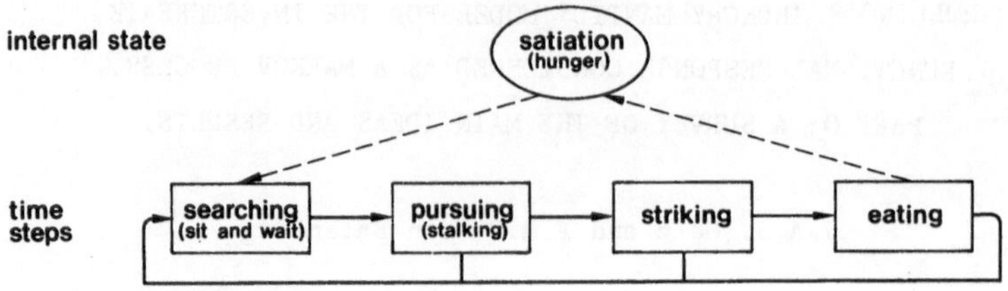

Fig 1: Decomposition of the prey catching process according to Holling.

For his preying mantid Holling found that

(1) satiation decreases exponentially during periods of fasting (fig 2),

(2) the form of the search field remains constant, but

(3) the size decreases linearly with satiation, except that it can never become negative (fig 3),

(4) pursuit occurs at a constant speed, independent of satiation,

(5) the prey, flies in Holling's experiment, escaped during pursuit by flying away at a constant rate,

(6) strike success is constant, independent of satiation,

(7) speed of eating is constant, independent of satiation, and so was the time needed to eat one fly, as fly size was kept rigorously constant.

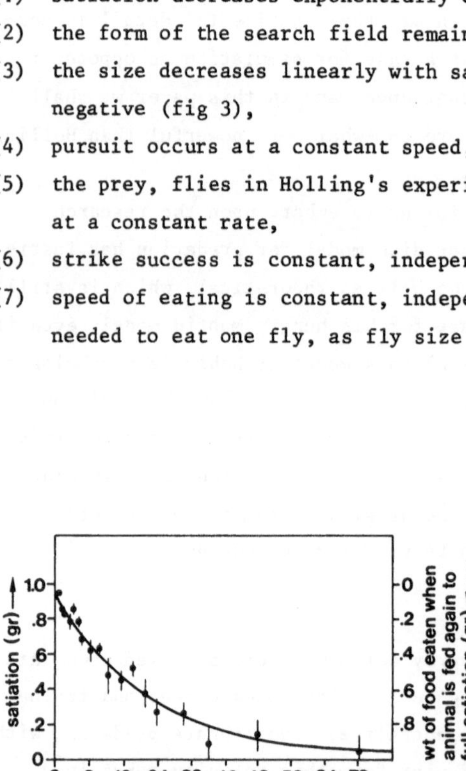

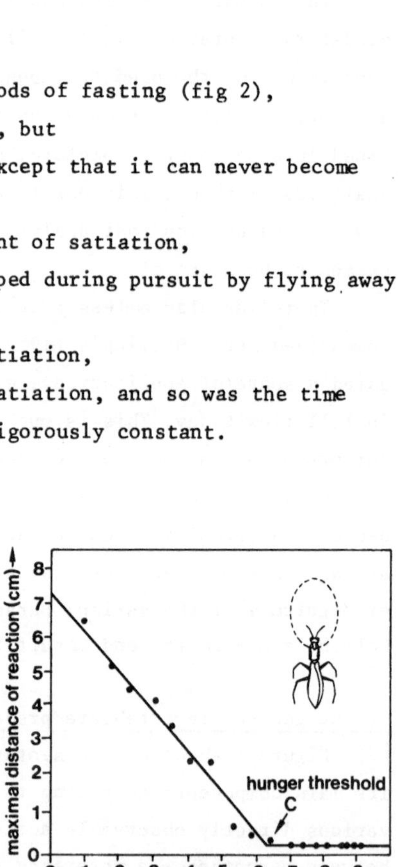

Fig 2: The decrease of satiation (≡ gut content) during a period of fasting in the mantid *Hierodula crassa*. Adapted from Holling (1966).

Fig 3: The size of the mantid's search field as a function of satiation. Adapted from Holling (1966).

To this list the following additional remarks have to be made:
(a) In Holling's experiments prey speed decreased with prey density. This will be accounted for here by referring to effective prey density, i.e. prey density multiplied by the speed reduction relative to the speed at zero density.
(b) The fact that the search field is not circular was corrected by Holling by the introduction of an equivalent circular field, based on the assumption that prey arrived from random directions. This is alright for the calculation of prey arrival rates, but not for e.g. the calculation of the probability of prey escaping during pursuit. For the sake of the exposition we shall not go into such minor refinements of Holling's original calculations, however. Moreover we shall neglect here the rare possibility that immediately after an unsuccesful pursuit or after completing eating a prey the predator spots a new prey inside its reactivated search field, as well as the very small addition to the search rate that derives from the increase of the search field in time.
(c) With Holling we shall equate satiation with gut content, and assume that during eating satiation changes due to the resultant of both ingestion and digestion, the digestion rate depending on satiation in the same manner as during periods of fasting.

If we interpret observation (2) combined with remarks (a) and (b) as implying that during search prey are encountered randomly at a rate equal to the effective prey density times the (maximal) prey velocity and the instantaneous width of the search field, then, given the various parameter values, the previous observations and remarks allow us to stochastically simulate the predator's behaviour, e.g. by successively constructing the (random) time intervals spent in the various activities. These assumptions therefore unambiguously define various stochastic processes like the satiation process S(t) or the total catch process N(t) by providing a means for constructing a probability measure on the set of sample functions.

Holling himself never went to a stochastic simulation. Instead he devised a calculation procedure which successively generated the time intervals spent in the various activities in a deterministic manner. To this end he immediately replaced any random quantity by its expectation, even if he had to deal with some nonlinear function of that same quantity later in his calculations. Still, the result of a fully stochastic simulation by the second author turned out to match pretty well that of Holling's deterministic one. The reason for this later proved to be that both the full stochastic process and Holling's deterministic version are very near to still another much simpler deterministic approximation of the full stochastic model, to be derived below by analytic means. The quality of Holling's predictions therefore seems to hinge on the special values of the parameters leading to the latter approximation. What happens in the general case we do not know yet. What we have done is construct various intermediate approximations which e.g. allow us to calculate numerically the asymptotic mean and variance of the catch for large times.

We are still in the process of checking the various approximations against each other for parameter values as well as functional relationships derived from other animals.

Most of the calculations below will be done in terms of general satiation dependent prey encounter and digestion rates. However, exponential decrease of satiation during fasting by now seems to be an almost universal law for invertebrate predators (and for many vertebrates as well). See e.g. Sabelis (1981) for a recent reference. Therefore we have without much ado sometimes made this simplifying assumption when it turned out to be convenient.

3. The behaviour of a predator represented as a Markov process

The main difference between Holling's secretary model and his mantid model is that in the former we have one unique Markovian searching state, during which the predator searches at a constant rate, whereas in the latter model, we have to deal with a continuum of searching states characterized by different values of the satiation. The Markov property of the searching activity in the secretary model leads us automatically to renewal theory as the source of appropriate tools. See e.g. Cox (1962). The fundamental role played by the Markov property in the analysis of behaviour sequences is discussed in Metz (1974) and Metz et al. (1983). In the mantid's case we have to develop other tools. One such tool box can be derived from the general theory of Markov processes.

To represent our hungry mantid as a Markov process we have to specify a state space, i.e. we have to look for a set of characteristics the value of which at any one time contains all information about past events relevant to the prediction of the animal's future behaviour. (This admittedly is somewhat of an oversimplification which will do, however, for the present purpose. A more detailed account of the problems involved in the construction of state spaces for deterministic as well as stochastic behavioural processes is given in Metz (1977, 1981).)

As long as the predator is searching life is easy: then the satiation S suffices as a summary of its prey catching history. As soon as a prey is sighted things change for the worse, however. The predator then starts pursuing. Pursuit may end either in the random event of prey escape or in a strike. When that strike will happen depends in a unique fashion on the distance the predator still has to pursue. During pursuit this observable distance therefore can be used as a second state variable augmenting satiation during this period. Its value is uniquely determined by past behavioural events as it equals the distance at which the prey was sighted (which in turn depends on the predator's satiation at the moment of first sighting) minus the distance already travelled in pursuit. If the prey escapes during pursuit the predator starts searching again and satiation alone provides a sufficient state description again. An unsuccesful strike has the same consequences as an unsuccesful pursuit. If the strike also is succesful the predator starts eating. It stops eating only when it has finished its meal. When this will happen depends in a unique fashion on how much of its meal still remains. So during eating we can use the remaining size of the meal as a second state variable. Our predator's state space finally is the union of three sets: the satiation axis, the product of the satiation axis times distances still to pursue, and the product of the satiation axis times remaining meal sizes. If we know where in this state space our predator resides at a particular moment, we know everything that can be known at that moment about its probabilistic prey catching future.

If we transform both the dimensions of the two additional state variables to time by dividing them respectively by the speed of pursuit and the speed of eating then these two new state variables can be combined simply by addition into one variable: maximum time still to be spent handling the prey. The predator's state space then is the product of the range of this new state variable and the satiation axis. This state space is depicted in figure 4, together with some segments of possible trajectories representing various behaviour sequences that may occur after

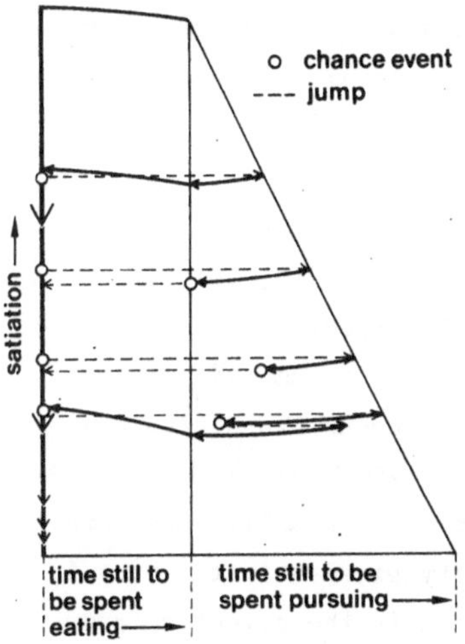

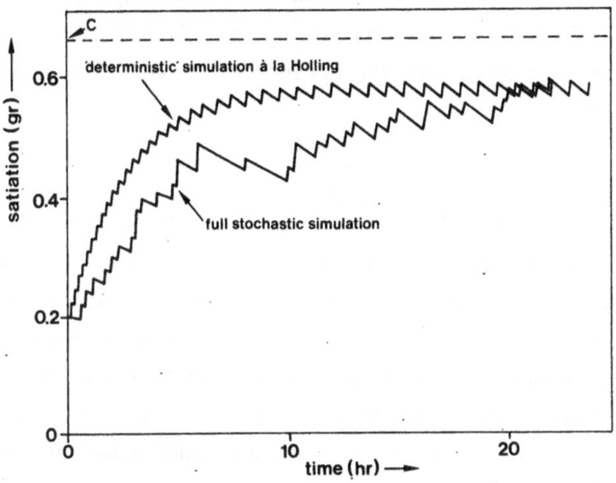

Fig 4: A Markovian state space for Holling's mantid. In the state space some segments of trajectories are depicted showing the various possible events that may happen after the sighting of a prey. The upper segment corresponds to a succesful prey capturing sequence. In the second segment the strike is unsuccesful. In the third segment the prey escapes during pursuit, and the fourth segment corresponds to the very rare event that during a unsuccesful pursuit a new prey has entered the visual field.

Fig 5: A simulated sample path of the satiation process together with the result of a deterministic simulation according to Holling's rules. The chosen sample path is not very representative as most sample paths kept much nearer to the path of the deterministic simulation.

a prey has entered the predator's visual field. In this figure we also have indicated the boundary lines delimiting the region that can be reached in the normal course of events. For example the rightmost boundary line is determined by the distance of first sighting a prey as a function of satiation.

Having arrived at a Markovian description we can write down a complicated set of partial integrodifferential equations for the probability distribution of the predator's state. These equations can not be solved explicitly, however, and for reasons of costs a Monte Carlo solution, i.e. a stochastic simulation, may well be preferable to a numerical solution to the integrodifferential equations. However, the parameter values of Holling's mantid, and probably of many other predators as well, are such that all movements off the satiation axis are much more rapid than the downward movement of satiation due to digestion. Moreover, since the jumps from the satiation axis are relatively infrequent compared to the return rate to the satiation axis, the main probability mass will be concentrated on the satiation axis itself. This effect is also demonstrated in figure 5, which shows a sample path of the satiation process, together with a deterministic simulation following Holling's rules. Here it can be seen that on catching a prey the satiation jumps almost instantaneously to its new level. Therefore we can turn to a simple approximating process in which all the horizontal movements in figure 4 are assumed to be instantaneous.

4. Negligable handling time

From now on we shall assume that 'handling time' is negligable, i.e. if a denotes the rate constant of digestion and $\&_s\tau$ denotes the expected time needed to return to the satiation axis after leaving this axis at s, we assume $a\&_s\tau \ll 1$. Under this assumption a prey capture results in an instantaneous transition $S \to S+w$, where w denotes prey weight. Between captures, S decreases according to the differential equation $dS/dt = f(S)$ where f equals minus the digestion rate. Finally, prey are captured at a rate $xg(S)$, i.e. the conditional probability given S that a prey is caught in an interval of length dt equals $xg(S)dt$, where x is the effective prey density and g is the satiation dependent rate constant of prey capture. On the basis of these observations we can write down the forward equation for the density p of S as

$$\frac{\partial p(s,t)}{\partial t} = - \frac{\partial f(s)p(s,t)}{\partial s} - xg(s)p(s,t) + xg(s-w)\, p(s-w,t) \tag{1a}$$

(see e.g. Cox & Miller (1965) p.237ff or Takács (1960) p.37). To make (1) well-defined we introduce the convention that gp equals zero for s negative. Moreover we have to add the (boundary) condition

$$p(s,t) = 0 \quad \text{for} \quad s \geq s_{max} \geq c+w , \tag{1b}$$

where c is the satiation threshold, i.e. the value of s for which g first becomes zero.

Apparently without great loss of biological generality we may write with Holling

$$f(s) = -as. \tag{2}$$

To calculate g we observe that

$$g(s) = g_1(s)g_2(s)g_3(s), \tag{3a}$$

where g_1 denotes the, satiation dependent, rate constant of prey encounter, and g_2 and g_3 denote respectively the probabilities of a succesful pursuit and strike. According to Holling for his mantid

$$g_1(s) = b(1-s/c)^+, \quad g_2(s) = \exp(-b'(1-s/c')^+), \quad g_3(s) = q, \tag{3b}$$

with $c' < c$, the $^+$ meaning that negative values are replaced by zero.

It is clear that our main interest is not in the distribution of S, but in the number of prey caught, N. Intuitively we write down immediately from 'law of mass action' considerations

$$\frac{d\mathcal{E}N}{dt} = x\int_0^\infty g(s)p(s,t)ds = x\mathcal{E}g(S). \tag{4}$$

A more formal derivation can be given by defining

$$p_n(s,t)ds \overset{def}{=\!=} P\{s < S(t) \leqslant s+ds,\ N(t)=n\}, \tag{5a}$$

to arrive at the 'generation expansion'

$$\frac{\partial p_n}{\partial t} = -\frac{\partial f p_n}{\partial s} - xgp_n + xg(s-w)p_{n-1}(s-w),$$

where $p_{-1}=0$ by convention, with initial and boundary conditions

$$p_0(s,0) = p(s,0), \tag{5b}$$

$$p_0(s,t) = 0 \text{ for } s > s_{max}, \quad p_n(s,t) = 0 \text{ for } s \geqslant c+w, \quad n=1,2,...$$

Multiplying the left and right hand sides of (5) with n, summing over n, and collecting terms gives us (4) again. By multiplying with n^2, summing and manipulating a bit more we arrive also at an equation which can be used to calculate var N:

$$\frac{d\ var(N)}{dt} = x\{2\ cov[N,g(S)] + \mathcal{E}g(S)\}, \tag{6a}$$

where [cov N,g(S)] can be calculated from

$$cov[N,g(S)] = \int_0^\infty g(s)\ z(s,t)\ ds, \tag{6b}$$

where z in turn can be calculated from

$$\frac{\partial z}{\partial t} = -\frac{\partial fz}{\partial s} - xgz + xg(s-w)\ z(s-w) + xg(s-w)\ p(s-w) - xp\mathcal{E}g(S), \tag{6c}$$

with initial and boundary conditions

$$z(s,0) = 0, \quad z(s,t) = 0 \text{ for } s > c+w. \tag{6d}$$

The quantity z itself also allows an interpretation as

$$z(s) = (\mathcal{E}_s N - \mathcal{E}N)p(s),$$

where $\mathcal{E}_s$ denotes the expectation conditional on S=s.

Heymans (in prep) has proven existence and uniqueness of the solutions to the equations for p and z, as well as the exponential convergence of those solutions to the stationary solutions $\hat{p}$ and $\hat{z}$ defined by

$$\begin{cases} 0 = -\dfrac{df\hat{p}}{ds} - xg\hat{p} + xg(s-w)\hat{p}(s-w) & (7a) \\ 1 = \displaystyle\int_0^\infty \hat{p}(s)ds & (7b) \end{cases}$$

$$\begin{cases} 0 = -\dfrac{df\hat{z}}{ds} - xg\hat{z} + xg(s-w)\hat{z}(s-w) + xg(s-w)\hat{p}(s-w) - x\hat{p}\displaystyle\int_0^\infty g(\sigma)\hat{p}(\sigma)d\sigma & (8a) \\ 0 = \displaystyle\int_0^\infty \hat{z}(s)ds. & (8b) \end{cases}$$

Moreover, from general central limit results for functions defined on Markov chains, like those in Grigorescu & Oprişan (1976) or Bolthausen (1982) it can be deduced with some slight effort that for $t\to\infty$

$$(N-\mathcal{E}N)/(\text{var } N)^{\frac{1}{2}} \xrightarrow{\frown} \text{Gaussian}(0,1). \tag{9}$$

To calculate $\hat{p}$ numerically we make use of the fact that (7a) reduces on (0,w] to an ordinary differential equation. So we can choose a starting value $\hat{p}(w) = \hat{p}_w > 0$ and integrate backwards from w to zero. Next we integrate from w to 2w, using the known values of $\hat{p}$ on (0,w], and so on. It can be shown that $\hat{p}$ stays positive on (0,c+w) and becomes exactly zero at c+w. Due to the linearity of (7a) we can normalize afterwards to conform to (7b). The only troublesome aspect may be the singular nature of (7a) at s=0, due to the fact that f(0)=0. As a result $\hat{p}$ may diverge for $s\downarrow 0$. However, near to s=0 we can easily derive an analytical approximation to $\hat{p}$ to replace the numerical one. The result of such a numerical calculation for Holling's mantid parameters and the lowest, highest and a middle prey density used by Holling in his final experiments is shown in figure 6.

The calculation of $\hat{z}$ proceeds in exactly the same manner as that of $\hat{p}$ except that instead of normalizing we add some multiple of $\hat{p}$ to satisfy (8

Finally if we know $\hat{p}$ we can calculate the functional response F as

$$F(x) = x\int_0^\infty g(s)\,\hat{p}_x(s)\,ds, \tag{10}$$

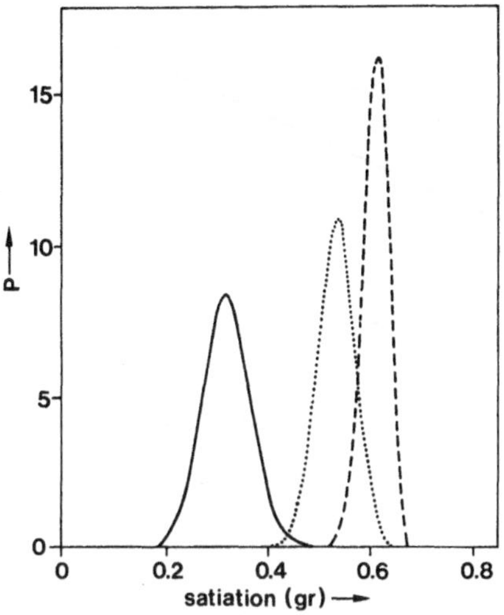

Fig 6: Calculated stationary distribution of the satiation for Holling's mantid for the three values of the effective prey density, corresponding to the lowest, the highest and one middle prey density used by Holling in his final experiments.

where the subscript x refers to the fact that $\hat{p}$ is dependent on the parameter x. For Holling's mantid the functional response is numerically found to be increasing and concave, but we do not yet have any proof of such properties. Direct probabilistic considerations tell us that

$$\frac{dF}{dx}(0) = g(0), \tag{11}$$

$$\lim_{x\to\infty} F(x) = \left(-\int_c^{c+w} (f(s))^{-1} ds\right)^{-1}. \tag{12}$$

To arrive at more manageable results we have to go on to the next approximation stage.

5. Small prey weights

If we look at figure 5 again we see that the prey weight is very small relative to the satiation threshold c, but the accumulated prey weight during a few hours of eating still leads to a considerable upwards shift of S despite the continuous digestion. Therefore we introduce the new parameter $\xi \stackrel{\text{def}}{=} x.w$, and look what happens to (1) if w becomes small, ξ remaining constant:

$$\frac{\partial p}{\partial t} = -\frac{\partial fp}{\partial s} - xgp + xg(s-w)p(s-w) \approx -\frac{\partial fp}{\partial s} - xgp + [xg(s)p(s) - xw\frac{\partial gp}{\partial s}(s)].$$

In the limit when $w\downarrow 0$ we get

$$\frac{\partial p}{\partial t} = -\frac{\partial(f+\xi g)p}{\partial s},$$

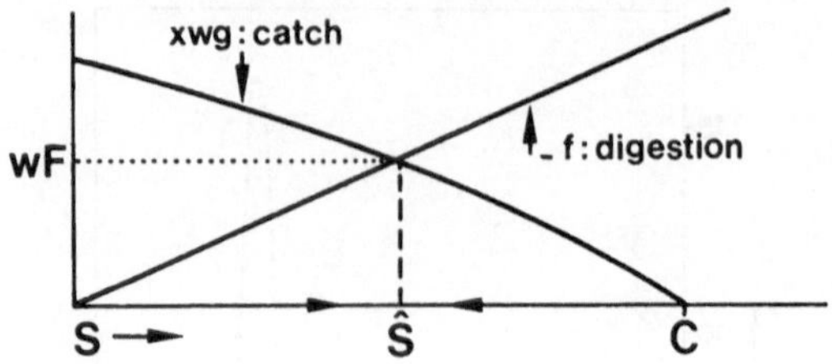

Fig 7: The processes contributing to the satiation in the deterministic limit model for one particular value of the effective prey density. Also indicated are the equilibrium level of the satiation, $\hat{s}$, and the corresponding value of the functional response F.

which is equivalent to

$$\frac{ds}{dt} = f(s) + \xi g(s), \tag{13}$$

i.e. the 'jump terms' have disappeared. So the prey catching process has become completely deterministic: the predator is slurping prey soup.
[Remark: The fact that we could make our approximation in two steps is based on the fact that for Holling's mantid $a\&_s \tau < 0.0027 \ll w/c = 0.037 \ll 1$.]

Figure 7 depicts what happens to our soup eating predator. Its satiation will quickly stabilize at some value $\hat{s}$, defined by

$$-f(\hat{s}) = xwg(\hat{s}) \tag{14}$$

and

$$F(x) \approx xg(\hat{s}). \tag{15}$$

Both $\hat{s}$ and F can be read off immediately in the figure. Changing x corresponds to multiplying the wg curve by different constants. From the construction we can easily prove that for f linear and g decrasing F will always increase; when g is in addition concave F will be concave too.

When pursuit is always succesful, i.e. if g is equal to $b(1-s/c)^+$, and if $f(s) = -as$ as usual, the model formulation derived here corresponds exactly to that of Rashevsky (1959).

6. Linear approximation

During the limiting process described in the previous section we have lost all information concerning the variance of the catch. To retrieve this information we may consider a diffusion approximation locally around the deterministic trajectory. It is easier, however, to derive the results we need by more direct means. To this end we consider first an 'almost linear' version of the satiation process, in which $f(s) = -as$ and $g(s) = b(1-s/c)^+$.

As a first step we observe that, on neglecting the probability mass situated between c and c+w,

$$\&g(S) \approx b(1-\&S/c).$$

To calculate &S we multiply both sides of (1) with s and integrate over s to obtain

$$\frac{d\&S}{dt} = -a\&S + xw\&g(S) \approx -(a+xwb/c)\&S + xwb.$$

Proceeding in this manner using (6) in combination with (1) we can also derive approximate differential equations for variances, covariances and so on. We shall not bother you here with the detailed calculations, but confine ourselves to giving the two most important final results:

$$\frac{d\&N}{dt}(\infty) = \frac{bx}{1+xwb/(ac)}, \qquad (16)$$

which corresponds to the deterministic result (15) under the same assumptions on f and g, and

$$\frac{d\ \mathrm{var}\ N}{dt}(\infty) = \frac{bx}{(1+xwn/(ac))^3}. \qquad (17)$$

As a 'worst case' check of the accuracy of (16) we can let $x\to\infty$, giving

$$\frac{d\&N}{dt}(\infty) = ac/w,$$

which should be compared to an exact value, calculated from (12), of

$$a/\ln(1+w/c) = (ac/w)(1+\tfrac{1}{2}w/c+o(w/c)).$$

So (16) is correct to first order in w/c even for p concentrated near the satiation threshold. For Holling's mantid w/c = 0.037.

For general g one should do a local linearization by replacing g with its tangent line in $\hat{x}$ in order to calculate the appropriate values of b and c to be substituted in (17). (If one substitutes these values for b and c in (16) one automatically gets back the deterministic result (15)).

7. Concluding remarks

In our opinion the most important outcome of the previous analysis is the emergence of a clearer and more complete picture of a predator's behaviour as envisioned by Holling. Moreover, we have brought Holling's model formulation within the analytic realm, thereby preparing it for the application of a host of ready made tools. Using this tool box we have developed procedures which allow us e.g. to numerically estimate the relative contribution of the various components of the predation process to the determination of the functional response.

As one of the results we have found a Rashevsky type deterministic approximation which is considerably more simple than Holling's own deterministic approximation procedure. The conditions for such an approximation to hold are relatively short handling time, $a\&_s\tau \ll 1$ for all s, and relatively small prey weight, $w/c \ll 1$. For Holling's mantid even the stronger assumption $a\&_s\tau \ll w/c \ll 1$ holds, which made it possible to derive a very simple approximate expression for the asymptotic variance of the total catch. If the handling time is small but the prey weight isn't, it is still possible to derive more complicated equations from which we can calculate numerically the functional response as well as the asymptotic variance of the total catch. Moreover 'small' does not necessarily mean very small: the various approximations seem to be pretty robust.

For the parameter values of Holling's own mantid our deterministic model essentially does the whole job: in forthcoming papers it will be shown that a functional response calculated from (14) and (15) does not deviate significantly from one calculated numerically from (7) and (10). Also the approximation to the variance of the total catch calculated from (17) compares very favourably with the variance calculated numerically from (6a,b) and (8).

Various extensions of the previous calculations are possible. One important possibility is to consider variable prey size. In that case we are not only interested in the variance of the number of prey caught, but also in the variance of the accumulated prey weight, as this effectively determines the probability of a predator dying due to random starvation. We may also consider different regions of the parameter space. For example, when $a\&_s\tau$ and w/c are both small but of the same order of magnitude, we get a different deterministic limit in which Holling's secretary model becomes merged into a Rashevsky type hunger model as deduced in section 5. These topics as well as a more detailed exposition of the preceding calculations will be the subject of forthcoming papers.

References

Bolthausen, E. (1982). The Berry-Esseén theorem for strongly mixing Harris recurrent Markov chains. Z. Wahrscheinlichkeitstheorie 60:283-289.

Cox, D.R. (1962). Renewal theory. Methuen, London.

Cox, D.R. and Miller, H.D. (1965). The theory of stochastic processes. Methuen, London.

Grigorescu, S. and Oprişan, G. (1976). Limit theorems for J-X processes with a general state space. Z. Wahrscheinlichkeitstheorie 35:65-73.

Heijmans, H.J.A.M. (in prep). Holling's hungry mantid model for the invertebrate functional response considered as a Markov process. Part III: Mathematical elaborations.

Holling, C.S. (1959). Some characteristics of simple types of predation and parasitism. Can. Entomol. 91:385-398.

Holling, C.S. (1966). The functional response of invertebrate predators to prey density. Mem. ent. Soc. Canada 48.

Metz, J.A.J. (1974). Stochastic models for the temporal fine structure of behaviour sequences. p.5-86 in D.J. McFarland (ed) Motivational control systems analysis. Acad. Press, London.

Metz, J.A.J. (1977). State space models for animal behaviour. Ann. Syst. Res. 6:65-109.

Metz, J.A.J. (1981). State space representations of animal behaviour; an expository survey. PhD Thesis Un. of Leiden, MC, Amsterdam.

Metz, J.A.J., Dienske, H., de Jonge, G. and Putters, F.A. (1983). Continuous time Markov chains as models for animal behaviour. Bull. Math. Biol.

Rashevsky, N. (1959). Some remarks on the mathematical theory of the nutrition of fishes. Bull. Math. Biol. 21:161-183.

Sabelis, M.W. (1981). Biological control of two-spotted spider mites using phytoseiid predators. Part I. Agr. Res. Rep. 910. PUDOC, Wageningen.

Takács. L. (1960). Stochastic processes; problems and solutions. Methuen, London.

THE EFFECT OF COMPETITION ON THE FLOWERING TIME OF ANNUAL PLANTS

by
Thomas L. Vincent
and
Joel S. Brown
University of Arizona
Tucson, Arizona 85721, USA

The optimal flowering time for a population of annual plants can be modelled as a continuous game. This paper utilizes Maynard Smith's ESS concept reformulated in terms of a continuous game to examine the optimal flowering time for annual plants under competition for resources. It is shown that a population of plants experiencing competition from each other should evolve to a flowering time later than that expected for an isolated plant. The results are general in that a class of functions is used for the plant model rather than a specific function.

1. INTRODUCTION

The results of artificial selection experiments demonstrate the availability of heritable variation. That is, most organisms show a phenotypic shift when subjected to artificial selection. This same heritable variation is available to natural selection, which operates through survival and fecundity differences to provide the most fit phenotypes. The phenotypic characteristic which results from this process will appear as though the individuals of a population have chosen an optimal strategy. However, the optimal strategy for a given individual depends upon the strategies of others, so there exist conflicts of interest between the individuals in a population and the optimizing process is actually a game (Von Neumann and Morgenstern, 1944).

The solution to the evolutionary game has been described by Maynard Smith and Price (1973) as the Evolutionarily Stable Strategy or ESS. An ESS is present when a rare mutant cannot establish itself successfully within a community. Maynard Smith's (1974) definition of an ESS is presented here for reference.

ESS Definition. A strategy I is an ESS for all strategies J if either $E(I,I) > E(J,I)$ or $E(I,I) = E(J,I)$ and $E(I,J) > E(J,J)$ where $E(I,J)$ denotes the expected payoff to an individual playing I against an opponent using strategy J, and both players draw from the same set of strategies.

This definition is directly applicable to matrix games. That is, games where there are a finite number of pure strategies and the payoffs between players are summarized as a matrix which matches all possible strategy pairs to a payoff for each player. The ESS for a matrix game is found by examining the payoff function and applying the definition directly.

The optimal flowering time for an annual plant is a phenotypic characteristic which may be thought of as a strategy. Since this strategy is defined over a continuous interval of flowering times (the growing season), the above ESS definition is not directly applicable. There are many other biological problems for which the set of evolutionary choices of phenotypes will be defined over a continuous set (Mirmimani and Oster, 1978; Vincent, 1977; and Lawlor and Maynard Smith, 1976). Since Maynard Smith's ESS definition has been formulated specifically from evolutionary stability requirements, we suggest that extending the ESS concept to continuous systems is a more satisfactory approach than simply using the similar Nash concept which is well defined and used in continuous game theory (Vincent and Grantham 1981). In the past, several authors have used the Nash solution as a substitute for the ESS (Mirmimani and Oster, 1978; Vincent, 1978; Auslander et al., 1978). This is acceptable under certain circumstances since an ESS will always be a Nash solution. However, the Nash solution does not impose all of the stability requirements required by the ESS.

2. THE BALANCED GAME

Here we will consider continuous games which are simple, parametric and balanced. A continuous parametric game is one in which the parameters (strategies) may be chosen over a continuous set. A simple game is one which is defined entirely in terms of the payoff criteria as functions of the strategies. A balanced game represents a new concept and will be developed in detail shortly.

Let $G_i(\cdot)$ represent the payoff to an individual i. This payoff is assumed to be a function of the strategies of all the individuals of the same population who are involved in "playing the game." Henceforth, we will refer to these individuals as the players and we will assume that each player can draw a single strategy from the same continuous set of strategies. The strategy associated with player i is designated by u_i, so that if there are n players, the i-th player's payoff may be represented by $G_i(u_1, \cdots, u_i, \cdots, u_n)$. The vector of strategies $[u_1, \cdots, u_i, \cdots, u_n]$ will be designated by $[u_i;v]$ or $[u_i,u_j;v]$ where it is understood that v is the vector of the remaining players' strategies (i.e., in the first case, v is every players' strategy but player i, in the second case, v is every players' strategy but player i and j). If the remaining players are using a common strategy s, then the notation $[u_i;s]$ means that every component of v has been replaced by s. The notation [s;s] or [s] means that every player is using the common strategy s.

Maynard Smith's ESS definition uses only one payoff function to define a two player game. This presupposes certain symmetry properties which are associated with a biological game. In particular, at an ESS point, each player must receive the same payoff. We will impose this stability requirement along with some others by restricting the class of $G_i(\cdot)$ functions. These restrictions will constitute what we call a balanced game.

Definition of a Balanced Game. Let $N = \{1, \cdots, n\}$ be the set of indicies representing n players. The set of payoff functions $G_i(u_i;v)$ for all $i \in N$ are said to constitute an n player balanced game if for any $i,j,k \in N$ and real numbers δ and $\bar{\delta}$ there exists an $\ell \in N$ such that

$$G_i(u_i, u_\ell;s) = G_j(u_j, u_k;s) \tag{2.1}$$

whenever $u_i = u_j = s + \delta$ and $u_\ell = u_k = s + \bar{\delta}$.

When n = 1, the "game" will always be balanced. However, a one player game is simply a scalar payoff optimization problem. When n > 1 the conditions imposed by (2.1) restrict the class of payoff functions in a number of ways. For example, the balanced game requires that a common strategy s will result in a common payoff for each player (Set $\delta = \bar{\delta} = 0$ in (2.1)). The payoff associated with a mutant player (a player with a phenotype different from the other players with a common phenotype) must be the same regardless of which player is the mutant (Set $\bar{\delta} = 0$ in (2.1)). There are additional restrictions imposed by (2.1) which are discussed in detail elsewhere (Vincent and Brown, 1983).

The definition of a balanced game given here is similar to the symmetric game of Von Neumann and Morgenstern (1944) and Nash (1951). However, there are important differences which makes the balanced game more generally applicable to evolutionary problems than the symmetric game.

A symmetric game is one in which all players have the same set of strategies to choose from and the payoffs depend only upon the choice of all players and not on the ordering of players. Given that all players in an n-person game have chosen a strategy the payoff to an individual is independent of whether it is the 1st, 2nd, or rth player. All players in the game using a given strategy will receive the same payoff independent of the ordering of the players. In the notation used here, this means that all players are faced with the same cost and control functions and in addition a given player's cost function, say $G_i(u_i;v)$ is not dependent upon the ordering of strategies in the vector v but only upon the values of the components of v.

An independence of cost functions from the ordering of players (the ordering of values in the vector v) is violated in natural systems. For example, consider a rock-hopper penguin in a nesting colony. It can make a big difference to its cost function whether its nearest neighbor is hyper-aggressive or whether a more remote neighbor is hyper-aggressive. In such a colony, it would be expected that each penguin is faced by the same cost and control functions, and that the ordering of strategies in each of the penguin's cost functions is important.

In comparing the symmetric game and the balanced game, it is noted that although each definition requires the cost and control functions to be identical for each player, the balanced game does not require the independence of the cost functions with respect to the order of strategies within the strategy vector v.

3. A BALANCED ESS

The authors have previously (Vincent and Brown, 1983) extended the ESS concept to a more complex parametric game than considered here. The results of that investigation may be readily applied to the case considered here. The following definition and necessary conditions are condensed from that study. Hence, the Theorem presented here is given without proof.

Balanced ESS Definition. Let $G_i(u_i;v)$ represent the payoff to the i-th player in an n player balanced game. A scalar strategy s is said to be a balanced ESS if and only if for any $i \in N$ and any nonzero scalar δ

$$G_i(s;s) \geq G_i(u_i;s) \tag{3.1}$$

where $u_i = s + \delta$. In the event that for some δ

$$G_i(s;s) = G_i(u_i;s) \tag{3.2}$$

then

$$G_i(s, u_j;s) > G_i(u_i, u_j;s) \tag{3.3}$$

for all such δ and all $j \in N$, $j \neq i$, where $u_j = s + \delta$.

Because of the balanced game assumption, if conditions (3.1) - (3.3) are satisfied by any one player, they will be satisfied by all players. This requirement along with the requirement that all the remaining players use the same strategy, distinguishes this definition from the Nash definition (Nash 1951). However, by virtue of condition (3.1), a balanced ESS will always be a Nash solution.

We now state some necessary conditions which are useful in finding candidates for a balanced ESS. We assume continuity of the payoff functions as required by the theorem.

Balanced ESS Theorem. Let the scalar strategy set S be open. If $s \in S$ is a balanced ESS, then it is necessary that for any $i \in N$

$$\frac{\partial G_i(s)}{\partial u_i} = 0 \tag{3.4}$$

and

$$\frac{\partial^2 G_i(s)}{\partial u_i^2} \leq 0. \tag{3.5}$$

In the event that there exists a scalar $\delta > 0$ such that for all u_i satisfying $s - \delta \leq u_i \leq s + \delta$ condition (3.2) holds, then it is necessary that

$$\frac{\partial^2 G_i(s)}{\partial u_i^2} = 0 \tag{3.6}$$

and

$$\frac{\partial^2 G_i(s)}{\partial u_i\ \partial u_j} \leq 0 \qquad (3.7)$$

for all $j \varepsilon N$.

With these necessary conditions, we may now analyze a general continuous model for annual plants which includes the effects of competition.

4. FLOWERING TIME FOR ANNUAL PLANTS UNDER COMPETITION

Cohen (1971) developed a model for predicting the flowering time for a single annual flowering plant. During the growing season of length T, the plant can devote energy to either vegetative growth or to seed production. The important variables in this model are

A = leaf biomass

R = net photosynthetic production per unit leaf mass

L = ratio of leaf mass to remaining vegetative mass

z = fraction of total plant growth allocated to reproduction

G = reproductive biomass.

Assuming that R and L are constants, the leaf growth and the reproductive growth of the plant are given by

$$\dot{A}(t) = RLA(t)[1-z(t)] \qquad (4.1)$$

$$\dot{G}(t) = RA(t)z(t) \qquad (4.2)$$

where the dot denotes differentiation with respect to time.

Under a growing season of fixed length T, Cohen demonstrated that the optimal strategy for maximizing seed production is to grow only vegetatively up to a time $u < T$ (i.e., $z(t) = 0$ for $t < u$) and then to cease vegetative growth and devote the remainder of the growing season to producing only seeds (i.e., $z(t) = 1$ for $t > u$). Integrating equation (4.1) from $t = 0$ yields

$$A(t) = A(o)\exp(RLt) \qquad (4.3)$$

Seed production as a function of u (the phenotype or strategy is identified by the value of u) can be obtained by substituting A(t) into equation (4.2) and then integrating from $t = u$ to $t = T$. This yields

$$G(u) = (T-u)RA(0)\exp(RLu) \tag{4.4}$$

Seed production is at a global maximum on the interval [0, T] when $u = u_1$ with u_1 defined by

$$u_1 = T - 1/RL \tag{4.5}$$

In general, of course, plants are growing in the company of others and competing with each other for water, nutrients, space and sunlight (Mirimani and Oster, 1978). The effect of this competition on an individual plant will depend on the size of other plants as well as their densities. We will assume here that the larger the plant the greater its competitive effect will be on its neighbors since its larger root system should draw resources more rapidly and its greater size should exert a greater shading effect. An individual plant's seed production is then not only going to be a function of its own size, as in the case of Cohen's model, but should also be a function of the size and number of neighboring plants. We can model the varying competitive effect of all neighboring plants on an individual as a continuous process and formulate the biological interaction as a game. We begin by developing a new qualitative model.

The inverse yield law (Kira et al., 1953) has been used successfully to relate plant density to yield in monoculture (Holliday, 1960). The relationship is expressed as

$$V = (a + b\ N)^{-1} \tag{4.6}$$

where V is the mean plant mass and N is plant density. Both a and b are constants and are dependent upon the particular plant species and the environmental conditions. Weiner (1982) has modified this expression to provide the yield of an individual as

$$G_i = \frac{S_i}{1+W} \tag{4.7}$$

where G_i is seed production of an individual identified here by the subscript i. The quantity S_i is the seed production of the i-th individual in the absence of competitors and W is the competitive effect of neighboring plants.

Equation (4.7) will be used to develop the payoff function for our model. Borrowing from Cohen's noncompetitive model, equation (4.4) suggests that S_i may be expressed more generally as

$$S_i = (T - u_i)f(u_i) \tag{4.8}$$

where u_i is the time when the i-th plant switches from vegetative to reproductive growth and $f(u_i)$ is a function which measures the plants ability to produce seeds given that flowering starts at time u_i. For Cohen's model, drop the subscript and it follows from equation (4.4) that $f(u) = RA(0)\exp(RLu)$. Equation (4.8) simply states that the product of the rate of seed production by the time available to produce seeds gives the expected number of seeds produced in the absence of competitors. Since $(T - u_i)$ measures the time remaining in the growing season to produce seeds, it follows that if $f(0)$ is small and $f(u_i)$ increases rapidly with u_i, the maximum value for S_i will occur between 0 and T.

Since $f(u_i)$ is a measure of the plant's size and photosynthetic capability, it is assumed that $f(\cdot)$ is a monotonic function of u_i, that is

$$\frac{df(u_i)}{du_i} > 0 \tag{4.9}$$

Note that we assume the function $f(\cdot)$ is the same for all plants. However, the argument u_i depends on the particular plant under consideration. The competitive effect W in equation (4.7) is not constant but depends on the control choices made by all the plants. That is, it depends on $[u_i;v]$ where v is the vector of switching times for all plants other than i. Combining equation (4.8) with (4.7) yields the i-th plant's payoff function

$$G_i(u_i;v) = \frac{(T-u_i)\ f(u_i)}{1 + W(u_i;v)} \tag{4.10}$$

Consider now some desirable properties of $W(u_i;v)$. If there are no other plants in the community, then there should be no effect of competitors. Furthermore a plant i should be able to diminish the effects of neighbors by flowering later and the competitive effect of neighbors should be intensified by postponing their flowering time. If each plant has the same flowering time, then the competitive effect should increase with an increase in this flowering time. We incorporate these properties with the following assumptions

$$W(u_i;v) = 0 \quad \text{if} \quad n = 1 \tag{4.11}$$

$$\frac{\partial W(s)}{\partial u_i} < 0 \tag{4.12}$$

$$\frac{dW(s)}{\partial s} > 0 \tag{4.13}$$

An example of a function possessing these properties which might also be useful to the problem at hand is

$$W(u_i;v) = \sum_{j=1}^{r} (T+u_j-u_i)g_j(u_j) \tag{4.14}$$

where $g_j(u_j)$ is an appropriate monotone increasing function of u_j with $g_j(u_j) = 0$ when $j = i$.

We will now examine flowering dates using model (4.10) for three different growing situations. The first situation is the same one considered before by Cohen, that of maximizing seed production for a single plant. We will then obtain a balanced ESS solution for a community of plants and compare both of these results with the optimum flowering date for a community of plants where community seed production is maximized. In what follows, it is convenient to define

$$E(s) \triangleq -f(s) + \frac{df(s)}{ds}(T-s) \tag{4.15}$$

With just a single plant ($n = 1$), we set $W(u_i;v) = 0$ in equation (4.10). If we then replace u_i by s, the necessary condition for the optimal flowering date is simply given by

$$E(s) = 0 \tag{4.16}$$

or alternatively

$$\frac{df(s)}{ds} = \frac{f(s)}{(T-s)} \tag{4.17}$$

Note that postponing the flowering date, increases the rate at which the plant can produce seeds but, at the same time, reduces the time left in the growing season to produce seeds. Equation (4.16) expresses the condition to maximize this product. This is similar to a well known result in economics associated with the cost of production. By producing more of a product, the average cost of production goes down but the marginal cost of production (i.e., instantaneous time rate of change of cost) goes up. Cost is minimized by producing at the point where marginal cost equals average cost. For plants, seed production is maximized when the marginal rate of increase in seed production equals the average rate of seed production taken with respect to the time remaining in the growing season. It is biologically reasonable to assume that at $s = 0$, the average is less than the margin, for otherwise the plant would devote no time to vegetative growth. With $f(\cdot)$ a finite monotonic function of s, and with the average approaching infinity as s approaches T, the solution to (4.17), say s^*, must satisfy $0 < s^* < T$.

Condition (2.1) with $\bar{\delta} = 0$ is satisfied by the model (4.10). To complete the requirements for a balanced game, we will assume the $W(u_i;v)$ function is such that condition (2.1) is satisfied with $\bar{\delta} \neq 0$.

The first derivative of (4.10) is given by

$$\frac{\partial G_i}{\partial u_i} = \frac{[1+W(u)][-f(u_i) + \frac{df(u_i)}{du_i}(T-u_i)] - (T-u_i)f(u_i)\frac{\partial W(u)}{\partial u_i}}{[1+W(u)]^2} \tag{4.18}$$

Substituting for each component of the vector u by the scalar s and setting the above expression equal to zero yields the following necessary condition

$$E(s) = \frac{\partial W(s)}{\partial u_i}\left[\frac{(T-s)\ f(s)}{1+W(s)}\right] \tag{4.19}$$

It follows from (4.12) that $E(s) < 0$ at a balanced ESS solution point. From the definition of $E(s)$, this requires

$$\frac{df(s)}{ds} < \frac{f(s)}{T-s} \tag{4.20}$$

Note that under our previous assumptions on $f(\cdot)$, equation (4.20) is satisfied only if a balanced ESS solution is later than s^*, the optimal flowering time for a single plant.

The second derivative condition (3.5) is also satisfied provided

$$\frac{dE(s)}{\partial u_i} \le G_i(s)\frac{\partial^2 W(s)}{\partial u_i^2}\ . \tag{4.21}$$

Clearly a sufficient condition for (4.21) to be satisfied is for the left-hand expression to be negative and the right-hand expression to be positive. If the curve describing $(T-u_i)f(u_i)$ is everywhere concave then $\frac{\partial E(s)}{\partial u_i} < 0$. We have already assumed $\frac{dW(s)}{du_i} < 0$. If the competitive effect of other plants with respect to u_i is reduced uniformly and asymptotically to a non-negative minimum, then $\frac{d^2W(s)}{du_i^2} > 0$.

It follows from (4.10) that the number of seeds produced per plant decreases with competition. This result is expected as the plants cannot avoid the interference of other plants through the term $W(u_i;v)$. The real question is how much of a penalty do plants pay under the balanced ESS solution as compared to a solution which maximizes the population's seed production? In order to obtain the solution which maximizes the total seed production for a community of plants, we first set each component of $[u_i;v]$ in equation (4.10) equal to s and then obtain the first derivative of $G_i(\cdot)$ with respect to s.

$$\frac{dG_i}{ds} = \frac{[1+W(s)][-f(s) + \frac{df(s)}{ds}(T-s)] - (T-s)f(s)\frac{dW(s)}{ds}}{[1+W(s)]^2} \tag{4.22}$$

Setting this derivative equal to zero, we obtain

$$E(s) = \frac{dW(s)}{ds}\left[\frac{(T-s)f(s)}{1+W(s)}\right] \tag{4.23}$$

It follows from (4.13) that $E(s) > 0$ at an optimal cooperative solution. From the definition of $E(s)$, this requires

$$\frac{df(s)}{ds} > \frac{f(s)}{T-s} \tag{4.24}$$

Equation (4.24) is satisfied at times earlier than s*, which is the optimal flowering time for a single plant. However, since we have obtained the maximum for $G_i(s)$, this solution will yield more seeds than the balanced ESS solution. Clearly this result is the desirable one in agriculture, however, it would have to be artificially maintained as this cooperative solution is not stable.

DISCUSSION

The analysis of the optimal flowering time presented here gives a general result and yields predictions which are qualitatively similar to those of Cohen (1971), Gadgil and Gadgil (1975) and Schaffer (1977). The balanced ESS solution is the point where the population can be expected to evolve in a natural enviornment. The result derived here is general in that a class of functions rather than individual functions are specified. We find that a population experiencing competition from neighbors should evolve to a flowering time later than that expected for an isolated plant. Furthermore, the flowering time which would maximize the seed production of the population is less than both the competitive and the isolated plant flowering times. Although the cooperative solution maximizes seed production, it is not evolutionarily stable and can be invaded by plants with a later flowering time. This may explain why certain hybrid agricultural grain crops are competitively inferior to wild forms.

REFERENCES

Auslander, D.J., Guckenheimer, J.M. and Oster, G. (1978). Random evolutionary stable strategies, *Theor. Pop. Biol.* 13, 276-293.

Cohen, D. (1971). Maximizing final yield when growth is limited by time or by limiting resources, *J. Theor. Biol.* 33, 299-307.

Gadgil, S.M. and Gadgil (1975). Can a single resource support many consumer species?, *J. Genet.*, 62, 33-47.

Holliday, R. (1960). Plant population and crop yield, *Nature*, 186, 22-24.

Kira, T., Ogawa, H. and Sakazaki, N. (1953). Intraspecific competition among higher plants. I. Competition yield-density interrelationships in regularly dispersed populations, *J. of the Institute of Polytechnics*, Osaka City University, Series D, 4, 1-26.

Lawlor, R.L. and Maynard Smith, J. (1976). The coevolution and stability of competing species, *Amer. Natur.* 110, 76-99.

Maynard Smith, J. and Price, G.R. (1973). The logic of animal conflicts, *Nature* 246, 15-18.

Maynard Smith, J. (1974). The theory of games and the evolution of animal conflicts, *J. Theor. Biol.* 47, 209-221.

Mirmimani, M. and Oster, G. (1978). Competition, kin selection and evolutionary stable strategies, *Theor. Pop. Biol.* 13, 304-339.

Nash, J.F. (1951). Non-cooperative games, *Ann. Math.* 54, 286-295.

Schaffer, W.M. (1977). Some observations on the evolution of reproductive rate and competitive ability in flowering plants, *Theor. Pop. Biol.* 11, 90-104.

Vincent, T.L. (1977). Environmental adaption by annual plants (An optimal controls/games viewpoint), *Lecture Notes in Control and Information Sciences*, 3, edited by P. Hagedorn, Springer-Verlag, Berlin.

Vincent, T.L. and Grantham, W.J. (1981). Optimality in Parametric Systems, John Wiley and Sons, Inc., New York.

Vincent, T.L. and Brown, J.S. (1983). Evolutionary stability using continuous parametric game theory, manuscript.

Von Neumann, J. and Morgenstern, O. (1944). Theory of Games and Economic Behavior, Princeton University Press, Princeton, NJ.

Weiner, J. (1982). A neighborhood model of annual-plant interference, *Ecology*, 63, 1237-1241.

EVOLUTIONARILY STABLE STRATEGIES FOR LARVAL DRAGONFLIES

Philip H. Crowley
T.H. Morgan School of Biological Sciences
University of Kentucky
Lexington KY 40506
U.S.A.

Many animals spend a considerable amount of time and effort watching, stalking, and attacking their neighbors even when it would seem to be mutually advantageous for them simply to ignore each other. A possible example of this paradoxical behavior, interference among dragonfly larvae, is analyzed from a game-theoretic viewpoint to see if such "strategies" appear to be evolutionarily stable. The results suggest that the ever-present possibility of ambush, in which the attacker has a significant chance of seriously injuring the victim, can culminate in "wars of attrition" or pre-emptive aggression by one or both neighbors. Testable hypotheses are presented, and the means of obtaining quantitative predictions from the theory are indicated.

INTRODUCTION

Much recent work by mathematical ecologists has attempted to identify the optimal morphological, physiological, or behavioral characteristics that organisms could have from some feasible character set. The resulting predictions can then be (and occasionally have been) compared with empirical observations as a test of the adequacy of the biological interpretation underpinning the optimization problem. The biological justification for expecting agreement between well-formulated predictions and empirical observations is generally that natural selection should tend to shift the population over evolutionary time toward those genotypes (and thus phenotypes) maximizing inclusive fitness. There are of course a number of possible reasons why the optimal solution may not be attainable, and at least some of these relate to the semantics of invoking the formal definition of "optimization". For instance, the game-theoretic outcome of two-player contests including those to be considered here is often non-optimal <u>in the mathematical sense</u>, even though each player is assumed to make the best possible response to an opponent. This is because the contingencies of the game can alter the stability topography and may shift the predicted behavior away from the high-fitness optimum for an unresponsive opponent toward an interactive solution that may lie on a lower fitness peak.

Though this general viewpoint offers no particular conceptual difficulty to evolutionary biologists, who are used to thinking in terms of interactive dynamics and relative rather than absolute fitness, its potential for generating testable explanations of apparently non-optimal biological characteristics is only just beginning to be realised. In particular, agonistic behavior sufficiently intense to sometimes cause serious injury (and related avoidance or placation behaviors) may be evolutionarily stable yet result in strikingly inefficient use of otherwise abundant resources.

The case in point, the agonistic interactions between larval dragonflies, may provide an example of this possibly widespread phenomenon. As a first step toward evaluating this interpretation, I postulate and analyze some plausible dragonfly strategies to look for evolutionary stability (cf. cases studied by Maynard Smith and Parker 1976, and by Maynard Smith 1982, p. 108ff; also see the example presented elsewhere in this volume by T.L. Vincent concerning the flowering phenology of annual plants). This presentation should also suggest some of the kinds of uses and difficulties to be found in the application of game theory to evolutionary biology.

Dragonflies (Odonata, including the damselflies) typically spend most of their lives as larvae in fresh-water habitats, passing through about twelve instars before emerging onto land, where mating occurs. During the larval period, individuals lengthen by more than an order of magnitude and increase in dry weight by more than three orders of magnitude (see Tillyard 1971, Corbet 1962, and Corbet 1980 for general information on the physiology and life-histories of dragonflies). Since dragonfly larvae are well known to be generalist predators on moving animals smaller or similar in size to themselves (Pearlstone 1973, Corbet 1980, Merrill 1980), the broad range of sizes reached during development suggests a high potential for inter- and intraspecific interference and predation.
The existence, though not the frequency, of intra-odonate predation is well documented (e.g. Corbet 1962, Macan 1964, Lawton 1970, Pearlstone 1973, Thompson 1978, Uttley 1980, and Merrill 1981). And recent field experiments (Benke 1978,

Benke et al 1982, Pierce et al in preparation, Johnson et al in preparation) indicate that larval dragonflies may inhibit each other's growth and survival both within and between species, an effect (apparently not attributable to prey depletion) that seems to implicate interference. But even where overtly aggressive interactions among odonate larvae are infrequent, the behavior of the animals can result in "passive interference" (Uttley 1980), in which prolonged staring and mutual avoidance can greatly restrict the feeding activities of the larvae. The analysis presented here bears directly on the question of why and when both active and passive interference would be expected to develop and persist under the influence of natural selection in populations of dragonflies. Since dragonflies are often significant and sometimes dominant predators in fresh-water littoral communities, understanding the mechanisms by which they interact may also provide considerable insight into the structure and dynamics of these communities.

In contrast to those of some terrestrial predators (e.g. spiders in Riechert 1978), the prey-hunting sites of dragonfly larvae are not well defined and characterized. The prevalent view in the literature seems to be that larval dragonflies have a limited number of fairly discrete "fishing sites" available within habitats, and that these tend to be occupied and actively defended by a relatively small proportion of the population (see Macan 1964, 1977). There appears, however, to be no direct evidence to support this "scarce, static, and discrete" characterization of hunting sites, and several lines of evidence indicate otherwise: the relative physical uniformity (fine grain) of microhabitats on the scale of an individual larva (e.g. bits of leaf material in a detrital deposit, or the regular geometry of stems in a rush bed); the difficulty of contriving discrete and static feeding sites in the laboratory (Baker 1980; J.H. Lawton, personal communication; and personal experience); and the notoriously high variances in time and space of invertebrate prey populations (Hassell 1978). Thus the kind of consistently recognizable owner-intruder asymmetry that gives rise to the "bourgeois strategies" considered by Maynard Smith (1982; see also

Maynard Smith and Parker 1976) seem inapplicable in the present study. And prey-hunting sites need not be assumed to be discrete entities, though this would apparently make little difference in the formulation.

So consider a number of dragonfly larvae moving occasionally and slowly within some fairly uniform habitat (see Crowley 1979, Savan 1979, Baker 1980; such behavior presumably permits slow dispersal while restricting exposure to other visual and tactile predators, notably vertebrates, and allowing moving prey to be ambushed). To keep the presentation straightforward, all larvae may be assumed conspecific, though relaxing this assumption introduces only minor complications, as noted later. When larvae are in the presence of abundant prey and few other odonates, movement is less frequent (Corbet 1962, Uttley 1980). But when prey are scarce or odonates abundant, dispersal becomes necessary. In either case, some encounters between odonate larvae will inevitably occur.

At least three kinds of encounters among these larvae can be usefully distinguished for present purposes: (1) confrontations (see Figure 1), initiated when two larvae orient toward each other at a safe distance outside the labial striking range of both larvae (each is assumed to establish the relative size of the opponent and to determine that it has been seen by the opponent); (2) ambushes, in which one larva strikes at the other while the other is not oriented toward its attacker, implying that only the attacker is able to estimate relative size; and (3) collisions, initiated when two larvae first orient toward each other within the striking distance of at least one of the larvae, and possibly triggering an attack in which neither individual has a good estimate of relative sizes. When light intensity and physical structure permit adequate visibility, confrontations will be relatively common. This kind of encounter involves the exchange of the most information between opponents, and the strategies employed are expected to be correspondingly more complex. Of particular interest for each confrontation strategy are its implications for ability and susceptibility to ambush. Thus most of the analysis focuses explicitly on confrontations, but ambushes and collisions are also taken into account.

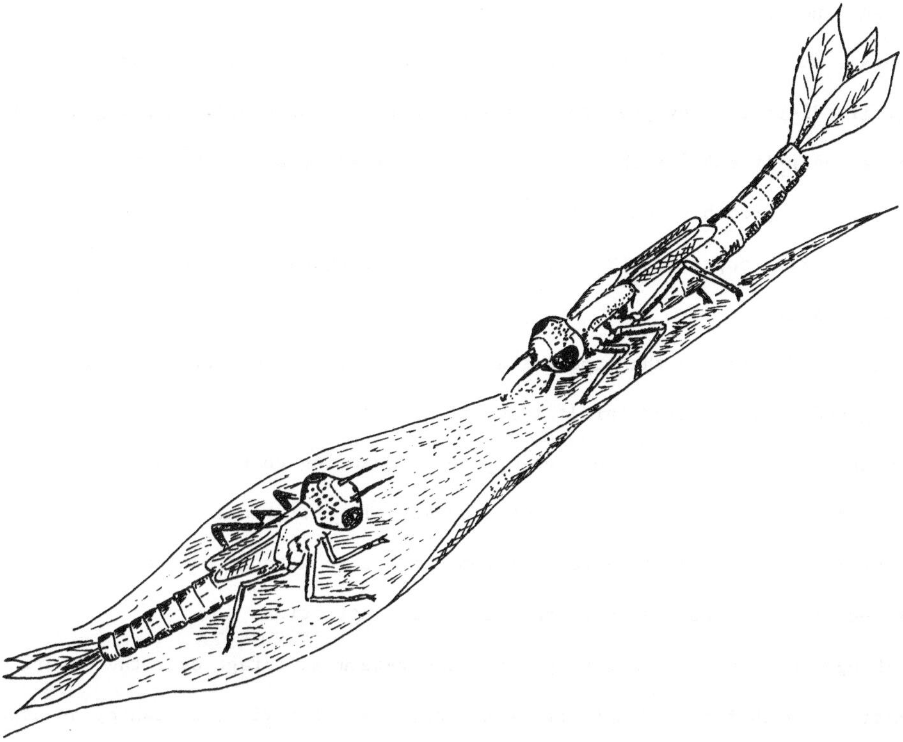

Figure 1. A confrontation between two coenagrinoid damselflies (Odonata: Zygoptera). The analysis presented here may apply best to this family, whose larvae have been observed most extensively in the laboratory; it may require some modification for rigorous application to the most actively foraging odonates, such as the lestids and aeshnids, but larvae from most of the other families should fit the assumptions and constraints of the model reasonably well.

SOME PLAUSIBLE STRATEGIES AND THEIR PAYOFFS

The two most obvious strategic possibilities for confrontations, equivalent to "hawk" and "dove" (Maynard Smith and Parker, 1976), are:

1. Escalate (E): Advance to attack; leave only if beaten in combat.

2. Retreat (R): Observe opponent without advancing to threaten; leave if opponent advances to threaten or if beaten in a war of attrition. (Wars of attrition are contests in which opponents compete by displaying or simply waiting for the other to leave. They typically result in negative-exponential frequency distributions of leaving times - see Maynard Smith (1982) for numerous examples and variations on this theme).

An alternative to retreating under pressure of an advancing opponent would be simply to leave as soon as the opponent is determined to be large enough to make winning unlikely. But while this strategy may offer better protection from the opponent and may initiate the search for a new site sooner, it cannot take advantage of the possible failure of the opponent to advance and threaten (e.g. if distracted by predators, other opponents, or prey). Leaving immediately also necessarily entails the risks and costs of additional movement, and the time saved is probably generally negligible relative to the expected time between encounters with opponents (T). Three other strategies that seem intuitively plausible are:

3. Ignore (O): Ignore the opponent and proceed with other activities (e.g. feeding and avoiding predators); leave if attacked.

4. Counter-escalate (C): Observe opponent advancing to threaten; counter-attack if attacked; leave if beaten in combat or war of attrition.

5. Intimidate (I): Advance to threaten but not to attack; counter-attack if attacked; leave if beaten in combat or war of attrition.

Though the ignore strategy has the short-term advantage over the others of permitting "high-fitness" activities to continue, it invites ambush by an opponent. It is assumed here that the best response to an ambush is an immediate withdrawal, since the victim has little or no information about the attacker (larger or smaller dragonfly? newt? fish?). Ambush thus involves minimal risk of counter-attack and probably increases the chance of damaging or consuming the victim. It follows that an opponent strictly adhering to the ignore strategy should eventually be ambushed by an opponent using any of the four strategies, even those that are otherwise non-aggressive. A non-negligible amount of time (T_1) may pass, however, before the ambush occurs.

The costs and gains associated with these five strategies are listed in Table 1; from these the payoff matrix in Table 2 has been constructed. It is immediately apparent from Table 2 that, though they are otherwise identical, "counter-escalate" (C) is inferior to "intimidate" (I) in confronting "retreat" (R). But the threatening implied by the I-strategy may entail some additional risk associated with movement that is not indicated in the table.

Table 1. Gains and costs associated with the strategies[1].

Gains

g_1, gain rate from successfully removing an opponent[2]

g_2, gain rate while ignoring an opponent[2]

G_3, incremental gain from successfully confronting an opponent[3]

G_4, incremental gain from successfully ambushing an opponent[3]

Costs

c_1, loss rate during time spent not feeding and/or not avoiding predators[2]

C_2, injury cost of losing in combat

C_3, injury cost of letting an opponent strike first in a confrontation

C_4, injury cost of letting an opponent strike first in an ambush

C_5, risk and energetic costs of moving to a new site

C_6, increased risk of ambush from ignoring an "ignore" strategist[4]

[1]All gains and costs are relative to the expected gain from other sites.

[2]When multiplied by the expected duration, these yield the cumulative component of gain or cost, such as that associated with feeding and avoiding predators, and are designated by lower case letters.

[3]Include gains from the chance of eating an opponent as a result of combat and should be decremented by any injury cost of winning in combat.

[4]Should be decremented by the gain from the chance of ambushing the opponent.

The evolutionary result could be a compromise between I and C involving an advance on the opponent just sufficiently menacing that an appropriately wary or rapid retreat by an R-strategist begins to be dangerous in itself. Therefore the C-strategy will hereafter be ignored in favor of a slightly conservative I-strategy. Table 2 also suggests that, except under rather special circumstances described in the next section, the "ignore" strategy will be inviable by virtue of its susceptibility to ambush.

Table 2. Payoff matrix for five strategies[1].

Superior Combatant

Inferior Combatant	E, escalate	I, intimidate	C, counter-escalate	O, ignore	R, retreat
E	g_1T+G_3 $-C_2-C_5$	$g_1T+G_3-C_3$ $-C_2-C_5$	$g_1T+G_3-C_3$ $-C_2-C_5$	$g_1T_1-C_2-C_4-C_5$ $g_1(T-T_1)+G_4-c_1T_1$	$-C_5$ g_1T
I	g_1T+G_3 $-C_2-C_3-C_5$	$\min(0,\frac{g_1T-C_5}{2})$	$\min(0,\frac{g_1T-C_5}{2})$	$g_1T_1-C_2-C_4-C_5$ $g_1(T-T_1)+G_4-c_1T_1$	$-C_5$ g_1T
C	g_1T+G_3 $-C_2-C_3-C_5$	$\min(0,\frac{g_1T-C_5}{2})$	$\min(0,\frac{g_1T-C_5}{2})$	$g_1T_1-C_2-C_4-C_5$ $g_1(T-T_1)+G_4-c_1T_1$	$\min(0,\frac{g_1T-C_5}{2})$
O	$g_1(T-T_1)+G_4-c_1T_1$ $g_1T_1-C_2-C_4-C_5$	$g_1(T-T_1)+G_4-c_1T_1$ $g_1T_1-C_2-C_4-C_5$	$g_1(T-T_1)+G_4-c_1T_1$ $g_1T_1-C_2-C_4-C_5$	g_2T-C_6	$g_1(T-T_1)+G_4-c_1T_1$ $g_1T_1-C_2-C_4-C_5$
R	g_1T $-C_5$	g_1T $-C_5$	$\min(0,\frac{g_1T-C_5}{2})$	$g_1T_1-C_2-C_4-C_5$ $g_1(T-T_1)+G_4-c_1T_1$	$\min(0,\frac{g_1T-C_5}{2})$

[1] T, expected time between encounters with opponents; T_1, expected time from beginning confrontation until successful ambush; the expected time to determine the outcome of a confrontation and the expected time to find a new site are assumed to be negligible relative to T. Other symbols are defined in Table 1. For double entries, the upper right expression applies to the superior combatant (i.e. the animal that would win if combat were to ensue) and the lower left to the inferior combatant; single entries apply to both. Entries are designated here by row-letter/column-letter, with subscripts for superior (s) or inferior(i) expressions where necessary. E/E_s is the gain for removing and possibly consuming the opponent; E/E_i entails both injury costs and the risks and costs of moving to a new site. E/I_s and E/C_s are the same as E/E_s, reduced by the cost of ceding the initiative to the opponent; E/I_i and E/C_i are identical to E/E_i. I/E_s and C/E_s (as for E/E_s) are the gain for removing and possibly consuming the opponent. I/E_i and C/E_i include the injury, initiative, and moving costs. For O/E, O/I, O/C, O/R, E/O, I/O, C/O, and R/O, the payoff for O is the feeding gain until the inevitable ambush, less the costs of injury, of allowing the opponent an ambush, and of moving to a new site; the payoff to the other strategy is the cumulative gain following ambush and the incremental gains from the ambush itself, less the cost of not feeding until the ambush. For R/E, R/I, E/R, and I/R, R bears the risks and costs of moving, and the other strategy gains from the removal of an opponent. The O/O payoff is a cumulative gain from feeding and predator-avoidance, less an increased cost of ambush risk. For I/I,I/C,C/I,C/C,C/R,R/C, and R/R, the outcome is assumed to be a war of attrition that lasts on average for the time t at which the expected payoff just falls to zero, i.e. $(g_1T-t)-C_5)/2-c_1t=0$, and $t = (g_1T-C_5)/(g_1+2c_1)$. When $g_1T\leq C_5$, the expected payoff is non-positive, and the outcome should be determined rapidly and randomly (a fast war of attrition?). Also note that if $g_1T\leq -C_5$, a new site should immediately be sought, whether or not an opponent is present.

Table 3. Abbreviated payoff matrix and expected payoffs for three strategies.[1]

Inferior Combatant	Superior combatant		
	E-escalate	I-intimidate	R-retreat
E	Q_6 Q_2	q_6 Q_2	Q_3 Q_5
I	Q_6 Q_1	Q_4	Q_3 Q_5
R	Q_5 Q_3	Q_5 Q_3	Q_4

$$Q_6 > q_6;\ Q_6 > Q_5 > Q_4 > Q_3 > Q_2 > Q_1$$

Expected Payoffs

$E_E(E) = pQ_6 + (1-p)Q_2$	$E_I(E) = pQ_6 + (1-p)Q_2$	$E_R(E) = Q_5$
$E_E(I) = pq_6 + (1-p)Q_1$	$E_I(I) = Q_4$	$E_R(I) = Q_5$
$E_E(R) = Q_3$	$E_I(R) = Q_3$	$E_R(R) = Q_4$

[1]The Q's represent the corresponding payoffs in Table 2. For double entries, the upper-right expression applies to the superior combatant and the lower left to the inferior combatant; single entries apply to both. The subscripts and cases of the letters indicate the relative magnitudes of the payoffs as shown. For the expected payoffs, the strategy in parentheses played against the subscripted strategy yields the indicated payoff, where p is the chance of being the superior combatant and depends on relative size.

An abbreviated payoff matrix and the expected payoffs for the remaining three strategies are presented in Table 3. The chance of winning in combat (p), along with the appropriate combinations of costs and gains, determines which strategy or strategies are evolutionarily stable in a particular case. (The critically important relationship between p and the relative sizes of the opponents is considered below). The specific combinations of costs and gains involved in establishing the ESS's are expressed by the lumped parameters δ and γ.

γ is the relative cost of injury in combat; and δ is the relative cost of losing in combat, including both the cost of injury and of risks associated with dispersal.

THE EVOLUTIONARILY STABLE STRATEGIES

An evolutionarily stable strategy (ESS) is a strategy from a set of alternatives that cannot be displaced by any initially rare alternative (cf. Maynard Smith and Price 1973); displacement is assumed to be an evolutionary process resulting from differences in fitnesses expressed here as payoffs. Thus from γ, δ, and the expected payoffs of Table 3, the ESS distributions on p-axes shown in Figure 2 have been obtained. The derivations are presented in the appendix.

The results can be summarized as follows: When the chance of winning in combat is high, the ESS is to escalate; and when the chance of winning in combat is low, the ESS is to retreat. For individuals having similar chances of winning in combat, if the benefits of winning in combat exceed the injury cost of losing, then the ESS is generally to engage in combat; but if the injury cost exceeds the benefits of winning, then the ESS is generally a waiting contest at close range. (cf. Maynard Smith 1982, p.16). Note in particular that the benefits of winning in combat will increase with hunger, which is thus of primary importance in determining the appropriate response to an opponent. Though more complex patterns involving combat and waiting could develop in certain circumstances (especially when the cost of moving is relatively high), the increased possibility of accidentally inappropriate behavior (see the caption of Figure 2) suggests that the simplest patterns will persist (Parker and Rubenstein 1981, Hammerstein and Parker 1982, Maynard Smith 1982).

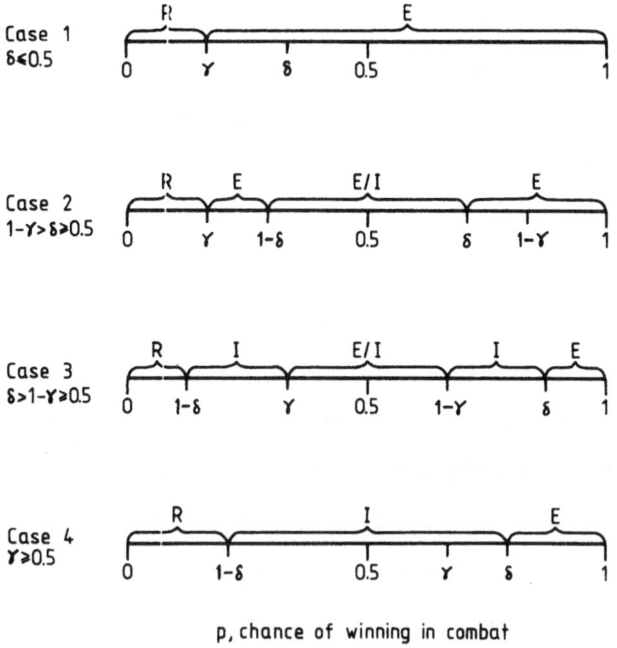

Figure 2. Evolutionarily stable strategies for the four possible cases. E is "escalate", I is "intimidate", and R is "retreat"; the ESS strategy depends on the chance of winning in combat p and on the magnitudes of the payoff parameters γ and δ. $\gamma = (Q_3-Q_2)/(Q_6-Q_2) = C_2/(g_1T+G_3+C_2+C_5)$; and $\delta = (Q_4-Q_2)/(Q_6-Q_2) = (C_2+C_5)/(g_1T+G_3+C_2+C_5)$, assuming $Q_4= 0$ (see Tables 1-3). The regions designated E/I are stable as any combination of E and I subregions in which the strategy at p is identical to the strategy at 1-p. The potentially complex patterns that could thus be found within these regions seem less plausible when the possibility of mistakes is taken into account. This view then suggests that the number of "switchovers" on the p-axis should be minimized in such cases, and thus that the central region will be exclusively E in the second case and exclusively I in the third. If so, then these four collapse to two distinct cases, in which the relative magnitudes of δ and 1-γ (or equivalently γ and 1-δ) are critical. In other words, if the injury cost of losing in combat (C_2) exceeds the gains of winning in combat (g_1T+G_3), then an "intimidation" strategy is stable for p~0.5 (i.e. opponents of similar size). but if the gains of winning exceed the cost of injury, then the stable behavior for similar-sized opponents is combat.

In order to postulate which of the four cases of Figure 2 are most likely to apply in nature, some idea of the relative magnitude of g_1T, G_3, C_2, and C_5 must be obtained. The gain rate g_1 exceeds zero to the extent that the site offers better than average feeding opportunities and predator protection. (Note that if prey are equally abundant or superabundant and predators equally dangerous at all sites, then g_1=0. High values of g_1 may result in low values of T, since a desirable site should attract intruders. G_3 will also shrink if prey are abundant, because the chance of consuming the opponent is of less consequence. The total cost of injury in combat and of being forced to disperse seems likely to be generally higher than the sum of the gains, largely because of the quite considerable chance that such injuries and the increased risk of predation can jeopardize survival or ultimate reproductive success. The most common situation in nature may therefore be $C_2 > C_5 + g_1 T + G_3$ (case 4), though it also seems probable that all of the other cases do occur in nature as well.

A considerably more cumbersome ESS analysis that includes the "ignore" strategy (0) can only be roughly summarized here: "Ignore" can at least replace "retreat" in the results of Figure 2 if $g_1 T_1 > C_2 + C_4 + C_5$; from the preceding discussion, however, this seems unlikely because the costs of losing in combat, particularly via ambush or vertebrate predation, are doubtless often severe. But if $g_1 T_1 < C_2 + C_4 + C_5$ and if $T_1(g_1 + c_1) > T(g_1 - g_2) + G_4 + C_6$, then both "ignore" and the results of Figure 2 are evolutionarily stable; if the second condition also fails then the 0-strategy cannot persist. The second condition suggests that 0 is most likely to be viable when encounters with opponents are frequent (low T, which reduces the payoff for ambushing an 0-strategist and may also imply that ambushes tend to be disrupted) and when prey are abundant (g_1 and g_2 are similar in magnitude, c_1 is large, and G_4 and C_6 are relatively small). But even if this condition is satisfied, it remains unclear that the 0-strategy could replace the more broadly applicable and similarly stable strategies of Figure 2.

It is the apparent failure of the "ignore" strategy to dominate the dangerous (E) and inefficient (I) strategies for individuals of comparable abilities in combat that can be considered "non-optimal" behavior in the sense discussed in the introduction. Thus the danger of ambush by a nearby individual results in ESS's with payoffs generally lower than they would be if the two individuals simply ignored each other. Though the possibility of local prey depletion (see Free et al. 1977 on "pseudointerference") may enhance the desirability of the ESS strategies, the general conclusion appears not to depend on this.

AMBUSHES AND COLLISIONS

Ambush apparently increases dramatically the attacker's chance of driving off or consuming an opponent over the chance of doing so in a confrontation (personal observations). It is primarily the skills at ambush of dragonflies and other inhabitants of the fresh-water littoral that can make moving around there so dangerous for insects and other small invertebrates. This is particularly true when visibility is adequate, since larval dragonflies can visually detect moving objects (such as advancing dragonflies) much more readily than stationary ones (such as another dragonfly waiting in ambush) (Corbet 1962). Thus many larval odonates confine most of their active searching activities associated with foraging and dispersal to the night (Corbet 1962). Nothing resembling a confrontation between larvae as defined here seems likely to occur in the dark. Though ambushes doubtless still occur in darkness, when increased activity may more or less compensate for the reduced reactive distance, any given ambush under these conditions seems somewhat less likely to result in injury or consumption of the victim. Collisions should become relatively common in the dark, but the lack of information conveyed about the chance of winning in combat may encourage a hasty disengagement. Overall then, most of the serious injuries caused by agnostic interactions between dragonfly larvae probably result from ambushes, a factor that looms large in determining the cost of losing in a confrontation. Combat is a credible ESS in confrontations only for the two cases (1 and 2) in which the injury costs of combat are exceeded by the gains of winning (see Figure 2).

RELATIVE SIZE AND OTHER ASYMMETRIES

Contests between two individuals capable of seriously injuring each other tend to be settled by asymmetries - discernable differences between the contestants - as long as the asymmetries can be mutually detected with high fidelity (Maynard Smith and Parker 1976). The fidelity requirement casts considerable doubt on the possible efficacy of an owner-intruder asymmetry for dragonfly larvae, as previously noted. But body size, as perhaps indicated by head width, head cross-sectional area, or eye size, seems certain to be involved in any visual assessment of an opponent, especially since size is generally a good indicator of resource-holding potential (see Maynard Smith 1982) and more specifically of the chance of winning in combat (see Uttley 1980). Relative size, expressed as the size ratio of the opponents, probably conveys almost as much information about the chance of winning as do the the absolute sizes, at least in many arthropods (see Uttley 1980 , Sigurjónsdóttir and Parker 1981). A logarithmic transformation is useful to linearize increments of relative size, particularly when these increments can roughly correspond to the geometrical progression of discrete life stages in arthropods known as instars. A sigmoid relationship between the chance of winning in combat (p) and the log size ratio (r) seems highly probable (Figure 3). Sufficient empirical observations to establish the shape of this curve and the magnitudes of the parameters γ and δ would permit quantitative predictions about the outcome of confrontations to be made from the results illustrated in Figure 2.

An important feature of this sigmoid curve also illustrated in Figure 3 is the tendency of uncertainties in the assessment of relative size to bias the p-estimate toward 0.5, equal chances of winning and losing. Thus when visibility is poor or perhaps in interactions between different species, any two opponents should tend to see themselves as being more similar in fighting ability. But when essentially no information is available about relative size (e.g. collisions in the dark), the best guess would come from the individual's accumulated information about its own size relative to the size-frequency distribution of the population. Though an opponent's size can probably be estimated quickly and accurately in most confrontations, an adequate assessment of an individual's own size (needed in any case to determine p) and of the population's size-frequency distribution must depend largely on experience. (Other clues to an individual's own size may also be available, including genetic, developmental, and physiological information; see Alcock, Jones and Buckman (1975) on size-related mating strategies of bees). There is reason to believe that such information and the characteristics of the ESS itself can generally be learned quite quickly even without the benefit of a sophisticated nervous system (Hartley 1981).

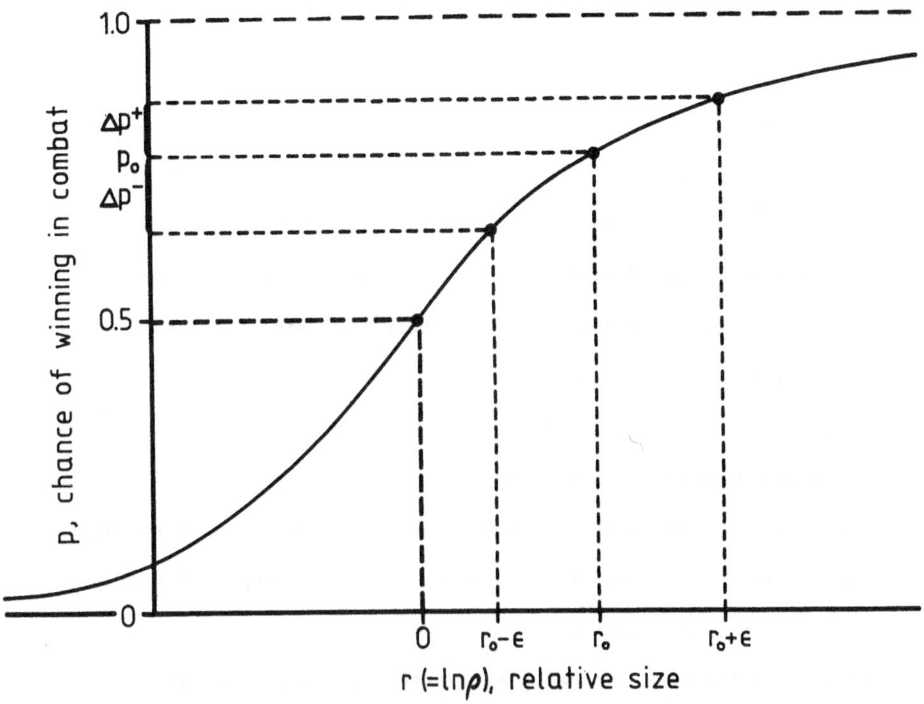

Figure 3. The relationship between the chance of winning in combat and the relative size of the combatant. Relative size r is the logarithm of the size ratio ρ of one combatant to the other and should be linearly related to differences in instar number for arthropods that increase in size geometrically with instars. The sigmoid curve results from the greater sensitivity of ρ to incremental changes in relative size in the region where sizes are most similar (see Jakobsson et al. 1979 for an extreme example with fish); where sizes differ considerably, the same incremental change in relative size can make little difference in p, since the curve presumably approaches p=0 and p=1 asymptotically. The symmetry of the limbs of the curve about r=0 and p=0.5 follows from the identity $p(\ln\rho) \equiv p(\ln(1/\rho))$.

Characteristics of dragonfly larvae other than size could also influence their estimates of the chance of winning in combat. Behavior or morphology conclusively associated with particular species, and the morphological changes associated with molting or emergence are possible examples. Certain other asymmetries that are likely to be at least of second-order importance - such as differences in hunger level or in information accumulated about the desirability of the site - do more than simply altering p, because the status of these asymmetries cannot be unequivocally known to both opponents. Some progress has been made in understanding the general implications of such asymmetries (see Maynard Smith 1982), though no attempt has been made to deal with them here.

SOME TESTABLE HYPOTHESES

In addition to the previously mentioned possibility of deriving quantitative predictions from estimates of γ, δ, and $p(r)$, several qualitative testable hypotheses emerge from the foregoing analysis and provide a coda:

1. When densities are sufficient that individuals are often close enough to detect each other, they should spend much of their time watching, stalking, and attacking each other rather than feeding.
2. Increasing uncertainty about relative size (e.g. when visibility is poor) biases the size estimate toward similarity.
3. Increasing the relative value of a site, decreasing the cost of losing in combat, or increasing the danger of attempting to go elsewhere all increase the range and frequency of aggression.
4. Small changes in the costs or gains associated with a particular site sufficient to shift the balance between the spoils of victory and the cost of defeat can shift behavior sharply over a wide range of sizes.

ACKNOWLEDGEMENTS

My thanks to Maureen Chapman for drafting Figures 2 and 3, to John Lawton for drawing Figure 1 and reading the manuscript, and to Sylvia Hogarth, Dan Johnson, Malcolm MacGarvin, and Mark McPeek for reading the manuscript. Special thanks to John Lawton, Mark Williamson, and the faculty and students in the Department of Biology at the University of York for their hospitality during a stimulating sabbatical year. This research was supported by NSF grant DEB 81-04424 and by a travel grant from the University of Kentucky Research Foundation.

LITERATURE CITED

Alcock, J., Jones, C.E. and Buckman, S.L. (1977). Male nesting strategies in the bee Centris pallida Fox (Anthophoridae: Hymenoptera). Am. Nat. 111: 145-155.

Baker, R.L. (1980). Use of space in relation to feeding areas by zygopteran nymphs in captivity. Can. J. Zool. 58: 1060-1065.

Benke, A.C. (1978). Interactions among co-existing predators - a field experiment with dragonfly larvae. J. Anim. Ecol. 47: 335-350.

Benke, A.C., Crowley, P.H. and Johnson, D.M. (1982). Interactions among co-existing larval Odonata: An in situ experiment using small enclosures. Hydrobiologia, in press.

Corbet, P.S. (1962). A Biology of Dragonflies. Quadrangle Books, Chicago.

Corbet, P.S. (1980). Biology of Odonata. Ann. Rev. Entomol. 25: 189-217.

Crowley, P.H. (1979). The behavior of zygopteran nymphs in a simulated weed bed. Odonatologica 8: 91-101.

Free, C.A., Beddington, J.R. and Lawton, J.H. (1977). On the inadequacy of simple models of mutual interference for parasitism and predation. J. Anim. Ecol. 46: 543-554.

Hammerstein, P. and Parker, G.A. (1982). The asymmetric war of attrition. J. Theor. Biol. 96: 647-682.

Hartley, C.B. (1981). Learning the evolutionarily stable strategy. J. Theor. Biol. 89: 611-633.

Hassell, M.P. (1978). The Dynamics of Arthropod Predator-Prey Systems. Princeton Univ. Press, Princeton.

Lawton, J.H. (1970). Feeding and assimilation in larvae of the damselfly Pyrrhosoma nymphula (Sulzer) (Odonata: Zygoptera). J. Anim. Ecol. 39: 669-689.

Macan, T.T. (1964). The Odonata of a moorland fishpond. Int. Revue ges. Hydrobiol. 49: 325-360.

Macan, T.T. (1977). The influence of predation on the composition of fresh-water animal communities. Biol. Rev. Cambridge Philos. Soc. 52: 45-70.

Maynard Smith, J. (1982). Evolution and the Theory of Games. Cambridge Univ. Press, Cambridge.

Maynard Smith, J. and Parker, G.A. (1976). The logic of asymmetric contests. Anim. Behav. 24: 159-175.

Maynard Smith, J. and Price, G.R. (1973). The logic of animal conflict. Nature, Lond. 246: 15-18.

Merrill, R. (1981). A comparison of the diets of dragonfly larvae (Odonata: Anisoptera) coexisting in an allochthonous detritus habitat. M.S. thesis, E. Tenn. State Univ., Johnson City.

Parker, G.A. and Rubenstein, D.I. (1981). Role assessment, reserve strategy, and acquisition of information in asymmetric animal conflicts. Anim. Behav. 29: 135-162.

Pearlstone, P.S.M. (1973). The food of damselfly larvae in Marion Lake, British Columbia. Syesis 6: 33-39.

Riechert, S.E. (1978). Games spiders play: behavioral variability in territorial disputes. Behav. Ecol. Sociobiol. 3: 135-162.

Savan, B.I. (1979). Studies on the foraging behavior of damselfly larvae (Odonata: Zygoptera). D. Phil. thesis, Univ. of London.

Sigurjónsdóttir, H. and Parker, G.A. (1981). Dung fly struggles: evidence for assessment strategy. Behav. Evol. Sociobiol. 8: 219-230.

Thompson, D.J. (1978). The natural prey of larvae of the damselfly, Ischnura elegans (Odonata: Zygoptera). Freshwat. Biol. 8: 377-384.

Tillyard, R.J. (1917). The Biology of Dragonflies (Odonata or Paraneuroptera). Cambridge Univ. Press, Cambridge.

Uttley, M.G. (1980). A laboratory study of mutual interference between freshwater invertebrate predators. D. Phil. thesis, Univ. of York.

APPENDIX

DERIVATIONS OF THE EVOLUTIONARILY STABLE STRATEGIES OF FIGURE 2, BASED ON THE PAYOFFS OF TABLE 3

Case 1. $\delta \leq 0.5$

$p>\delta$[1]

$E_E(E)>E_E(I)\sim E_E(R)$

$E_I(E)>E_I(I)>E_I(R)$

$E_R(E)=E_R(I)>E_R(R)$

E is the ESS because it yields the largest payoff regardless of the opponent's strategy[2].

[1] $p = \delta$, $p = \gamma$, and $\delta = 1-\gamma$, since they are of negligible biological significance, are omitted to simplify the presentation. The matrix

$$\begin{bmatrix} p & p & \delta^+ \\ p^- & \delta & \delta^+ \\ \gamma & \gamma & \delta \end{bmatrix}$$

, where $\delta^+ = (Q_5-Q_2)/(Q_6-Q_2)$ and $p^- = (pq_6+(1-p)Q_1-Q_2)/(Q_6-Q_2)$, is obtained from the expected payoffs in Table 3 by subtracting Q_2 and then dividing by Q_6-Q_2. This alternative form simplifies the algebra somewhat.

[2] Because $E_R(E)=E_R(I)$ and R is favored for $p<\gamma$, I could appear for $p>\delta$ and increase in frequency by drift - but this of course still produces the same result in an encounter with R. If I for $p>\delta$ were to become frequent enough to favor I at $p<\delta$, E at $p>\delta$ would then increase in frequency and restore R at $p<\gamma$. This kind of "wobble" in the ESS may allow I to persist at low frequency for $p<\gamma$ and result in occasional wars of attrition between individuals with quite different chances of winning in combat. There are clearly several other ESS's described here with comparable potential for "wobble".

$\gamma<p<\delta$	$E_E(E)>E_E(I)\sim E_E(R)$	E is the ESS - only the payoffs against E are relevant, since the ESS for the 1-p opponent is E, as shown above.
$p<\gamma$	$E_E(R)>E_E(E)>E_E(I)$	R is the ESS, since the opponent's ESS is E

Case 2. $1-\gamma>\delta\geq 0.5$[1]

$p>\delta$	$E_E(E)>E_E(I)\sim E_E(R)$ $E_I(E)>E_I(I)>E_I(R)$ $E_R(E)=E_R(I)>E_E(R)$	E is the ESS, as for Case 1.
$\delta>p>1-\delta$	$E_E(E)>E_E(R)\sim E_E(I)$ $E_I(I)>E_I(E)>E_I(R)$ $E_R(E)=E_R(I)>E_R(R)$	E and I are both ESS's and could evolve any pattern such that the strategies at p and 1-p are the same (but see the text). $E_I(E)>E_E(I)$ implies that E has a larger domain of attraction (Maynard Smith 1982), but $E_I(I)>E_E(E)$ indicates that I is a stronger attractor.
$1-\delta>p>\gamma$	$E_E(E)>E_E(R)\sim E_E(I)$	E is the ESS, as for $\gamma<p<\delta$ in Case 1.
$p<\gamma$	$E_E(R)>E_E(E)>E_E(I)$	R is the ESS (see Case 1).

Case 3. $\delta>1-\gamma\geq 0.5$

$p>\delta$	$E_E(E)>E_E(I)\sim E_E(R)$ $E_I(E)>E_I(I)>E_I(R)$ $E_R(E)=E_R(I)>E_R(R)$	E is the ESS, as for Cases 1 and 2.
$\delta>p>1-\gamma$	$E_E(E)>E_E(R)\sim E_E(I)$ $E_I(I)>E_I(E)>E_I(R)$ $E_R(E)=E_R(I)>E_R(R)$	I is the ESS by the following argument (refer to $\gamma>p>1-\delta$ below): Suppose E were fixed here and for $\gamma>p>1-\delta$. Then R would invade and take over for $\gamma>p>1-\delta$, since $E_E(R)>E_E(E)$. I could then drift into $\delta>p>1-\gamma$, favoring I to begin a takeover for $\gamma>p>1-\delta$. I is the inevitable result in both regions, as can also be shown for any other initial frequencies.

$1-\gamma>p>\gamma$	$E_E(E)>E_E(R)\sim E_E(I)$ $E_I(I)>E_I(E)>E_I(R)$ $E_R(E)=E_R(I)>E_R(R)$	E and I are both ESS's, just as for $\delta>p>1-\delta$ in Case 2.
$\gamma>p>1-\delta$	$E_E(R)>E_E(E)>E_E(I)$ $E_I(I)>E_I(R)>E_I(E)$ $E_R(E)=E_R(I)>E_R(R)$	I is the ESS - see $\delta>p>1-\gamma$ above.
$p<1-\delta$	$E_E(R)>E_E(E)>E_E(I)$	R is the ESS, as for $p<\delta$ in Cases 1 and 2.

Case 4. $\gamma\geq 0.5$

$p>\delta$	$E_E(E)>E_E(I)\sim E_E(R)$ $E_I(E)>E_I(I)>E_I(R)$ $E_R(E)=E_R(E)>E_R(R)$	E is the ESS - see Cases 1 - 3.
$\delta>p>\gamma$	$E_E(E)>E_E(R)\sim E_E(I)$ $E_E(I)>E_I(E)>E_I(R)$ $E_R(E)=E_R(I)>E_R(R)$	I is the ESS - see $\delta>p>1-\gamma$ in Case 3.
$\gamma>p>1-\gamma$	$E_E(R)>E_E(E)>E_E(I)$ $E_I(I)>E_I(R)>E_I(E)$ $E_R(R)=E_R(I)>E_R(R)$	I is the ESS, since it is the only strategy with a higher payoff than the others against itself.
$1-\gamma>p>1-\delta$	$E_E(R)>E_E(E)>E_E(I)$ $E_I(I)>E_I(R)>E_I(E)$ $E_R(E)=E_R(I)>E_R(R)$	I is the ESS - see $\delta>p>1-\gamma$ in Case 3.
$p<1-\delta$	$E_E(R)>E_E(E)>E_E(I)$	R is the ESS, as for Case 3.

PART II
POPULATION BIOLOGY

THE STORAGE EFFECT IN STOCHASTIC POPULATION MODELS

Peter L. Chesson
Department of Zoology
The Ohio State University
1735 Neil Avenue
Columbus, Ohio 43210 U.S.A.

Introduction

Many populations occur in environments that vary substantially in time. Some of this variation is regular, for example, seasonal variation, and some of it is stochastic, i.e., has a strong element of randomness. Although I shall focus mainly on this stochastic variation there are many parallels between the effects of regular and stochastic variation and some of the mathematics below holds equally well for both kinds of temporal environmental variation.

The presence of significant stochastic variation suggests that stochastic models, not deterministic models, should be used to describe population dynamics. Yet they rarely are. The lack of emphasis on stochastic models seems in part due to the difficulty in analyzing stochastic models and in part due to the feeling that either stochasticity is mostly noise obscuring a deterministic signal or its effect is one that destabilizes systems. Discussions of these views are to be found in Goh (1976), Turelli (1978a), Murdoch (1979), Chesson and Warner (1981), and Chesson (1982). The perception that environmental variability is destabilizing might well have encouraged the use of stochastic models; instead it has lead to an emphasis on features in deterministic models that ought to prevent destabilization by stochasticity (Beddington et al. 1976, Goh 1976). However, if destabilization is equated with the likely extinction of one or more species, then the conclusion that stochasticity is generally destabilizing is not correct. Recently there have appeared a number of models in which the effect of stochasticity is to promote coexistence of competing species (Chesson and Warner 1981, Chesson 1982, Chesson, in press, Ellner, in press, Shmida and Ellner, in press, Abrams, ms). Indeed, stochastic models have a variety of interesting and important behaviors that cannot be guessed from their deterministic counterparts. Thus no longer is there any good justification for the low emphasis on stochastic models.

There is a particular broad class of models where a stochastic environment seems especially likely to promote coexistence of competing species. Models in this class contain a feature that has been called the storage effect (Chesson, in press, Warner and Chesson, ms).

The Storage Effect

For many organisms the life cycle is naturally divisible into pre-reproductive individuals (juveniles) and reproductive individuals (adults). Maturation to an adult is called recruitment and the per capita number of new adults appearing in a unit of time is called the recruitment rate. The environment can affect adults and juveniles quite differently. In many organisms, highly variable birth rates and juvenile survivorship rates lead to highly variable recruitment rates. On the other hand, adult survivorship may be relatively constant and much higher than juvenile survivorship (Chesson, in press, Warner and Chesson, ms). These features are especially well documented for fish populations (Gulland 1982) and perennial plant populations (Grubb 1977, Harper 1977, Hubbell 1980).

To model the consequences of variable recruitment rates and low, less variable, adult death rates, let $X_i(t)$ be the adult population size of species i at time t, δ_i the death rate of adults, and $R_i(t)$ the recruitment rate. With these definitions

$$X_i(t+1) = (1-\delta_i)X_i(t) + R_i(t)X_i(t). \tag{1}$$

The recruitment rate is assumed to take the form

$$R_i(t) = f_i(\xi(t), X_1(t), X_2(t), \ldots, X_n(t)) \tag{2}$$

where f_i is some function of the randomly varying environment, $\xi(t)$, and the adult densities of the n species in the system. For simplicity, the death rate is constant in this system. However, provided adult death rates are small, the results of Chesson and Warner (1981) and Chesson (1982) extend to show that moderate adult death rate variation does not qualitatively affect the results we obtain below. For definiteness we shall generally assume that the environment process $\xi(0)$, $\xi(1)$, ... is an independent and identically distributed (i.i.d.) sequence, but many features of our analysis hold more generally.

To investigate the consequences of this model for population growth, we define $\rho_i(t) = R_i(t)/\delta_i$ so that

$$X_i(t+1)/X_i(t) = 1 + \delta_i(\rho_i(t)-1). \quad (3)$$

Thus species i increases or decreases depending on whether $\rho_i(t) > 1$ or < 1. Such situations of population increase and decrease will be referred to respectively as favorable and unfavorable periods. According to the invasibility criterion (Turelli 1978b) species i will persist in the system if

$$E \log\{1 + \delta_i(\rho_i(t) - 1)\} > 0 \quad (4)$$

when this expression is evaluated at 0 density for species i. (It is usually assumed that the system has a unique stationary distribution in the absence of species i and (4) is evaluated for this stationary distribution.)

The LHS of (4) is essentially the mean instantaneous growth rate of species i at low density and it has some quite remarkable properties. For example, a sufficient, but by no means necessary condition for (4), and hence for persistence is

$$E\left[\log \rho_i | \rho_i > 1 \right] > c(\delta_i)P(\rho_i < 1)/P(\rho_i > 1) \quad (5)$$

(Chesson, in press), where $c(\delta_i) = -\delta_i^{-1}\log(1-\delta_i) \simeq 1$ for small δ_i. This sufficient condition involves only the magnitude of ρ_i during favorable periods: no account is taken of what happens during unfavorable periods. In other words, for a given frequency of occurrence of favorable periods, a species can persist provided sufficient benefit in terms of recruitment is derived during favorable periods, independently of the magnitude of the recruitment costs that are incurred during unfavorable periods. This is not to say that the costs during unfavorable periods are unimportant, but simply that they are less important than the benefits gained during favorable periods.

Some idea of the effect of unfavorable periods can be gained from the approximation

$$E \log\{1 + \delta_i(\rho_i(t) - 1)\} \simeq \delta_i\{E\rho_i - 1\} \quad (6)$$

which holds for small δ_i or situations where ρ_i takes only values near 1. Recalling that $E[X;A] = E[X|A]P(A)$, the persistence criterion becomes

$$E[\rho_i; \rho_i < 1] + E[\rho_i; \rho_i > 1] > P(\rho_i \neq 1). \tag{7}$$

Clearly both favorable and unfavorable periods contribute to this inequality. In particular, values of ρ_i near 1 during unfavorable periods mean that favorable periods do not have to be strongly favorable for the persistence criterion to be satisfied. However, it is equally clear that favorable periods have a potentially larger effect: while 0 is a lower limit to ρ_i, so that $E[\rho; \rho_i < 1] \geq 0$, in general there is no reason for any particular upper limit to ρ_i. Thus $E[\rho_i; \rho_i > 1]$ has the potential to be large, and will be large if strong recruitments are not too infrequent.

What we have seen is that in populations with small adult death rates and variable recruitment rates, there is a definite asymmetry in the effects of favorable versus unfavorable periods. Strong recruitments contribute quite significantly to population growth while poor recruitments can be made arbitrarily poor without causing much decline in the mean growth rate of the population. This asymmetry between the effects of favorable versus unfavorable periods is called the storage effect because it comes about by summation or "storage" of recruitment in the adult population: the adult population consists of the sum of past recruitments, each being discounted every time unit by the adult death rate. The fact that the adult population is essentially a sum over a number of periods of recruitment diminishes the harm that occurs whenever recruitment fails. The storage effect is present whenever generations are overlapping, i.e., whenever $\delta_i < 1$, but it is strongest when adult death rates are small, for then the effects of favorable periods of recruitment persist for a long time.

The Storage Effect as a Mechanism of Coexistence

Consider a group of species that compete quite strongly, but assume that this competition mostly affects recruitment of juveniles to the adult population. Then if adults are competing, the number and vigor of their offspring are affected by competition, while competition among juveniles affects their chances of surviving to adulthood. If competitive dominance varies through time from species to species, each species is able to have periods of strong recruitment and the asymmetry imparted by the storage effect means that periods of poor recruitment, when a species is competitively inferior, need not cancel out the effects of these favorable periods. In this way the storage effect favors coexistence.

To demonstrate the potential role of the storage effect in promoting coexistence in a variety of competition models we take a conservative approach using

$$R_i'(t) = \inf_{\underset{\sim}{X} \mid X_i = 0} f_i(\mathfrak{E}(t), \underset{\sim}{X}) \tag{8}$$

which is the minimum value of the recruitment rate that will occur for a given state of the environment with species i at 0 density. We are considering a worst case situation for the possible effects of the other species.

If $R_i'(t)$ can be made arbitrarily large by a suitable choice of values of the environment $\mathfrak{E}(t)$, then $\rho_i(t)$ can take on arbitrarily large values, and if there are such values of $\mathfrak{E}(t)$ for each species, then there are probability distributions for $\mathfrak{E}(t)$ such that the species coexist. For example, if the δ_i are small, condition (7) says that coexistence will occur if $P(\rho_i > n) = 1/n$ for each species. However, since the storage effect diminishes the harm to population growth during unfavorable periods, there is clearly a broad variety of probability distributions for $\mathfrak{E}(t)$ that permit coexistence.

This analysis gives merely sufficient conditions for coexistence, and because of its conservatism these sufficient conditions themselves may not be very useful. Thus the analysis is best viewed as demonstrating the trend toward coexistence as variation in the environment is increased in certain broad directions. It gives little indication of the actual magnitude of variation necessary for coexistence. However, the strength of the analysis is its generality as illustrated by the examples below. Moreover, the analysis above is not restricted to the case where the environment is i.i.d. Indeed it demonstrates invasibility with very general kinds of environment processes including both stationary ergodic processes and regular environment processes in which $\mathfrak{E}(t)$ is a periodic function of time. On the other hand, retaining the assumption that the environment is i.i.d. allows us to conclude that each species persists in the strong sense of stochastic boundedness (Chesson, in press).

Example 1. The multispecies lottery model.

This model involves competition for space. Each adult holds a unit of space and thus death of adults releases $\Sigma\delta_j X_j(t)$ units of space in the time interval $(t, t+1)$. The $\beta_i(t)X_i(t)$ offspring of species i compete for this space with the $\Sigma\beta_j(t)X_j(t)$ offspring from all species. Under the assumption that space is allocated at random, the proportion of available space taken by species i is $\beta_i(t)X_i(t) \,/\, \Sigma\,\beta_j(t)X_j(t)$. This involves also the assumption that the number of offspring always exceeds adult deaths, so that space is always in short supply. It now follows that

$$R_i(t) = (\Sigma\delta_j X_j(t)) \frac{\beta_i(t)}{\Sigma\,\beta_j(t)X_j(t)} \quad . \tag{9}$$

In the simplest case the "birth rates" $\beta_i(t)$ (which include density independent juvenile mortality) are simply functions of the environment and we deduce

$$R_i'(t) = \delta_i \frac{\beta_i(t)/\delta_i}{\max\limits_{j\neq i} \beta_j(t)/\delta_j} \tag{10}$$

The ratio $\beta_i(t)/\delta_i$ is a natural measure of the competitive ability of a species. Thus expression (10) says that all species can coexist in the system if each species experiences periods when it is sufficiently competitively superior to all of the other species. This will be achieved with appropriate variation in the environment.

This simplest form of the lottery model does not permit coexistence in a constant environment, nor in an environment with non-overlapping generations (Chesson and Warner 1981). It follows that the storage effect is essential to coexistence.

The storage effect can also be shown to promote coexistence in a variety of more complex versions of the lottery model (Chesson, in press, in prep.). These models allow for such features as density dependence in the $\beta_i(t)$, spatial heterogeneity, and the possibility that fecundity is not always sufficiently high to fill all of the available space.

Example 2. Generalized Lotka-Volterra Competition.

To obtain examples that are analogous with the more usual sorts of competition models, consider first the case of just two competing species and let $L_i(t)$ be the number of juveniles of species i produced in time (t, t+1). The model for $L_i(t)$ is

$$L_i(t) = B_i(t)X_i(t)f_i(X_i(t), X_j(t)) \tag{11}$$

where the environmentally varying parameter $B_i(t)$ is the per capita birth rate in the absence of competition, and the function f_i represents the proportionate reduction in the birth rate due to competition among adults. The function f_i is thus assumed decreasing in both arguments.

Competition may also occur among juveniles affecting their survivorship to adulthood. Thus we have

$$R_i(t)X_i(t) = \Theta_i L_i(t)g_i(L_i(t), L_j(t)), \tag{12}$$

where Θ_i is survivorship without competition and g_i is the reduction in survivorship due to competition; g_i also is decreasing in both arguments.

Equations (11) and (12) may represent discrete forms of Lotka-Volterra competition as suggested by Chesson (in press) or they may represent some arbitrary generalization of Lotka-Volterra competition.

Under the assumption that $\ell g_i(\ell, 0)$ has a finite maximum M_i we see that

$$X_i(t+1) \leq (1-\delta_i)X_i(t) + \Theta_i M_i \tag{13}$$

and so $X_i(t)$ is bounded above by $\kappa_i = \Theta_i M_i/\delta_i$ if it starts below this value. Thus we obtain

$$R_i'(t) \geq \Theta_i B_i(t)\ g_i(0, B_j(t)\kappa_j)f_i(0, \kappa_j). \tag{14}$$

The monotonicity of g_i now implies that $R_i'(t)$ will be large whenever $B_i(t)$ is large provided that $B_j(t)$ is not simultaneously large. It follows that the two species will coexist provided only that the density independent components of their birth rates (the $B_i(t)$) take on sufficiently large values at different times. This particular example generalizes quite trivially to cover the case of an arbitrary number of species. Coexistence occurs by the storage effect in the multispecies case if each species has large values of $B_i(t)$ while the values for the other species are small or moderate.

Without specifying the g_i and f_i, it is quite possible that coexistence occurs without the storage effect, either without a stochastic environment or without overlapping generations, but the important point is that the storage effect can lead to coexistence regardless of the specific form of the f_i and g_i, and so there is no doubt that it broadens the range of situations in which coexistence can occur.

Comparison with a spatial model

We have seen that the storage effect promotes coexistence in a variety of circumstances but the analysis gives little indication of how much variation is necessary for coexistence to occur nor does it say how effective the storage mechanism is relative to other mechanisms of coexistence. Some idea of the necessary amount of variation can be obtained from the two-species lottery model where this has been well-documented (Chesson and Warner 1981). A comparison with spatial heterogeneity lets us judge the relative efficacy of the storage mechanism. This is especially interesting because spatial heterogeneity is commonly regarded as a strong promoter of coexistence.

To make the comparison we construct an analogous spatial model which applies to a planktonic larval situation. Local populations of adults are assumed to exist on discrete patches, their offspring enter a pool of plankton which are then redistributed to the patches where the larvae may or may not mature as adults depending on the outcome of larval competitive interactions. The number of patches in the system, k, is assumed to be effectively infinite, and this assumption is justified by the usual convergence of the dynamics of finite systems of patches to those of infinite systems (Chesson 1981). The equation describing the dynamics of species i on patch j is

$$X_{ij}(t+1) = (1-\delta_i)X_{ij}(t) + R_{ij}(t)\overline{X}_i(t) \quad . \tag{15}$$

In this equation $X_{ij}(t)$ is the number of adults of species i on patch j, $\overline{X}_i(t) = \frac{1}{k}\Sigma_{j=1}^{k} X_{ij}(t)$ (the spatial average of adult numbers), and $R_{ij}(t)\,\overline{X}_i(t)$ is the number of new recruits to the adult population on patch j. Adults do not migrate: the only connection between patches is through the larval pool. The local recruitment rate $R_{ij}(t)$ takes the form

$$R_{ij}(t) = f_i(\mathcal{E}_j(t), \overline{X}_1(t), \ldots, \overline{X}_n(t)) \qquad (16)$$

which essentially embodies the idea that the total density of adults of each species in the system determines the size of the larval pool of each species. The local environment $\mathcal{E}_j(t)$ affects both the relative rates of migration of larvae to different patches, and the outcomes of the interactions among larvae on individual patches. Certainly more complex spatial models are possible but then they must be compared with models more complex than (1) for the storage effect. However for spatial versions of the lottery model it is best to have

$$R_{ij}(t) = (\Sigma_\ell \delta_\ell X_{\ell j}(t)) \frac{\beta_{ij}(t)\overline{X}_i(t)}{\Sigma_\ell \beta_{\ell j}(t)\overline{X}_\ell(t)} \qquad (17)$$

so that local recruitment depends on the local amount of space becoming available. But, because the $X_{\ell j}(t)$ enter (17) linearly, $\overline{X}_\ell(t)$ can be substituted for $X_{\ell j}(t)$ without altering the results we obtain (Chesson, in prep.). Thus it is sufficient to work with the general form (16).

To model spatial heterogeneity it is assumed that $\mathcal{E}_1(t), \mathcal{E}_2(t), \ldots$ are i.i.d. for fixed t, and that the distribution of the $\mathcal{E}_j(t)$ does not depend on t. These assumptions provide spatial variation, and permit temporal variation locally in space, but do not allow any temporal variation that is correlated over all patches.

With these assumptions the dynamics of the spatial averages of population numbers are given by the equation

$$\overline{X}_i(t+1) = (1-\delta_i)\overline{X}_i(t) + \phi_i(\overline{X}_1(t), \ldots, \overline{X}_n(t))\overline{X}_i(t) \qquad (18)$$

where

$$\phi_i(x_1, \ldots, x_n) = Ef_i(\mathcal{E}_j(t), x_1, \ldots, x_n). \tag{19}$$

Note that (18) is a simple difference equation and so the $\overline{X}_i$ will behave deterministically but this deterministic behavior depends on spatial heterogeneity through its effects on the ϕ_i.

If $\mathcal{E}(t)$ is substituted for $\mathcal{E}_j(t)$ in (16), the subdivision of the total population into local populations becomes irrelevant and equation (18) no longer holds, at least not exactly; instead, the $\overline{X}_i(t)$ satisfy the storage model (1). With this observation the assumption that $\mathcal{E}(t)$ and $\mathcal{E}_j(t)$ have the same probability distribution establishes a one-one correspondence between the storage model and the spatial model. However, the correspondence goes beyond a mere formal relationship for the two models actually converge to each other numerically as the death rates are made small. To see how this happens let $\delta_i = h\delta_i'$, $f_i = hf_i'$, $\phi_i = h\phi_i'$. Decreasing h decreases death rates and lengthens the lives of individuals. The recruitment rate also decreases with h. In general this is necessary to prevent unbounded growth in the total reproductive output of an individual in its lifetime, but in the lottery models it is an automatic consequence of decreasing the δ_i with h.

Defining $\overline{\underset{\sim}{X}}_h(t) = (\overline{X}_1(t/h), \ldots, \overline{X}_n(t/h))$ for the spatial process, and $\underset{\sim}{X}_h(t) = (X_1(t/h), \ldots, X_n(t/h))$ for the storage model, we can view these models, appropriately, on a time scale commensurate with the life expectancy of an adult. A general theorem of M.F. Norman applies. In the presence of mild regularity conditions (Norman 1975) the difference $\underset{\sim}{X}_h(t)-\overline{\underset{\sim}{X}}_h(t)$ converges in probability to $\underset{\sim}{0}$, provided $\underset{\sim}{X}_h(0) = \overline{\underset{\sim}{X}}_h(0)$. Thus the storage model and the spatial model are essentially indistinguishable for long-lived organisms: at least for long-lived organisms, the storage effect promotes coexistence just as effectively as does spatial heterogeneity.

The results of Norman (1975) are actually stronger than this simple convergence in probability for he shows that the rate of convergence is of order $h^{\frac{1}{2}}$: specifically,

$$h^{-\frac{1}{2}}[\, \underset{\sim}{X}_h(.) - \overline{\underset{\sim}{X}}_h(.)\,] \tag{20}$$

converges weakly in distribution to a diffusion process $\underset{\sim}{Z}(t)$. If $\overline{\underset{\sim}{X}}_h(0) = \underset{\sim}{x}^*$ is an equilibrium point for the spatial process (such points do not depend on h), then $\underset{\sim}{Z}(t)$ has an especially simple structure. It is a multivariate Ornstein-Uhlenbeck process and the distribution of $\underset{\sim}{Z}(t+s)$ given $\underset{\sim}{Z}(t)$ is

$$N(e^{As}Z(t),\ \int_0^s e^{Au}\Sigma_0 e^{A^T u}du) \tag{21}$$

where $N(\underset{\sim}{\mu}, \Sigma)$ means multinormal with mean $\underset{\sim}{\mu}$ and variance matrix Σ,

$A = \left(x_i^* \frac{\partial\phi_i'}{\partial x_j}(\underset{\sim}{x}^*)\right)$, $\Sigma_0 = \left(x_i^* x_j^* \mathcal{C}\{\, f_i'(\boldsymbol{\xi}(t), \underset{\sim}{x}^*),\ f_j'(\boldsymbol{\xi}(t), \underset{\sim}{x}^*)\}\right)$ and $\mathcal{C}$ means covariance.

In the event that the eigenvalues of A have negative real parts so that the spatial model is locally stable at $\underset{\sim}{x}^*$ for small h, the process $\underset{\sim}{Z}(t)$ converges as $t \to \infty$ to a stationary stochastic process with mean $\underset{\sim}{0}$ and variance $\Sigma = \int_0^\infty \exp(Au)\Sigma_0 \exp(A^T u)du$. Thus for large t and small h, $\underset{\sim}{X}_h(t)$ will be approximately a stationary process with distribution

$$N(\underset{\sim}{x}^*,\ h\Sigma). \tag{22}$$

This means that for long-lived organisms the storage model will give us small fluctuations about the spatial equilibrium, with the variance of these fluctuations being proportional to the adult death rates.

On the other hand, if any eigenvalue of A has a positive real part, the spatial model will be locally unstable at $\underset{\sim}{x}^*$ and the storage model will show increasingly severe fluctuations about the value $\underset{\sim}{x}^*$. This close link between stability in the two sorts of model further strengthens our conclusion that the stabilizing effects of spatial heterogeneity are matched in the analogous models of the storage effect.

<u>Example</u>: The multispecies lottery model.

The spatial version of the lottery model (Chesson, in prep.) has

$$\phi_i(\underset{\sim}{x}) = (\Sigma\delta_\ell x_\ell)\ E\,\frac{\beta_{ij}}{\Sigma_\ell \beta_{\ell j} x_\ell}\quad . \tag{23}$$

Equilibrium points satisfy the equation

$$E \frac{\Gamma_i}{\Sigma \Gamma_\ell u_\ell^*} = 1 \tag{24}$$

where $\Gamma_i = \beta_{ij}/\delta_i$ and $u_i^* = x_i^* \delta_i^* / \Sigma_\ell \delta_\ell x_\ell^*$. In the lottery model, A has a 0 eigenvalue because $\Sigma X_i(t)$ = a constant (the total amount of space in the system). However if the distribution of $(\Gamma_1, \ldots, \Gamma_n)$ is n-dimensional, i.e., cannot be supported by a linear space of fewer than n dimensions, then 0 is a simple root of the characteristic equation of A while the other eigenvalues have negative real parts. It follows that small h and feasibility of the solution to (24) give stability of the spatial model at $\tilde{x}^*$, and they give small fluctuations about this value in the storage model.

An instructive special case is obtained by assuming that $(\Gamma_1, \ldots, \Gamma_n)$ has an exchangeable distribution, i.e., $(\Gamma_{\pi_1}, \ldots, \Gamma_{\pi_n})$ has the same joint distribution for all permutations $(\pi_1, \ldots, \pi_n)$ of $(1, \ldots, n)$. An exchangeable distribution gives a model of similar species, and for this model (24) has the feasible solution $x_i^* = \delta_i^{-1}/\Sigma_\ell \delta_\ell^{-1}$. If the distribution of $(\Gamma_1, \ldots, \Gamma_n)$ is also n-dimensional, the storage effect leads to coexistence with small fluctuations about $\tilde{x}^*$, for small h. An n-dimensional and exchangeable distribution for $(\Gamma_1, \ldots, \Gamma_n)$ is consistent with arbitrarily small but positive variances for the Γ_i. Thus this particular example provides an important complement to our previous results for it shows that in the lottery model similar species can coexist by the storage effect with arbitrarily small amounts of temporal environmental variability.

Summary

In many organisms the product of the birth rate and the juvenile mortality rate (the recruitment rate) is highly variable while the adult death rate is low and relatively less variable. These conditions lead to an asymmetry between the contributions of favorable and unfavorable periods to population growth. This asymmetry is called the storage effect and it can permit a species to persist provided favorable periods convey sufficient benefit regardless of the costs incurred during unfavorable periods. The storage effect is a stochastic mechanism of coexistence capable of acting in a broad variety of situations. For long-lived organisms its efficacy appears comparable to that of spatial heterogeneity.

Acknowledgements

I am grateful for comments on the manuscript by Jerry Downhower, Stephen Ellner and Michael Turelli.

Literature Cited

Abrams, P.A. (manuscript). Variability in resource consumption rates and the coexistence of competing species.

Beddington, J.R., Free, C.A. and Lawton, J.H. (1976). Concepts of stability and resilience in predator-prey models. J. Anim. Ecol. 45:791-816.

Chesson, P.L. (1981). Models for spatially distributed populations: the effect of within-patch variability. Theoret. Pop. Biol. 19:288-325.

Chesson, P.L. (1982). The stabilizing effect of a random environment. J. Math. Biol. 15:1-36.

Chesson, P.L. (in press). Coexistence of competitors in a stochastic environment: the storage effect. Proc. International Conference on Population Biology. Edmonton, Alberta 1982. Lecture Notes in Biomathematics.

Chesson, P.L. (in prep.). A stochastic model of competition in a patchy environment.

Chesson, P.L. and Warner, R.R. (1981). Environmental variability promotes coexistence in lottery competitive systems. Am. Nat. 117:923-943.

Ellner, S.P. (in press). Asymptotic behavior of some stochastic difference equation population models. J. Math. Biol.

Goh, B.S. (1976). Nonvulnerability of ecosystems in unpredictable environments. Theoret. Pop. Biol. 10:83-95.

Grubb, P.J. (1977). The maintenance of species richness in plant communities: the importance of the regeneration niche. Biol. Rev. 52:107-145.

Gulland, J.A. (1982). Why do fish numbers vary? J. Theoret. Biol. 97:69-75.

Harper, J.L. (1977). Population Biology of Plants. Academic Press, New York. 892pp.

Hubbell, S.P. (1980). Seed predation and the coexistence of tree species in tropical forests. Oikos 35:214-229.

Murdoch, W.W. (1979). Predation and the dynamics of prey populations. Fortschr. Zool. 25:245-310.

Norman, M.F. (1975). Approximation of stochastic processes by Gaussian diffusions, and applications to Wright-Fisher genetic models. SIAM J. Appl. Math. 29:225-242.

Shmida, A. and Ellner, S.P. (in press). Coexistence of plant species with similar niches. Vegetatio.

Turelli, M. (1978a). A reexamiantion of stability in randomly varying versus deterministic environments with comments on the stochastic theory of limiting similarity. Theoret. Pop. Biol. 13:244-266.

Turelli, M. (1978b). Does environmental variability limit niche overlap? Proc. Natl. Acad. Sci. USA. 75:5085-5089.

Warner, R.R. and Chesson, P.L. (manuscript). Coexistence mediated by environmental fluctuations: a field guide to the storage effect.

THE STABLE SIZE DISTRIBUTION: AN EXAMPLE IN STRUCTURED POPULATION DYNAMICS.

Odo Diekmann

Mathematisch Centrum, Kruislaan 413
1098 SJ Amsterdam, the Netherlands.

1. INTRODUCTION

If some characteristic of the individuals is essential for describing the dynamics of a population properly, one has to distinguish the individuals from each other according to this characteristic. As an example of such a trait, which can take a continuum of values, we shall consider "size" (denoted by the symbol x), by which we mean any relevant quantity satisfying a physical conservation law. *)

Then, to begin with, one has to specify the dynamics of the individuals. The basic processes fall into two categories:

I *Change:* the size of each individual changes continuously (according to some law which has to be specified) when nothing special happens:
$\frac{dx}{dt} = g$ = growth rate = prescribed function of x and, possibly, other variables.

II *Chance:* some individuals undergo spectacular processes, while others do not. One has to specify the chances (per unit of time) that this will happen as a function of x and .... For example,
$\mu = \mu(x) = \mu(x,\ldots)$ = chance to die as a function of x,...,
$b = b(x) = b(x,\ldots)$ = chance to split into two identical parts as a function of x,... .

(Although we use the word "chance", we shall deal with deterministic models which are based on the assumption of large numbers).

In the second step, one introduces a *density function* n to describe the state of the population and one derives an equation for n by drawing up the *balance* of I and II. For a species which reproduces by binary fission one obtains:

*) e.g., weight, N-, or P- content, but not age, since there is no conservation of age in the fission process.

$$(1)\qquad \underbrace{\frac{\partial n}{\partial t} + \frac{\partial}{\partial x}(gn)}_{\text{growth}} = \underbrace{-\ \mu(x)n(t,x)}_{\text{death}}\ \underbrace{-\ b(x)n(t,x)}_{\text{reproduction sink}}\ \underbrace{+\ 4b(2x)n(t,2x)}_{\text{reproduction source}} ,$$

where $\int_{x_1}^{x_2} n(t,\xi)d\xi$ = number of individuals with size between x_1 and x_2 at time t. (Exercise: explain the factor 4. Hint: check conservation of mass during fission). This is a special case of Sinko & Streifer's (1971) mathematical model for organisms reproducing by fission.

The year 1967 showed a remarkable outburst of papers formulating similar models for the dynamics of structured populations: Bell & Anderson (1967), Fredrickson, Ramkrishna & Tsuchiya (1967), Sinko & Streifer (1967). Although there has been some follow up (see, for instance, Streifer (1974), Oster (1977), Nisbet (this volume) and the references therein), we can conclude today, fifteen years later, that the mathematical theory is still in its infancy (possibly with the exception of age-dependent population growth).

From a mathematical point of view, the theory is concerned with first order partial differential equations with non-local terms (transformed arguments, integrals,...) which are nonlinear as soon as interaction is taken into account. From a biological point of view, the aim is to use information about the behaviour and the physiology of individuals to describe, understand and predict the dynamics of the population as a whole (see Streifer's (1974) excellent survey paper for an elaborate presentation of the main ideas). In practice one frequently encounters the inverse problem: how to use measurements of the density function to derive conclusions about the dynamics of individuals (see, e.g., Bell & Anderson, 1967).

The above observations form the basis for a recently started research project in the Netherlands (at the Mathematical Centre), which aims at analysing specific examples in this category of equations and models with an eye for a general theory. This note is a progress report, based on work of T. Aldenberg, H.J.A.M. Heijmans, H.A. Lauwerier, J.A.J. Metz, H. Thieme (Heidelberg), and the author. We shall deal with two topics:

i) linear equations: convergence towards a stable distribution,

ii) nonlinear equations: interaction via the growth function (a feedback mechanism which admits a clear biological interpretation).

2. THE STABLE SIZE DISTRIBUTION

In this section we assume that g, μ and b are functions of x only. As a further specification of the model we require:

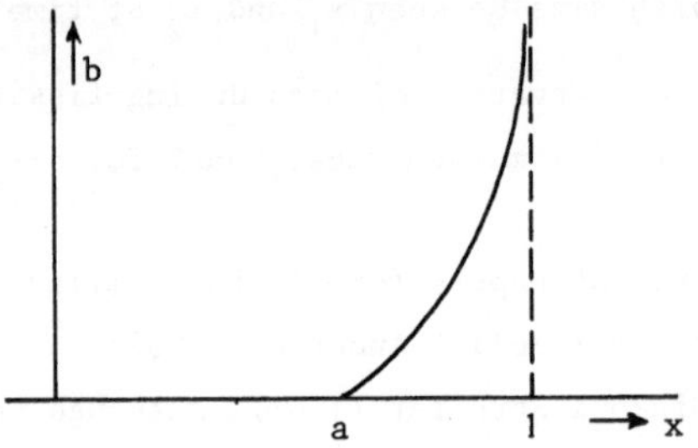

for some $a \in (0,1)$:

$b(x) = 0$, for $x \in (0,a)$

$b(x) > 0$, for $x \in (a,1)$, b continuous,

$\lim_{x \uparrow 1} \int_a^x b(\xi)d\xi = +\infty$.

These are mathematical counterparts of the following biological assumptions:

i) there is a minimal size, called a, which an organism should have in order to have some chance to undergo fission.

ii) there is a maximum size, normalized to be 1, which an organism can reach (note that the chance to grow from a to x without splitting is given by

$$\exp - \int_{T(a)}^{T(x)} b(\xi(t))dt = \exp - \int_a^x \frac{b(\xi)}{g(\xi)} d\xi,$$

in case μ equals zero; here $T(x)$ denotes clock time when the organism has size x and $\xi(t)$ the size as a function of time).

On account of i) we supplement (1) with the boundary condition

(2) $$n(t,\tfrac{1}{2}a) = 0$$

which expresses that organisms with size less than $\frac{1}{2}a$ do not exist. In (1) we interpret the term $4b(2x)n(t,2x)$ as zero for $x > \frac{1}{2}$. The functions μ and g are assumed to be continuous functions on $[\frac{1}{2}a,1]$, with μ nonnegative and g strictly positive. Finally, we assume that the situation at $t = 0$ is known:

(3) $$n(0,x) = \phi(x), \quad x \in [\tfrac{1}{2}a,1], \quad \phi \geq 0.$$

Question (by analogy with Lotka's celebrated result for unlimited age dependent population growth).

Is it true that

(4) $$n(t,x,\phi) \sim C(\phi)e^{\lambda_0 t} n_0(x), \quad t \to +\infty,$$

where λ_0 is a real number (the Malthusian parameter or intrinsic rate of natural increase) and $n_0(x) \geq 0$ is a stable size distribution?

Answer yes if $g(2x) < 2g(x)$ (or $g(2x)>2g(x)$),
no if $g(2x) = 2g(x)$.

Elucidation: Consider two organisms A and B with equal size. A splits into a and a. During some time interval a, a and B grow. Then B splits into b and b. How do the sizes of a and b compare? If $g(2x) = 2g(x)$ they are identical and the initial condition is, apart from multiplication, copied again and again. This merry-go-round character implies that all properties of the initial condition remain manifest for all times. In sharp contrast, when $g(2x) < 2g(x)$, only a one-dimensional projection (the constant C) of the initial condition influences the asymptotic behaviour.

We refer to Diekmann, Heijmans & Thieme (in preparation) for a precise mathematical formulation and a proof (in addition this paper will contain extensions to periodic environments, like in Thieme (preprint 1982)). The following *mathematical techniques* are used:

i) eigenvalue problem ⇒ integral operator equation ⇒ positive operator theory ⇒ dominant eigenvalue (Heijmans, preprint 1982).
ii) evolution equation ⇒ integral operator equation ⇒ existence and uniqueness of a solution ⇒ definition of a semigroup.
iii) semigroup + compactness + dominant eigenvalue ⇒ asymptotic behaviour for $t \to +\infty$ (it is remarkable that the condition on g is used only to get compactness of the semigroup after finite time).

Moreover, it is possible to derive a transcendental equation for λ_0 (and the other eigenvalues; Heijmans (preprint 1982)), which in the case $a \geq \frac{1}{2}$ takes the form

$$2 \int_a^1 \frac{b(\xi)}{g(\xi)} \exp\left(- \int_{\frac{1}{2}\xi}^{\xi} \frac{b(\eta)+\mu(\eta)+\lambda}{g(\eta)}\, d\eta\right) d\xi = 1.$$

Here the left hand side with $\lambda = 0$ has the usual interpretation: it is the offspring of the average individual (with x=a taken as the reference point). Similarly, n_0 and $C(\phi)$ are quite computable. So, although the proof uses abstract machinery, the outcome is rather concrete.

3. THE LIMITED WORLD

How does a population of, say, unicellular organisms, react upon a given, limited, supply of nutrients? This question immediately leads to another one: how do the organisms use nutrients for growth and reproduction? The main advantage of structured models is that one can use submodels for processes within the individuals and combine these to obtain an overall population model (Streifer, 1974).

Sinko and Streifer (1971) made a detailed model for a population of the planarian worm Dugesia tigrina, starting from the assumption that the important physiological characteristics can be described by their mass alone. They specified how the available food was distributed among the individuals, how the consumed food was used for maintenance and growth and how the "birth" function was influenced by food shortage. Moreover, they solved the resulting equations numerically and compared the outcome with available data.

In addition to the detailed modelling of real populations, one can try to enlarge understanding and intuition by analysing relatively simple idealized mathematical models. That is the approach taken here.

So assume that $g = g(x,c)$ and $b = b(x,c)$, where c describes the concentration of some important chemical substance. In a chemostat we would have

$$(5) \qquad \frac{dc}{dt} = \underbrace{\gamma}_{\text{inflow}} - \underbrace{\int_{\frac{1}{2}a}^{1} h(x,c)n(t,x)dx}_{\text{uptake by the population}} - \underbrace{\mu c}_{\text{outflow}}$$

for some function h. (If we are dealing with a structural chemical, as is assumed below, we may set h equal to αg, for some constant α).

Questions: 1) Do we still obtain a stable size distribution?

2) If so, how does the time-dependent factor (the amplitude) behave?

We don't know (yet) the answers in general. However, in the very special case that (abusing notation)

$$(6) \qquad \begin{cases} \text{i)} \quad b(x,c) = g(x,c)b(x), \\ \text{ii)} \quad g(x,c) = \beta(c)\, g(x), \qquad g(2x) < 2g(x), \\ \text{iii)} \ \mu \text{ independent of } x, \end{cases}$$

we have the following

Answers 1) Yes.

2) The asymptotic time dependence is described completely by a computable system of autonomous o.d.e.'s.

First we comment on the assumptions. In Diekmann et al. (preprint 1983) it is shown how (i) arises in a variant of the previous linear model, where one postulates a stochastic division threshold (the chance to undergo fission is determined by the size gained, independent of the time needed to realize this size increase). When energy (from food) is involved in c (ii) is certainly unrealistic, since it ignores the basic metabolism. However, it might apply to phosphate or nitrate limitation since these chemicals are used for building material. Assumptions (i) and (ii) imply that fission stops completely immediately after exhaustion of the substrate and, in principle, this consequence can be tested experimentally. However, a practical complication is formed by the fact that the fission process of each cell takes time (and that it will complete once started) and that, consequently, the instant at which fission stops is difficult to define or measure exactly. (Anyhow, we admit that (ii) is suggested by the fact that it makes mathematical life easy). Finally, (iii) is appropriate in a chemostat. That explains why we took the same μ in (5).

Next, we sketch the analysis of (1) & (5) under assumption (6). Abstractly, we can write the equation for n as

$$\frac{dn}{dt} = - \mu n + \beta(c)An,$$

where A is a linear operator. Let λ_o be the dominant eigenvalue of A and n_o the corresponding eigenfunction. Substitute

(7) $$n(t,x) = \rho(t)\ \{n_o(x) + n_1(t,x)\},$$

where n_1 is in the appropriate complementary subspace. By a trick (based on time scaling; note that under our assumptions growth and division scale in the same way) one can prove that $n_1(t,x) \to 0$ as $t \to +\infty$. Hence we can take limits in the equations for ρ and c to obtain:

(8) $$\begin{cases} \rho' = \rho(\lambda_o\beta(c)-\mu) \\ c' = \gamma - H(c)\rho - \mu c \end{cases}$$

where by definition

$$H(c) = \int_{\frac{1}{2}a}^{1} h(x,c)n_o(x)dx.$$

Note that both λ_0 and H are amenable to numerical calculation. We refer to Diekmann et al. (preprint 1983) for the details and for other feedback mechanisms which can be modelled and analysed in a similar manner.

So, under some rather special assumptions, these complicated models yield o.d.e. systems which can be analysed in all detail. This certainly is encouraging. Theoretically at least, one can relate in this way parameters in an o.d.e. total population model like (8) to (observable?) properties of individuals like growth and fission rates. Whether or not this has any practical significance remains to be seen.

LITERATURE CITED

Diekmann, O., Lauwerier, H.A., Aldenberg, T. and Metz, J.A.J., (preprint 1983). Growth, fission and the stable size distribution, MC Report TW 235, Amsterdam.

Bell, G.I. and Anderson E.C. (1967). Cell growth and division. I. A mathematical model with applications to cell volume distributions in mammalian suspension cultures. Biophys. J. 7: 329-351.

Diekmann, O., Heijmans, H.J.A.M., and Thieme, H. (in preparation). On the stability of the size distribution.

Fredrickson, A.G., Ramkrishna, D. and Tsuchiya, H.M. (1967). Statistics and dynamics of procaryotic cell populations. Math. Biosci. 1: 327-374.

Heijmans, H.J.A.M. (preprint 1982). A linear eigenvalue problem related to cell growth, MC Report TW 229, Amsterdam.

Nisbet, R.M. and Gurney, W.S.C., "Stage-structure" models of uniform larval competition, this volume.

Oster, G. (1977). Lectures in Population Dynamics. In: DiPrima, R.C. (ed.). Modern Modeling of Continuum Phenomena. Lectures in Applied Math. Vol. 16: 149-190.

Sinko, J.W. and Streifer, W. (1967). A new model for age-size structure of a population. Ecology 48: 910-918.

Sinko, J.W. and Streifer, W. (1971). A model for populations reproducing by fission. Ecology 52: 330-335.

Streifer, W. (1974). Realistic models in population ecology. In: Mac Fadyen, A. (ed.). Advances in Ecological Research 8: 199-266.

Thieme, H. (preprint 1982). Renewal theorems for linear periodic Volterra integral equations, Preprint 152, SFB 123, Heidelberg.

"STAGE-STRUCTURE" MODELS OF UNIFORM LARVAL COMPETITION

R.M. Nisbet and W.S.C. Gurney

1. Introduction

The investigation in this paper is motivated by the observation that laboratory insect populations, regulated by the availability of larval food, may exhibit "quasi-cyclic" fluctuations of two distinct types:

(a) cycles with period "two and a bit" times the maturation time as in Nicholson's blowflies (Fig. 1),

(b) cycles with period "one and a bit" times the maturation time as in Lawton's experiments on the Indian meal moth *Plodia interpunctella* (Fig. 2).

Any model constructed to describe such fluctuations (where the time scales of interest are comparable with the maturation time of an individual) clearly must include age-structure. Just as vital however is that the final mathematical description should be sufficiently simple to yield some "gut feeling" for the demographic consequences of larval competition and not just a million computer simulations.

We adopt the approach we have advocated in several recent papers (Gurney and Nisbet 1983; Gurney, Nisbet and Lawton 1983; Nisbet and Gurney 1983a; Blythe, Gurney and Nisbet 1983) and in the course of lectures preceding this symposium (Nisbet and Gurney 1983b). In these papers we developed methods for simplifying single-species age-structure models in situations where we can meaningfully subdivide the life history into developmental classes of arbitrary (and possibly varying) duration, and then follow the dynamics of the "subpopulations" of the various classes. By assuming that all individuals at the same developmental stage had the same vital rates, we were able to reduce the normal equations describing an age-structured population to a set of coupled delay-differential equations. In this paper we demonstrate that the formalism is capable of earning its keep; we do this by constructing a family of models of larval competition and demonstrating that they exhibit testable differences in dynamical behaviour. In particular, we are able to distinguish mechanisms producing cycles with periods equal to "one and a bit" and "two and a bit" times the maturation time.

2. Formulating Stage-Structure Models

We idealise the life history of our insect and subdivide the lifetime of an individual into two stages - "larvae" and "adults", the transition from the "larval" to the "adult" stages being called "maturation" or "pupation". We define:

$\tau(t)$ = time which was spent as larvae by individuals maturing at time t,

$L(t)$ = total number of larvae at time t

$\Delta(t)$ = per capita death rate of larvae at time t (assumed identical for all individuals irrespective of age)

$P(t)$ = proportion of larvae born at time $t - \tau(t)$ who survive to age $\tau(t)$ at time t.

$\alpha(t)$ = probability of successful pupation at time t

$R(t)$ = rate of adult recruitment at time t

$A(t)$ = total number of adults at time t

$B(t)$ = rate of production of viable eggs at time t

$\delta(t)$ = per capita death rate of adults at time t (assumed identical for all individuals irrespective of age).

The "t" dependence in the quantities: $\tau(\cdot)$, $\Delta(\cdot)$, $P(\cdot)$, $\alpha(\cdot)$, $R(\cdot)$, $B(\cdot)$ and $\delta(\cdot)$ includes both explicit time-dependence and implicit time-dependence (via density-dependence).

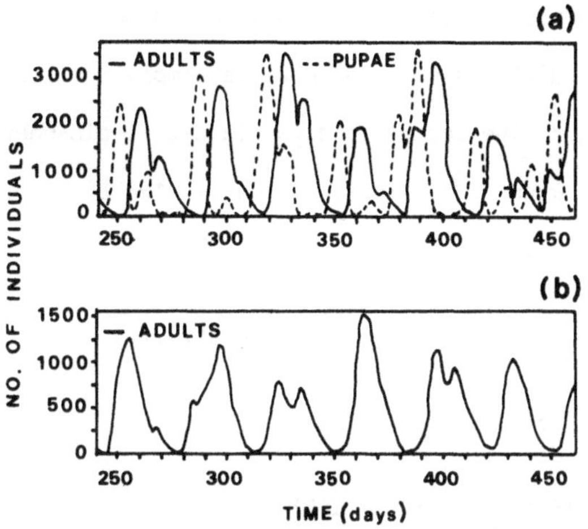

Fig. 1: Quasi-cyclic fluctuations in Nicholson's blowfly populations regulated by larval food supply:
(a) 50g per day, (b) 25g per day.
The egg to adult maturation time is around 15 days. (Nicholson, 1954).

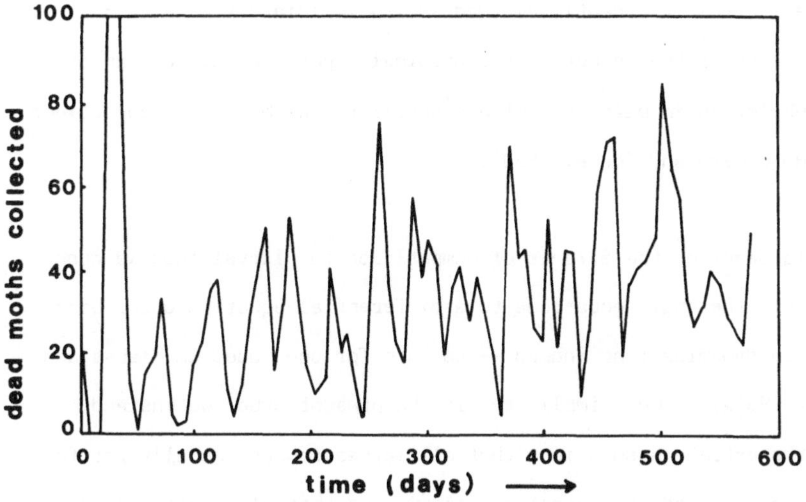

Fig. 2 Weekly counts of adult dead moths in one of Lawton's long-term cultures of Plodia interpunctella (from Gurney, Nisbet and Lawton, 1983). The egg to adult maturation time is around 30 days.

With these definitions writing down equations describing the changes in the subpopulations L(t) and A(t) is simply a matter of careful book-keeping. The recipe is given in detail elsewhere (Gurney, Nisbet and Lawton 1983; Nisbet and Gurney 1983a); for the present model the result is

$$\dot{L}(t) = B(t) - B(t-\tau)P(t)\,\{1 - \dot{\tau}(t)\} - \Delta(t)L(t) \qquad (1)$$

$$\dot{A}(t) = R(t) - \delta(t)A(t) \qquad (2)$$

where $R(t) = B(t-\tau)P(t)\,\{1 - \dot{\tau}(t)\}\,\alpha(t)$ (3)

and $P(t) = \exp\,\{-\int_{t-\tau(t)}^{t} \Delta(x)dx\}$ (4)

These equations presuppose an empty (i.e. A(t) = L(t) = 0) system for t < 0, which is "triggered" by the addition of a few newly hatched larvae or recently emerged adults immediately after t = 0. Again we refer the reader to the original papers for discussion of the subtleties in particular those associated with initial conditions (cf. Busenberg and Cooke, 1980).

To describe the effects of competition for larval food we ought to couple the above equations to a differential equation describing the food dynamics (and indeed we do this for one model in Nisbet and Gurney 1983a). For simplicity, in the present study we instead make the extreme assumptions that all larvae compete equally for food which is supplied at a constant rate Φ, and that all uneaten food decays rapidly or is removed rapidly. The available ration of food per individual per day is thus

$$\rho(t) = \Phi/L(t) \qquad (5)$$

We further assume that the per capita food consumption rate I(t) for the larvae has a hyperbolic (Michaelis-Menten) dependence on $\rho(t)$, implying

$$I(t) = \frac{I_{max}\rho(t)}{K_\rho + \rho(t)} = \frac{I_{max}}{1 + L(t)/L_o} \qquad (6)$$

where the characteristic larval population size (L_o) is ΦK_ρ^{-1}.

The models to be developed in the rest of the paper involve different ways of coupling I(t) to the various quantities in equations (1) - (4).

3. Immediate Effects of Larval Competition

Eating promotes growth and prevents death! Thus the immediate manifestation of larval competition will be density dependence of larval growth and/or death rates.

Growth

There is rather convincing experimental evidence (reviewed by Beddington, Hassell and Lawton, 1976) that the relationship between the growth rate of an individual arthropod larva, g(t), and its food uptake rate, I(t), is given by

$$g(t) = \begin{cases} \varepsilon(I(t) - I_o) & \text{for } I > I_o \\ 0 & \text{for } I < I_o \end{cases} \qquad (7)$$

There are of course exceptions; for some species the dependence on $I(t) - I_o$ is close to logarithmic (e.g. damselfly larvae: Lawton, Thompson and Thompson 1980), while John Lawton has told us of a spider that performs successive moults in the absence of food thereby effecting sustained negative growth! However equation (7) is commonly valid and we stick with it, combining it with equation (6) to obtain

$$g(t) = \begin{cases} \dfrac{g_m}{1 + L(t)/L_o} - \Gamma & \text{for } L(t) < L_o\left(\dfrac{g_m}{\Gamma} - 1\right) \\ 0 & \text{otherwise} \end{cases} \quad (8)$$

where $\Gamma(=\varepsilon I_o)$ represents a maintenance cost, and $g_m(=\varepsilon I_{max})$ has the property that $g_m - \Gamma$ represents the maximum possible individual growth rate.

Death

Beddington, Hassell and Lawton (1976) show that, for a number of species, through-instar survival probabilities rise monotonically from zero when $I = I_o$ to a maximum value (<1) which is achieved when food is available in excess of all requirements. Unless this reflects concentrated mortality at pupation or is the indirect result of changes in instar duration (both discussed later), then the results are consistent with a dependence of the per capita larval death rate on the feeding rate, of the form

$$\Delta(t) = \begin{cases} \dfrac{\mu}{I(t) - I_o} & \text{for } I > I_o \\ \infty & \text{for } I < I_o\,, \end{cases} \quad (9)$$

which when combined with equation (6) yields

$$\Delta(t) = \begin{cases} \dfrac{X(1+L(t)/L_o)}{1-Y-YL(t)/L_o} & \text{for } L < L_o Y^{-1}(1-Y) \\ \infty & \text{otherwise} \end{cases} \quad (10)$$

where $$X = \mu/I_{max}, \quad Y = I_o/I_{max} \quad (11)$$

4. Indirect Effects of Larval Competition

Since larvae are reproductively inactive, the growth rate changes discussed in the previous section can have no direct effect on overall population dynamics. They can, however have indirect effects in many ways from which we have selected three for detailed study:

(a) where the transition from the larval to adult stages is size-triggered, slower growth implies delayed maturation and hence lower egg-to-adult survival,

(b) where maturation to the adult stage is age-triggered, but its success rate is dependent on larval size, slower growth implies smaller mature larvae and hence increased mortality at pupation,

(c) if the adults do not feed (as in *Plodia*) or if the adult females derive much of their protein needs from food consumed as larvae, and if the maturation to the adult stage is triggered by age, then slower growth implies smaller mature larvae and hence less fecund adults.

We shall consider each of these possibilities in turn, and to assist us in the analysis we define $W(t)$ to be the weight of a larva which matures to adulthood at time t. If we assume that all individuals have the same (negligibly small) weight W_0 on hatching, then

$$W(t) = W_0 + \int_{t-\tau(t)}^{t} g(x)dx \simeq \int_{t-\tau(t)}^{t} g(x)dx \qquad (12)$$

Case (a) - Variable Maturation Time

If the maturation from larva to adult occurs at a critical size denoted by W^+, then we can calculate the time spent growing to that size by an individual who matures at time t by noting that

$$W(t) = \int_{t-\tau(t)}^{t} g(x)dx = W^{+} = \text{constant} \tag{13}$$

Then, following procedures detailed in Nisbet and Gurney (1983a), we note that this integral condition can be replaced by a differential equation describing the temporal development of $\tau(t)$, namely

$$\dot{\tau}(t) = 1 - \frac{g(t)}{g(t - \tau(t))} , \tag{14}$$

and an initial condition

$$\tau(o) = W^{+}/(g_m - \Gamma). \tag{15}$$

(We assume that $g(t)$ has its maximum value, $g_m - \Gamma$, for all $t < 0$ at which times the system is empty.)

Case (b) - Variable Pupation Success

If, as Beddington, Hassell and Lawton (1976) suggest, inadequately developed larvae are more prone than well developed individuals to die during moults or pupation, then we might expect a relationship between the weight at maturation, $W(t)$, and the probability of successful pupation $\alpha(t)$ which could be approximated by the offset hyperbolic form

$$\alpha(t) = \begin{cases} \dfrac{W(t) - W_1}{K_W + (W(t) - W_1)} & \text{if } W(t) > W_1 \\ 0 & \text{otherwise} \end{cases} \tag{16}$$

This can be incorporated into the basic population equations (1) - (4) provided we know the temporal evolution of $W(t)$. In principle this is obtained from equation (12), but it is computationally advantageous to play the well-tried game of replacing this integral relationship with the combination of the delay-differential equation

$$\dot{W}(t) = g(t) - g(t - \tau) \tag{17}$$

(obtained by differentiating equation (12)), and the initial condition

$$W(o) = \tau(g_m - \Gamma). \tag{18}$$

Case (c) - Variable Adult Fecundity

Although the existence of a relationship between adult fecundity and weight at maturation appears to be well documented, we know of no experimental evidence to guide us in the choice of functional forms to use in the analysis. The primary reason for this is that the effect is most likely to operate through a reduction in the total egg load produced by an adult; predicting the effect on fecundity then involves making assumptions about adult mortality (which may be weight-dependent) and about the details of the egg-laying schedule. Given this near-total ignorance of the facts, we make the simplest possible, defensible assumption, namely that an adult recruited into the population at time t_R has a fecundity proportional to its weight at recruitment (i.e. $W(t_R)$). We further assume that this fecundity is retained for the duration of its reproductively active life. It is then possible to calculate the instantaneous rate of egg production (again as an integral over time). We do not display the resulting expression, but again (perhaps not to the reader's total surprise) comment that it can be differentiated, and yields a differential equation, namely

$$\dot{B}(t) = qR(t)W(t) - \delta(t)B(t) \tag{19}$$

where q is the proportionality constant relating fecundity and weight at pupation.

5. Models

So far we have highlighted four ways (one direct, three indirect) in which larval food limitation may influence population dynamics. In order to elucidate their effects we have studied four models each incorporating one of the effects on its own. The assumptions defining each model are now listed.

(a) Model LD - Density Dependence of Larval Death Rate Only

We assume:
- (i) constant maturation time: $\tau(t) = \tau =$ constant,
- (ii) constant pupation success: $\alpha(t) = 1$ at all times,
- (iii) constant adult fecundity: $B(t) = eA(t)$,
- (iv) constant adult death rate: $\delta(t) = \delta =$ constant,
- (v) larval death rate given by equation (10).

(b) Model MT - Density Dependence of Maturation Time Only

We assume:
- (i) constant larval death rate: $\Delta(t) = \Delta =$ constant
- (ii) constant adult death rate: $\delta(t) = \delta =$ constant
- (iii) constant adult fecundity: $B(t) = eA(t)$
- (iv) constant pupation success $\alpha(t) = 1$ at all times
- (v) density-dependent growth described by equation (8) implying density dependence of maturation time through equations (14) and (15).

(c) Model PS - Density Dependence of Pupation Success Rate Only

We assume:
- (i) constant maturation time: $\tau(t) = \tau =$ constant
- (ii) zero larval death rate: $\Delta(t) = 0$
- (iii) constant adult fecundity: $B(t) = eA(t)$
- (iv) constant adult death rate: $\delta(t) = \delta =$ constant
- (v) density-dependent growth described by equation (8) implying density-dependence of pupation success via equations (16) - (18).

(d) Model AF - Density Dependence of Adult Fecundity Only

We assume: (i) constant maturation time: $\tau(t) = \tau$ = constant

(ii) zero larval death rate: $\Delta(t) = 0$

(iii) constant adult death rate: $\delta(t) = \delta$ = constant

(iv) constant pupation success: $\alpha(t) = 1$

(v) density-dependent growth described by equation (8) implying density-dependent fecundity via equation (19).

The dynamic equations representing each of these models are listed in the Appendix. For each model, provided certain obvious biological constraints on the parameters are obeyed, there is a single non-trivial, positive, steady state, the local stability of which can be investigated by studying the roots of the relevant characteristic equations (also listed in the Appendix). A glance at these equations suggests that models PS and AF might behave rather differently from the other two because of the terms involving $e^{-2\lambda\tau}$ in their characteristic equations. We have confirmed this by making extensive numerical studies (avoiding the traps elucidated by Busenberg and Cooke, 1980!) of the models, using where possible parameters appropriate to Plodia interpunctella. We indeed find that two very distinct patterns of dynamical behaviour occur.

In models LD and MT, it is rather common to find oscillations with a period slightly greater than τ (or τ^*, the steady state delay in the case of MT). Depending on the magnitude of the maintenance cost, these oscillations may be damped or may grow into limit cycles; however the observed periods are very insensitive to this transition and we can fairly describe the behaviour as single generation cycles.

By contrast in models AF and PS, while "single-generation" oscillations are commonly present early in the transients, the long-term behaviour of the population is dominated by oscillations (which again may be damped or growing) with a period slightly exceeding twice the maturation time. This behaviour is reminiscent of the "delayed feedback" mechanism which we believe to be responsible for the well-studied cycles in Nicholson's blowflies under conditions of adult food limitation (Auslander et al 1974, Oster 1976, Oster and Ipaktchi 1978, Gurney Blythe and Nisbet 1980; Nisbet and Gurney 1982, chapter 8).

This sharp contrast in the dynamics is reflected in the roots of the characteristic equations for the various models. As is evident from the Appendix, these equations are transcendental and distinctly unlovely. However one cannot but be struck by the structural similarities between equations (A4) and (A13) (for models LD and MT) and between equations (A19) and (A25) (for models PS and AF). The significance of these similarities is confirmed by numerical computations of the roots. If we again use "Plodia-like" parameters where possible and scan the remaining parameter groups we find that the roots behave in a manner consistent with the dynamics just described; for models LD and MT the dominant roots have imaginary parts a bit below $2\pi/\tau$ while for the other two models (AF and PS) the imaginary parts are close to π/τ (reflecting a period of 2τ).

Details of the computations and a more formal investigation of the location of the characteristic equation roots will be published later.

6. Discussion

We know that age-structure has the effect of introducing delays into population dynamics, and furthermore that the effects of such delays can be extremely subtle and delicate (see for example Hastings 1983 Cushing 1980). However, the present study does point to some rather robust generalisations concerning the effects of competition at the larval stage.

We first notice that in the short term, all the models are capable of producing "single-generation" oscillations in response to a sharp innoculation of individuals of a particular age. In two models (LD and MT), if the longer-term behaviour of the system is at all oscillatory, it is these single-generation cycles which dominate, a feature which reflects the fact that in both models the effects on adult recruitment of a perturbation in larval population is immediate. These single-generation oscillations are simply a manifestation of the "discrete generations" so commonly assumed in insect population models (e.g. Varley, Gradwell and Hassell 1973), but it is important to note that they are likely to be damped as successive generations overlap to an increasing extent. Lawton's experiments on Plodia illustrate that this "dephasing" of generations also occurs in the real world (of which his laboratory is a part!); consequently although mathematically unfamiliar in structure, our models probably constitute the minimal defensible representation of such laboratory populations.

Secondly, we note that in the other two models (AF and PS), the dominant cycles are likely to be of the delayed feedback type which produces a period in excess of twice the delay. The lovers of discrete generation models will recognise "two-point" cycles as for instance were found in the experiments of Fujii (1967, 1968) and described as the effects of "overcompensation". This overcompensation is due to the delayed influence on adult recruitment of an effect in larval population. Again however, the cycles may be damped - a second clear pointer to the need for simple continuous time models that can incorporate the overlap of generations.

There are of course many aspects of larval competition for food which we have not considered. Another suite of models emerges if we consider contest competition, a third if we assume that competition only occurs among individuals of the same age or size (cohort competition: Gurney, Nisbet and Lawton 1983). Also of potential importance is asymmetry, a property of proven significance in interspecific competition (Lawton and Hassell, 1981) but arguably as likely to occur between larval instars in a single-species population. Notwithstanding these difficulties, we believe that there is often sufficient experimental data in existence to justify a "component-by-component" assembly of arthropod population models. Furthermore the order in which various density-dependent factors act is known to have marked effects on population dynamics (cf. May et al 1981). There is thus a clear need for simple, interpretable models that do not rely on the fiction of discrete generations, and we propose that the "stage-structure" models used in this paper exploit a methodology which matches this need.

Acknowledgments

This work was motivated by John Lawton's systematic study of the population dynamics of *Plodia interpunctella*, and helped along by many discussions with him. Don Ludwig (at Trieste!) coined the phrase "stage-structure" models. Stephen Blythe contributed many useful comments on the first draft of this paper.

LITERATURE CITED

Auslander, D.M., Oster, G.F., and Huffaker, C.B. (1974). Dynamics of interacting populations. J. Franklin Institute 297, 345-376.

Beddington, J.R. Hassell, M.P., and Lawton, J.H. (1976). The components of arthropod predation: II. The predator rate of increase. J. Anim. Ecol. 45, 165-185.

Blythe, S.P., Nisbet, R.M., and Gurney, W.S.C. (1983). Formulating population models with differential aging. Proceedings of International Conference on Population Biology (H.I. Freedman, Editor) Springer-Verlag.

Cushing, J.M. (1980). Model stability and instability in age structured populations. J. Theor. Biol. 86, 709-730.

Busenberg, S. and Cooke, K.L. (1980). The effect of integral conditions in certain equations modelling epidemics and population growth. J. Math. Biol. 10, 13-32.

Fujii, K. (1976). Studies on interspecies competition between the azuki bean weevil Callosobruchus chinensis and the southern cowpea weevil C. maculatus: II. Competition under different environmental conditions. Research in Population Ecology 9, 192-200.

Fujii, K. (1968). Studies on interspecies competition between the azuki bean weevil and the southern cowpea weevil: III. Some characteristics of strains of two species. Research in Population Ecology 10, 87-98.

Gurney, W.S.C., Blythe, S.P. and Nisbet, R.M. (1980). Nicholson's blowflies revisited. Nature, London 287, 17-21.

Gurney, W.S.C. and Nisbet, R.M. (1983). The systematic formulation of delay-differential models of age or size structured population. Proceedings of International Conference on Population Biology (H.I. Freedman, Editor). Springer-Verlag.

Gurney, W.S.C., Nisbet, R.M. and Lawton, J.H. (1983). The systematic formulation of tractable single species models incorporating age-structure. J. Anim. Ecol. (in press).

Hastings, A. (1983). Age dependent predation is not a simple process - this volume.

Lawton, J.H. and Hassell, M.P. (1981). Asymmetrical Competition in Insects. Nature, London, 289, 793-795.

Lawton, J.H., Thompson, B.A., and Thompson, D.J. (1980). The effects of Prey Density on Survival and Growth of Damselfly Larvae. Ecol. Entomology 5, 39-51.

May, R.M., Hassell, M.P., Anderson, R.M., and Tonkyn, D.W. (1981). Density Dependence in Host-Parasitoid Models. J. Anim. Ecol. 50, 855-866.

Nicholson, A.J., (1954). An Outline of the Dynamics of Animal Population, Aust. J. Zool 2, 9-65.

Nisbet, R.M. and Gurney, W.S.C. (1982). Modelling Fluctuating Populations, John Wiley & Sons Ltd., Chichester.

Nisbet, R.M. and Gurney, W.S.C. (1983a). The systematic formulation of population models for insects with dynamically varying instar duration. Theor. Pop. Biol.

Nisbet, R.M. and Gurney, W.S.C. (1983b). The formulation of age structure models in Lecture Notes on Mathematical Ecology (T. Hallam, Ed.) Springer-Verlag.

Oster, G.F. and Ipaktchi (1978). Population Cycles in Periodicities in Chemistry and Biology, Vol. 4. (H. Eyring, Editor) pp 111-132. Academic Press, New York.

Oster, G.F. (1976). Lectures in Population Dynamics in Modern Modelling of Continuum Phenomena. (R.C.Di Prima, Editor) pp 149-190. American Mathematical Society, Providence, R.I.

Varley, G.C., Gradwell, G.R., and Hassell, M.P. (1973). Insect Population Ecology: An Analytical Approach. Blackwell Scientific Publications, Oxford.

APPENDIX - The Model Equations

Model LD

$$\dot{L}(t) = B(t) - B(t-\tau)P(t) - \Delta(t)L(t) \qquad (A1)$$

$$\dot{A}(t) = R(t) - \delta A(t) \qquad (A2)$$

$$\dot{P}(t) = P(t)\,\{\Delta(t-\tau) - \Delta(t)\} \qquad (A3)$$

with $R(t) = B(t-\tau)\,P(t)$, $B(t) = eA(t)$, and $\Delta(t)$ defined by equation (10). The characteristic equation determining local stability is (in dimensionless form)

$$\lambda^2(\lambda+\Theta) - \delta\tau(e^{-\lambda}-1)(\lambda^2+z_1\lambda+z_2) = 0 \qquad (A4)$$

with

$$\Theta = \left[\Delta^* + L^*\left(\frac{d\Delta}{DL}\right)^*\right]\tau \qquad (A5)$$

$$z_1 = \Theta - A^*\left(\frac{d\Delta}{dL}\right)^*\tau \qquad (A6)$$

$$z_2 = A^*\delta\left(\frac{d\Delta}{DL}\right)^*\tau^2\,\{\frac{e}{\delta} - 1\}, \qquad (A7)$$

in which the asterisk denotes the steady state value of a quantity. For stability, all values of λ satisfying (A4) must have negative real parts.

Model MT

$$\dot{L}(t) = B(t)-B(t-\tau(t))\,\exp\{-\Delta\tau(t)\}\frac{g(t)}{g(t-\tau(t))} - \Delta L(t) \qquad (A8)$$

$$\dot{A}(t) = R(t) - \delta A(t) \qquad (A9)$$

$$\dot{\tau}(t) = 1 - g(t)/g(t-\tau(t)) \qquad (A10)$$

with

$$B(t) = eA(t) \qquad (A11)$$

and

$$R(t) = B(t-\tau)\exp\{-\Delta\tau(t)\}g(t)/g(t-\tau(t)) \qquad (A12)$$

The characteristic equation is

$$\lambda^2 - (e^{-\lambda}-1)(z_1\lambda + z_2) = 0 \qquad (A13)$$

in which $z_1 = \delta\tau^* + (g^*)^{-1}\left(\frac{dg}{dL}\right)^* \delta\tau^* A^*$

$$z_2 = (g^*)^{-1}\left(\frac{dg}{dL}\right)^* \delta(\delta-e)(\tau^*)^2 A^*$$

Model PS

$$\dot{L}(t) = B(t) - B(t-\tau) \qquad (A16)$$

$$\dot{A}(t) = B(t-\tau)\alpha(t) - \delta A(t) \qquad (A17)$$

$$\dot{W}(t) = g(t) - g(t-\tau) \qquad (A18)$$

in which $\alpha(t)$, $g(t)$ are defined by equations (16) and (8) respectively.

The characteristic equation is

$$\lambda^3 - \delta\tau\lambda^2 (e^{-\lambda}-1) + z_1(e^{-\lambda}-1)^2 \qquad (A19)$$

with $z_1 = -e^2\tau^3 A^*(1-\delta/e)^2(dg/dL)^* K_W^{-1}$ (A20)

Model AF

$$\dot{L}(t) = B(t) - B(t-\tau) \qquad (A21)$$

$$\dot{A}(t) = B(t-\tau) - \delta A(t) \qquad (A22)$$

$$\dot{W}(t) = g(t) - g(t-\tau) \qquad (A23)$$

$$\dot{B}(t) = qB(t-\tau)W(t) - \delta B(t) \qquad (A24)$$

in which $g(t)$ is given by equation (8). The characteristic equation is

$$\lambda^3 - \delta\tau\lambda^2(e^{-\lambda}-1) + K(e^{-\lambda}-1)^2 = 0 \qquad (A25)$$

with $K = qB^*\left(\frac{dg}{dL}\right)^* \tau^3$ (A26)

SIMPLE MODELS FOR AGE DEPENDENT PREDATION

Alan Hastings
Department of Mathematics
University of California
Davis, California 95616

The study of predator-prey systems has had a long history in theoretical ecology, dating from the time of Lotka and Volterra. More recently the effects of age structure on the dynamics of interacting predator-prey systems has been considered. In this note I will summarize results described in more detail elsewhere on the stability and dynamics of a particular class of interactions between age structured prey and predators, namely age dependent predation. Numerous examples ranging from molluscs to insects to fish illustrate this phenomenon.

Some of the best documented cases of age dependent predation are from fish. Nielsen (1980) discusses the interaction between walleye and yellow perch, where the major diet item for adult walleye is juvenile yellow perch (see also Le Cren et al. 1977 and references in these papers). Other examples from other taxa are discussed in more detail in Hastings (1983a).

Predation only on juveniles or only on adults has been treated in a number of continuous time models. Particularly relevant to the discussion here are the following: For the simple interaction terms used by Lotka and Volterra, predation on only juveniles or on only adults has been shown to act as a stabilizing influence. In fact, the model, which is originally neutrally stable, becomes stable for all values of the parameters (May, 1974 and Smith and Mead, 1974), for particular choices of maturation functions. However, results for another model, where the predator eats only "eggs", seem to indicate that age dependent predation cannot be stabilizing, and in fact may be destabilizing (Gurtin and Levine, 1979).

In the current paper, I will summarize results on age dependent predation in continuous and discrete time models which include the possibility of an arbitrary functional response by the predator. (More details can be found in Hastings 1983a, 1983b). Since many if not most reasonable functional responses are destabilizing (Murdoch and Oaten, 1975), a primary question of whether age dependent predation can overcome the destabilizing effect and lead to a model with a stable equilibrium is important. The results will help to reconcile the disparate results described above.

CONTINUOUS TIME MODELS

I will first describe the set of models for the case where predation is only on adults. Let H(t) denote the number of adult prey at time t, and let P(t) denote the number of predators at time t. When there is a predator with a functional response, the equation for the prey dynamics will be assumed to have the form

$$(1) \qquad dH/dt = r\int_0^\infty H(t-s)G(s)ds - DH - Pf(H)$$

where G(z) is the probability that an individual survives to age z and matures from juvenile to adult at age z, r is the per capita birth rate, D is the death rate of adults and f(H) is the functional response of predators to prey (see e.g. Murdoch and Oaten, 1975).

The predator population will obey the following equation:

$$(2) \qquad dP/dt = cPf(H) - kP.$$

The model will be completely specified by a choice of the function G(s).

The different choices I will employ for G(s) are easier to express in terms of

$$(3) \qquad g(s) = 1 - \int_0^s G(x)dx,$$

the proportion of individuals remaining juveniles at age s, ignoring the effects of mortality. I will investigate three possibilities, namely:

$$(4) \qquad g_1(s) = 1, \quad s \leq \tau; \quad 0, \; s \geq \tau$$

$$(5) \qquad g_2(s) = e^{-\alpha s}$$

$$(6) \qquad g_3(s) = (1+s)e^{-\alpha s}$$

where the parameters τ and α measure the length of the juvenile period.

The most important difference among the three models is in the variance in the length of the juvenile period. No variance is represented by (4), with the variance in (6) intermediate between that of (4) and (5).

Analogous models can be formulated for the case where only juveniles are predated upon, although it is extraordinarily difficult to write these models down in such generality (Hastings, 1983a).

DISCRETE TIME MODELS

The general form of the discrete time models is the following: They are based on a Nicholson-Bailey host-parasite formulation with a Leslie type age structure in the prey. Various predator-prey models have been derived from the Nicholson-Bailey model, which is then a special case. Since the special case has fewer parameters it is the one that I will investigate first. I will first present a model where only juveniles (young of the year) are preyed upon (parasitized) by the predator (parasite). Let $H_i(t)$ be the number of prey (hosts) in age class i at time t, and let $P(t)$ be the number of predators (parasites). Then a general model is:

(7) $$H_0(t+1) = \left\{\sum_{i=0}^{\infty} B_i H_i(t)\right\} e^{-aP(t)} s_0$$

(8) $$H_i(t+1) = H_{i-1}(t) s_i; \quad i \geq 1,$$

(9) $$P(t+1) = \left\{\sum_{i=0}^{\infty} B_i H_i(t)\right\}\left\{1-e^{-aP(t)}\right\}$$

where B_i is the age specific birth rate and s_i is the age specific survival rate for the prey and a is the "attack area" for the predator. Note that the order of events in the life cycle of the prey is census, births, predation, survival, census. (Hence the birth rate may include some survival factors.) Similar models can be formulated incorporating other forms of age specific predation.

By special choices of the parameters the model can be simplified. In particular, I report results below for the following choices:

(10) $$B_0 = 0; \quad B_i = B, \quad i \geq 1$$

and either

(11) $$s_i = s$$

for all i, or else the case where (11) holds only up to some greatest age, with no survival after that.

RESULTS

I will merely outline the results of the analysis of these models here. For more detail the reader is referred to Hastings (1983a and 1983b) where the analysis is described. I will first summarize the results for the continuous time models, beginning with predation only on adults. Unless otherwise noted, I will be considering only destabilizing functional responses. For the continuous time models, the major results are a local stability analysis of the unique nontrivial equilibrium point with both predator and prey present, which exists whenever the predator's death rate is small enough relative to the predation term. Numerical work indicates that the local stability analysis is reflected in the global behavior. With the fixed juvenile period (4), and predation only on adults, one can show, using results of Cooke and Grossman (1982), that as the length of the juvenile period (τ) is varied, there are 'switches' in stability. For short enough juvenile periods, the model is unstable, then becomes stable as the juvenile period is increased and then unstable again. There can be (depending on the parameters) any number of switches, with the final switch being to unstable. One can also show that such switching behavior can occur with respect to any of the parameters in the model. Additionally, the unstable lengths of the juvenile periods correspond to the length of the predator-prey oscillation, as determined by linearizing the model with no juvenile period.

For the other maturation forms (5) and (6) the results are simpler. With (5), there is a single switch from instability to stability as the length of the juvenile period is increased. With (6), which has a smaller variance than (5), one complete set of switches (as in (4)) is possible.

For predation only on juveniles, the form (4) is too difficult to analyze. With the form (5) again only a single switch in stability is possible as the length of the juvenile period is increased. However, there are functional responses for which the model is unstable no matter how short the juvenile period is. Finally, quite complex behavior is possible with the form (6), even with the Lotka-Volterra functional response. Here, the model can be stable only for short juvenile periods, or only for intermediate juvenile periods, or for both.

For the discrete time models, I will consider only the two cases described above. Here, age dependent predation has a stabilizing influence, but global behavior is not well reflected by local behavior. The Nicholson-Bailey model without age structure is always unstable, and trajectories quickly wind out to the axes. For the age structured model described by (7)-(11) with no maximum age, the model always has a stable periodic attractor whenever the nontrivial equilibrium point (both predator and prey present) exists and is unstable. This periodic attractor remains quite far from the axes and is quickly approached from a very wide range of initial conditions. On the other hand if the maximum age for the prey is 10, or even 20, the range of initial values which are attracted to the periodic attractor (or the equilibrium) is so greatly reduced as to make persistence of the predator-prey system unlikely.

CONCLUSIONS

One of the primary conclusions is that age dependent predation is a complex phenomenon. It can act as a stabilizing influence; but whether or not it is stabilizing may depend on a complex interaction between time delays due to juvenile periods and inherent oscillations in the model, as illustrated by predation only on adults with the maturation form (4). Another general conclusion is that time delays and increases in time delays are not necessarily destabilizing (cf. Cooke and Grossman, 1982 and Cushing and Saleem, 1982). The variance in the length of the juvenile period seems to play an important role, with larger variances in general leading to greater stability. This observation, if it carries over to other models, may help us to explain and understand the importance of the large variance in juvenile period for many amphibians (Wilbur, 1980).

For the discrete time models, the important conclusion I wish to stress here is that changes in the model which are very subtle, such as removing all individuals greater than ten years of age, may have profound consequences, altering the persistence of the system. This can be true even if the survival rate is quite low, less than .5 per year, in which case individuals older than 10 represent an extremely small fraction of the total population. This suggests the fragility of some natural systems and the importance of rare events.

ACKNOWLEDGMENT. Supported by NSF Grant DEB-8002593.

LITERATURE CITED

Cooke, K. and Grossman, Z. 1982. Discrete delay, distributed delay and stability switches. Jour. Math. Anal. Appl. 86:592-627.

Cushing, J. and Saleem, M. 1982. A predator-prey model with age structure. J. Math. Biol. 14:231-250.

Gurtin, M.E. and Levine, D.S. 1979. On predator-prey interactions with predation dependent on age of prey. Math. Biosci. 47:207-219.

Hastings, A. 1983a. Age dependent predation is not a simple process. 1. Continuous time models. Theo. Pop. Biol. (in press).

Hastings, A. 1983b. Age dependent predation is not a simple process. 2. Discrete time models. Manuscript.

Le Cren, E.D., Kipling, C. and McCormack, J.C. 1977. A study of the numbers, biomass and year-class strengths of perch (Perca fluviatilis L.) in Windemere from 1941 to 1966. J. Anim. Ecol. 46:281-307.

May, R. 1974. "Stability and Complexity in Model Ecosystems, 2nd ed.", Princeton U.P., Princeton.

Murdoch, W. and Oaten, A. 1975. Predation and population stability. Adv. Ecol. Res. Res. 9:1-125.

Nielsen, L. 1980. Effect of Walleye (Stizostedion vitreum vitreum) predation on juvenile mortality and recruitment of yellow perch (Perca flavescens) in Oneida Lake, New York. Can. J. Fish. Aquat. Sci. 37:11-19.

Smith, R.H. and Mead, R. Age structure and stability in models of predator-prey systems. Theor. Pop. Biol. 6:308-322.

Wilbur, H. 1980. Complex life cycles. Annu. Rev. Ecol. Syst. 11:67-93.

A MODEL OF NATICID GASTROPOD PREDATOR-PREY COEVOLUTION[1]

D. L. DeAngelis,[2] J. A. Kitchell,[3] W. M. Post,[2] and C. C. Travis[4]

Introduction

This paper presents the first in a series of increasingly complex models of naticid gastropod predator-prey coevolution. This model is restricted to one predator and one prey and to coevolutionary feedback in size change. Subsequent models will involve multiple species and evolving morphologies. Gastropods of the family Naticidae are predators on other molluscs, both gastropods and bivalves, which they attack by drilling through the shells of their prey. The unique characteristics of the naticid gastropod system for testing aspects of coevolutionary theory in the fossil record were discussed by Kitchell et al. (1981) and Kitchell (1982).

Recent studies (Kitchell et al. 1981) indicated that prey selection by naticid gastropods is consistent with a strategy of maximizing utilization of energy and time. Relevant predator parameters, such as prey handling time, drilling rate, borehole geometry, limits of the ability of predators to manipulate potential prey, and probability of drilling success, each as functions of predator size, were obtained by laboratory studies on the naticid gastropod *Polinices duplicatus*. Relevant prey parameters, such as shell thickness at the standard site of drilling and biomass energy value, as functions of prey size, were also obtained for preferred species of bivalve prey and the cannibalistic interaction (see Kitchell et al. 1981). A cost/benefit parameter, time spent drilling per unit energy gained, was calculated for preferred prey species. These curves for the bivalve prey *Mercenaria mercenaria* as functions of prey length for two predator sizes are shown in Fig. 1. The descending limbs of the curves represent increasing energetic profitability to a predator due to energy of the prey increasing at a faster rate than drilling time, while the abruptly ascending limbs represent the sharply increasing difficulty of successful drilling of prey above a certain prey:predator size ratio. Using these curves, Kitchell et al. (1981) showed that rankings of prey energetic profitability predicted prey selection rankings reported from selection experiments.

---

[1]Research sponsored in part by the National Science Foundation's Ecosystems Studies Program under Interagency Agreement No. DEB80-21024 with the U.S. Department of Energy under Contract W-7405-eng-26 with Union Carbide Corporation, and in part by the National Science Foundation's Ecology Program Grant DEB81-09914 with the University of Wisconsin. Publication No. 2149, Environmental Sciences Division, ORNL.

[2]Environmental Sciences Division, Oak Ridge National Laboratory, Oak Ridge, Tennessee 37830

[3]Department of Geology and Geophysics, University of Wisconsin, Madison, Wisconsin 53706.

[4]Health and Safety Research Division, Oak Ridge National Laboratory, Oak Ridge, Tennessee 37830.

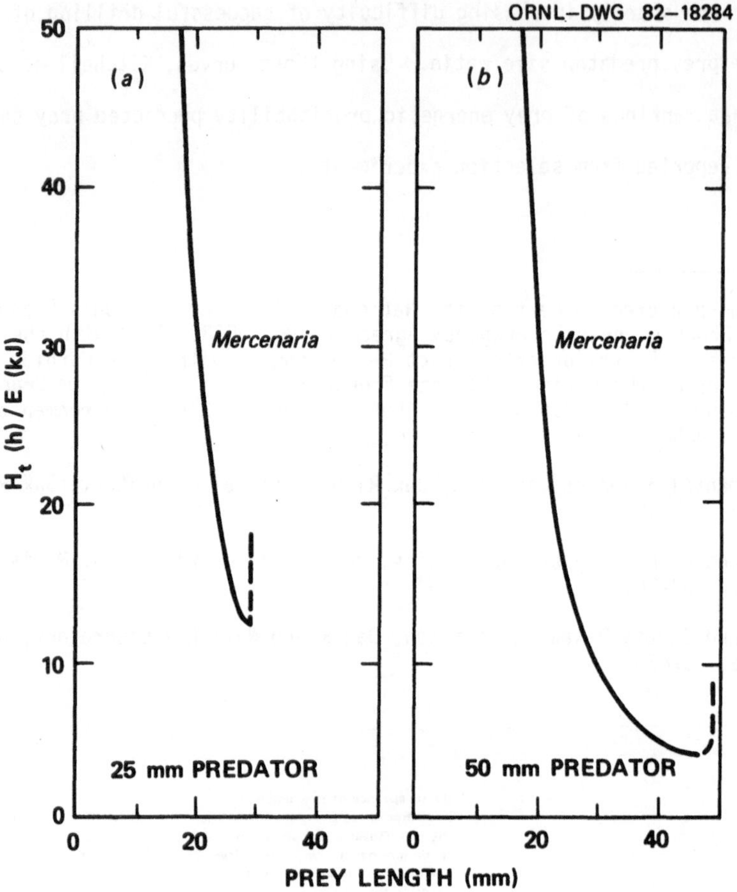

Figure 1. Cost benefit curves for Polinices duplicatus feeding on Mercenaria mercenaria; (a) for a predator length of 25 mm, and (b) for a predator length of 50 mm. H_t is prey handling time (in hours) and E is the energy (in kilojoules) obtained from prey of a particular size.

From Fig. 1 it is apparent that the predator's cost/benefit ratio from feeding is a function not only of its choice of prey but also of the predator's size. Larger predators are able to exploit a wider range of prey sizes. All other things being equal, there would be a selective pressure in favor of larger predator size. By contrast, the prey have two evolutionary options regarding size: (1) reducing the growth rate may keep prey sizes in the range of higher predator cost/benefit ratios, thus reducing prey desirability to the predator; (2) increasing the growth rate may increase the probability of prey reaching the critical escape size, denoted by the sharply ascending limbs of the cost/benefit curves (Fig. 1).

Consequently, a model of the energetics of predatory naticids and their prey was developed to determine whether or not coevolution in sizes generated by the predator-prey interaction is likely to occur in this system. In such a model, the advantages of faster growth in either facilitating predation (for the predator) or avoiding it (for the prey) are weighed against the increased costs of respiration due to diverting energy to growth and the alternative option of diverting energy to reproduction. The model contains no genetics - it is simply assumed that average predator and prey adult sizes can change through evolutionary time in response to selective pressure. Our model permits both the predator and prey simultaneously to seek their optimal sizes, given arbitrary sub-optimal starting conditions.

The coevolution model

A difficulty in relating evolutionary theory to its ecological context is the difference in optimization criteria (see Roughgarden 1978). In evolutionary theory one maximizes individual fitness whereas in ecological theory it is more common to consider maximization of population size or efficiency of use of a resource by a population as optimization criteria. Roughgarden showed that these two criteria often do not coincide. We make a deliberate choice of individual fitness as the optimization criterion. This choice is prompted by the available data, which tell us much about the evolving characteristics of individuals but little about population sizes.

The choice of which specific indicators of individual fitness to use is a second difficulty. The effective reproductive contribution over a lifetime,

$$J = \int_0^\infty \ell(x) b(x) e^{-rx} dx \quad , \qquad (1)$$

is a reasonable, convenient representation of individual fitness (e.g., Goodman 1982). In Eq. (1) x is the age of an individual, $\ell(x)$ is survival to age x, b(x) is reproduction (for our purposes the energy devoted to reproduction, which can be translated into the number of offspring times the energetic investment per offspring) at age x, and r is the growth rate of the population. If r is set to zero, which is reasonable for a stable population size, J is the investment in offspring that an individual produces in its lifetime.

This definition of fitness must necessarily incorporate all adaptations to the environment that influence survival and reproduction, many of which cannot be quantified. In developing a model of coevolution, we simplified the problem by ignoring the influence on evolution of aspects of prey fitness not related to energetics or naticid predation (i.e., effects of other predators, parasites, or competitors, and abiotic factors).

Over evolutionary time the prey is allowed to change the percentage of its energy that it devotes to growth at different ages. Energy diverted into growth at a given age is not available for reproduction at that age. The combination of age-dependent survival and reproduction that maximizes J for the prey is sought.

An equation equivalent to Eq. (1) could also be used for the predator. However, we make the simplifying assumption that the amount of energy a predator can devote to reproduction is the surplus, or is proportional to the surplus, of the rate at which the predator extracts energy from the prey population minus the predator's own respiration losses. We also assume that predator survival is affected only through its size-dependent ability to drill prey successfully. Hence, predator fitness can be represented by its net rate of energy intake. This assumption may not be too unrealistic. Lawton et al. (1975) showed that the reproductive output of several arthropods is approximately proportional to their energy intake.

In both the prey and the predator the size and fitness (prey reproductive value and predator rate of energy intake) form feedback loops that can be either positive or negative. The coevolutionary connection comes through the effect of prey size on predator energy intake and the effect of predator size on prey reproductive potential. Prey and predator population densities are not included as variables in the model. They occur in the model only as constant parameters whose values are approximated from present-day data.

<u>Predator energy intake</u> - We begin with a standard model for the energy intake of a predator feeding on a single prey species and specialize this model for the naticid predator-prey interaction.

Let us define the fractions of time a predator spends searching for and handling prey by T_s and T_h, respectively. Also, define by T_r the fraction of time (assumed fixed here) spent by the predator on all other activities. Then

$$T_s + T_h + T_r = 1 \quad . \tag{2}$$

The number of predator "attacks" on prey per unit time per predator for prey of type j (i.e., age class j) is

$$N_j = a_j s_j n_j T_s \quad , \tag{3}$$

where

a_j = coefficient of predator-prey encounter rate,

s_j = selection coefficient; i.e., the probability that the predator will attack a prey individual of type j given an encounter, and

n_j = population number density (per unit area) of prey.

The total predator handling time is

$$T_h = \sum_j N_j(t_{d,j} + p_j t_{c,j}) \quad , \tag{4}$$

where

$t_{d,j}$ = average time spent drilling a prey individual of size j

$t_{c,j}$ = average time spent consuming a prey individual that has been successfully drilled, and

p_j = probability of successful "capture" (drilling) of a prey, given an attack.

From Eqs. (2), (3), and (4), we can solve for T_s;

$$T_s = \frac{(1 - T_r)}{1 + \sum_j a_j s_j n_j (t_{d,j} + p_j t_{c,j})} \quad , \tag{5}$$

from which N_j is easily computed using Eq. (3);

$$N_j = \frac{a_j s_j n_j (1 - T_r)}{1 + \sum_j a_j s_j n_j (t_{d,j} + p_j t_{c,j})} . \quad (6)$$

The net energy flow to the predator is

E_t = (energy gained from the prey) - (energy respired during consumption) - (energy respired during attack) - (energy respired during searching) - (energy respired during other activities), i.e.,

$$E_t = \sum_j \{p_j \alpha_j E_j N_j - p_j e_{c,j} t_{c,j} N_j - e_{d,j} t_{d,j} N_j\} - \xi_s T_s - \xi_r T_r , \quad (7)$$

where

α_j = assimilation coefficient,

E_j = energy value of prey of age class j,

$e_{d,j}$ = energy expended per unit time in drilling prey of type j,

$e_{c,j}$ = energy expended per unit time in consuming prey of type j,

ξ_s = energy expended per unit time during searching, and

ξ_r = energy expended per unit time during other activities.

Finally, E_t can be written as

$$E_t = \frac{(1-T_r) \sum_j a_j s_j n_j \{p_j(\alpha_j E_j - e_{c,j} t_{c,j}) - e_{d,j} t_{d,j}\} - \xi_s}{1 + \sum_j a_j s_j n_j (t_{d,j} + p_j t_{c,j})} - \xi_r T_r . \quad (8)$$

The energy intake, E_t, can be expressed in terms of predator length, L_g, by expressing ξ_s, ξ_r, $e_{d,j}$, $e_{c,j}$, and p_j in terms of L_g. This is done for the first four of these parameters in Table 1. The probability of successful capture of prey depends on the length of the prey's shell, L_b, and on the predator length. Laboratory studies (Kitchell et al. 1981) show that there is a sharp break between size ratios for which predation can and cannot occur. For Mercenaria prey (see Fig. 1) we define the function as

Table 1. Empirical regressions and parameter values relevant to P. *duplicatus*

Measurement	Regression	Parameter values
Weight[a]	$W_g = 10^{-\alpha_1+\alpha_2 L_g}$ g	$\alpha_1 = 0.885$ $\alpha_2 = 0.052$
Respiration[a]	$\xi_s = \xi_t = \beta_3 10^{\beta_1} W_g^{\beta_2}$ kJ/year $e_{c,j} = e_{d,j} = \beta_4 \xi_s$	$\beta_1 = 2.43$ $\beta_2 = 0.536$ $\beta_3 = (24)(365)(2.02)$ $\beta_4 = 1.2$
Time of drilling[b]	$t_{d,j} = S_t/D_r$	$D_r = 0.0223$ mm/h S_t = see Table 2
Time of consumption[b]	$t_{c,j} = t_{d,j}$	
Assimilation coefficient[a]		$\alpha_g = 0.7$

[a]Huebner and Edwards 1981.

[b]Kitchell et al. 1981.

$$p_j = \begin{cases} 0 & (L_b > L_g) \\ 1 & (L_b \leq L_g) \end{cases} . \tag{9}$$

Prey reproductive contribution - The reproductive contribution, J, for a steady-state population with (as assumed for convenience here) discrete age classes, can be written

$$J = \sum_j \ell_j b_j \quad , \tag{10}$$

where ℓ_j is survival to age j and b_j is the amount of reproduction at age j. We take b_j to represent the amount of energy put into reproduction at age j, with no specification as to how the energy is apportioned into separate offspring.

The amount of surplus energy that is available for either growth or reproduction is given by

$$\text{Growth + Reproduction = Consumption - Feces - Excretion - Respiration.} \tag{11}$$

To model an "abstract" bivalve we use, as a starting point, data on energy intake, excretion and respiration for M. mercenaria. From this information it is possible to write a differential equation for growth of the bivalve if the fraction of energy diverted to reproduction is known, or can be estimated, for each age class. We have assumed that the fraction of energy diverted to reproduction will increase monotonically, with the general form of the family of curves in Fig. 2a. Typical growth curves of the prey will have the form shown in Fig. 2b. These resemble real bivalve growth curves (e.g., Valentine 1973), although precise description of known growth curves is not our aim. For computation of J, prey survival to age class j, ℓ_j, must also be computed. Prey mortality consists of predation mortality caused by the gastropod and background mortality from all other causes (see Table 2).

Finally, the predator selection coefficient, s_j, measures the decision of whether or not to attack a prey of age class j (s_j = 0 means no, s_j = 1 means yes). The ideal predator will not attack a prey individual whose relative size is such that p_j = 0, and it will not choose a prey individual so small that the energy balance is negative; i.e., that

$$p_j(\alpha_j E_j - e_{c,j} t_{c,j}) - e_{d,j} t_{d,j} < 0 \quad . \tag{12}$$

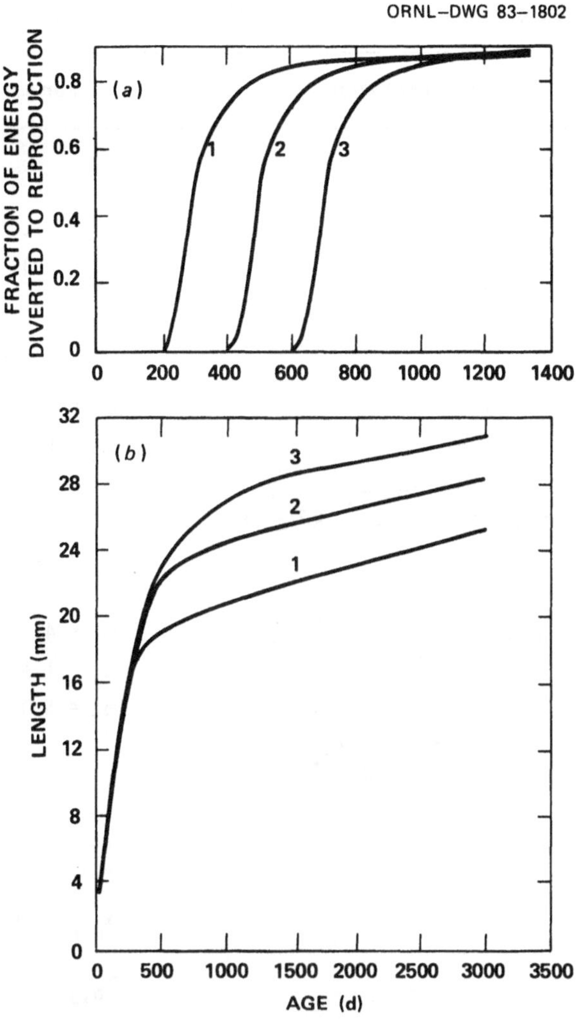

Figure 2. (a) A series of three hypothetical prey strategies of allocation of energy into reproduction (for different values of a_3; see Table 2); (b) the corresponding growth through time.

Model simulations

To determine whether maximum predator length is to be increased or decreased in a given coevolutionary time step, the current energy intake of the adult predator is compared to the net energy intake of the preceding time step. If the net energy intake has increased and at the same time the predator length has been increased, then a further increase in predator length is prescribed; a negative

Table 2. Empirical regressions and parameter values relevant to *M. mercenaria*

	Regression	Parameter values
Weight[a]	$W_b = 10^{\gamma_1} L_b^{\gamma_2}$ g	γ_1 = -5.544 γ_2 = 3.279 γ_3 = 21.17
Energy value[a]	$E_1 = \gamma_3 W_b$ kJ	
Shell thickness[b]	$S_t = 10^{\sigma_1 + \sigma_2 L_b}$ mm	σ_1 = -0.389 σ_2 = 0.0126
Energy intake[a]	$E_{prey} = \eta_1 10^{\eta_2} L_b^{\eta_3}$ kJ/year	η_1 = 8.6 η_3 = 1.0 η_2 = 1.388
Respiration[a]	$R_{prey} = \omega_1 10^{-\omega_2} L_b^{\omega_3}$ kJ/year	ω_1 = 4.83 ω_2 = 1.73 ω_3 = 1.016
Survival[a]	$\ell_j = e^{-\sum_j m_{b,j} - \sum_j m_{g,j}}$	
Predation mortality[c]	$m_{g,j} = m'_g s_j$	
Assimilation		α_{prey} = 0.29
Background mortality[c]		$m_{b,j}$ = 0.2 year^{-1}
Predation mortality coefficient[c]		m'_g = 5.0 year^{-1}
Energy allocation to reproduction[c]	$E_{repro} = \dfrac{a_4(\alpha_{prey} E_{prey} - R_{prey})}{1 + a_1(t - a_3)^{-a_2}}$ (for $t > a_3$; E_{repro} = 0 otherwise)	a_1 = 900.0 a_2 = 1.6 a_3 = variable a_4 = 0.97

[a]Hibbert 1977.

[b]Kitchell et al. 1981.

[c]Arbitrary estimates.

correlation would cause a change in predator length in the opposite direction on the next step. In either case, the predator length is changed by an amount between 0 and 1 mm chosen by means of a uniform pseudo-random number generator.

The prey switching strategy (from growth to reproduction) is altered evolutionarily in exactly the same way. Next the reproductive value, J, is assessed. If a change in the age of switching improved J on the last time step, the same change is made again in the next time step. Computer simulations were performed with the coevolutionary model (available from the authors on request) using the best possible estimates of those parameters for which data were available. Reasonable guesses had to be made for some parameter values (see Table 2).

Before we simulated coevolution, we allowed the prey to evolve in the absence of the predator. Regardless of which starting conditions were used, the prey evolved to a stable equilibrium with a strategy of switching from growth to reproduction at about 0.6 years. This corresponds to a prey size of about 40 mm at maturation. Evidently, above a size of about 30 mm, natural mortality and the increase in prey respiration more than compensate for the enhanced reproduction due to increased adult size.

To determine which direction the coevolution of predator length and prey reproductive strategy would take under a variety of initial conditions, several runs were made with different initial sizes of the adult predator and switching ages of the prey. Sixteen combinations of the initial four prey switching ages and four predator lengths were tested. The seed for the pseudo-random number generator was changed for each set of 16 combinations. The results of the sets of runs were approximately the same, indicating that stochasticity had only a slight effect on the coevolutionary directions of change. The results for 16 combinations are shown in Fig. 3.

There are two attractors evident in Fig. 3. One occurs at a predator length of zero and prey maturation at 220 days and attracts trajectories for predator lengths that start too small. The other attractor occurs at a predator size of about 30 to 35 mm and a prey adult maturation time of approximately 330 days.

ORNL–DWG 83–1940

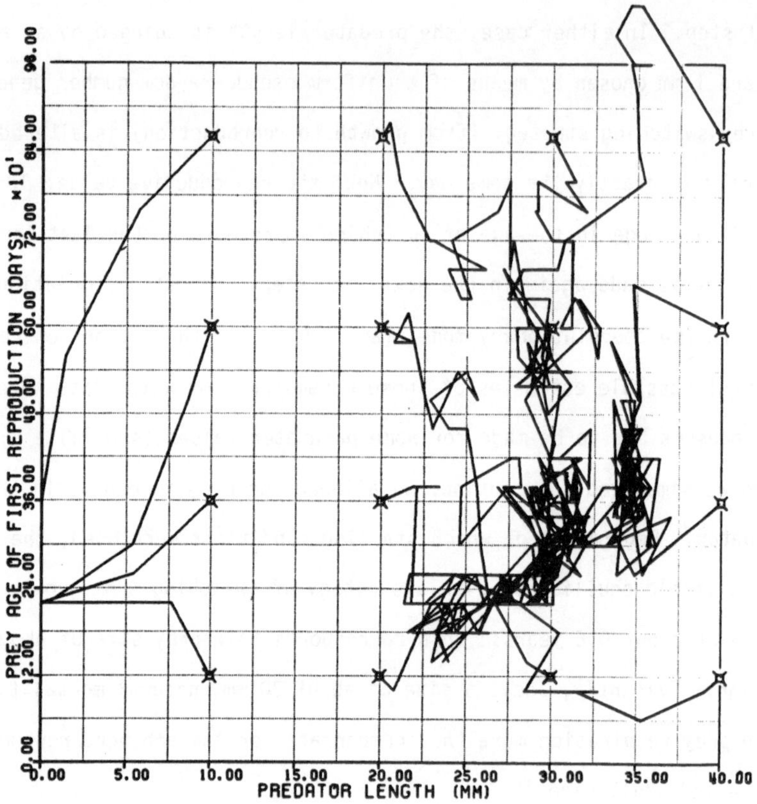

Figure 3. Coevolution trajectories for different prey and predator length starting values.

Because the model is a discrete-time step model, there is considerable fluctuation about this second attractor, especially by the prey population.

Approach to a particular attractor, as well as the rate of approach through evolutionary time, depends on the interplay between the sizes of the prey and the predator. The jerkiness of the trajectories probably reflects the discrete size increments allowed the predators and prey. For short initial predator sizes ($\leq$10 mm) the predator has little effect on the prey switching strategy. The prey evolves as if the predator was not present. The predator, when initially small, tends to specialize on smaller classes of prey by becoming increasingly smaller itself, until, at least according to the model, it vanishes. This vanishing is quite probably an artifact, caused in part by the fact that important factors may have been omitted. These omitted factors would no doubt reduce the fitness of the predators before they become too small and, therefore, would act to prevent further evolution towards smaller lengths.

For larger predator sizes (20 to 40 mm) the coevolutionary dynamics are dominated by changes in predator size. Predators move fairly rapidly toward their optimum energy efficiency, while prey maturation time changes more slowly. But note from Fig. 3 that when the predator is present the prey delays maturation beyond its optimal value (in the absence of predators). The presence of the predator reduces the fitness of prey by decreasing prey survival. The prey respond by delaying their age of switching to reproduction so that there is greater probability that prey attain an adult size that is safe from predator attack. The predator is apparently limited by increasing respiration from continuing a coevolutionary "chase" beyond about 35 mm.

Discussion

Phyletic size increase is a very common phenomenon among fossil invertebrates (e.g., Newell 1949, Gould 1977), and evolutionary size increase among bivalves has been frequently reported (e.g., Hallam 1975). Kitchell (in prep.) analyzed change in size among naticid gastropods within the genera Polinices, Natica, and Neverita (data from Marincovich 1977). A pronounced mean size increase over Cenozoic time has occurred in all three genera of predatory naticids. In this study, we restrict our analysis to size change over time between two interacting species, a predatory naticid gastropod and its bivalve prey. These species, though patterned on actual, modern species, are intended to be abstract "models" of species that might have existed at any time during the Cenozoic. The coevolutionary model is designed to study the propensity of system behavior, not to make detailed predictions.

We show that two simultaneous, maximizing algorithms (the predator maximizes energy intake; the prey maximizes reproductive output) result in an endogenous, coevolutionary size change, to a stable attracting point. In particular, we show that selection for delayed reproduction in a predator-prey system that is highly size-selective due to the predatory strategy of cost/benefit prey selection, coupled with the relative allometries of cost (prey shell thickness) and benefit (prey biomass) with prey size and the highly size-dependent probability of successful predation, lead to a coevolutionary size change for both predator and prey, to a limit condition dictated by predatory respiration costs. In the absence of predation, the prey species begins reproducing at an earlier age, resulting in smaller sizes for given ages than when predation is present. Addition of the predator results in a delay in the timing of reproduction by the prey, thereby facilitating a size response.

Acknowledgements

We appreciate helpful comments on earlier versions of this paper by J. E. Breck, J. L. Elmore, R. V. O'Neill, R. I. Van Hook and three anonymous reviewers.

LITERATURE CITED

Goodman, D. (1982). Optimal life histories, optimal notation, and the value of reproductive value. Am. Nat. 119:803-823.

Gould, S. J. (1977). Ontogeny and Phylogeny. Belknap Press, Harvard University, Cambridge, Massachusetts.

Hallam, A. (1975). Evolutionary size increase and longevity in Jurassic bivalve and ammonites. Nature 258:493-496.

Hibbert, J. C. (1977). Growth and survivorship in a tidal-flat population of the bivalve Mercenaria mercenaria from Southhampton. Water Mar. Biol. 44:71-76.

Huebner, J. D. and Edwards, D. C. (1981). Energy budget of the predatory marine gastropod Polinices duplicatus. Mar. Biol. 61:221-226.

Kitchell, J. A. (1982). Coevolution in a predator-prey system. Proc., Third No. Amer. Paleontol. Conv., Vol. 2:301-305.

Kitchell, J. A., Boggs, C. H., Kitchell, J. F. and Rice, J. F. (1981). Prey selection by naticid gastropods: Experimental tests and application to the fossil record. Paleobiology 7(4):533-552.

Lawton, J. H., M. P. Hassell, and J. R. Beddington. 1975. Prey death rates and rate of increase of arthropod predator populations. Nature 255:60-62.

Marincovich, L., Jr. (1977). Cenozoic Naticidae (Mollusca:Gastropods) of the northeastern Pacific. Bull. Am. Paleontol. 70:165-494.

Newell, N. D. (1949). Phyletic size increase, an important trend illustrated in fossil invertebrates. Evolution 3:103-124.

Roughgarden, J. (1978). Coevolution in ecological systems III. Coadaptation and equilibrium population size. pp. 27-48. IN P. F. Brussard (ed.), Ecological Genetics: The Interface. Springer-Verlag, New York.

Valentine, J. 1973. Evolutionary Paleoecology of the Marine Biosphere. Prentice-Hall, Inc., New Jersey.

LISTS OF FIGURES

Figure 1. Cost benefit curves for Polinices duplicatus feeding on Mercenaria mercenaria; (a) for a predator length of 25 mm, and (b) for a predator length of 50 mm. H_t is prey handling time (in hours) and E is the energy (in kilojoules) obtained from prey of a particular size.

Figure 2. (a) A series of three hypothetical prey strategies of allocation of energy into reproduction (for different values of a_3; see Table 2); (b) the corresponding growth through time.

Figure 3. Coevolution trajectories for different prey and predator length starting values.

A THEORETICAL MODEL FOR THE COEVOLUTION OF A HOST AND ITS PARASITE

S.D.Jayakar

Department of Genetics and Microbiology, University of Pavia, Pavia, Italy.

Introduction

It is well known that parasites are adapted to their hosts. This adaptation is a result of evolution on the part of the parasitic species. On the other hand, it can, except perhaps in exceptional circumstances, be of no advantage to the host species that the parasite be so well adapted to it. One would expect therefore that the host species show a tendency to evolve away from this adaptation. This would result then in a situation in which there would be continuous "coevolution" of the two species, the host tending to "get away" from the parasite and the parasite tending to "catch up". The precise biological changes implied in this coevolution depend of course on the kind of organisms involved.

There are two fields of study of the interaction between hosts and parasites which have been fairly extensively studied, namely: (a) their population dynamics, dealing purely with the relative numbers of the species involved in the interaction (see e.g. Scudo and Ziegler 1978) and (b) changes in the genetic composition of the species (see e.g. Flor 1955, Jayakar 1970, Leonard 1977, and Lewis 1981). Slatkin and Maynard Smith (1979), in their review of coevolutionary theory referred to two families of models. The first concerned host-parasite interaction and dealt only with gene frequency changes; the second concerned prey-predator relationships and dealt only with changes in population densities without any reference to genetic variation. It would be more realistic to consider the two aspects together, and in this paper we intend to develop some simple models of the type which take into account both population densities and their genetic compositions. This has been done by Levin and Udovic (1977) for two interacting populations (species). They discussed the general aspects of such a model. Although this model was in discrete time, they compared the analysis of their model with the equivalent model in continuous time. In particular they analysed the stability of the boundary equilibria. Gillespie (1975) analysed a host-parasite model in which genetic variation is considered in the host, but for the parasite only the demographic response is investigated. We will deal with a rather general situation involving parasites which infect a host individual with a given probability, this infection resulting in some sort of loss of fitness of the host, and in reproduction on the part of the parasite. Although the formulation of these models will be of a general nature, after the general presentation we will provide an example of an attempt at applying the model to a particular problem.

Specific genetic (single gene) changes in a host which alter the capacity of the parasite to infect it are known, for example, in bacteria-bacteriophage systems. In plants, genetic polymorphisms in hosts and parasites (rusts) are known which govern the resistance of the hosts and the capacities of the parasites to infect the host. More complicated genetic systems have also been worked out in some plant-parasite systems. In animals also there are some well understood examples; for example Drosophila species and the cynipid wasp parasite Pseudeucoila bochei in which the genetic bases of host resistance and of overcoming this resistance by the parasite have also been studied. In higher animals, immunogenetics has begun to show the importance of specific genetic differences in resistance to several kinds of infections. Clearly the latter relationships are extremely complicated multilocus systems and modelling them is beyond our scope here.

Although we refer throughout to the host-parasite relationship, the models developed here would apply equally well to the prey-predator situation. Formally, there would be no difference in the models. The interpretation of the parameters used would of course be entirely different.

Before discussing the models involving both species densities and genetic compositions, let us briefly review the purely frequency models.

Frequency models

Although we do not wish to discuss the purely frequency models here in any detail, it is nevertheless instructive to describe two of them briefly in order to introduce the type of biological parameter that is being contemplated.

Jayakar (1970) considered a host species (η) and a parasite species (π). The genetically simplest form of the model was that in which both these species were assumed to be haploid. A host genotype (η_1) was assumed to be susceptible to the parasite genotype π_1. However, a mutant η_2 was assumed to be resistant to π_1, but not to π_2 which was capable of infecting both η_1 and η_2. The only parameter in the model was the probability that a given host individual is infected if it is susceptible. Several basic variations of this basic theme were also considered and analysed.

The second model we will describe is that studied by Leonard (1971). The model is very similar in structure to the Jayakar model. The parameters he considered were: (1) the cost of virulence, in other words the loss in fitness incurred by parasites of a given genotype which had an increased infective capacity; (2) the effectiveness of host resistance in suppressing pathogen reproduction; (3) the advantage to the pathogen in having a gene which breaks down the resistance of the host; (4) the loss of fitness of the susceptible host attacked by the avirulent parasite genotype; and (5) the cost of resistance on the part of the host.

Various generalisations of this basic type of model have been studied by various authors - a general theory has been published by Lewis (1981).

A fairly general model could be set out as follows. If we have a set of host genotypes η_i, $i=1,\ldots,m$ and a set of parasite genotypes π_j, $j=1,\ldots,n$, one could consider then the following interaction properties.

κ_{ij} = Prob {contact between an individual of η_i and one of π_j results in infection} - one assumption implicit here is that it is very unlikely that a host individual encounters more than one parasite during the susceptible stage of its life history.

μ_{ij} = Prob {individual of η_i dies as a result of infection by a π_j, given that it has been infected}.

α_{ij} = average number of offspring produced by π_j on η_i if η_i dies.

β_{ij} = average number of offspring produced by π_j on η_i if η_i survives.

These parameters can however be reduced for our purposes to two for each host-parasite pair, namely,

average reproduction of the parasite =

$$\nu_{ij} = \kappa_{ij}\,\{\mu_{ij}\alpha_{ij} + (1-\mu_{ij})\beta_{ij}\}$$

and average host mortality =

$$\lambda_{ij} = \kappa_{ij}\mu_{ij}.$$

An analysis of such a model shows that in the absence of fitness differences between the genotypes due to causes other than the parasitic interaction, only one host genotype and one parasite genotype can survive - which particular pair survives depends on the relative values of the ν_{ij} and the λ_{ij}. Clearly such models are unrealistic and very limited from the biological point of view. Indeed, any realistic model must consider population densities as well, and this we will do in the rest of this paper.

Density frequency models

Although it is mathematically more convenient and less restrictive in most cases to model evolutionary processes in continuous time, this becomes impossible for the study of genetical changes in diploid organisms. Since we wish to develop fairly general genetic models, this requires us to set them up as discrete processes. One has to be careful however since in this case some variables of the system can at some stage assume negative values inadmissible as they would be biologically meaningless.

Before we introduce population densities into the models it is essential to make some assumptions regarding the regulation of the population densities of the species involved. In the simple models developed here we will assume that in the absence of parasites each host population is logistically controlled; in other words, for each of m host species, if H_i represents the density of the ith host, the one generation difference equation is of the type

$$H_i' = H_i(1 + r_i - \frac{r_i H_i}{K_i}) \; ; \quad i=1,\ldots,m, \tag{1}$$

where r_i and K_i are the standard parameters of such a model, namely those which are usually referred to as the rate of natural increase and the carrying capacity respectively. It is implicit in the above formulation that we are assuming no interaction of the competitive type between host species. Each host has, in other words, its own independent resources.

The parasite species on the other hand are assumed to be completely dependent on the given set of hosts for survival. We assume in other words that in the absence of these species each parasite species would independently of the others die out exponentially; if there are n parasite species and P_j denotes the density of the jth, we have

$$P_j = P_j t_j \; ; \quad j=1,\ldots,n, \tag{2}$$

where t_j is the proportion surviving after one generation.

In this framework we now introduce the consequences of parasitisation. Since the combination (i,j) of host-parasite implies a mortality of λ_{ij} host/individual host/individual parasite, the total host mortality due to this combination will be $\lambda_{ij}H_iP_j$. The parasite yield on the other hand will be $\nu_{ij}H_iP_j$. Equations (1) and (2) must then be corrected for these quantities and they become

$$H_i' = H_i(1 + r_i - \frac{r_i H_i}{K_i} - \sum_j P_j \lambda_{ij}) \qquad (3)$$

and

$$P_j' = P_j \sum_i H_i \nu_{ij} t_j \, . \qquad (4)$$

Several variations of detail in the assumptions of this basic model can be considered but would not make any difference to the conclusions.

The equilibria of this model can be calculated by solving the equations obtained by putting

$$H_i' = H_i = \hat{H}_i, \quad i= ,\dots,m; \quad P_j' = P_j = \hat{P}_j, \quad j= ,\dots,n,$$

where $\hat{H}_i$ and $\hat{P}_j$ represent the equilibrium densities. This gives a system of $(m+n)$ equations:

$$\sum_j \hat{P}_j \lambda_{ij} = r_i(1 - \frac{\hat{H}_i}{K_j}) \, , \quad i=1,\dots,m; \qquad (5)$$

$$\sum_i \hat{H}_i \nu_{ij} = 1/t_j \, , \quad j=1,\dots,n. \qquad (6)$$

For these equilibrium values to make sense they must evidently be positive quantities. Otherwise the species with negative values must be excluded from the system.

With this general formulation of the model, our intention is next to see the consequences of analysing the model for limited numbers of species so that fairly simple explicit solutions can be written down. Let us then consider the case $m = n = 2$.

Analysis of the case $m = n = 2$

In this situation it is possible to write down explicit equilibria for the system (3), (4). These are:

$$\hat{H}_1 = \frac{\nu_{22}/t_1 - \nu_{21}/t_2}{\nu_{11}\nu_{22} - \nu_{12}\nu_{21}} \, ; \qquad \hat{H}_2 = \frac{\nu_{11}/t_2 - \nu_{12}/t_1}{\nu_{11}\nu_{22} - \nu_{12}\nu_{21}} \, ; \qquad (7.1)$$

$$\hat{P}_1 = \frac{\lambda_{22}\phi_1 - \lambda_{12}\phi_2}{\lambda_{11}\lambda_{22} - \lambda_{12}\lambda_{21}} \, ; \qquad \hat{P}_2 = \frac{\lambda_{11}\phi_2 - \lambda_{21}\phi_1}{\lambda_{11}\lambda_{22} - \lambda_{12}\lambda_{21}} \, ; \qquad (7.2)$$

where $\phi_i = r_i(1 - \frac{\hat{H}_i}{K_i})$.

It is interesting in this case to look for conditions under which particular combinations of these four species can coexist. The strategy we will follow in order to do this will be that of testing for the stability of the boundary equilibria, i.e. those with incomplete subsets of these species. Suppose for example that η_1 and π_1 are present in the community. The equilibrium between them is maintained at the densities:

$$\hat{H}_1 = 1/\nu_{11}t_1 \; ; \quad \hat{P}_1 = \phi_1/\delta_{11} = \frac{r_1}{\lambda_{11}}(1 - 1/\nu_{11}t_1K_1).$$

For this to be biologically meaningful, we must have $\hat{P}_1$ positive.

In order to examine now whether the species η_2 can be successfully introduced into this community, we need to evaluate the matrix of partial derivates of the system with the three species η_1, η_2 and π_1 and find the conditions under which we have an eigen value greater than 1. In other words if η_2 is introduced at a small density, it can also establish itself if and only if we have an eigenvalue greater than 1. It turns out that at the (η_1,π_1) equilibrium the derivatives $\partial H_2'/\partial H_1$ and $\partial H_2'/\partial P_1$ are zero, and consequently it is sufficient to check whether the partial derivative $\partial H_2'/\partial H_2$ of the three species system calculated at the equilibrium given above is greater than 1.

But

$$\left.\frac{\partial H_2'}{\partial H_2}\right|_{(\hat{H}_1,\hat{P}_1)} = 1 + r_2 - \hat{P}_1\lambda_{21}$$

$$= 1 + r_2 - \frac{\lambda_{21}}{\lambda_{11}} r_1(1 - 1/\nu_{11}t_1K_1).$$

Thus η_2 can establish itself iff.

$$\frac{r_2\lambda_{11}}{r_1\lambda_{21}} \geqq 1 - 1/(\nu_{11}t_1K_1). \quad (8)$$

By symmetry, the condition for η_1 to be able to invade the equilibrium with η_2 and π_1 present is given by:

$$\frac{r_1\lambda_{21}}{r_2\lambda_{11}} \geq 1 - 1/(\nu_{21}t_1K_2). \qquad (9)$$

If (8) and (9) are both satisfied then both η_1 and η_2 can coexist with π_1. What is more important is that one of these species can be eliminated if one of these conditions is violated. Remember however that these two species, under the assumptions of the model, are not in competition and interact only through their common parasite. The fact of having a parasite in common has thus introduced some sort of competition between the host species independently of resource utilisation.

A similar analysis shows that π_2 can invade the (η_1,π_1) equilibrium iff.

$$\left.\frac{\partial P_2'}{\partial P_2}\right|_{(\hat{H}_1,\hat{P}_1)} = \hat{H}_1\nu_{12}t_2 \geq 1,$$

i.e. iff. $\nu_{11}t_1 \leq \nu_{12}t_2.$ (10)

This is a reciprocal relationship which implies that one host can only support one parasite species at a time. This is characteristic of models of this type, and can be a limitation if we wish to examine situations which are contrary to the one predicted.

Further by evaluating $\partial P_2'/\partial P_2$ in the situation (η_1,η_2,π_1) one can examine the possibility of establishing a complete four species community. The condition turns out to be fairly simple but tedious to write down, namely:

$$\frac{1}{t_1\{\frac{\nu_{12}K_1\lambda_{11}}{r_1} + \frac{\nu_{22}K_2\lambda_{21}}{r_2}\}} - \frac{1}{t_2\{\frac{\nu_{11}K_1\lambda_{11}}{r_1} + \frac{\nu_{21}K_2\lambda_{21}}{r_2}\}} <$$

$$\frac{\nu_{12}K_1 + \nu_{22}K_2}{\frac{\nu_{12}K_1\lambda_{11}}{r_1} + \frac{\nu_{22}K_2\lambda_{21}}{r_2}} - \frac{\nu_{11}K_1 + \nu_{21}K_2}{\frac{\nu_{11}K_1\lambda_{11}}{r_1} + \frac{\nu_{21}K_2\lambda_{21}}{r_2}}.$$

A similar condition holds for the introduction of π_1 in the (η_1,η_2,π_2) community. A suitable choice of numerical values for the parameters shows that it is possible to establish a stable situation with two hosts and two parasite species.

Some numerical examples

Let us consider the following set of parameters:

$r_1 = 0.2 \quad K_1 = 2000 \quad t_1 = 0.5$

$r_2 = 0.4 \quad K_2 = 500 \quad t_2 = 0.5$

$\lambda_{11} = 0.001 \quad \lambda_{12} = 0.001 \quad \nu_{11} = 0.005 \quad \nu_{12} = 0.005$

$\lambda_{21} = 0.001 \quad \lambda_{22} = 0.002 \quad \nu_{21} = 0.005 \quad \nu_{22} = 0.010.$

Then the (η_1,π_1) equilibrium is given by:

$\hat{H} = 400; \quad \hat{P} = 160.$

Similarly the (η_2,π_2) equilibrium is given by:

$\hat{H} = 200; \quad \hat{P} = 120.$

Clearly it is not the absolute values of the parameters that are important but their values relative to each other.

Since in the above example

$$\frac{r_1\lambda_{22}}{r_2\lambda_{12}} \geqq 1 - 1/(\nu_{22}t_2K_2),$$

η_1 can be successfully introduced into the community (η_2,π_2) which then reaches the new equilibrium:

$\hat{H}_1 = 800/3; \quad \hat{H}_2 = 200/3; \quad \hat{P}_2 = 520/3.$

However the equation corresponding to eqn. (11) is not satisfied so that π_1 will never be able to invade the community. A change in parameters, for example $t_1 = 2/3$, will however make it possible to establish a four species community.

As an example of a situation in which one species will replace another in the presence of a parasite, let us consider the following parameter set:

$r_1 = 0.2 \quad K_1 = 2000 \quad t_1 = 0.5 \quad \nu_{11} = 0.002$

$r_2 = 0.4 \quad K_2 = 2000 \quad t_2 = 0.5 \quad \nu_{21} = 0.002$

$\lambda_{11} = 0.001 \quad \lambda_{21} = 0.001.$

In Fig. 1, which shows the effect of introducing π_2 into the (η_1,π_1) equilibrium, we can see the oscillatory approach to the new equilibrium typical of all such models, both of the pure density type and of the pure frequency type.

Genetic models

The model proposed above concerns separate species of hosts and of parasites. They would also be valid with minor modifications for different genotypes of the same haploid species, except for the small probabilities of mutation from one type to another. It would perhaps be preferable to modify the structure of competition between the types since different genotypes of the same species are likely to com-

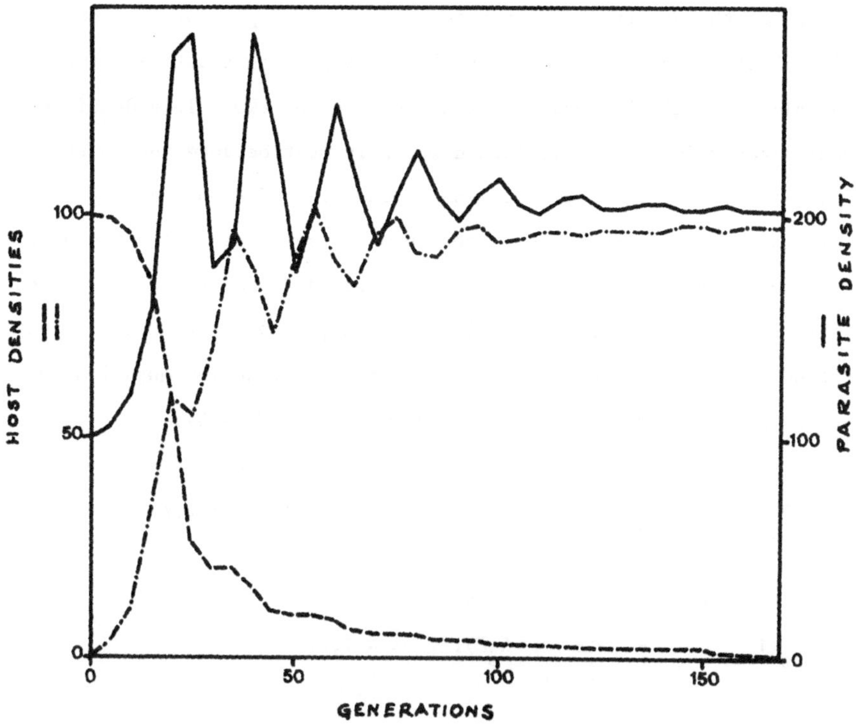

Figure 1. The dynamics of the replacement of one host by another due solely to their having a common parasite.

pete for resources. In this case eqn. (3) would have to be modified as follows:

$$H_i' = H_i\{1 + r_i - r_i \frac{\sum_j H_j c_{ij}}{K_i} - \sum_j P_j \lambda_{ij}\} \qquad (12)$$

where c_{ij} is a measure of the competition of the jth host type on the ith. If we consider all the c_{ij} to be identical as a first approximation we can then incorporate them into the K_i and thus obtain the following modified equation:

$$H_i' = H_i\{1 + r_i - r_i \frac{\sum_j H_j}{K_i} - \sum_j P_j \lambda_{ij}\}. \qquad (13)$$

Adequate adjustments must be made for the models concerning diploid organisms. Consider for example one diploid host and one monomorphic parasite species. If one considers a single genetic locus in the host genome with two alleles A_1 and A_2, one has to consider the relative frequencies of the three genotypes. Let us refer to the product of the genotypic frequencies and the host population density as the "densities" of the genotypes, and to the corresponding quantities for the alleles as allele "densities". Thus, if H_1 and H_2 are the two allele densities, the one generation recursion can be written down. This must be done in two stages. In the first stage, starting from the gene frequencies of A_1 and A_2 (p_1 and p_2, $p_1 + p_2 = 1$) in any given generation, one can write down the zygote frequencies assuming random mating and no selection differences at this stage. We will then have the three genotypes in the proportions p_1^2, $2p_1p_2$, p_2^2. After the effects of competition and mortality due to the parasites, and ignoring mutation, we will have equations analogous to eqn. (3), namely (with $H_1 + H_2 = H$):

$$H_{ij}' = H_{ij}(1 + r_{ij} - r_{ij}\frac{H}{K_{ij}} - \lambda_{ij}P); \qquad i,j = 1,2; i \leqq j. \tag{14}$$

The recursion for the allele densities will be:

$$H_i' = H_i(1 + R_i - S_iH - \Lambda_iP); \qquad i = 1,2, \tag{15}$$

where

$$R_1 = \frac{H_{11}r_{11} + H_{12}r_{12}/2}{H_1}; \qquad R_2 = \frac{H_{22}r_{22} + H_{12}r_{12}/2}{H_2};$$

$$S_1 = \frac{H_{11}r_{11}/K_{11} + H_{12}r_{12}/2K_{12}}{H_1}; \qquad S_2 = \frac{H_{22}r_{22}/K_{22} + H_{12}r_{12}/2K_{12}}{H_2};$$

$$\Lambda_1 = \frac{H_{11}\lambda_{11} + H_{12}\lambda_{12}/2}{H_1}; \qquad \Lambda_2 = \frac{H_{22}\lambda_{22} + H_{12}\lambda_{12}/2}{H_2}.$$

Let us assume for simplicity however that the genotypes have identical values of r and K so that this rather complicated system of recursions reduces to:

$$H_i' = H_i(1 + r - \frac{rH}{K} - \Lambda_iP). \tag{16}$$

The Λ_i are however functions of the Hs and are therefore not constant in time. An alternative way of expressing this system is in the two variables H and $p = p_1$. It then becomes:

$$H' = H(1 + r - rH/K - \Lambda P); \tag{17.1}$$

$$p' = p(1 + r - rH/K - \Lambda_1P)/(1 + r - rH/K - \Lambda P). \tag{17.2}$$

where $\Lambda = \Lambda_1H_1 + \Lambda_2H_2$.

Thus at equilibrium we will have

$$\Lambda_1P = \Lambda P = r(1 - H/K). \tag{18}$$

In addition, from the recursion for the parasite density, we have:

$$P' = Pt\Lambda,$$

and at equilibrium,

$$\Lambda = 1/t. \quad (19)$$

From (18) and (19) one can calculate the equilibrium values and their stability conditions.

To introduce more complicated genetical situations into this scheme is merely a matter of writing down the systems that are so generated. This kind of system can be analysed in any particular problem which one wishes to investigate and the genetical consequences, such as conditions for the maintenance of genetic polymorphisms, could be explored. It would not be instructive however to carry the general analysis any further here.

A particular problem

Let us consider the following evolutionary problem. It has been observed in several host species, for example in barley (Hordeum vulgare) and flax, that when the host carries two or more genes resistant to different pathogens, they commonly belong to the same linkage group and often are allelic to each other. The reasons for this may be physiological in that it may be more likely that genes resistant to similar pathogens can be more easily found as mutants at a locus which already carries resistant alleles, or there may be a few such loci available. The reason may however lie in the particular mathematical structure of the genetic system. This problem suggests the following model.

Consider a host species with a gene resistant to parasite π_1 at locus A. Assume now that a second parasite π_2 is introduced, and that the host has a choice of two resistance genes, one situated at the same locus and one at another locus B. Let us denote by A_0 and A_1 the susceptible and (dominant) resistance genes to π_1. Let A_2 be the gene resistant to π_2 at the A locus and B_2 that at the B locus where the susceptible gene is B_0. There are then six possible chromosomes (see Table 1) with their relative frequencies. Let us assume that the amount of resistance conferred against the two species of parasites is σ_1 and σ_2 respectively. The other parameters of the model are listed in Table 2.

It is now a fairly straightforward matter to write down the recursion equations relative to the six host chromosomal "densities" and the two parasite densities (Table 3). This system of 8 independent equations is however extremely difficult to

Table 1. Genetic variation of resistance.

		Locus A		
		Susc.	Res. 1	Res. 2
Locus B	Susc.	A_0B_0	A_1B_0	A_2B_0
	Res. 2	A_0B_2	A_1B_2	A_2B_2

Frequency of recombination between A and B = θ.

Table 2. Parameters and the values used in the numerical example, and variables.

Parameters

K	=	carrying capacity common to all host genotypes	= 500
r	=	rate of natural increase of all host genotypes	= 0.2
t_1, t_2	=	intrinsic survival rates of the parasitic species	= 0.5, 0.5
λ_1, λ_2	=	probability of death of host due to infection	= 0.002, 0.002
ν_1, ν_2	=	reproduction rate of parasites on host / infection	= 5, 6
σ_1, σ_2	=	degrees of resistance conferred by the genes relative to the two parasites	= 0.5, 0.5
θ	=	frequency of recombination between the two loci	

Variables

p_{ij}	=	frequency of chromosome A_iB_j
N	=	host density
n_i	=	density of parasite i

solve analytically. In order to obtain some idea of the equilibria of the system we have resorted to a numerical investigation with a given set of parameters (see Table 2). Note that an important assumption of the model is that the two genes resistant to π_2 confer the same degree of resistance. Clearly one can modify the model to remove this restriction.

It emerges from the numerical investigation that for any value of θ there are at least three equilibria present, all of which are boundary equilibria in the sense that one of the chromosome frequencies is zero. The three equilibria are shown in Table 4. Two of these equilibrium types are such however that the first

Table 3. System of recursion equations.

$$N'p_{00}' = N\Big[p_{00}(1+\rho)-n_1\lambda_1\{p_{00}(1-p_{10}-p_{12})-D_0\theta\}-n_1\sigma_1\lambda_1\{p_{00}(p_{10}+p_{12})-D_1\theta\}$$
$$-n_2\lambda_2 p_{00}(p_{00}+p_{10})-n_2\sigma_2\lambda_2\{p_{00}(1-p_{00}-p_{10})-D_0\theta-D_1\theta\}\Big]$$

$$N'p_{10}' = N\Big[p_{10}(1+\rho)-n_1\sigma_1\lambda_1(p_{10}+D_1\theta+D_2\theta)$$
$$-n_2\lambda_2 p_{10}(p_{00}+p_{10})-n_2\sigma_2\lambda_2\{p_{10}(1-p_{00}-p_{10})+D_1\theta+D_2\theta\}\Big]$$

$$N'p_{20}' = N\Big[p_{20}(1+\rho)-n_1\lambda_1\{p_{20}(1-p_{10}-p_{12})+D_0\theta\}$$
$$-n_1\sigma_1\lambda_1\{p_{20}(p_{10}+p_{12})-D_2\theta\}-n_2\sigma_2\lambda_2(p_{20}+D_0\theta-D_1\theta\}\Big]$$

$$N'p_{02}' = N\Big[p_{02}(1+\rho)-n_1\lambda_1\{p_{02}(1-p_{10}-p_{12})+D_0\theta\}$$
$$-n_1\sigma_1\lambda_1\{p_{02}(p_{10}+p_{12})+D_1\theta\}-n_2\sigma_2\lambda_2(p_{02}+D_0\theta+D_1\theta)\Big]$$

$$N'p_{12}' = N\Big[p_{12}(1+\rho)-(n_1\sigma_1\lambda_1+n_2\sigma_2\lambda_2)(p_{12}-D_1\theta-D_2\theta)\Big]$$

$$N'p_{22}' = N\Big[p_{22}(1+\rho)-n_1\lambda_1\{p_{22}(1-p_{10}-p_{12})-D_0\theta\}$$
$$-n_1\sigma_1\lambda_1\{p_{22}(p_{10}+p_{12})+D_2\theta\}-n_2\sigma_2\lambda_2(p_{22}-D_0\theta+D_2\theta)\Big]$$

$$n_1' = Nn_1t_1\nu_1\lambda_1\{\sigma_1+(1-\sigma_1)(1-p_{10}-p_{12})^2\}$$

$$n_2' = Nn_2t_2\nu_2\lambda_2\{\sigma_2+(1-\sigma_2)(p_{00}+p_{10})^2\}$$

where

$$\rho = r(1-N/K)$$
$$D_0 = p_{00}p_{22} - p_{02}p_{20}$$
$$D_1 = p_{00}p_{12} - p_{02}p_{10}$$
$$D_2 = p_{12}p_{20} - p_{10}p_{22}.$$

parasite π_1 is eliminated from the system (types A and B). In type C, where both parasites are maintained at the stable equilibrium, the chromosomal types which are present in non-zero frequencies are A_1B_0, A_2B_0 and A_0B_2. In other words, both resistant genes are retained. It is interesting to note also that the chromosomes which are retained in the B equilibrium are complementary to those retained at the C equilibrium.

It would of course be interesting to know what the regions of attraction of these equilibria are. In general this task is immensely difficult. For particular parameter sets, one could obtain some idea of their extent by numerical trials. That these regions depend on the recombination frequency θ is evident from the fact that in our numerical investigations, we obtained convergence to different

Table 4. The stable equilibria.

A. $\hat{n}_1 = \hat{p}_{00} = \hat{p}_{10} = \hat{p}_{20} = 0$

$\hat{N} = \{t_2\nu_2\lambda_2\sigma_2\}^{-1}$

$\hat{n}_2 = \{\rho\lambda_2\sigma_2\}^{-1}$

the values of $\hat{p}_{02}$, $\hat{p}_{12}$, and $\hat{p}_{22}$ depend on initial conditions.

B. $\hat{n}_1 = \hat{p}_{10} = \hat{p}_{20} = \hat{p}_{02} = 0$

$$\hat{N} = \left[\sigma_2 + \frac{\sigma_2{}^2\theta_2(1-\sigma_2)}{\{1-\sigma_2(1-2\theta)\}^2}\right]^{-1} / t_2\nu_2\sigma_2$$

$\hat{p}_{00} = \sigma_2\theta/\{1-\sigma_2(1-2\theta)\}$

$$\hat{n}_2 = \frac{\rho\{1-\sigma_2(1-2\theta)\}}{\lambda_2\sigma_2\{1-\sigma_2(1-\theta)\}^2}$$

the values of $\hat{p}_{12}$ and $\hat{p}_{22}$ depend on initial conditions.

C. $\hat{p}_{00} = \hat{p}_{12} = \hat{p}_{22} = 0.$

none of the equilibrium values depend on initial conditions. $\hat{N}$ and $\hat{p}_{10}$ are independent of θ but the equilibrium values of the other variables depend on θ (see Fig. 2).

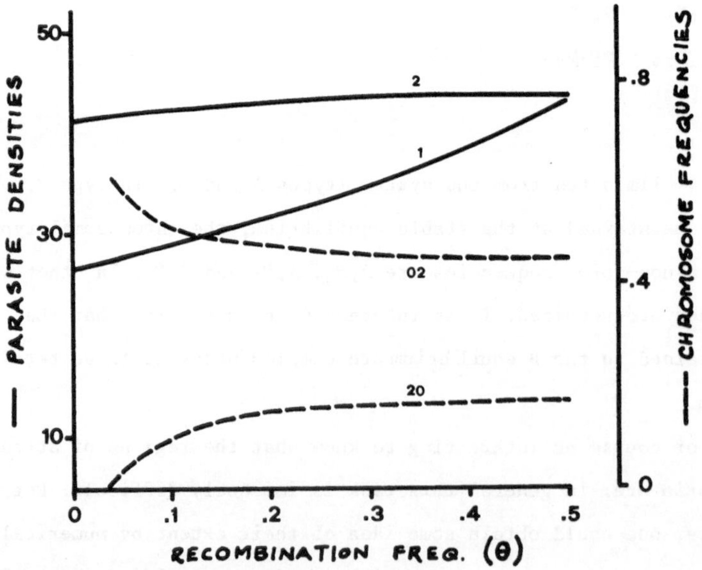

Figure 2. Variation of equilibrium C with the recombination frequency θ. There is a threshold value of θ (≅ 0.04) below which $\hat{p}_{20}$ becomes zero.

equilibria from the same initial conditions for different values of θ. In fact θ is likely to be a key parameter in the outcome of such a model.

Discussion

We have suggested here an extension of the standard population genetics techniques to the question of the coevolution of a set of host species (or genotypes) and a set of parasite species (or genotypes). We feel that such a formulation is essential for answering questions regarding the genetic strategies which hosts and parasites would have to adopt for survival.

Though the model in its complete form is complicated, to answer specific questions regarding evolutionary problems it can be suitably simplified and rendered more amenable to analysis. Thus, of the several parameters considered above relative to single species characteristics and interaction properties, one could investigate the evolutionary path of a species by considering the fate of genetic variants for a single characteristic at a time.

The advantages of these models over previous models are that they consider both the densities and the genetic compositions of the host and the parasite species. We feel that this kind of approach is essential in host-parasite evolutionary models.

Though we have referred throughout this paper to a host-parasite system, as we have mentioned in the Introduction, the same kind of approach is valid also for prey-predator systems, or indeed for any system which considers coevolution at two trophic levels, and all the considerations of the model would hold after suitable reinterpretation of the parameters. As we have stressed while developing the model, the changes in fine detail, for example the choice of the density regulation of the host species, would not require extensive changes in its formulation. Our intention here is to present a theoretical approach to the study of the coevolution of hosts and parasites, or in general at two different trophic levels.

Biologically, in the course of coevolution, there would be quantitative changes in the interaction parameters. These changes can also be modelled without considering their genetics (see e.g. Levin, this volume), but it is probably more informative from the evolutionary point of view to consider more specifically genetic models. A one locus model is probably not realistic, but it helps in getting a feel for the kind of strategy (change in parameters) a species is likely to adopt in a given situation, and as such could help us to make predictions regarding the changes which species interactions will undergo, in other words regarding coevolution.

Acknowledgements

This work was supported by the National Research Council (C.N.R.), Italy.

LITERATURE CITED

Flor, H.H. (1955). Host-parasite interaction in flax rust - its genetic and other implications. Phytopathology 45:680-685.

Gillespie, J.H. (1975). Natural selection for resistance to epidemics. Ecology 56: 493-495.

Jayakar, S.D. (1970). A mathematical model for the interaction of gene frequencies in a parasite and its host. Theor. Pop. Biol. 1:140-161.

Leonard, K.J. (1977). Selection pressures and plant pathogens. Ann. N. Y. Acad. Sc. 287:201-222.

Levin, S.A. and Udovic, J.D. (1977). A mathematical model of coevolving populations. Amer. Nat. 111:657-675.

Lewis, J.W. (1981). On the coevolution of pathogen and host: I. General theory of discrete time coevolution; II. Selfing hosts and haploid pathogens. J. Theor. Biol. 93:927-985.

Scudo, F.M. and Ziegler, J.R. (1978). The Golden Age of Theoretical Ecology: 1923-1940. Lecture Notes in Biomathematics, 22. Springer-Verlag, Berlin, Heidelberg, New York.

Slatkin, M. and Maynard Smith, J. (1979). Models of coevolution. Quart. Rev. Biol. 54:233-263.

PART III
COMMUNITY AND ECOSYSTEM THEORY

PARTICLE SIZE SPECTRA IN ECOLOGY

William Silvert

Marine Ecology Laboratory
Bedford Institute of Oceanography
P. O. Box 1006
Dartmouth, Nova Scotia B2Y 4A2
CANADA

INTRODUCTION

The use of size distributions as descriptors of ecosystems goes back at least to Elton (1927), although quantitative applications have appeared only during the past decade. Elton noted that there was a characteristic difference between terrestrial and aquatic ecosystems; on land the organisms in the lower trophic levels tend to be larger than the organisms which feed on them (e.g. trees, giraffes, tigers), while in the sea the reverse is true (phytoplankton, zooplankton, fish). The value of this observation was dramatically emphasized by the advent of automated particle size measurement devices and their use in marine biology, which occurred in the 1960's (Sheldon and Parsons 1967). A Coulter counter was used on the 1970 circumnavigation of the western hemisphere by the Canadian research vessel Hudson and this provided the first opportunity to compare particle size distributions from widely distributed pelagic marine environments, including both polar regions and the tropical waters of the Atlantic and Pacific oceans (Sheldon *et al*. 1972). This work greatly extended Elton's observation by showing that there are also patterns in the size distributions from temperate and polar regions which are characteristically different from those found in tropical waters.

The results of the Hudson '70 cruise led to a flurry of interest in the nature of these patterns and how they arose, and it became evident that bioenergetic models based on the conservation of energy could explain the observed regularities (Kerr 1974, Sheldon *et al*. 1977, Platt and Denman 1977, 1978, Silvert and Platt 1978, 1980). Furthermore these models predicted that similar patterns should be found in freshwater systems, and size patterns have indeed been observed, although they are not as smooth as those found in offshore environments (Sprules and Holtby 1979). Since these models involve assumptions about the size structure of predator-prey relations it has usually been assumed that they would not apply in systems with different feeding patterns; for example, filter-feeding is an important size-dependent feeding strategy characteristic of aquatic ecosystems, and the trophic structure of benthic fauna seems different from what is found in the water column. Therefore even though patterns of size distribution had also been observed among bottom fauna (Ursin 1973, Thiel 1975), these were not usually expected to be as general or significant as those found in pelagic systems. It therefore came as a surprise to many workers in the field when Schwinghamer (1981) showed that benthic ecosystems exhibit regularities at least as pronounced as those found in pelagic systems, and that the same patterns are found in spatially separated systems with different community structure and bottom type. Although it was certainly recognized that other factors could be as important as food web structure in determining the size distributions of ecosystems (Smith 1976), there is no quantitative theory of how this occurs, and the absence of such a theory has left many ecologists with the feeling that such patterns must be vague and qualitative. Since these benthic results indicate that the patterns are pronounced and predictable, it seems that theoretical ecologists will have to address the problem of developing quantitative theories of particle size distributions in communities for which size-dependent feeding is not the dominant organizing force.

THE PARTICLE SIZE SPECTRUM

In its simplest terms a particle size spectrum is a histogram describing how many particles fall into each of several different size ranges. Translated into operational terms there are several refinements and complications which arise. The size classes are usually defined so as to be equal on a logarithmic scale, and automated counting systems usually use classes which increase by a factor of two in volume or mass (Sheldon and Parsons 1967). There are generally many more particles in the smaller size classes than there are in the larger ones, so total volume or biomass is a more convenient variable than number. The definition of the size of a particle is a difficult question which must often be settled in a less than satisfactory way by operational considerations such as the idiosyncrasies of the measuring apparatus; for living particles the appropriate measure might be dry weight or surface area (Harding 1977, Brodie 1982), but often something less satisfactory like a linear dimension is used. The issue, although important, is not necessarily of fundamental concern, since they are all highly correlated; however, the use of a convenient proxy measurement like wet weight to describe the energy content of an organism is bound to increase the scatter of the data around any model which is basically correct (Cammen 1980).

Given all these difficulties, it is indeed surprising to find that particle size distributions are fairly regular. The most general and striking characteristic of many of the marine distributions is that the total biomass in logarithmically equal size intervals is roughly comparable, even though the size range covered is enormous (from a few micrometers to several dozen meters in length, meaning a factor of more than 10^{20} in mass or volume, and a correspondingly great factor in the number of particles). This is one of several patterns which are characteristic of different environments; pelagic distributions tend to be smoothest, while coastal distributions have a regular pattern of bumps, and benthic distributions have two pronounced and characteristic minima. Thus there are two types of theoretical questions to be dealt with; one is to understand the basic regularity, and the other is to understand how these different patterns arise.

BIOENERGETIC MODELS

Theoretical work so far has focussed on explaining the regularity of pelagic distributions, using conservation of energy as the underlying assumption. Kerr (1974) developed the first equilibrium model for a size-structured food chain, while Platt and Denman (1977, 1978) looked at the steady state formulation of a continuous model incorporating growth and respiration. The Platt-Denman model was developed into a dynamic model by Silvert and Platt (1978), and in a later form of the model they incorporated spatial heterogeneity and some degree of taxonomic disaggregation (Silvert and Platt 1980).

The Silvert-Platt model is basically in the form of a von Foerster equation, which in turn is based on the flow equations of hydrodynamics; the chief difference between this model and previous ecological applications of the von Foerster equation (Sinko and Streifer 1967, Rubinow 1973, Streifer 1974, Levin 1976) is in the way in which the biomass or energy is transferred from one part of the size spectrum to another. Hydrodynamic flows are continuous, since molecules of fluid must follow continuous trajectories, but this is not true in the modelling of size distributions. One can treat size as a fourth dimension, and while flows due to normal growth are continuous (i.e., when an organism grows from one size to another it passes through all intermediate sizes), the same is not true of energy transfers due to predation or reproduction, or indeed to growth involving metamorphosis. Furthermore, some of these processes, predation for example, are nonlinear, which is a substantial mathematical complication. It appears that these nonlinearities may destabilize the system and be responsible for the strength of the seasonal cycles which are observed in temperate marine ecosystems (Silvert and Platt 1980).

The basic form of the continuous equations is:

$$D\beta/Dt+\mu\beta = S$$

where β is the biomass density as a function of size w and the position vector, μ is a loss term incorporating respiration and non-grazing mortality, and the four-dimensional flow derivative D/Dt is defined by

$$D\beta/Dt = \partial\beta/\partial t+\partial(\beta g)/\partial w+\mathrm{div}(\beta v)$$

where g=dw/dt is growth. This is a generalization of the standard hydrodynamic flow derivative and represents the change in biomass density as one moves along with the particle; in hydrodynamics this corresponds to following the advective flow of the fluid in which the particle is embedded, and we have added a term to represent the change in size of the particle as well. In other words, the flow derivative represents the difference between the biomass density for particles of a given size at a specific point, and the density for slightly larger particles located downstream at a later time. The source term S incorporates discontinuous flows such as the transfer of energy from large particles to the smaller sizes by reproduction, and the transfers due to predation which usually go from smaller organisms to larger ones in aquatic systems. Growth plays the role of velocity in the fourth (size) dimension in this formulation; however, the situation is complicated by the essential nonlinearities introduced by predation, since not only are the sources and sinks on the right-hand of the equation affected by this, but also the growth rate g depends on ingestion.

A similar formalism has been used in other applications of the von Foerster equation mentioned previously, with age or patch size corresponding to the "fourth dimension", but the discontinuous flows represented by the source term S represents a major difference.

A serious difficulty associated with the existence of flows which are discontinuous in the size "dimension" is that boundary conditions must be specified over an interval, rather than simply at the boundary of the region. In biological terms this means that for a continuous model, such as the original Silvert-Platt (1978) formulation, the phytoplankton are represented by particles of a single size which drive the entire system, but in the later approach an entire size spectrum of primary production must be specified (Silvert and Platt 1980).

Because of these mathematical difficulties, detailed solutions of the model equations have not yet been carried out. Before doing so it seems essential to ensure that all aspects of the model formulation are correct, and there are grounds for believing that the predation submodel used may not be realistic. Silvert and Platt (1980) used the quadratic interaction characteristic of Lotka-Volterra models, but when the size range of particles grazed by a given size of predator is large this may lead to spurious instabilities of the type associated with omnivory in Lotka-Volterra food web models (Silvert 1983). This can be avoided by using more realistic submodels, such as one based on size-selective feeding (Silvert, in preparation).

THE ROLE OF DIMENSIONALITY

The bioenergetic approach, although satisfactory for the study of pelagic ecosystems, does not seem to be justified in physically structured environments such as those associated with benthic communities. The regularity of the distributions found for these communities (Schwinghamer 1981) suggests that conservation of energy is only one of the factors affecting the underlying size structure of ecosystems (Smith 1976). The existence of a general topology for size structures in different environments indicates that there may be a fundamental link between these environments and the kinds of particle size distributions they can support, regardless of the evolutionary history of the particular region.

An obvious feature which distinguishes pelagic communities from benthic or terrestrial ones is dimensionality, in that pelagic marine organisms live in an essentially three-dimensional environment, while terrestrial organisms are restricted to two dimensions and soft bottom benthic environments seem to fall somewhere in between. While this idea requires further quantification (dimensionality is not normally considered a continuous variable, even though the use of fractional dimensions has proved useful in the study of phase transitions), some of the ways in which dimensionality affects size structure are quite clear. The existence of a two-dimensional substrate is essential for the growth of macrophytes, and thus pelagic marine ecosystems are distinguished by the fact that primary production is restricted to unicellular algae. Because of physiological constraints on feeding behaviour this means that herbivores must also be small, which leads to the close connection between size and trophic level which must be an important contributing factor to the relatively uniform size distributions found in these systems. Another contributing factor is the fact that in water the maximum speed of a swimming organism is closely related to its size, so that larger raptors can chase their prey more rapidly. However, there is no signficant dependence of speed on size in the terrestrial environment (Gunther and Morgado 1982), so frequently predators are smaller than their prey. Dimensionality and the importance of gravity on land also play a role here, since a small active raptor often catches prey by crippling it and bringing it down, while in the water the predator more often needs to engulf the prey completely to capture it, or at least to attach itself securely to the prey without being able to pin it against a substrate. Furthermore, terrestrial raptors often operate in groups and surround prey, which is practical in two dimensions but is virtually impossible in three.

Since primary production on land is mainly by macrophytes, the size of herbivores is not constrained in the same way as in the water. A more important factor appears to be seasonal changes in plant biomass which make mobility essential. Herbivores tend to be much more mobile on land than in the water, even though the range of sizes (from small insects to elephants) is great. Thus we find that primary production consists of macrophytes ranging in size from grasses to trees, but the herbivores almost always include large vertebrates such as ruminants, and the carnivores are of an intermediate size.

The smaller terrestrial animals, notably insects, may have an airborne phase, and thus they occupy an environment of dimensionality intermediate between land and water. This may be why the marine pattern of larger eating smaller is more prevalent among this group, since the capture process and dependence of speed on size among flying carnivores is similar to that found in the sea. Similar remarks may apply to animals which live within the soil.

Another aspect of dimensionality is that it is easier to search on a plane than in three dimensions, and thus spatial dispersion plays a different role in the two environments. This may also be important in determining the size structure of predators, since a small randomly searching predator has a better chance of finding prey in two dimensions than in three.

The benthic community is intermediate in many respects, particularly since some macrobenthic organisms have pelagic phases. There are many factors which make the benthic environment unique. Predator-prey size ratios are not as regular as they tend to be in other environments, and often this is directly related to the nature of the substrate; deposit feeders, for example, can scrape diatoms and bacteria off the surface of sand grains in a way which permits even large benthic grazers to feed on these microscopic organisms. It is likely that studies of size structure in benthic communities will play a major role in helping us understand the relative importance of a two-dimensional substrate in a three-dimensional living universe.

DISCUSSION

The above arguments are highly speculative, since it is one thing to show that we can build plausible models of particle size distributions in different types of ecosystems but quite another to demonstrate that these models lead to fundamental insights about how these different ecosystems are organized. Certainly the existence and nature of a substrate is a major determinant of ecosystem organization, and Marcotte (1978) has pointed out that dimensional differences in the feeding environment are an important factor in niche differentiation. Particle size distributions are likely to prove the best way to investigate such questions, since a highly aggregated description seems most appropriate for a problem of this degree of generality, and studies of system aggregation have shown that aggregation by body size, which is basically the same as aggregation by turnover time (Humphreys 1979, Banse and Mosher 1980), is usually superior to other approaches (Cale and Odell 1979, O'Neill and Rust 1979). The use of allometric relations to estimate physiological rates for system models of this type makes it possible to use particle size distributions for different kinds of ecological investigations (Platt and Silvert 1981), so models of this type can be general despite the simplified level of system description. Particle size distributions have proven useful in the study of bioenergetic flows in pelagic ecosystems, and they may well prove of even greater value in the study of more general and perhaps fundamental questions in ecology.

REFERENCES

Banse, K., and Mosher S. (1980). Adult body mass and annual production/biomass relationships of field populations. Ecol. Monogr. 50: 355-379.

Brodie, P.F. (1982). A surface area or thermal index for marine mammal energetic studies. Proc. 3rd Theriological Congress, Helsinki. Acta Zool. Fennica 169 (4): 000-000.

Cale, W.G. Jr., and Odell, P.L. (1979). Concerning aggregation in ecosystem modelling. In Theoretical systems ecology (E. Halfon, ed.). Academic Press.

Cammen, L.M. (1980). Ingestion rate: an empirical model for aquatic deposit feeders and detritovores. Oecologia 44: 303-310.

Elton, C.S. (1927). Animal ecology. Sidgewick and Jackson, London.

Gunther, B., and Morgado E. (1982). Theory of biological similarity revisited. J. Theor. Biol. 96:543-559.

Harding, G.C.H. (1977). Surface area of the euphausiid Thysanoessa raschii and its relation to body length, weight, and respiration. J. Fish. Res. Bd. Canada 34:225-231.

Humphreys, W.F. (1979). Production and respiration in animal populations. J. Anim. Ecol. 48:427-453.

Kerr, S.R. (1974). Theory of size distribution in ecological communities. J. Fish. Res. Bd. Canada 31:1859-1862.

Levin, S.A. (1976). Population dynamic models in heterogeneous environments. Ann. Rev. Ecol. Syst. 7:287-310.

Marcotte, B.M. (1978). The ecology of meiobenthic harpacticoids (Crustacea: Copepoda) in West Lawrencetown, Nova Scotia. Ph. D. Thesis, Dalhousie University. 212 p.

McKendrick, A.G. (1926). Applications of mathematics to medical problems. Proc. Edin. Math. Soc. 44: 98-130.

O'Neill, R.V., and Rust, B. (1979). Aggregation error in ecological models. Ecol. Modelling 7:91-105.

Platt, T., and Denman, K. (1977). Organization in the pelagic ecosystem. Helgol. Wiss. Meeresunters. 30:575-581.

Platt, T. and Denman, K. (1978). The structure of pelagic marine ecosystems. Rapp. P. -V. Reun. Cons. Int. Explor. Mer 173:60-65.

Platt, T., and Silvert, W. (1981). Ecology, physiology, allometry and dimensionality. J. Theor. Biol 93:855-860.

Rubinow, S.I. (1973). Mathematical problems in the biological sciences. SIAM, Philadelphia. 90 p.

Schwinghamer, P. (1981). Characteristic size distributions of integral benthic communities. Can. J. Fish. Aquat. Sci. 38:1255-1263.

Sheldon, R.W., and Parsons, T.R. (1967). A continuous size spectrum for particulate matter in the sea. J. Fish. Res. Bd. Canada 24:909-915.

Sheldon, R.W., Prakash, A., and Sutcliffe, W.H. Jr. (1972). The size distribution of particles in the ocean. Limnol. Oceanogr. 17:327-340.

Sheldon, R.W., Sutcliffe, W.H. Jr. and Paranjape, M.A. (1977). The structure of the pelagic food chain and the relationship between plankton and fish production. J. Fish. Res. Bd. Canada 34:2344-2353.

Silvert, W. (1983). Is dynamical systems theory the best way to understand ecosystem stability? Proc. Int. Conf. Pop. Biol. (In press).

Silvert, W. and Platt, T. (1978). Energy flux in the pelagic ecosystem: a time-dependent equation. Limnol. Oceanogr. 23:813-816.

Silvert, W., and Platt, T. (1980). Dynamic energy-flow model of the particle size distribution in pelagic ecosystems. In Evolution and ecology of zooplankton communities (W.C. Kerfoot, ed.). Univ. Press of New England.

Sinko, J.W., and Streifer, W. (1967). A new model for age-size structure of a population. Ecology 48:910-918.

Smith, F. (1976). Ecosystems and evolution. Bull. Ecol. Soc. America. Spring 1976:2-6.

Sprules, W.G., and Holtby, L.B. (1979). Body size and feeding ecology as alternatives to taxonomy for the study of limnetic zooplankton community structure. J. Fish. Res. Bd. Canada 36:1354-1363.

Streifer, W. (1974). Realistic models in population ecology. Adv. Ecol. Res. 8: 199-266.

Thiel, H. (1975). The size structure of the deep-sea benthos. Int. Rev. Gesamten Hydrobiol. 60:575-606.

Ursin, E. (1973). On the prey size preferences of cod and dab. Medd. Dan. Fisk. Havunders. 7:85-98.

SPECIES-ABUNDANCE RELATION AND DIVERSITY

Ei Teramoto, Nanako Shigesada and *Kohkichi Kawasaki

Department of Biophysics, Kyoto University,
Kyoto, Japan

*Science and Engineering Research Institute,
Doshisha University, Kyoto, Japan

1. Introduction

The scenic affluence of nature is attributable to the fact that most ecological communities contain a rich variety of biological species with widely differing abundances. Ecologists long have been interested in the distribution of abundances of species within a given area. Commonly the "species-abundance" curves obtained from the observed data for particular taxonomic groups are studied. Considering data from several different areas, one then seeks a "best" functional form for the frequency distribution, so that by adjusting a small number of parameters one can fit closely the data from the majority of observed communities. These parameters then serve to characterize the individual communities.

Several types of frequency distribution laws have been proposed (see for example, Fisher, Corbet and Williams (1943); Preston (1948); Brian (1953)). Among forms suggested have been the logarithmic series distribution, the lognormal distribution, and the negative binomial distribution. In general, the distribution which works best for one community is not necessarily the best choice for others; thus one cannot easily answer the question: which is the optimum distribution as a general law?

However, roughly speaking, it is a common qualitative feature of species-abundance relations found in the majority of natural communities that there exist a few species with large population sizes and a large number of rare species with small numbers of individuals. In other words, singleton species are common; doubleton and tripleton species are less so; and so on. Thus, if we plot number of species as a function of the population size, we always obtain a monotone decreasing concave curve.

The question immediately arises: why do natural communities nave such a common qualitative pattern of "species-abundance" relations? An early proposed explanation was due to MacArthur (1957) who put forth the well-known "broken stick model" by assuming random occupation of nonoverlapping niches and that population sizes of species were proportional to the sizes of the species' niches. Alternatively, Utida (1943) derived a geometric series distribution by using a very simple inter-specific competition model.

A third and quite simple relation, "Zipf's law", states that, in a collection of many subjects of the same item, if we give the rank x (integer) to each subject on the basis of the order of size y, then, in many cases, we find empirically a rank-size relation $xy = \text{const}$. For example, English words ranked by the frequency of usage (the first rank "the", the second rank "and" and so on), urban communities ranked by population size (the first rank "Tokyo", the second rank "Osaka" and so on), and rivers ranked by length show approximately the Zipf relation, $xy = \text{const}$. If we consider the frequency distribution of size, $f(y)$, it is easily shown that the rank x of a subject with size y is given by $x = \int_y^\infty f(y')dy'$. Therefore, in terms of frequency distributions, Zipf's rank-size relation can be expressed by $f(y) = c/y^2$.

Yule (1924) also discussed the size (number of species) distribution of genera and derived a hyper-geometric series distribution by taking into account "specific mutation" and "generic mutation" as a stochastic birth process. He showed that the resultant frequency distribution has a form $f(y) \propto y^{-(1+1/\rho)}$ (where $\rho<1$) for large values of y. The result was compared with the observed data used by Willis (1922) in his discussion of evolution. As has been discussed by Rapoport (1978), if this type of the distribution is commonly found in a wide variety of objects, there may be a possibility that the distribution can commonly be characterized by some special probabilistic model, just as the normal distribution is appropriate to the sum of independent random variables. Studies on species-abundance relations in ecological communities, which are also characterized by similar monotone decreasing concave curves, are also of interest from this point of view.

2. Geometric Series Distribution and Competition Model

In Japan, Motomura (1932) proposed the geometric series distribution to fit the data of the bottom fauna of Japanese lakes obtained by Miyadi. He found a good fit to the data by introducing the rank-size relation

$$\log y + ax = b,$$

where y is the population size of the species of rank x. Here all species are ranked in the order of their population sizes. Obviously this expresses a geometric progression with the ratio exp(-a), and gives the size distribution $f(y) \propto 1/y$, instead of $1/y^2$ (Zipf's law). Typical examples are shown in Figs. 1 and 2. In order to explain the geometric series distribution, Utida (1943) considered a very

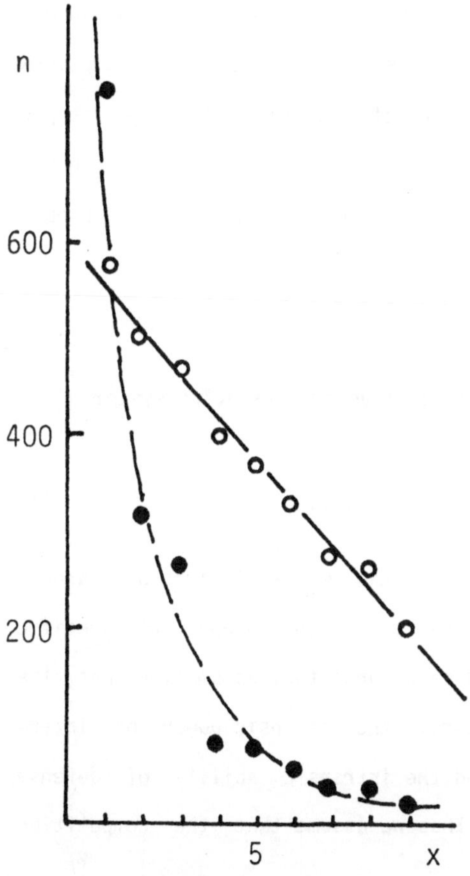

Fig.1. Bottom fauna of Lake Aoki (Miyadi, 1931)

simple competition model. Consider an area which consists of M compartments and n individuals of each species A,B,... randomly occupying these compartments. Here we assume that more than two individuals of the same species cannot occupy the same compartment simultaneously (this assumption was removed later by Motomura). Then we have compartments with different species (e.g. (A,D), (A,B,D), (B,C) and so on.) Furthermore we suppose that these species have a relative superiority for interspecific competition in the order A,B,... and that only one individual of the strongest species can survive in each compartment. Then the above example is reduced to (A), (A(, (B),.... Under these assumptions, all n individuals of species A can obviously survive occupying n compartments; among n individuals of species B, on the average, n(1-n/M) individuals (which entered A-absent compartments) will survive. Similarly, on the average $n\{1-n/M - \frac{n}{M}(1-n/M)\} = n(1-n/M)^2$ individuals of species C survive. In this way, it is easily shown that the population size distribution becomes a geometric series with the ratio (1-n/M). One may criticize the assumptions underlying this simple model. However, the effects of such competition seem to provide a possible rationale for the common type of species abundance relations in ecological communities. Thus, in the following sections, we shall discuss this problem from a population dynamical standpoint.

3. Competitive Multi-Species Model

Consider the Lotka-Volterra model of a competitive multi-species system

$$\frac{d}{dt}x_i = (\varepsilon_i - \sum_j \mu_{ij}x_j)\, x_i\,, \qquad i = 1,2,\ldots,N \tag{1}$$

where x_i is the population density of the ith species, ε_i is its intrinsic growth rate. μ_{ii} and μ_{ij} $(i \neq j)$ are the coefficients of intra- and interspecific competition, respectively. Here we introduce a conceptual assumption that the competitive interaction consists of two factors, the intrinsic power of interference (or attack) to other individuals and the intrinsic ability of defense against the attack of other individuals. Thus we assume that the competition coefficients can be written in the form

$$\begin{aligned} \mu_{ij} &= \beta_i \gamma_j \qquad \text{for } i \neq j, \\ &= \beta_i \alpha_i \qquad \text{for } i = j, \end{aligned} \tag{2}$$

where γ_j is an intrinsic factor of interference of an individual of the jth species, and its effect on the ith species is reduced, due to the defense ability of the ith species, by a factor β_i ($\beta_i < 1$); intraspecific competition is distinguished by using a different factor of interference α_i. Hereafter in our discussion we assume that

$$\alpha_i > \gamma_i \qquad i = 1,2,\ldots,N \tag{3}$$

This assumption is not so unrealistic, because intraspecific interference in the same ecological niche should be severest, due to dependence on the same resource.

Rewriting Eq.(1), we have

$$\begin{aligned} \frac{d}{dt} x_i &= (\varepsilon_i - \beta_i \alpha_i x_i - \sum_{j \neq i} \beta_i \gamma_j x_j)\, x_i \,, \\ &\equiv f_i(x) x_i \,, \qquad i = 1,2,\ldots,N \end{aligned} \tag{4}$$

where x denotes the set of variables $x_1, x_2, \ldots, x_N$; $\alpha_i > \gamma_i$ for all i; and species are so ordered that

$$\frac{\varepsilon_1}{\beta_1} > \frac{\varepsilon_2}{\beta_2} > \cdots > \frac{\varepsilon_N}{\beta_N} \,. \tag{5}$$

4. Dynamical Properties

Here we shall study the final state of the system given by the solution of (4) when starting from some initial state x^0, with all positive x_i^0. Eq.(4) in general has 2^N critical points in the whole state space ($-\infty < x_i < \infty$ for all i), including degenerate ones. First of all we shall consider the condition for the existence of a positive critical point x^* ($x_i^* > 0$ for all i). The critical point $x^* = (x_1^*, x_2^*, \ldots, x_N^*)$ which satisfies the equations

$$f_i(x) = \varepsilon_i - \beta_i\alpha_i x_i - \beta_i \sum_{j\neq i} \gamma_j x_j = 0 , \qquad i = 1,2,\ldots,N \tag{6}$$

can be obtained as

$$x_i^* = \frac{\xi_i}{\gamma_i}\{ \frac{-\sum_{k=1}^{N} \frac{\varepsilon_k}{\beta_k} \xi_k}{1 + \sum_{k=1}^{N} \xi_k} + \frac{\varepsilon_i}{\beta_i}\} , \qquad i = 1,2,\ldots,N \tag{7}$$

where

$$\xi_i = \frac{\gamma_i}{\alpha_i - \gamma_i} \quad (>0) .$$

Thus, using the assumptions (3) and (5), we can easily derive the following Lemma:

Lemma

The values of x_i^* ($i=1,2,\ldots,N$) at the critical point (7) satisfy the relation

$$(\alpha_1 - \gamma_1) x_1^* > (\alpha_2 - \gamma_2) x_2^* > \cdots > (\alpha_N - \gamma_N) x_N^* , \tag{8}$$

and this gives the positive critical point ($x_i^* > 0$ for all i) if and only if the parameters satisfy the condition

$$-\sum_{k=1}^{N} (\frac{\varepsilon_k}{\beta_k} - \frac{\varepsilon_N}{\beta_N}) \xi_k + \frac{\varepsilon_N}{\beta_N} > 0 . \tag{9}$$

As we can see in the next section, this positive critical point is globally stable. However, the relation (9) may be a very severe condition especially for a many species system (large N). Actually the condition (9) scarcely holds, except for such special cases as

(i) $\frac{\varepsilon_i}{\beta_i} - \frac{\varepsilon_N}{\beta_N} \ll 1$, for all i,

or

(ii) $\xi_i \ll 1$ $(\alpha_i \gg \gamma_i)$.

Thus it has been shown that competitive multi-species populations will not in general coexist; this is a manifestation of "Gause's principle of competitive exclusion". Therefore, as the next step, we shall study more natural cases.

In order to proceed with our discussion, here we shall introduce the concept "sector stability" defined by Goh (1980).

Definition

We start from a dynamical model of a multi-species system

$$\frac{d}{dt}x_i = f_i(x)x_i \ , \qquad i = 1,2,\ldots,N \qquad (10)$$

where all $f_i(x)$ have continuous partial derivatives. Considering a nonnegative critical point $x^* = (x_1^*, x_2^*, \ldots, x_N^*)$, we define the subsets P and Q of $I = \{1,2,\ldots,N\}$ by

$$P = \{ i \mid x_i^* > 0 \} \quad \text{and} \quad Q = \{ j \mid x_j^* = 0 \} , \qquad (11)$$

and let Ω be the subspace

$$\Omega = \{ x \mid x_i > 0 \text{ for } i \in P; \ x_j \geq 0 \text{ for } j \in Q \}.$$

The nonnegative critical point x^* is *globally sector stable* if every solution of (10) which starts from Ω remains in Ω for all finite time t and converges to x^* as $t \to \infty$.

If we use this prescription in our model (4), we can prove the following theorem (Proof in Appendix).

Theorem

In the system (4), by choosing an arbitrary number $n \in \{1,2,\ldots,N\}$, we consider a set of equations

$$\begin{aligned} f_i(x) &= 0 \qquad \text{for } i = 1,2,\ldots,n, \\ \text{and} \qquad x_i &= 0 \qquad \text{for } i = n+1, n+2, \ldots, N. \end{aligned} \qquad (12)$$

Let

$$x^*(n) = (x_1^*(n), x_2^*(n), \ldots, x_n^*(n), 0, 0, \ldots, 0)$$

be the critical point given by (12) for each $n = 1,2,\ldots,N$ and let s be the maximum of n which satisfies the condition

$$x_1^*(n) > 0,\ x_2^*(n) > 0,\ \ldots,\ x_n^*(n) > 0.$$

Then $x^*(s) = (x_1^*(s), x_2^*(s), \ldots, x_s^*(s), 0, \ldots, 0)$, where

$$x_i^*(s) = \frac{\xi_i}{\gamma_i}\left\{\frac{-\sum_{k=1}^{s}\frac{\varepsilon_k}{\beta_k}\xi_k}{1+\sum_{k=1}^{s}\xi_k} + \frac{\varepsilon_i}{\beta_i}\right\}, \qquad i = 1,2,\ldots,s \tag{13}$$

is a globally sector stable critical point of the system (4) and the relations

$$(\alpha_1 - \gamma_1)\, x_1^*(s) > (\alpha_2 - \gamma_2)\, x_2^*(s) > \cdots > (\alpha_s - \gamma_s) x_s^*(s) > 0\ . \tag{14}$$

are satisfied.

This theorem presents the criterion for the survival of species in competitive multispecies communities. Among N species ranked in the order of values ε_i/β_i, starting from arbitrary initial population densities $x_i^0 > 0$ $(i=1,2,\ldots,N)$, only the species of rank up to s can survive asymptotically approaching the finite stationary population densities $x_i^*(s)$ $(i=1,2,\ldots,s)$; other species of ranks from s+1 to N tend to extinction. Here it should be noted that the population sizes realized at the final state do not necessarily follow the order of ranks, but satisfy the relation (14).

Here we shall consider the invasion of new species in some chosen area where already several former occupant species are living with the stationary population densities. At the stage of invasion, we can apply our theorem to the system of the community including the invader species. Then it is clear that if the rank of the invader species in terms of ε_i/β_i is higher than that of the former occupant species with the lowest rank, the invasion will succeed and some of the former occupant species with the rank lower than the invader species become extinct unless the criterion of survival given by the theorem is fulfilled again for this new system. Therefore, if the N species considered in our theorem invade the given area one after another in random order, the final stable community will be established with the s surviving species satisfying the conditions of the theorem. Other species cannot invade, or if present tend to extinction. Therefore, it is concluded that the theorem can represent also the results of ecological succession; i.e., the successive invasion of new species.

5. Environmental Heterogeneity and Species-Abundance Relation

Generally for any area within which we are interested in the ecological community, the environmental conditions are not uniform; instead, there is a complex heterogeneous structure consisting of many different ecological niches. The structure of the ecological niche may continuously change from place to place in the given area. However, in order to take into account this complexity, we shall assume that the niche space can be divided into M patches. In each patch there are both intra- and interspecific competition among the invader species, but no competition between those of different patches.

It seems to be natural to assume that the parameters in Eq.(4) for a given species i have different values ε_i^μ, α_i^μ, β_i^μ, γ_i^μ $(\alpha_i^\mu > \gamma_i^\mu)$ depending on the patches $\mu = 1,2,\ldots,M$. Then we have the equations

$$\frac{d}{dt}x_i^\mu = \Big(\varepsilon_i^\mu - \beta_i^\mu \alpha_i^\mu x_i^\mu - \beta_i^\mu \sum_{j \neq i} \gamma_j^\mu x_j^\mu \Big) x_i^\mu - D_i^\mu x_i^\mu + \sum_{\mu' \neq \mu} D_i^{\mu\mu'} x_i^{\mu'} , \qquad \begin{array}{l} \mu = 1,2,\ldots,M \\ i = 1,2,\ldots,N \end{array} \tag{15}$$

where we take migration into consideration by introducing the rate of migration $D_i^{\mu\mu'}$ from the patch μ' to μ and $D_i^\mu = \sum_{\mu' \neq \mu} D^{\mu'\mu}$.

When migration is low and all patches can be regarded as isolated systems, we can apply our theorem directly to each set of equations for patch $\mu = 1,2,\ldots,M$. Then we have the set of globally sector stable solutions

$$x^{*\mu} = (x_1^{*\mu}, x_2^{*\mu}, \ldots, x_{s_\mu}^{*\mu}, 0, \ldots, 0) \qquad \mu = 1,2,\ldots,M \tag{16}$$

where at the final stable state, the patches obviously have different numbers of species s_μ. As for the problem of how this stable solution is affected by the presence of migration terms, Levin (1976) and Goh (1980) have already discussed the same problem for more general systems. By using the theorem given by Levin, we can show that if all migration rates are sufficiently small, there exists a sector stable solution $\{x^{*\mu}(D)\}$, and it continuously approaches the solution (16) as the migration rates tend to zero. Therefore, we can say that, so far as the migration rates are sufficiently small, the number of species and their population sizes can be approximately given by the stable solution (16). Thus the species-abundance relation as the result of our competition model can be obtained by calculating the populations of surviving species in the whole area

$$x_i^* = \sum_{\mu} x_i^{*\mu}, \qquad i = 1,2,\ldots,N \tag{17}$$

where obviously $x_i^{*\mu} = 0$ for $i > s_\mu$.

6. Numerical Simulation of Results

In order to see the qualitative features of the species-abundance relation, we assumed, in our computer calculations, that the parameters except ε_i^μ $(\mu = 1,2,\ldots,M)$ have the same values, independent of i, and used the values $\mu_{ii} = \beta_i\alpha_i = 1$ and $\mu_{ij} = \beta_i\gamma_j = 0.5$ for all i and j. For particular values of the parameters ε_i^μ, we considered a frequency distribution of the values of ε_i^μ $(\mu = 1,2,\ldots,M)$,

$$\mathrm{Prob}(\varepsilon < \varepsilon_i^\mu < \varepsilon + d\varepsilon) = \rho_i(\varepsilon)d\varepsilon, \qquad i = 1,2,\ldots,N$$

and actually used a box type distribution

$$\begin{aligned} \rho_i(\varepsilon) &= 1/\sigma \qquad \text{for } E_i - \sigma/2 < \varepsilon_i < E_i + \sigma/2, \\ &= 0 \qquad \text{otherwise} . \end{aligned} \tag{18}$$

N real numbers randomly chosen from the interval (5 10) were assigned to the mean values E_i in the order of their magnitude. Then the values ε_i^μ $(\mu = 1,2,\ldots,M)$ were randomly selected according to the frequency distribution (18) for each species, respectively. Using these parameter values and $N = 200$, $\sigma = 2.5$, stable populations at every patch were calculated by the procedure stated in our theorem. Fig.3 shows our result for the species-abundance relation which expresses the population sizes of surviving species given by (17) as a function of their ranks. Here all surviving species are ranked in the order of their population sizes, so it should

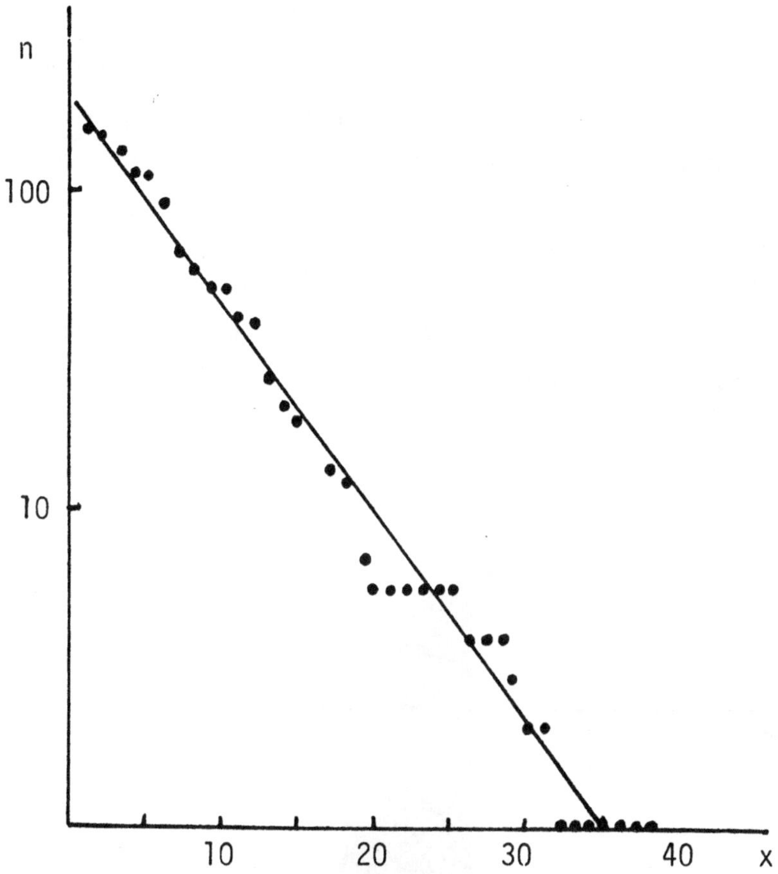

Fig.2. Crabs at Amakusa Bay (Ono,1961)

be noticed that the rank r does not necessarily coincide with the suffix i which specifies the species.

We can conclude from the result shown in these figures that when the area consists of many patches (M>10), the logarithm of the population size shows a linear relation with the species rank; namely, the geometric series distribution becomes a plausible approximation over a wide range of ranks. This qualitative feature of the distribution was not altered by different choices of parameter values. Similar results were also obtained by Yodzis (1978) in his discussion of the successional processes of a system of competitive species, in which random values were assigned to the coefficients of interspecific competitions.

In conclusion, we can say that the common qualitative feature of most species-abundance relations - that is, that they consist of a few species with large population sizes and many rare species - can be expected wherever the niche space within a given area has a complex patchy structure.

Acknowledgment: This work was supported by a Grant-in-Aid for Special Project Research on Biological Aspects of Optimal Strategy and Social Structure from the Japan Ministry of Education, Science and Culture.

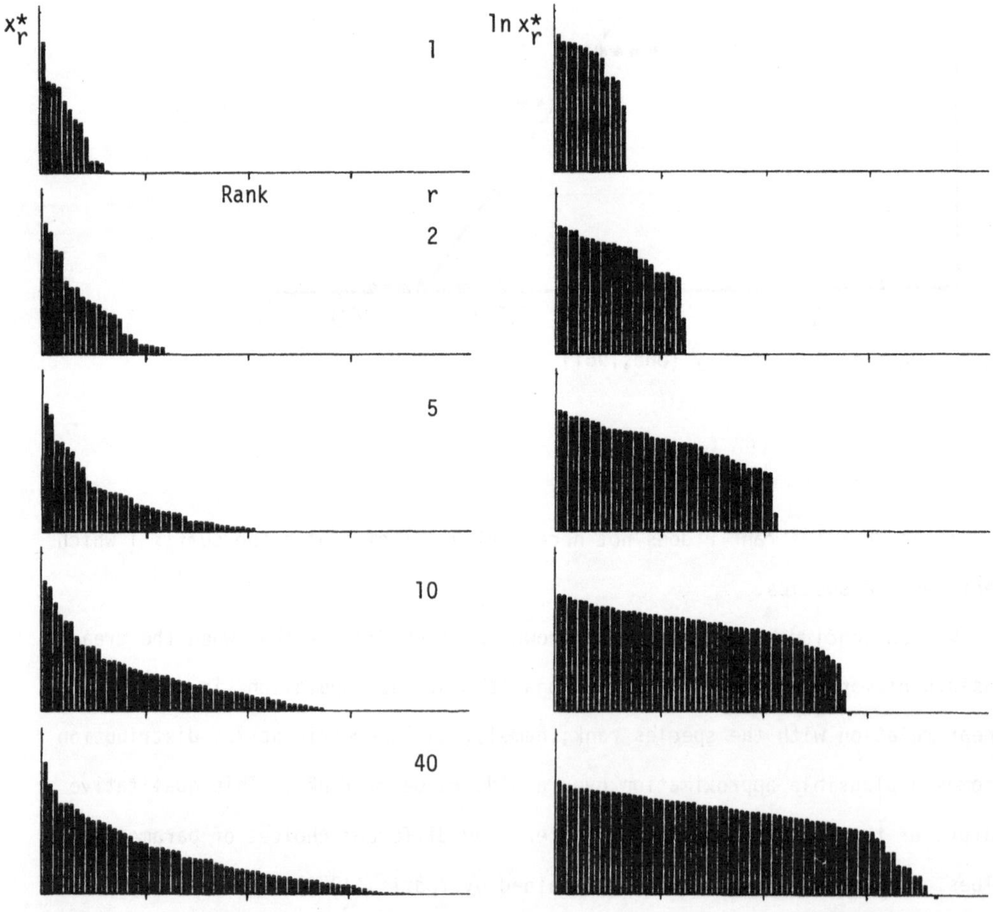

Fig.3. Species-abundance relations for the cases M=1, 2, 5, 10 and 40.

References

Brian, M.V. (1953). Species frequencies in random samples from animal populations. J. Anim. Ecol. 22:57-64.

Fisher, R.A., Corbet, A.S. and Williams, C.B. (1943). The relation between the number of species and the number of individuals in a random sample of an animal population. J. Anim. Ecol. 12:42-58.

Goh, B.S. (1980). Management and Analysis of Biological Population. Elsevier Scientific Publishing Company.

Levin, S.A. (1976). Spatial patterning and the structure of ecological communities. In Some Mathematical Questions in Biology, 7. Lectures on mathematics in the life science. Levin, S.A.,(ed.), Vol.8,1-35, Providence, R.I., Amer. Math. Soc.

MacArthur, R.J. (1957). On the relative abundance of bird species. National Academy of Science, Proceeding, 43:293-295.

Motomura, I. (1932). Statistical treatment of ecological communities. J. Zool. Soc. Japan 44:379-383 (in Japanese).

Preston, F.W. (1948). The commonness, and rarity, of species. Ecol. 29:254-283.

Rapoport, A. (1978). Rank-size relations. International Encyclopedia of Statistics (eds., William, H.K., and Judith, M.T.). The Free Press.

Utida, T. (1943). A theoretical consideration on Motomura's geometric series law of species-abundance relation in animal communities. Report on Ecology 9:173-178 (in Japanese).

Willis, C.J. (1922). Age and Area. Cambridge University Press.

Yodzis, P. (1978). Competition for space and the structure of ecological communities. Lecture Notes in Biomathematics. Vol.25 Springer-Verlag.

Yule, G.U. (1924). A mathematical theory of evolution, based on the conclusions of Dr. J.C. Willis, F.R.S. Phil. Trans. Roy. Soc. Series B, 213:21-81

Appendix (Proof of Theorem)

Let $P=\{1,2,\dots,s\}$ and $Q=\{s+1,s+2,\dots,N\}$ where s is defined in the Theorem, and consider a domain defined as $\Omega=\{x|x_i>0 \text{ for } i\in P;\ x_j\geq 0 \text{ for } j\in Q\}$. It is evident that the solution of Eq.(4) which begins in the set Ω remains in Ω for all finite values of t, since $\frac{d}{dt}x_i=0$ at $x_i=0$ for $i=1,2,\dots,N$.

In order to prove the stability of the solution of Eq.(4) confined in Ω, we will show that any solution of Eq.(4) starting from Ω converges to $x^*(s)$ as $t\to\infty$. To this end, we propose the following Lyapunov function

$$V(x)=\sum_{i\in P}\frac{\gamma_i}{2\beta_i}\{x_i-x_i^*(s)-x_i^*(s)\ln x_i/x_i^*(s)\}+\sum_{j\in Q}\frac{\gamma_j}{2\beta_j}|x_j|\geq 0, \tag{A-1}$$

where the equality sign holds only at the point $x=x^*(s)\equiv(x_1^*(s),x_2^*(s),\dots,x_s^*(s),$

0,...,0) in Ω, which is given by

$$x_i^*(s) = \frac{\frac{\xi_i}{\gamma_i}\{\frac{\varepsilon_i}{\beta_i} + \sum_{k\in P}(\frac{\varepsilon_i}{\beta_i} - \frac{\varepsilon_k}{\beta_k})\xi_k\}}{1 + \sum_{k\in P}\xi_k} \qquad \text{for } i = 1,2,\ldots,s. \qquad (A-2)$$

The time derivative of Eq.(A-1) is calculated as

$$\frac{d}{dt}V(x) = -\frac{1}{2}\{\sum_{i\in I}\gamma_i(x_i - x_i^*(s))\}^2 - \frac{1}{2}\sum_{i\in I}(\alpha_i - \gamma_i)\gamma_i(x_i - x_i^*(s))^2$$

$$+ \sum_{j\in Q}\frac{\gamma_j}{2\beta_j}\{\varepsilon_j - \beta_j\sum_{i\in P}\gamma_i x_i^*(s)\}x_i(s) , \qquad (A-3)$$

where $I = P + Q \equiv \{1,2,\ldots,N\}$. Thus if

$$f_l(x^*(s)) \equiv \varepsilon_l - \beta_l\sum_{i\in P}\gamma_i x_i^*(s) \le 0 \qquad \text{for all } l\in Q, \qquad (A-4)$$

then

$$\frac{d}{dt}V(x) \le 0 ,$$

where equality holds only at $x=x^*(s)$, and hence $x^*(s)$ becomes globally sector stable.

Hereafter we will prove the relation (A-4). Substituting (A-2) into (A-4), we obtain

$$f_l(x^*(s)) = \varepsilon_l - \beta_l\frac{\sum_{k\in P}\frac{\varepsilon_k}{\beta_k}\xi_k}{1 + \sum_{k\in P}\xi_k}$$

$$= \frac{\beta_l\{-\sum_{k\in\{P,l\}}(\frac{\varepsilon_k}{\beta_k} - \frac{\varepsilon_l}{\beta_l})\xi_k + \frac{\varepsilon_l}{\beta_l}\}}{1 + \sum_{k\in P}\xi_k} \qquad \text{for } l\in Q. \qquad (A-5)$$

To describe this equation more concisely, let us introduce the following set of equations for $l\in Q$,

$$f_i(x) \equiv \varepsilon_i - \beta_i\alpha_i x_i - \sum_{j\neq i}\beta_i\gamma_j x_j = 0 \qquad \text{for } i = 1,2,\ldots,s \text{ and } l,$$

$$x_j = 0 \qquad \text{for } j = s+1,s+2,\ldots,N \text{ except } l. \qquad (A-6)$$

The solution of (A-6), $x^*(s|l) = (x_1^*(s|l),x_2^*(s|l),\ldots,x_s^*(s|l),0,\ldots,0,x^*(s|l),$ $0,\ldots,0)$, is given by

$$x_i^*(s|l) = \frac{\frac{\xi_i}{\gamma_i}\{-\sum_{k\in\{P,l\}}(\frac{\varepsilon_k}{\beta_k}-\frac{\varepsilon_i}{\beta_i})\xi_k+\frac{\varepsilon_i}{\beta_i}\}}{1+\sum_{k\in\{P,l\}}\xi_k} \qquad \text{for } i\in\{P,l\}. \tag{A-7}$$

Especially, the l-th element $x_l^*(s|l)$ is written as

$$x_l^*(s|l) = \frac{\frac{\xi_l}{\gamma_l}\{-\sum_{k\in\{P,l\}}(\frac{\varepsilon_k}{\beta_k}-\frac{\varepsilon_l}{\beta_l})\xi_k+\frac{\varepsilon_l}{\beta_l}\}}{1+\sum_{k\in\{P,l\}}\xi_k}. \qquad \text{for } l\in Q. \tag{A-8}$$

Note that the parenthesized term in the numerator of (A-8) is exactly the same as that of (A-5). Thus $f_l(x^*(s))$ can be rewritten as

$$f_l(x^*(s)) = \frac{\gamma_l\beta_l(1+\sum_{k\in\{P,l\}}\xi_k)}{\xi_l(1+\sum_{k\in P}\xi_k)}x_l^*(s|l) \qquad \text{for } l\in Q. \tag{A-9}$$

Now we will show that

$$x_l^*(s|l) \le 0 \qquad \text{for all } l\in Q.$$

which assure $f_l(x^*(s)) \le 0$ for all $l\in Q$. By using Eq.(A-8) and the relation (5), we find the following equation,

$$x_{s+1}^*(s|s+1)\times\frac{\gamma_{s+1}}{\xi_{s+1}}\times(1+\sum_{k\in\{P,s+1\}}\xi_k) - x_l^*(s|l)\times\frac{\gamma_l}{\xi_l}\times(1+\sum_{k\in\{P,l\}}\xi_k)$$

$$=(\frac{\varepsilon_{s+1}}{\beta_{s+1}}-\frac{\varepsilon_l}{\beta_l})(1+\sum_{k\in P}\xi_k) \ge 0 \qquad \text{for } l\in Q. \tag{A-10}$$

Comparing the set of equations of (A-6) with that of (12) in the text, we find that (A-6) for the case of $l=s+1$ is identical to (12) for the case of $n=s+1$. Thus we have the following equation

$$x_{s+1}^*(s|s+1) = x_{s+1}^*(s+1).$$

Furthermore, recalling the definition of s and the relation (8) in the Lemma in the text, we have

$$x_{s+1}^*(s+1) \le 0,$$

and hence from (A-10), we can conclude

$$x_l^*(s|l) \le 0 \qquad \text{for all } l\in Q.$$

A COMPETITION MODEL WITH AGE STRUCTURE

J. M. Cushing and M. Saleem

The University of Arizona
Department of Mathematics and
Program on Applied Mathematics
Building # 89
Tucson, Arizona 85721
USA

1. Introduction. In recent years there has been a great deal of interest in modeling and analyzing the growth and interactions of populations with age structure. A large portion of the recent literature was stimulated by some seminal papers of Gurtin and MacCamy [7-9] in which the growth of a single age-structured population was modeled by a first order partial differential equation (which has come to be called "McKendrick's equation) for the population density. More recently papers have appeared in which systems of such equations have been used to describe interactions between age-structured populations [2,4-6,11,12,15,16] nearly all of which have dealt with predator-prey interactions. As pointed out in [2,10] the presence of age structure within one or more of the interacting populations can significantly change the dynamics of the interaction and in fact can lead to violations of often held basic tenets in theoretical population dynamics.

In this paper we will derive and investigate a model for the dynamics of several competing age-structured populations in order to see to what extent some of the fundamental notions of the theory of competition (e.g. competitive exclusion, ecological niche and limiting similarity) can be affected by the presence of age structure within the populations. The model equations will be derived from the McKendrick equations and will serve as a generalization of the classical Lotka-Volterra equations. We will derive the system parameters as is done in the MacArthur-Levins theory [1,13], but with allowance for dependence of the various parameters in this theory on age. Although the MacArthur-Levins theory is perhaps naive and a bit out-dated, we felt that given the influence which it has had on the thinking in theoretical ecology, it would be both interesting and reasonable in this, a preliminary study of age-structure and competition, to model the derivation on (and in fact generalize) this theory. Regardless of how the "coefficients" in the equations are derived, however, it will be seen to what extent the usual dynamics and concepts in competition theory generalize to age-structured populations and what new dynamics can possibly result.

2. Derivation of the Model Equations. If it is assumed that n populations are described by age-specific density functions $\rho_i(t,a)$, $i = 1$ to n, (where t is time and a is age) so that $\int_{a_2}^{a_1} \rho_i(t,a)da$ is the total population between ages a_1 and a_2, then the McKendrick equations which describe the growth dynamics are

$$\partial\rho_i/\partial t + \partial\rho_i/\partial a + \mu_i\rho_i = 0 \tag{2.1}$$

$$\rho_i(t,0) = \int_{a=0}^{\infty} f_i\rho_i(t,a)\,da\,, \quad i = 1,2,\ldots n\,. \tag{2.2}$$

Here μ_i and f_i are respectively the per unit density death and fecundity rates and hence (2.1) simply accounts for the removals from the ith population by death and (2.2) accounts for the additions to the ith population at age $a = 0$ made by births from all contributing age classes. In general μ_i and f_i are functions of t and a and for populations which exhibit density dependent self-regulation and/or mutual interactions they are functionals of the densities ρ_j, $j = 1$ to n, as well.

In this paper these vital rates μ_i, f_i will be assumed explicitly independent of time t so that (2.1)-(2.2) is an autonomous set of equations. We are interested in competing species so that the functional dependencies of μ_i and f_i on population densities will have the properties that increased densities cannot increase fecundity nor decrease the death rate. In fact, the case when the effect of competition for a common resource is predominantly felt in reduced fecundity (rather than increased death rate) will be considered and hence it will be assumed that μ_i is independent of the densities ρ_j. Finally, it will be assumed throughout that μ_i is also independent of age a. Although the results of this paper remain valid if this assumption is not made and $\mu_i = \mu_i(a)$ is an integrable function of a, there is a great simplification in notation and details if we assume $\mu_i = \text{constant} \geq 0$. This is a frequently made assumption in age-structured population dynamics and corresponds to an exponentially descreasing survivorship curve.

We consider now, as in the classical MacArthur-Levins theory, the case of competition for a one-dimensional resource as measured by a real parameter r. If

$R_i(r,a)dr$ = the amount of resource of type r to $r + dr$ available to age class a ;

$u_i(r,a)$ = the age-specific per unit density resource consumption rate of resource r by species ρ_i ;

$\beta_i(r,a)$ = the age-specific conversion of per unit resource r to number of off-spring of species ρ_i (per unit density) ;

then

$$f_i = \int_{-\infty}^{+\infty} \beta_i(r,a)u_i(r,a)R_i(r,a)dr . \tag{2.3}$$

If it assumed that the amount of resource r which would be available to species ρ_i but is instead consumed by species ρ_j is proportional to the total consumption rate of resource r by species ρ_j then

$$R_i(r,a) = A_i(r,a) - \Sigma_{j=1}^{n} c_{ij}(r,a)\int_{s=0}^{\infty} u_j(r,s)\rho_j(t,s)ds . \tag{2.4}$$

Here n is the number of competing species and $A_i(r,a)$ is the amount of resource r made available to age class a of species ρ_i in its habitat. If the total birth rate of species ρ_i is denoted by $B_i(t) := \rho_i(t,0)$, then the linear partial differential equation (2.1) implies that $\rho_i(t,a) = B_i(t-a)\exp(-\mu_i a)$. When substituted into the birth equation (2.2) this expression, together with (2.3)-(2.4), results in a system of integral equations to be solved for $B_i(t)$:

$$B_i(t) = \int_0^{\infty} \int_{-\infty}^{+\infty} \beta_i u_i [A_i - \Sigma_{j=1}^{n} c_{ij} \int_0^{\infty} u_j B_j(t-s)\exp(-\mu_j s)ds]dr \; B_i(t-a)\exp(-\mu_i a)da. \tag{2.5}$$

Our primary interest is with the existence and stability of equilibrium solutions of (2.5). A positive equilibrium $B_i(t) \equiv B_i^0 > 0$ is a solution of the algebraic equations

$$\Sigma_{j=1}^{n} \int_{a=0}^{\infty}\int_{-\infty}^{+\infty} \beta_i u_i c_{ij} \int_{s=0}^{\infty} u_j \exp(-\mu_j s)ds \, \exp(-\mu_i a)da \, B_j^0 = \int_{a=0}^{\infty}\int_{-\infty}^{\infty} \beta_i u_i A_i dr \, \exp(-\mu_i a)da - 1. \tag{2.6}$$

The first term on the right hand side is the inherent per capita net reproductive

<u>rate</u> n_i of species ρ_i and we arrive at the conclusion that (2.5) can have a positive equilibrium only if this inherent net reproductive rate is larger than one, a not too surprising result because $n_i = 1$ means exact replacement.

In order to make this system more tractable, we will make several further simplifying assumptions (using the MacArthur-Levins theory as a quide). First of all it will be assumed that the resource availability A_i is the same for all age classes: $A_i(r,a) = A_i(r)$ and that the constants of proportionality c_{ij} are independent of a and r: $c_{ij}(r,a) = c_{ij} > 0$. Secondly we write

$$u_i(r,a) = n_i w_i(a) p_i(r) \tag{2.7}$$

where $p_i(r)$ is a probability density function (referred to as the resource "picking" or "preference" function) for which $p_i(r)dr$ represents the probability of choosing resource in the range r to $r + dr$ (or it is the fraction of resource in this interval consumed) and $n_i w_i(a)$ is the <u>age-specific resource consumption rate</u>. Here the age-specific distribution $w_i(a)$ is normalized so that

$$A_i \int_{a=0}^{\infty} \beta_i(a) w_i(a) \exp(-\mu_i a) da = 1 \, , \quad A_i := \int_{-\infty}^{\infty} p_i(r) A_i(r) dr \tag{2.8}$$

which has the effect of introducing the inherent net reproductive rates n_i explicitly into the equations and analysis through (2.7). Finally, it will be assumed that a given species' resource consumption rate effects all other species' resource availabilities equally: $c_{ij} = c_j$. By rescaling the units used to measure population densities, one can assume that $c_{ii} = 1$. Thus in (2.5) we will take $c_{ij} = 1$.

With all these assumptions in place, (2.5) reduces to

$$B_i(t) = [A_i - \Sigma_{j=1}^{n} n_j p_{ij} \int_0^{\infty} w_j(a) B_j(t-a) \exp(-\mu_j a) da] \int_0^{\infty} \beta_i(a) n_i w_i(a) B_i(t-a) \exp(-\mu_i a) da \tag{2.9}$$

where $p_{ij} := \int_{-\infty}^{\infty} p_i(r) p_j(r) dr$. In the following sections the resource preference functions will be taken to be Gaussian with mean r_i and standard deviation W_i

(the niche "location" and "width" respectively). If it is assumed that the resource niche widths are equal for all species: $W_i = W > 0$ and that the niches are equally spaced: $|r_i - r_j| = d|i-j|$, $d > 0$, then

$$p_{ij} = \gamma\alpha^{(i-j)^2}, \quad \alpha := \exp(-d^2/4W^2), \quad \gamma := (4\pi W^2)^{1/2} \tag{2.10}$$

in (2.9). Note that $0 < \alpha < 1$.

The technical requirements on the functions $w_i(a)$, $A_i(r)$ and $p_i(r)$ needed throughout are that these are continuous, non-negative functions for which the integrals $\int_0^\infty w_i(a)\exp(-\mu_i a)da$, $\int_0^\infty a w_i(a)\exp(-\mu_i a)da$ and $\int_{-\infty}^{\infty} A_i(r)p_i(r)dr$ are finite.

If all parameters are independent of age a: $\beta_i(a) \equiv$ constant and $w_i(a) \equiv$ constant, then it is possible to show that the above model reduces to the classical Lotka-Volterra system. To see this, integrate (2.1) from $a = 0$ to $a = +\infty$, introduce the total population sizes $P_i(t) := \int_0^\infty \rho_i(t,a)da = \int_0^\infty B_i(t-a)\exp(-\mu_i a)da$ and use (2.9) in the result. This will yield the Lotka-Volterra system of ordinary equations for the P_i.

3. Equilibria. An equilibrium $B_i(t) \equiv B_i^0$ = constant > 0 of equation (2.9) yields a steady-state population density $\rho_i = B_i^0\exp(-\mu_i a)$. Such an equilibrium is a positive solution of the algebraic equations (2.6), which under the added assumptions made at the end of Section 2, become

$$\Sigma_{j=1}^n \alpha^{(i-j)^2}\Gamma_j = A_i(n_i-1)/n_i \tag{3.1}$$

where $\Gamma_j := n_j\gamma w_j^*(\mu_j)B_j^0$ and "*" denotes Laplace transform. A conclusion which can be drawn from (3.1) is that the classical principle of "limiting similarity" remains valid for this more general age-structured model. For, if it is assumed that the species are similar in reproductive output $n_i = N > 0$ and are supplied with equal amounts of resource $A_i = A > 0$, then the positive equilibrium equations (3.1) can be written

$$\Sigma_{j=1}^n \alpha^{(i-j)^2}\Omega_j = 1 \tag{3.2}$$

where $\Omega_j := N\Gamma_j/A(N-1)$, which are identical to those for the classical Lotka-Volterra-MacArthur-Levins equations [1,13]. It is known that (3.2) has a positive equilibrium for all niche separation to width ratios d/W if $n = 2$, but that if $n \geq 3$ then there is a positive lower bound for d/W of order of magnitude one below which no positive equilibrium exists and above which one does exist and is stable. Thus, this principle of limiting similarity remains intact (as far as the existence of a positive equilibrium is concerned) for the age-structure model considered here.

Note that complete symmetry is not required for this result. The death rates μ_i and the age-specific resource consumption rates $Nw_i(a)$ are not necessarily identical for all species.

There is, as will be seen in the following sections, a crucial difference, however, between the age-structure model above and the classical Lotka-Volterra model. In the classical model a positive equilibrium, when it exists, is always (asymptotically) stable [1]. Positive equilibria of the equations (2.9) can on the other hand be unstable, even in the case $n_i = N$, $A_i = A$ which is analogous to the classical case.

In the next section the stability of equilibrium solutions of (2.9) is studied under certain restrictions for the case of two species $n = 2$.

4. Stability of Equilibria. The stability (local, asymptotic) of an equilibrium solution of (2.9)-(2.10) can be investigated analytically by the usual procedure of investigating the roots of the characteristic equation of the linearization [14]. If $x_i = B_i - B_i^0$ and all terms higher order in x_i are ignored, then (2.9) takes the (vector) form

$$x(t) = \int_0^\infty K(a)x(t-a)da$$

whose characteristic equation is given by

$$D := \det(I - K^*(z)) = 0 \tag{4.1}$$

If there are no roots z for which $\operatorname{Re} z \geq 0$, then B_i^0 is (locally, asymptotically) stable while if there exists a root $\operatorname{Re} z > 0$ then B_i^0 is unstable.

In general (4.1) is a transcendental equation in z and an analysis of the roots is very complicated. One case for which we have managed to carry out a rather complete analysis is the case of two species $n = 2$ under the further restrictions

$$\mu_1 = \mu_2 = \mu > 0, \quad \beta_1(a) = \beta_2(a) \equiv \text{constant} > 0, \quad w_1(a) = w_2(a) = w(a) \qquad (4.2)$$
$$A_1 = A_2 = A > 0 \; .$$

Thus, it is assumed that the two species have identical death rates μ_i, resource-to-offspring conversion factors β_i and age-specific resource rate age distributions $w_i(a)$. We have also assumed here that the β_i are non-age dependent and consequently that the only age-specific parameter in the model is the resource consumption rate $n_i w(a)$. Also the total resource availability A_i is the same for both species. Note that the inherent net reproductive rates n_1, n_2 are not assumed identical, however.

Under these simplifying assumptions, the characteristic equation (4.1) in the case $n = 2$ reduces, for an arbitrary equilibrium (B_1^0, B_2^0), to

$$D := \det\begin{pmatrix} 1-n_1\beta[A-2n_1\gamma w^*(\mu)B_1^0-n_2\gamma\alpha w^*(\mu)B_2^0]w^*(z+\mu) & n_1 n_2\gamma\alpha\beta w^*(\mu)B_1^0 w^*(z+\mu) \\ n_1 n_2\gamma\alpha\beta w^*(\mu)B_2^0 w^*(z+\mu) & 1-n_2\beta[A-n_1\gamma\alpha w^*(\mu)B_1^0-2n_2\gamma w^*(\mu)B_2^0]w^*(z+\mu) \end{pmatrix} =$$

The normalization (2.8) of $w(a)$ implies that

$$A\beta w^*(\mu) = 1 \, . \qquad (4.3)$$

Note that any z for which $w^*(z+\mu) = 0$ cannot be a root of this characteristic equation. Thus we assume throughout that $w^*(z+\mu) \neq 0$.

The equilibria $(B_1^0, B_2^0) = E$ are given by

$$E_0 := (0,0), \quad E_1 := (\beta A^2(n_1-1)/n_1^2\gamma, 0), \quad E_2 := (0, \beta A^2(n_2-1)/n_2^2\gamma)$$

$$E_3 = (\beta A^2[1-n_1^{-1}-\alpha(1-n_2^{-1})]/\gamma n_1(1-\alpha^2),\ \beta A^2[1\ n_2^{-1}-\alpha(1-n_1^{-1})]/\gamma n_2(1-\alpha^2)) .$$

What makes the analysis of (4.1) tractable for any of these equilibria is that although the characteristic equation is transcendental in z, it turns out to be a polynomial (of degree two or less) in the ratio

$$\eta(z) := w^*(\mu)/w^*(z+\mu) .$$

This fact, together with $|\eta(z)| \geq 1$ for $\text{Re } z \geq 0$, allows us to obtain stability for those parameter values for which all roots of this polynomial in η lie in the unit circle $|\eta| < 1$. We have used this approach to obtain those regions in the n_1,n_2-plane where the individual equilibria E_i above are stable. Instability, at least for n_1,n_2 near the boundaries of these regions, can be proved by an implicit function theorem argument which yields a root with $\text{Re } z > 0$. A simple example is detailed below. FIGURE 1 contains a summary of these general results.

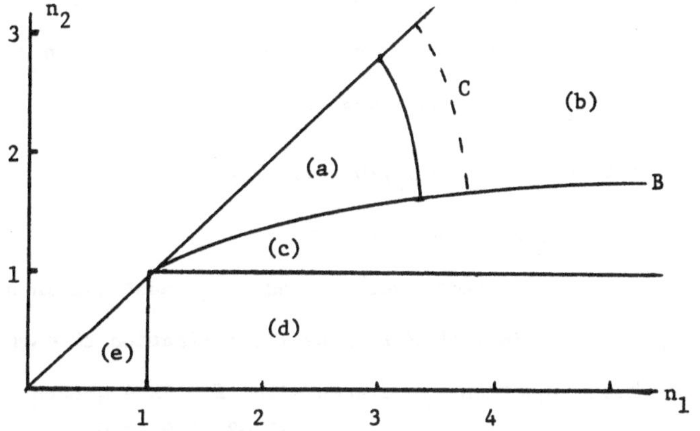

FIGURE 1: A positive equilibrium E_3 of (2.9) exists for $n_1 < n_2$ only in the regions (a) and (b) bounded by the line $n_1 = n_2$ and the curve B which is asymptotic to the line $n_2 = 1/(1-\alpha)$. It is stable in at least region (a) whose boundary is the solid line ($\eta_- = -1$). The dashed line is the actual stability boundary for the example (4.4), $m \geq 2$. In (c), E_1 and E_2 are both nonnegative, but only E_1 is stable (at least for $n_1 < 3$). In (d) only E_1 is nonnegative (it is stable at least for $n_1 < 3$). In (e) only E_0 is non-negative and it is stable (E_0 is unstable outside (e)).

We give here the details only for the simplest case of E_0. The remaining equilibria are analysed in a similar manner, although the details are in some cases very tedious. For E_0 the characteristic equation reduces to the factorable quadratic equation in η

$$D := (1 - n_1/\eta)(1 - n_2/\eta) = 0$$

whose roots lie inside the unit circle $|\eta| \leq 1$ if and only if n_1 and $n_2 < 1$. Thus E_0 is stable if both $n_i < 1$. This manner of reasoning yields sufficient conditions for stability. It does not follow that the equilibrium is necessarily unstable if an η root lies outside the unit circle. To establish that E_0 is unstable outside the square $0 < n_i < 1$ (at least near the boundaries $n_i = 1$) we treat D as a function of z and one of the n_i, say n_1 (holding n_2 fixed, the opposite case is symmetric):

$$D = D(z,n_1), \quad D(0,1) = 0, \quad D_z(0,1) = \beta A\int_0^\infty aw(a)\exp(-\mu a)da(1-n_2) > 0.$$

The implicit function theorem implies the existence of a root $z = z(n_1)$, $z(1) = 0$, for $n_1 \sim 1$. An implicit differentiation gives further that

$$\text{Re } z'(0) = (1 - n_2)/D_z(0,1) > 0$$

and hence $\text{Re } z(n_1) > 0$ for $n_1 > 1$ near 1.

For the equilibrium E_3 to be positive, n_1 and n_2 must lie in a region R of the shape of (a) plus (b) in FIGURE 1 (plus its reflection through the line $n_1 = n_2$). The characteristic equation associated with E_3 is a quadratic equation $\eta^2 + c_1\eta + c_2 = 0$ in the ratio η where $c_1 := \gamma\beta^{-1}A^{-2}(n_1^2B_1^0+n_2^2B_2^0)-2$ and $c_2 := 1-\gamma\beta^{-1}A^{-2}(n_1^2B_1^0+n_2^2B_2^0)+(1-\alpha^2)\gamma^2\beta^{-2}A^{-4}n_1^2n_2^2B_1^0B_2^0$. It is not difficult to show that this quadratic has two real roots $\eta_- < \eta_+ < 1$ and hence the characteristic roots are the roots of the two equations $\eta(z) = \eta_-$, $\eta(z) = \eta_+$. Thus $\eta_- > -1$ implies stability and $\eta_- = -1$ <u>defines the boundary of a region of stability of the positive equilibrium</u> <u>E_3</u>. This subregion of R in FIGURE 1 consists of (a) and its reflection through the line $n_1 = n_2$ and contains the point $(n_1,n_2) = (3,3)$ on its boundary. <u>This</u>

<u>stability region is not necessarily maximal as the example below shows, but it is the largest possible in general (i.e. for arbitrary</u> $w(a)$) as the example below also shows.

It is curious to note the importance of the integers 1,2 and 3 for this competition system of equations. For all other values of the other system parameters and for all age-specific resource consumption distributions $w(a)$, 1 is the minimal value of the net reproductive rates n_i for the survival of both species, 2 is the value of the n_i which maximizes the equilibrium E_3 (i.e. $\max(|B_1^0| + |B_2^0|)$ occurs at $n_1 = n_2 = 2$) and 3 is the maximum value of n_1, n_2 for which the stability of the positive equilibrium can be guaranteed in general.

As an example set

$$w(a) = La^m \exp(-am/T) \tag{4.4}$$

$T > 0$, $m = 1,2,\ldots$ and $L = (\mu T+m)^{m+1}/m!A\beta T^{m+1}$ is chosen so that the normalization (4.3) holds. For this case $\eta(z) = (zT+\mu T+m)^{m+1}/(\mu T+m)^{m+1}$. The characteristic roots are $z = (\mu T+m)T^{-1}(-1+|\eta_\pm|^{1/(m+1)}r_j)$ where r_j are the $(m+1)^{st}$ roots of $+1$ if $\eta_\pm > 0$ and -1 if $\eta_\pm < 0$. Thus, the positive equilibrium is always stable if $m = 1$, but is unstable for $m \geq 2$ when η_- or $\eta_+ < 0$ and $-1+|\eta_\pm|^{1/(m+1)}\cos(\pi/(m+1)) > 0$. Thus the region of stability is defined by the curve C: $\eta_- = -\sec^{m+1}(\pi/(m+1))$ in the n_1,n_2-plane. Note that this curve approaches the curve $\eta_- = -1$ as $m \to +\infty$ (i.e. as the age distribution of the resource consumption rate narrows around the mean $T > 0$).

5. <u>Oscillations</u>. With the onset of instability occuring as one leaves a region of stability in the n_1,n_2-plane, the possibility of nonconstant oscillations arises. Rigorous theorems and bifurcation techniques for integral equations of the form (2.9) can be used to prove the existence of nonconstant periodic solutions for n_1,n_2 near these boundaries [3]. We will describe this phenonemnon only for the example given in Section 4, our purpose being restricted here to demonstrating clearly the possibility of nonconstant (stable) periodicities in an age-structured competition model (a possibility which does not occur in non-age-structured models) and to illustrate

some interesting features of these oscillations.

Bifurcation of nonconstant periodic solutions from the positive equilibrium E_3 occurs at a parameter point $(n_1,n_2) = (n_1^0,n_2^0)$ at which the characteristic equation has a root $z = i\omega$, $\omega \neq 0$. For $w(a)$ as in the example in Section 4 above, this occurs for (n_1^0,n_2^0) on the boundary line C when $m \geq 2$ in FIGURE 1. Thus $\omega = (\mu T+m)^{-1}\tan(\pi/(m+1))$. The bifurcation theory in [3] implies the existence of a path emanating from (n_1^0,n_2^0) in the n_1,n_2-plane along which there exist nonconstant solutions of (2.9) of fixed period

$$p = 2\pi/\omega = 2\pi(\mu T+m)^{-1}T\cot(\pi/(m+1)) . \tag{5.1}$$

Note that this period p is independent of the critical point (n_1^0,n_2^0) and hence we have an unusual case where all bifurcating periodic solutions near the critical boundary curve C have the same period (usually the period changes with the parameters (n_1,n_2)). This is corroborated by numerically computed solutions of (2.9) which we have carried out.

The solutions graphed in FIGURES 2-5 were computed with $n_1 = n_2 = N$, $m = 17$ and parameter values $\mu = W = \beta = A = 1.0$ and $T = 0.5$. Equal initial data was used so that $B_1(t) \equiv B_2(t)$ in these graphs. FIGURE 6 shows a case with $n_1 \neq n_2$ so that $B_1(t) \not\equiv B_2(t)$. Note that the species oscillate in phase. FIGURE 2 shows both species' birth rates tending to zero for $N < 1$ and approaching an equilibrium for $1 < N < 3$ with the maximum equilibrium occuring at $N = 2$ and damped oscillations occuring for $N = 3$. FIGURE 3 shows a sinusoidal-like periodicity for N above criticality with period $p \sim 1.1$ which is consistent with (5.1). The periodicities in FIGURES 4 and 5 exhibit an interesting period doubling (to approximately 2.2 in FIGURE 4 and to approximately 4.4 in FIGURE 5 for the slowly varying outer harmonic) as N is increased further. This is reminiscent of the same phenomenon which is well-known to occur in difference equations.

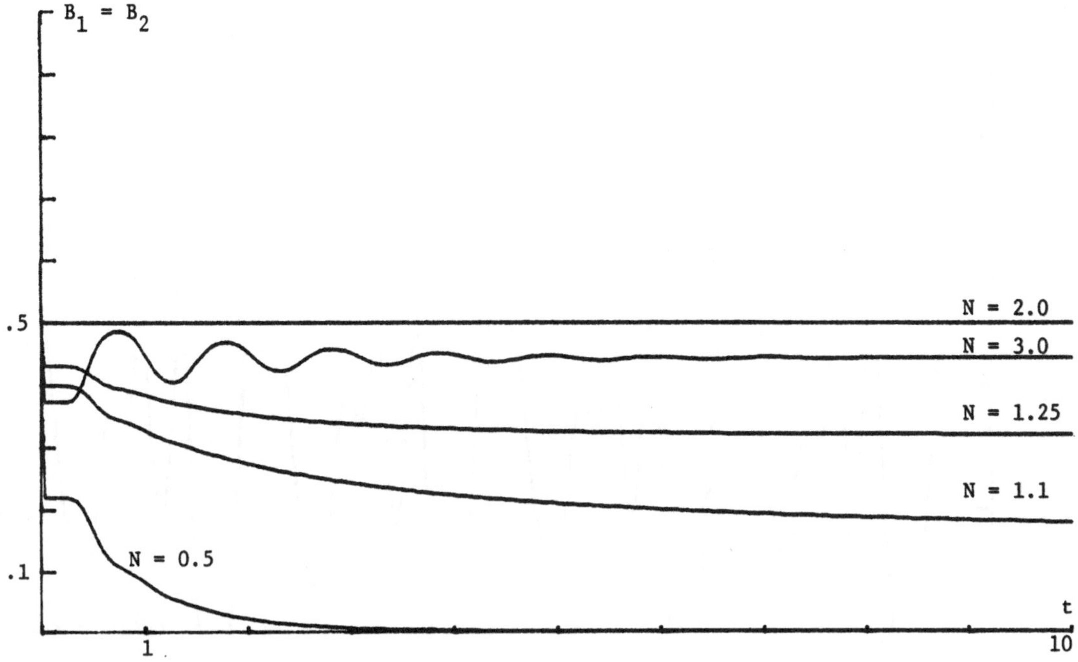

FIGURE 2

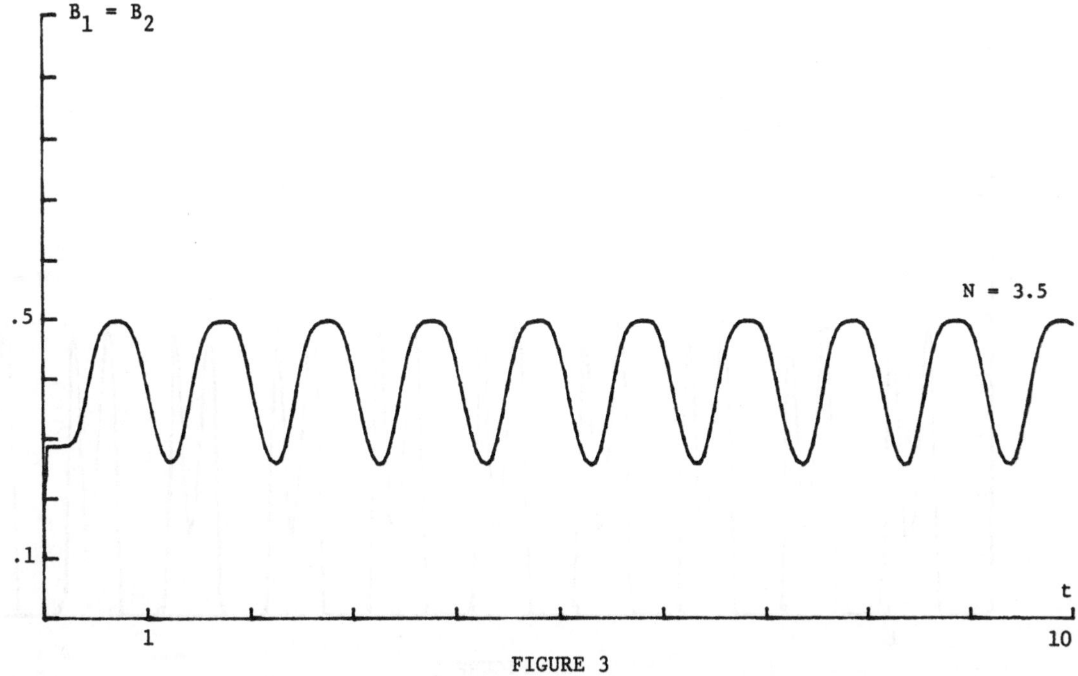

FIGURE 3

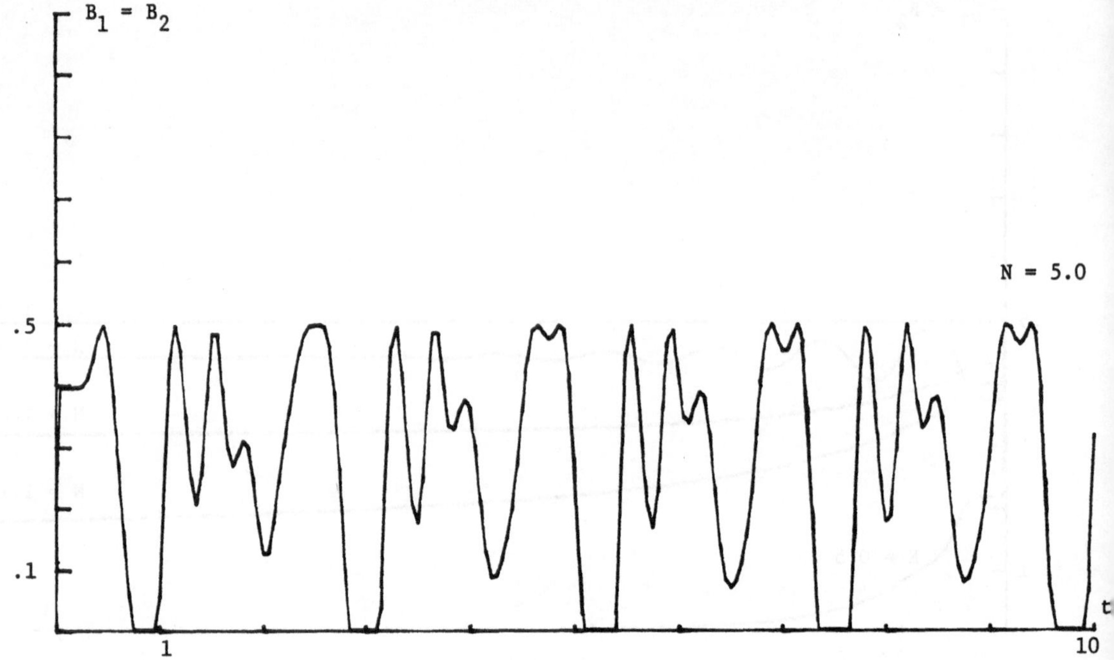

FIGURE 4

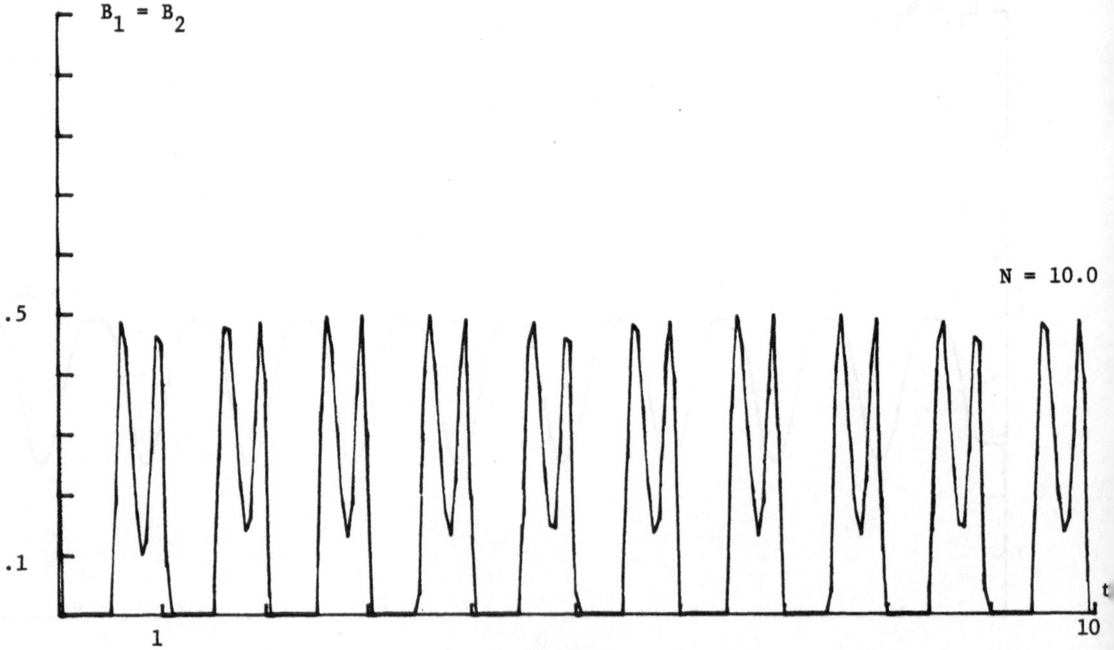

FIGURE 5

A connection to such difference equations can be made by formally passing $m \to +\infty$ in the model equations (in which case $w(a)$ in (4.4) becomes a Dirac function) which then reduce (with $n_1 = n_2 = N$ so that $B_1 \equiv B_2 \equiv B$) to the difference equation

$$B(t) = (A - N(1+\alpha^2)\exp(-\mu T)B(t-T))N\beta\exp(-\mu T)B(t-T).$$

If we let $t = jT$, $j = 0,1,2,\ldots$, $B_j := B(jT)$ and $d_1 := NA\beta\exp(-\mu T)$, $d_2 := N^2\mu(1+\alpha^2)\exp(-2\mu T)(1+\alpha^2)$ then we find that

$$B_{j+1} = B_j(d_1 - d_2 B_j) .$$

This is a well studied difference equation which is known to have rich and sometimes exotic dynamical behavior, including cascading period doubling periodicities and "chaos".

It would have been interesting to have found "chaotic" solutions for the integral equations (perhaps to be expected for large m and/or n_i), but as can be seen from FIGURES 4 and 5 the equations become quite "stiff", a fact which prevented us from carrying out further numerical solutions for larger $n_i = N$ or larger m.

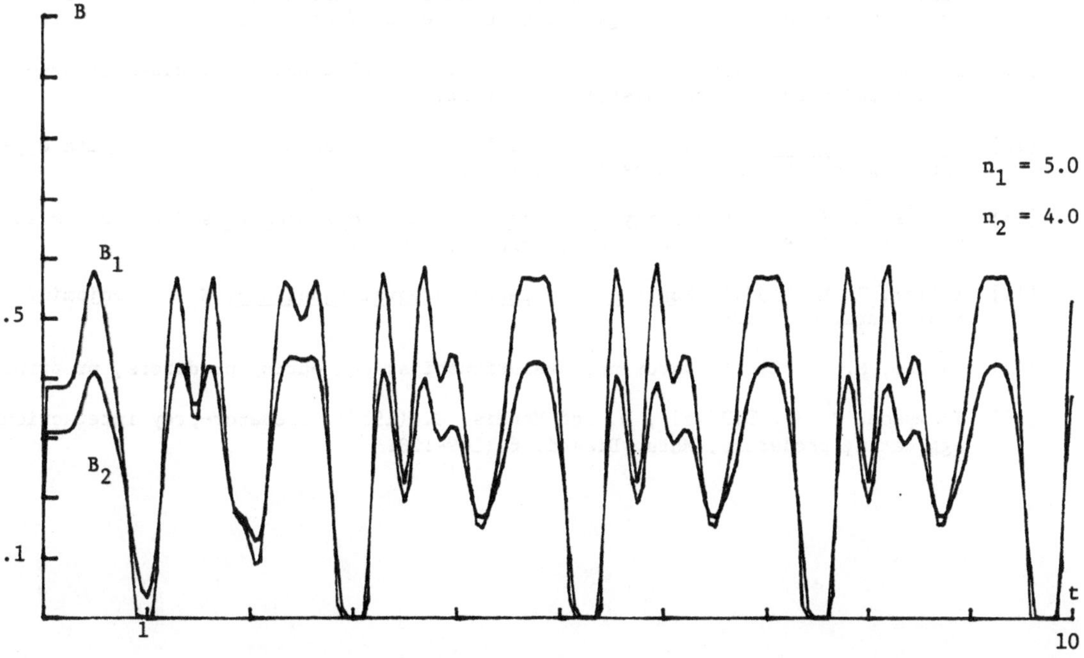

FIGURE 6

REFERENCES

[1] Christiansen, F. B. and Fenchel, T. M. (1977). Theories of Populations in Biological Communities, Ecol. Studies 20, Springer, Berlin.

[2] Cushing, J. M. and Saleem, M. (1982). A predator-prey model with age structure, J. Math. Biol. 14:231-250.

[3] Cushing, J. M. and Simmes, S. D. (1980). Bifurcation of asymptotically periodic solutions of Volterra integral equations, J. Integral Equations 4:339-361.

[4] Gopalsamy, K. (1982). Time lags and density dependence in age dependent two species competition, Bull. Aust. Math. Soc.

[5] Gurtin, M. E. and Levine, D. S. (1979). On predator-prey interactions with predation dependent on age of prey. Math. Biosci. 47:207-219.

[6] ______________ (1982). On populations that cannibalize their young, SIAM J. Appl. Math. 42:94-108.

[7] Gurtin, M. E. and MacCamy, R. C. (1974). Non-linear age-dependent population dynamics, Arch. Rat. Mech. 3:281-300.

[8] ______________ (1979). Population dynamics with age dependence, in Nonlinear Analysis and Mechanics, Heriot-Watt Symposium Vol. 3 (R. J. Knops, ed.), Pitman, London :1-35.

[9] ______________ (1979). Some simple models for nonlinear age-dependent population dynamics, Math. Biosci. 43:199-211.

[10] Hastings, A. (1982). Age dependent predation is not a simple process, Proc. Autumn Course on Math. Ecology, Int. Centre for Theo. Physics, Trieste, Italy.

[11] Levine, D. S. (1982). Bifurcating periodic solutions for a class of age-structured predator-prey systems, preprint.

[12] ______________ (1981). On the stability of a predator-prey system with egg-eating predators, Math. Biosci. 56:27-46.

[13] May, R. M. (1974). Stability and Complexity in Model Ecosytems, Monographs in Pop. Biol. 6, Princeton U. Press, Princeton, N. J.

[14] Miller, R. K. (1971). Nonlinear Volterra Integral Equations, W. A. Benjamin, Menlo Park, California.

[15] Saleem, M. (1982). Predator-prey relationships: egg eating predators, preprint.

[16] Thompson, R. W., DiBiasio, D. and Mendis, C. (1982). Predator-prey interactions: egg eating predators, Math. Biosci. 60:109-120.

STABILITY VS.COMPLEXITY IN MODEL COMPETITION COMMUNITIES

Pavel Kindlmann

INTRODUCTION

The question of the so-called stability vs. complexity relationship, i.e. the question whether a complex ecosystem tends to be more or less stable than a simple one, was broadly discussed in recent years. In the 'fifties, ecologists put forward the hypothesis that complex biological communities are more stable than simple ones (Mac Arthur, 1955; Elton, 1958; Hutchinson, 1959), but recent theoretical investigations have demonstrated that stability is not a simple mathematical consequence of complexity. The contrary frequently seems to be true (Gardner and Ashby, 1970; May, 1972; Hastings, 1982a, 1982b). The general context in which these studies are of biological interest has recently been well surveyed by Pimm (1982).

The mathematical analysis of the stability vs. complexity relationship is based on the assumption that the population dynamics of a system of m interacting species can be described by a system of m (in general nonlinear) first order differential equations of the following form:

$$dn_i(t)/dt = F_i(n_1(t), n_2(t), \ldots, n_m(t)); \; i = 1, 2, \ldots, m, \tag{1}$$

where $n_i(t)$ represents the population of i-th species at time t. This system is said to be stable if the corresponding linearized system

$$dx/dt = Ax \tag{2}$$

is locally asymptotically stable, i.e., if all the eigenvalues of the so-called interaction or community matrix in (2) have negative real parts (or lie within a unit circle if, instead of (1), the corresponding discrete model is analysed).

The interaction matrix A of a great ecosystem is considered as being a random matrix consisting of the elements

$$a_{ij} \begin{cases} = 0 \text{ with probability } 1-C, \\ \neq 0 \text{ with probability } C, \text{ where } a_{ij}\text{'s are chosen independently from distributions with mean } = 0, \text{ standard deviation } = s, \end{cases} \text{ for } i \neq j \quad (3)$$
$$= -1 \text{ for } i=j,$$

where, in the ecological sense, s represents the average interaction strength near the equilibrium, m is the number of species in the community, and C is connectance (i.e., the fraction of non-zero off-diagonal elements in the matrix A). The simulative and analytical investigations of several authors (Gardner and Ashby, 1970; May, 1972; Mc Murtrie, 1975; Hastings, 1982a, 1982b) have led to the conclusion that for a very wide range of matrices defined as in (3), their complex eigenvalue distribution tends, for $m \to \infty$, to a uniform density disc in the complex plane with the centre -1 (due to the choice of the diagonal elements) and radius $s\sqrt{mC}$. The fraction of real eigenvalues in the eigenvalue spectrum is scaling like $1/\sqrt{m}$. If $m \gg 1$, this fraction becomes negligible. Therefore, the probability of stability of (2), P_{stab}, tends to 0 if $s\sqrt{mC} > 1$ and P_{stab} approaches 1 if $s\sqrt{mC} < 1$, provided that the matrix A is constructed as in (3). An example of the eigenvalue spectrum of such matrices is given in Fig. 1.

A broad discussion arose with the aim to clarify the discrepancy between May's and empirical results. One of the arguments in this discussion was that matrices generated entirely randomly do not present ecologically plausible systems in most cases. The length of trophic chains is not limited and such phenomena as "plant eating a carnivore" are not excluded in a random construction. This problem becomes irrelevant if communities with only one trophic level are taken into account. In the following it will be shown that matrices which

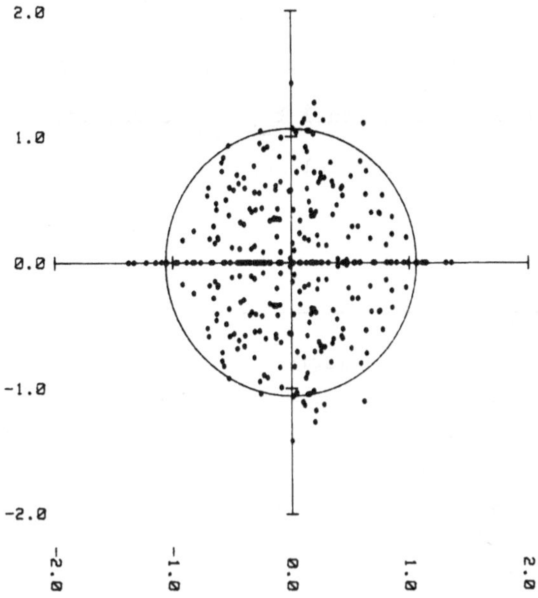

Fig. 1. The eigenvalue distribution of 50 matrices generated randomly with the elements chosen as in (3). Here, m = 10, s = 0.4, C = 0.7 and the diagonal is equal to 0.

represent such communities exhibit more complex behaviour than matrices generated entirely randomly and that in such matrices stability may even increase with increasing complexity in some cases.

THE EFFECT OF CONNECTANCE ON STABILITY FOR COMPETITIVE SYSTEMS

Competitive communities, i.e., such systems in which between each pair of species there is either no interaction or the influence on each other is negative, represent an example of communities with only one trophic level. We have constructed (Rejmánek, Kindlmann and Lepš, 1983) random matrices the elements of which were set equal to:

$$b_{ij} = \begin{cases} -|\mathrm{RAND}| & \text{for } i \neq j, \\ -1 & \text{for } i = j, \end{cases} \tag{4}$$

where the values of RAND were chosen from a normal distribution with the mean = 0 and standard deviation = s. The connectance C (≤ 1) had been introduced by zeroing a certain number of randomly chosen off-diagonal b_{ij}-b_{ji} pairs. Such "competitive matrices" represent models of quite general multispecies competitive communities.

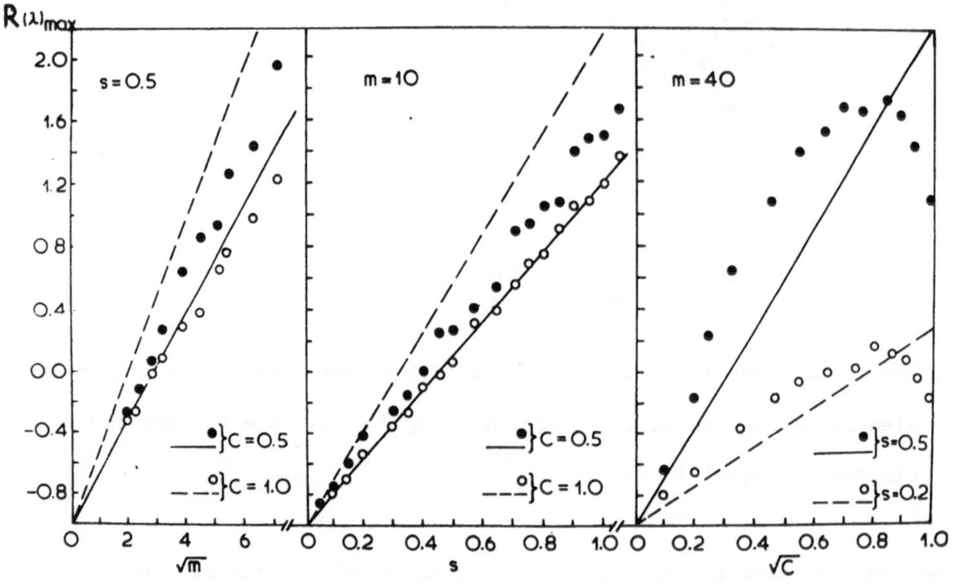

Fig. 2. The mean largest real parts, $R(\lambda)_{max}$, of eigenvalues of competitive matrices (see (4) for construction) against $\sqrt{m}$, s and $\sqrt{C}$. The circles are results of our simulations, the lines are May's prediction.

We have determined the largest real parts, $R(\lambda)_{max}$, of eigenvalues of interaction matrices B consisting of the elements defined in (4). Samples of 200 matrices were generated for each selected combination of m, s and C values. Results of this analysis are presented in Fig. 2. The dependence of mean $R(\lambda)_{max}$ on $\sqrt{m}$ and on s is positive and linear, differing more or less from May's prediction, depending on connectance. On the contrary, the relationship between mean $R(\lambda)_{max}$ and $\sqrt{C}$ is conspicuously non-linear and partly in direct contradiction to May's rule. Not only does it indicate decrease of stability until connectance $\sqrt{C} \doteq 0.8$ is achieved, but it also indicates an increase in stability (decrease of mean $R(\lambda)_{max}$) with the subsequent increase in C.

In the same way as Gardner and Ashby (1970), we have expressed the proportion of stable systems A and B (estimate of P_{stab} of A and B) against C for three different m values (see Fig. 3). The extent of the interval of C within which P_{stab} increases with C depends on m, s and b_{ii}.

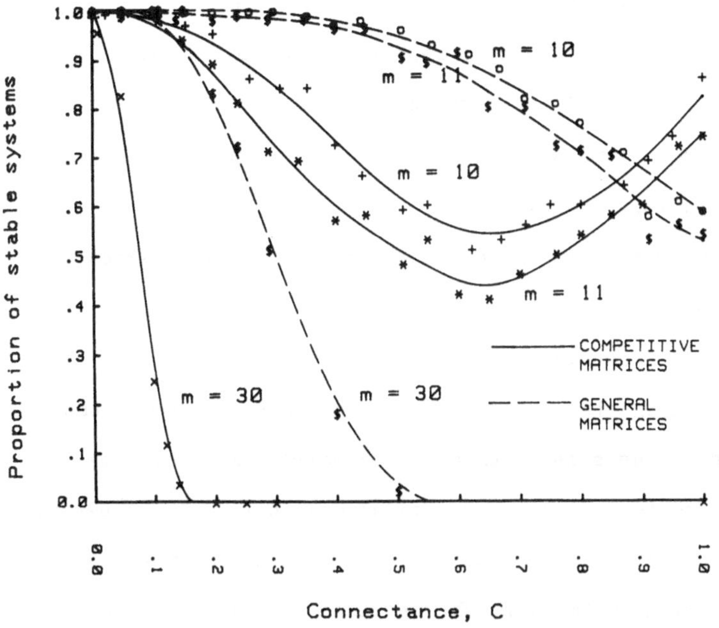

Fig. 3. Proportion of stable systems against connectance for general and competitive matrices (see (3) and (4) for construction). Each point represents the proportion of stable matrices in a sample of 200 matrices. Here, s = 0.4, b_{ii} = -1.

For m = 10 and for m = 11 the increase of stability with increasing connectance for higher values of connectance is obvious. For m = 30 and higher values of connectance, the probability of stability is zero. There also occurs decrease of $R(\lambda)_{max}$ in higher values of connectance, but $R(\lambda)_{max}$ remains positive. But if suitable s or b_{ii} were chosen, the same phenomenon could be observed for m = 30 or for any other m. Furthermore, from Fig. 3 it is evident that a more complex competitive system, i.e., a more connected system having more species (m = 11 and C = 1 in our case) may be more stable than a simpler one (i.e., the system with m = 10 and C being about 0.6).

THE EFFECT OF AMENSALISTIC INTERACTIONS

Many natural systems are organized in such a way that the zeros are not distributed symmetrically along the main diagonal of the interaction matrix, i.e. that not only competitive but also amensalistic interactions occur in the system. In this case, our computer simulations resulted in a similar qualitative behaviour as that shown in Figs. 2 and 3: increase of stability with complexity in some cases. Moreover, such systems are a better subject to analytical investigation than pure competitive ones.

I shall restrict my analysis only to the critical case of the transition from stable to unstable systems. As I will show, this transition occurs for the values Cms^2 not being very large in the competitive/amensalistic systems, too. Moreover, it is clearly seen that if for given C, m and s the corresponding matrix is almost certain to be unstable, then a random matrix with the same rank and connectance and standard deviation $s' > s$ is almost certain to be unstable, too. Therefore, I shall assume that there exists such a constant K, $\infty > K > 1$ that for each matrix under question $Cms^2 < K$. The matrices excluded are almost certain to be unstable.

Let the elements b'_{ij} of the "competitive/amensalistic" matrix B' be chosen in the same way as the elements of B with the exception that the zeros are distributed entirely randomly and asymmetrically along the main diagonal:

$$b'_{ij} \begin{cases} \left.\begin{array}{l} = 0 \text{ with the probability } 1 - C, \\ \neq 0 \text{ with the probability } C, \text{ then} \\ \quad b'_{ij} = -|X|,\ X \sim N(0, s^2), \end{array}\right\} \text{ for } i \neq j, \\ = -1 \text{ for } i = j. \end{cases} \tag{5}$$

The mean of the elements b'_{ij} is $-\mu = -Cs\sqrt{2/\pi}$ for $i \neq j$ and -1 for $i=j$. The variance of the elements b'_{ij} is $\sigma^2 = Cs^2(1 - 2C^2/\pi)$ for $i \neq j$ and 0 for $i=j$. We shall denote the eigenvalues of B' as $\lambda'_1, \lambda'_2, \ldots, \lambda'_m$. Let $A' = B' + \mu I_m$, where $-\mu$ is the mean of non-diagonal elements of B' and I_m is a matrix of rank m, all elements of which are equal to 1. The mean of the elements a'_{ij} of the matrix A' is 0 for $i \neq j$ and $\mu - 1$ for $i=j$. The variance of the elements a'_{ij} is the same as that of the elements b'_{ij}. If we denote the eigenvalues of A' as $\lambda_1, \lambda_2, \ldots, \lambda_m$, we may assume (on the basis of McMurtrie's and our simulations) that the eigenvalues of A' lie, for $m \gg 1$, in a uniform-density disc with the centre $\mu - 1$ and radius $\sigma\sqrt{m} = s\sqrt{mC(1 - 2C^2/\pi)}$.

Matrix	Elements	
D_1	${}_1d_{ij}$	$= b'_{ij} = a'_{ij} - \mu$, for $i,j = 1,2,\ldots,m$, $i \neq j$
	${}_1d_{ii}$	$= b'_{ii} + 1 - \mu = a'_{ii} + 1 - 2\mu$, for $i = 1,2,\ldots,m$
	${}_1d_{m+1,j}$	$= 0$, for $j = 1,2,\ldots,m+1$
	${}_1d_{i,m+1}$	$= 1$, for $i = 1,2,\ldots,m$
D_2	${}_2d_{ij}$	$= a'_{ij}$, for $i,j = 1,2,\ldots,m$, $i \neq j$
	${}_2d_{ii}$	$= a'_{ii} + 1 - \mu$, for $i = 1,2,\ldots,m$
	${}_2d_{m+1,j}$	= defined by (6), for $j = 1,2,\ldots,m$
	${}_2d_{i,m+1}$	$= 1$, for $i = 1,2,\ldots,m$
	${}_2d_{m+1,m+1}$	$= -\mu m$
D_3	${}_3d_{ij}$	$= a'_{ij}$, for $i,j = 1,2,\ldots,m$, $i \neq j$
	${}_3d_{ii}$	$= a'_{ii} + 1 - \mu$, for $i = 1,2,\ldots,m$
	${}_3d_{m+1,j}$	$= 0$, for $j = 1,2,\ldots,m$
	${}_3d_{i,m+1}$	$= 1$, for $i = 1,2,\ldots,m$
	${}_3d_{m+1,m+1}$	$= -\mu m$

Table 1. Definitions of auxiliary matrices D_1, D_2 and D_3.

In order to elucidate the relationship between eigenvalue distributions of A' and B' I shall introduce auxiliary matrices D_1, D_2, D_3 (see Table 1 for definitions). Evidently, the eigenvalues of D_1 are $\lambda'_1 + 1 - \mu$, $\lambda'_2 + 1 - \mu$, $\ldots$, $\lambda'_m + 1 - \mu$, 0 and those of D_3 are $\lambda_1 + 1 - \mu$, $\lambda_2 + 1 - \mu, \ldots, \lambda_m + 1 - \mu$, $-\mu m$. The characteristic matrix of D_1, $D_1 - \lambda E_m$, where E_m is the unit matrix of rank m, is equivalent to the characteristic matrix of D_2, $D_2 - \lambda E_m$, which may be obtained from $D_1 - \lambda E_m$ by the addition of μ-multiple of the last column to each of the remaining columns and subsequent subtraction of μ-multiple of all the rows except the last one from the last row. It might be easily shown that the mean of each of the random variables

$$\mu \left(\sum_{i=1}^{m} a'_{ij} + 1 - \mu \right) \qquad (6)$$

in the last row of $D_2 - \lambda E_m$ is zero and the variance of each of them is $2mC^3s^4(1 - 2C^2/\pi)/\pi$. If Cms^2 is bounded by a constant K, $0 < C \leqq 1$, and $m \to \infty$, this variance tends to zero. (The term $2C(1 - 2C^2/\pi)/\pi$ is bounded by a constant $2/\pi$, $(Cms^2)^2 < K^2$). The matrices D_2 and D_3 differ only in the elements ${}_2d_{m+1,j}$, ${}_3d_{m+1,j}$ for $j = 1,2,\ldots,m$. The mean of each of these elements is zero and the variance approaches zero if $m \to \infty$. It follows from the continuous dependence of the eigenvalues of a matrix on its elements that, for $m \gg 1$, the eigenvalues of D_2 differ only little from those of the matrix D_3. The eigenvalues $\lambda_i + 1 - \mu$ of D_3 lie in a uniform-density disc with the centre 0 and radius $\sigma\sqrt{m}$; the eigenvalue $-\mu m$ is real and lies outside the disc for $C > 0.5(-m + \sqrt{m^2 + 2\pi})$, i.e. for nearly all possible connectances, which may be derived from the expressions for μ and $\sigma\sqrt{m}$. The eigenvalues of D_1 differ only little from those of D_3 for $m \gg 1$; therefore we may assume that the asymptotic eigenvalue distribution of D_1 is, for $m \gg 1$, the same as that of D_3. Finally, it follows from these considerations that the asymptotic eigenvalue distribution of B' is, for $m \gg 1$, a uniform-density disc in the complex plane with the centre $\mu - 1$ and radius $\sigma\sqrt{m}$. For $C > 0.5(-m + \sqrt{m^2 + 2\pi})$ there appears one isolated eigenvalue outside this disc of the magnitude about $-\mu m + \mu - 1$.

The estimate of the eigenvalue distribution of the competitive/amensalistic matrix B' was in quite a good agreement with the results of our simulations even for m being quite small (see Fig. 4). The flattening of the eigenvalue distribution of competitive matrices is not an artefact which has occured in

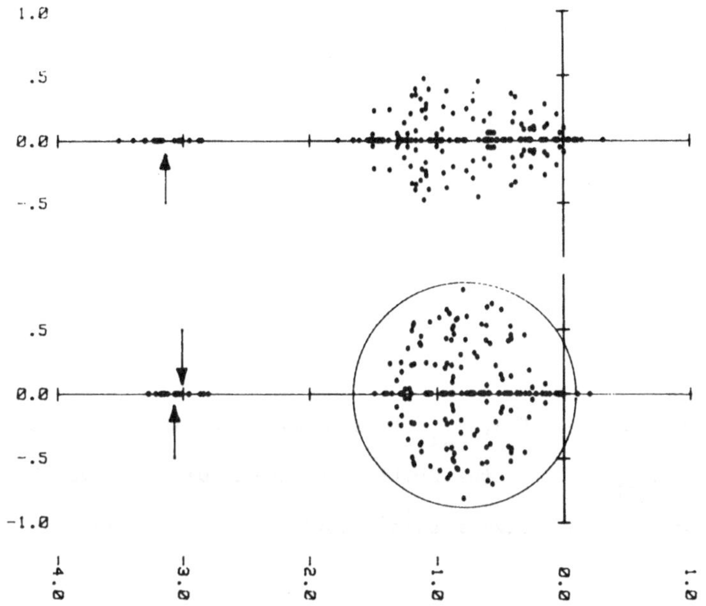

Fig. 4. The eigenvalue distribution of 20 competitive (upper graph) and 20 competitive/amensalistic (lower graph) matrices. In both samples, the rank, standard deviation and connectance were chosen m = 10, s = 0.4 and C = 0.7. The diagonal elements were equal to -1. The circle indicates our estimate of the eigenvalue distribution; ↓ is our estimate of the position of the isolated eigenvalue, ↑ are the actual means of the isolated eigenvalue.

Fig. 4, but a general phenomenon. It may be attributed to a certain symmetry of competitive matrices - to the symmetrical distribution of their zero elements - which "pushes" the eigenvalues in the direction towards the real axis.

By using the expression for μ we obtain an expression for λ_m in the original terms C, m and s:

$$\lambda_m \doteq -\mu m + \mu - 1 = (1 - m)\sqrt{2/\pi}\, sC - 1. \tag{7}$$

Further, all the complex numbers lying inside the disc with the centre $\mu - 1$ and radius $\sigma\sqrt{m}$, as well as the number $-\mu m$, have their real parts less than $\mu - 1 + \sigma\sqrt{m}$. Therefore, we may state an inequality for the greatest real part of the eigenvalues of the competitive/amensalistic matrix B', for $R(\lambda)_{max}$:

$$R(\lambda)_{max} < \mu - 1 + \sigma\sqrt{m}. \tag{8}$$

By using the expressions for μ and σ we obtain that

$$R(\lambda)_{max} < s(C\sqrt{2/\pi} + \sqrt{Cm(1 - 2C^2/\pi)}) - 1. \tag{9}$$

For a given m, the actual expectation of $R(\lambda)_{max}$ is somewhat less than the just mentioned upper limit for $R(\lambda)_{max}$, due to the finite number (m) of the eigenvalues. By using the uniformity of the eigenvalues distribution in the above-

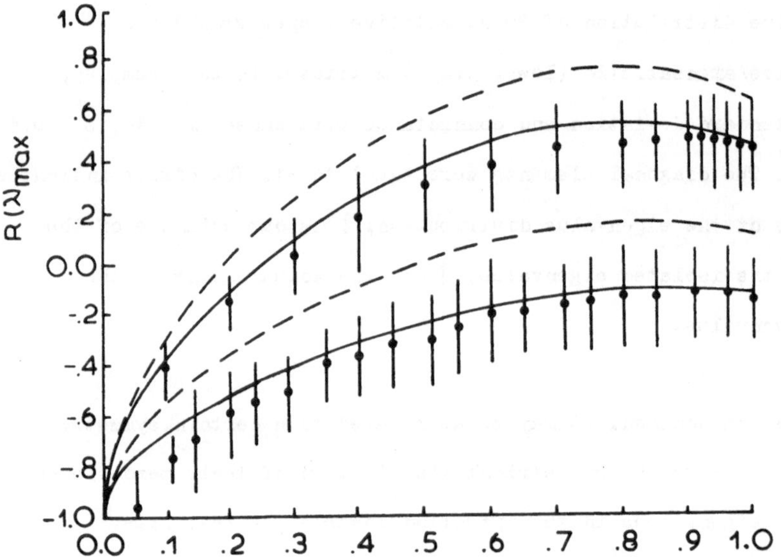

Fig. 5. Maximum real part of eigenvalues, $R(\lambda)_{max}$, against connectance, C, for competitive/amensalistic matrices; s = 0.4, m = 11 and m = 30. Expectation of $R(\lambda)_{max}$ (eq. (10)) is represented by solid curves, upper bound for $R(\lambda)_{max}$ (eq. (9)) by broken curves, points indicate mean values and vertical bars represent standard deviations of $R(\lambda)_{max}$ obtained by simulations. Each point corresponds to 200 randomly filled matrices.

-mentioned disc a relation for the expectation of $R(\lambda)_{max}$ may be derived:

$$E(\ R(\lambda)_{max}(B')\ ) = \int_{-R}^{R} x \frac{d}{dx}\left[\frac{2}{\pi R^2} \int_{-R}^{x} \sqrt{R^2 - y^2}\, dy\right]^{m-1} dx - 1, \qquad (10)$$

where $R = s\sqrt{mC(1 - 2C^2/\pi)}$.

In Fig. 5 the comparison of the expectation of $R(\lambda)_{max}$ and of the upper bound for $R(\lambda)_{max}$ with the result of our numerical simulations is shown. The decrease of $R(\lambda)_{max}$ for higher values of connectance is obvious. The same phenomenon as in May's and McMurtrie's figures of the dependence of $R(\lambda)_{max}$ on C occurs in this figure: the eigenvalues obtained by the simulations are slightly smaller than those predicted by the expression for $R(\lambda)_{max}$. It is a consequence of the fact that the expectation of $R(\lambda)_{max}$ is only an asymptotical estimate valid for $m \gg 1$.

THE CONSEQUENCES OF USING FINITE DIFFERENCE EQUATIONS

Another interesting fact follows from this analysis: the existence of a real negative isolated eigenvalue with a great magnitude of about $(1 - m)\sqrt{2/\pi}\ sC - 1$. The existence of a real negative eigenvalue with its absolute value equal to the spectral radius of a matrix with negative off-diagonal elements and zeros on its diagonal is predicted also by the Perron-Frobenius theorem. Our findings are in accordance with this theorem, but in our case the absolute value of this eigenvalue is much greater than the absolute values of the remaining m - 1 eigenvalues in most cases. This phenomenon demonstrates that between discrete competitive or competitive/amensalistic systems and their continuous counterparts there is a much greater difference in stability than between discrete and continuous systems generated entirely randomly (as by McMurtrie and May). In such competitive or competitive/amensalistic discrete systems an increase of stability within the interval of higher values of connectance does not occur because of the existence of the above-mentioned eigenvalue (see Figs. 6 and 7). For example, the shape of the curve for discrete model and m = 15 in Fig. 6

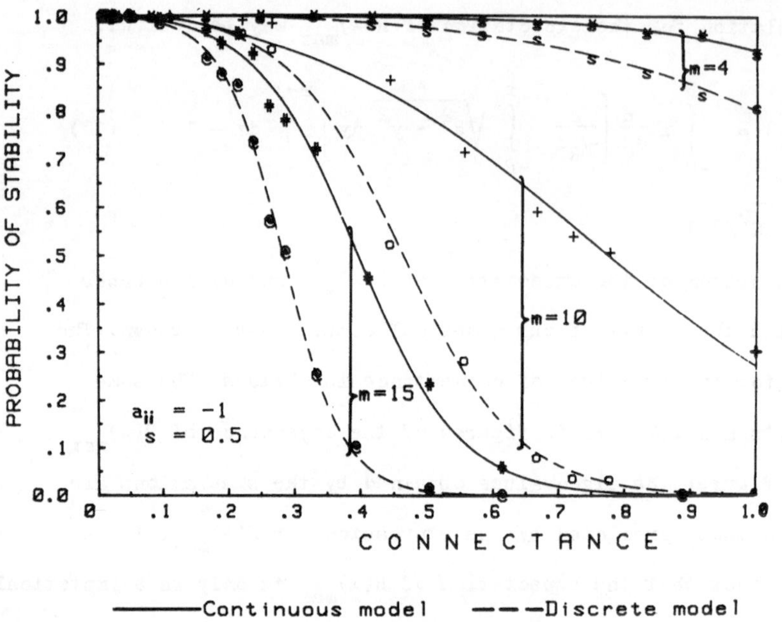

Fig. 6. Proportion of stable systems against connectance for continuous (solid curves) and discrete (broken curves) general systems (see (3) for construction). Each point corresponds to 200 randomly filled matrices. The value of s was chosen s = 0.5.

is nearly the same as the shape of the curve for discrete model and m = 10 in Fig. 7, i.e. for smaller number of species, though the value of s is smaller, too. On the contrary, the shapes of the corresponding curves for continuous models are quite different, especially for higher values of connectance.

DISCUSSION

The equilibria of our systems were not checked for positivity. But following the way of Goh and Jennings (1977) step by step, it may be easily shown that also the subset of competitive or competitive/amensalistic Lotka-Volterra systems, each of which has a feasible equilibrium, has the same stability property as a set of linear competitive or competitive/amensalistic systems which is assembled randomly in the same manner.

Moreover, these findings provide a new insight into the results of stability analysis of 40 real food webs presented recently by Yodzis (1981). He concluded that simulative increase of the proportion of interspecific competitive interactions "usually exerts a destabilizing influence". Five exceptions from this trend (webs nos. 2, 12, 13, 14 and 15 in his table 3) may result from the stabilizing influence of increasing competitive connectance. Only the species richness and structure of these five webs lie within the interval in which the stabilizing influence of increasing competitive connectance can manifest itself.

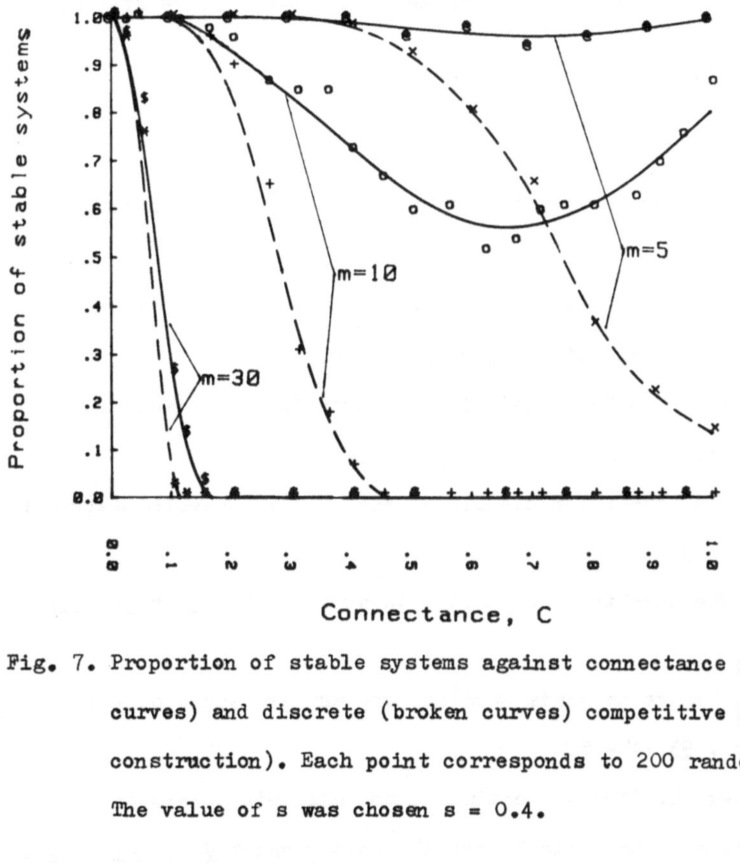

Fig. 7. Proportion of stable systems against connectance for continuous (solid curves) and discrete (broken curves) competitive systems (see (4) for construction). Each point corresponds to 200 randomly filled matrices. The value of s was chosen s = 0.4.

COMPARISON OF THE KNOWN STABILITY CRITERIA

For randomly generated systems, the following stability criteria are known:

1. For general continuous systems there holds the well-known May's criterion $Cms^2 < 1$. May's simulations indicate that it is a rather conservative estimate.

2. For general discrete systems, May's criterion is rather optimistic. Kindlmann and Rejmánek (1982) have developed another, very conservative criterion for discrete general systems: $Cm(m - 1)^3 s^2 < 1$.

3. In this paper there were derived rather conservative stability criteria for discrete and continuous competitive/amensalistic systems (see also Rejmánek, Kindlmann and Lepš, 1983). Our simulations have shown that the competitive systems are less stable than the competitive/amensalistic ones, but the difference between them is very small. Therefore, the criteria for competitive/amensalistic systems may approximately be applied to the competitive systems, too.

LITERATURE CITED

Elton, C.S. (1958). The ecology of invasion of animals and plants. London: Methuen.

Gardner, M.R. and Ashby, W.R. (1970). Connectance of large dynamic (cybernetic) systems: critical values for stability. Nature 228: 784.

Goh, B.S. and Jennings, L.S. (1977). Feasibility and stability in randomly assembled Lotka-Volterra models. Ecol. Modelling 3:63-71.

Hastings, H.M. (1982a). The May-Wigner stability theorem for connected matrices. Bull. Am. Math. Soc. 7:387-388.

Hastings, H.M. (1982b). The May-Wigner stability theorem. J. Theor. Biol. 97: 155-166.

Hutchinson, G.E. (1959). Homage to Santa Rosalia; or, why are there so many kinds of animals? Am. Nat. 93:145-159.

Kindlmann, P. and Rejmánek, M. (1982). Continuous vs. discrete models of multispecies systems: How much less stable are the latter ones? J. Theor. Biol. 94:989-993.

MacArthur, R.H. (1955). Fluctuations of animal populations and a measure of community stability. Ecology 36:533-536.

May, R.M. (1972). Will a large complex system be stable? Nature 238:413-414.

McMurtrie, R.E. (1975). Determinants of stability of large randomly connected systems. J. Theor. Biol. 50:1-11.

Pimm, S.L. (1982). Food webs. London and New York: Chapman and Hall.

Rejmánek, M., Kindlmann, P. and Lepš, J. (1983). Increase of stability with complexity in model competition communities. J. Theor. Biol. 101:649-656.

Yodzis, P. (1981). The stability of real ecosystems. Nature 289:674-676.

Persistence in Food Webs

T.C. Gard
Department of Mathematics
University of Georgia, Athens, Georgia 30602

1. Introduction. Of the stability concepts associated with population dynamics models, the notion of persistence has emerged as one of the most important to ecologists. (See, for example, Holling (1973), Botkin and Sobel (1974), Innis (1974), Maynard Smith (1974).) Persistence, generally, refers to that quality of such models whereby population density levels remain within certain acceptable bounds despite perturbations of model parameters or initial values. Various specific mathematical formulations of persistence have been given by Innis (1974), Botkin and Sobel (1974), Wu (1974), Freedman and Waltman (1977), McGehee and Armstrong (1977), and Harrison (1979a) for models taking the form of systems of ordinary differential equations. All of these definitions of persistence are closely related to the dynamical system concept of flow-invariance. Here this relationship will be discussed, and an extension to food webs of Freedman and Waltman's definition of food chain persistence will be exhibited for the Lotka-Volterra case. In particular, a sufficient condition for top predator persistence in terms of model parameters will be given. Although the discussion will be restricted to ordinary differential equation models, some brief remarks about persistence in dynamical models of other types are in order first.

For stochastic systems, in particular systems subject to random perturbations, Ludwig (1975) has suggested the exit time from a specified set as a measure of system persistence. The exit time is a random variable whose statistics can be estimated using perturbation techniques in this case. Tier and Hanson (1981) have carried out such a program, for example, for a single species population undergoing demographic as well as environmental random fluctuations. Allen (1981) has extended the Freedman-Waltman definition of persistence to systems incorporating spatial effects via different diffusion mechanisms. She has studied both discrete (patch type), and continuous (reaction-diffusion) models for prey-predator, competition, and mutualism systems. The importance of taking into account random fluctuations and spatial heterogeneity in ecosystem models, along with the development of mathematical tools for the analysis of stochastic and reaction-diffusion models, makes this a promising area for future research.

2. Persistence and flow-invariance. All of the persistence definitions cited above for models involving ordinary differential equations require or have as an immediate consequence the existence of a flow-invariant set in state space. To be precise, in this case the model takes the form

$$\frac{dx(t)}{dt} = f(x(t),a(t)) \tag{1}$$

where for each t,

$$x(t) = \{x_i(t)\} \in R_+^n = \{x = \{x_i\} \in R^n | x_i \geq 0,\ 1 \leq i \leq n\} \quad ,$$

representing the population densities of n species at time t, and a belongs to some admissible class $\mathcal{A}$ so that $a(t) = \{a_j(t)\} \in R^m$ gives the state of the environment at time t; the function $f = \{f_i\}: R_+^n \times R^m \rightarrow R^n$ denotes the species' net growth rates. Under mild assumptions on f and $\mathcal{A}$, given an $a \in \mathcal{A}$ and $x_0 \in R_+^n$ there exists a unique solution $x(t)$ of (1) which satisfies $x(t_0) = x_0$ and represents the evolution of species' population densities on some time interval. A set $M \subseteq R_+^n$ is flow-invariant with respect to (1) if each solution $x(t)$ of (1) having initial value $x(0) \in M$ satisfies $x(t) \in M$ for all $t > 0$. Equilibrium points and periodic trajectories are examples of flow-invariant sets. Redheffer and Walter (1975), Séifert (1976), and Gard (1980) review the mathematical literature on flow-invariance. Most of the results give criteria for checking whether or not a given set is flow-invariant. Generally, determining flow-invariant sets for a particular model is a difficult task.

Harrison (1979a) defines a subset M of

$$R_+^{n,o} = \{x = \{x_i\} | x_i > 0,\ 1 \leq i \leq n\} \quad ,$$

(sometimes referred to as the feasible region), as persistent with respect to (1) and an admissible class $\mathcal{A}$ provided M is flow-invariant with respect to (1) for each $a \in \mathcal{A}$; he also allows M to vary with time (i.e., require $x(t) \in M(t)$, if $x(0) \in M(0)$) and points out that the persistence definitions given by Innis (1974), Botkin and Sobel (1974), and Wu (1974) are special cases of his definition. Furthermore Harrison (1979b) has shown how persistent sets can be determined from Lyapunov functions; the sets so obtained are Lyapunov-stable. It seems reasonable to require the asserted flow-invariant set to be stable for persistence; otherwise, for example, a system which preserves, under various environments, an equilibrium, even an unstable one, would be ruled persistent. The existence of a stable flow-invariant set in $R_+^{n,o}$ corresponds to the definition of persistence used by McGehee and Armstrong (1977) in their treatment of competition systems.

Freedman and Waltman (1977) take a different approach in their analysis of food chain models. Persistence here means that for each solution $x(t) = \{x_i(t)\}$ of (1) with initial value $x(0) \in R_+^{n,o}$, and maximal interval of existence $[0,T)$,

$$\limsup_{t \to \tau} x_i(t) > 0 \tag{2}$$

for each i, $1 \leq i \leq n$, and each $\tau \in (0,T]$. That is, persistence means that no solution having all components positive initially experiences any component tending to zero in finite or infinite time. This is a stronger requirement than the definition given previously; for autonomous systems having bounded trajectories in $R_+^{n,o}$, it implies the existence of one or more stable flow-invariant sets in $R_+^{n,o}$, toward which all trajectories must move as $t \to \infty$. It is easy to see, for food chain models, that this definition of persistence is equivalent to persistence of the top predator; that is, if x_n denotes the density of the top predator, (2) is equivalent to

$$\limsup_{t \to \tau} x_n(t) > 0 \ , \tag{3}$$

for each $\tau \in (0,T]$, for each solution $x(t) = \{x_i(t)\}$ with initial value $x(0) \in R_+^{n,o}$ and maximum interval of existence $[0,T)$. For food web models, a natural extension of this persistence definition is, then, that given any population with all species initially present, at least some top predator survives indefinitely, which one survives being possibly dependent on the initial population configuration. This notion of persistence asserts the preservation of the web's trophic structure rather than all species in the web, which may be significant from the point of view of assessing ecological effects. (Paine (1966), in his studies of intertidal communities, has shown that removal of a top predator can drastically reduce community structure.) A disadvantage of this approach to persistence is that, although the existence of flow-invariant sets is asserted, the location of such sets in the feasible region is not addressed. Indeed, for food webs, these sets will not necessarily be situated in $R_+^{n,o}$; however, at least one species from each trophic level will be represented in each such set. The situation is somewhat mitigated by the improved mathematical tractability of the problem which is demonstrated in the next section where a criterion in terms of model parameters is deduced for Lotka-Volterra food webs. It is emphasized that these models are primarily of theoretical, as opposed to predictive, value at least at their current stage of development. Freedman (1980) has given a detailed mathematical treatment of the basic properties of Lotka-Volterra models.

3. Persistence in Lotka-Volterra food webs. In this section, the Lotka-Volterra food web, represented by the system of ordinary differential equations

$$\begin{aligned} \frac{dx_i}{dt} &= x_i(a_i - \sum_{j=1}^{k} b_{ij}x_j - \sum_{j=1}^{m} c_{ij}y_j), && 1 \le i \le k \\ \frac{dy_i}{dt} &= y_i(-d_i + \sum_{j=1}^{k} e_{ij}x_j - \sum_{j=1}^{p} f_{ij}z_j), && 1 \le i \le m \qquad (4) \\ \frac{dz_i}{dt} &= z_i(-g_i + \sum_{j=1}^{m} h_{ij}y_j), && 1 \le i \le p \end{aligned}$$

is considered. In this model $x_i(t)$, $y_i(t)$, and $z_i(t)$ denote the population densities at time t of the i th prey, intermediate predator, and top predator respectively; the a_i, b_{ij}, c_{ij}, d_i, e_{ij}, f_{ij}, g_i, and h_{ij} are positive constants representing the various intrinsic growth and interaction rates. Solutions of (4) with initial values in

$$R_+^n = \{(x_1,\dots,x_k,y_1,\dots,y_m,z_1,\dots,z_p) \mid x_i \geq 0,\ y_i \geq 0,\ z_i \geq 0,\ \text{each } i\}\ ,$$

$$n = k + m + p\ ,$$

are unique, bounded, and remain in R_+^n on their entire interval of existence. (These facts are easily deduced from the basic theory of ordinary differential equations.) Of primary interest here are solutions of (4) with initial values in

$$R_+^{n,o} = \{(x_1,\dots,x_k,y_1,\dots,y_m,z_1,\dots,z_p \mid x_i > 0,\ y_i > 0,\ z_i > 0,\ \text{each } i\}\ .$$

Top predator persistence means that for any solution

$$\varphi(t) = (x_1(t),\dots,x_k(t),y_1(t),\dots,y_m(t),z_1(t),\dots,z_p(t))$$

with initial value $\varphi(0) \in R_+^{n,o}$

$$\limsup_{t\to\tau} z_i(t) > 0 \tag{5}$$

for any $\tau \in (0,T]$, the maximal interval of existence of $\varphi(t)$, and for some index i, $1 \leq i \leq p$, which may depend on choice of $\varphi(0)$. Similarly to the corresponding proof in Gard and Hallam (1979) for food chain models, it can be shown that any such solution $\varphi(t)$ is defined on the entire interval $[0,\infty)$ and remains in $R_+^{n,o}$ for all finite time. Thus top predator persistence (5) becomes, for some i,

$$\limsup_{t\to\infty} z_i(t) > 0\ . \tag{6}$$

The basic problem addressed here is to obtain a criterion in terms of the model parameters which will guarantee (6).

A procedure for obtaining such a criterion involves the construction of a Lyapunov type function

$$\rho(x_1,\dots,x_k,y_1,\dots,y_m,z_1,\dots,z_p)$$

which is positive on $R_+^{n,o}$ and satisfies

$$\rho \to 0 \quad \text{if any} \quad z_i \to 0 \tag{7}$$

Assuming, by way of contradiction, that there is a solution $\varphi(t)$ of (4) with initial value $\varphi(0) \in R_+^{n,o}$ which exhibits $z_i(t) \to 0$, as $t \to \infty$, for all i, $1 \leq i \leq p$, one considers the function

$$\rho(\varphi(t)) = \rho(x_1(t),\ldots,x_k(t),y_1(t),\ldots,y_m(t),z_1(t),\ldots,z_p(t)) .$$

If a differential inequality of the form

$$\dot{\rho} = \frac{d\rho(\varphi(t))}{dt} = \nabla\rho \cdot \left(\frac{dx_1}{dt},\ldots,\frac{dx_k}{dt},\frac{dy_1}{dt},\ldots,\frac{dy_m}{dt},\frac{dz_1}{dt},\ldots,\frac{dz_p}{dt}\right) \geq \lambda\rho \quad , \tag{8}$$

for some positive constant λ, can be established for sufficiently large t, then the required contradiction is obtained. Indeed, inequality (8) holding for sufficiently large t implies $\rho(\varphi(t))$ does not tend to 0 as $t \to \infty$ which itself contradicts that $z_i(t) \to 0$ as $t \to \infty$.

For the food web model (4) one chooses ρ of the form

$$\rho = \prod_{i=1}^{k} x_i \prod_{i=1}^{m} y_i^{r_i} \prod_{i=1}^{p} z_i^{s_i}$$

where the constants r_i and s_i are to be determined. Then

$$\begin{aligned}\dot{\rho} = \rho\{ &\sum_{i=1}^{k} (a_i - \sum_{j=1}^{k} b_{ij}x_j - \sum_{j=1}^{m} c_{ij}y_j) \\ &+ \sum_{i=1}^{m} r_i(-d_i + \sum_{j=1}^{k} e_{ij}x_j - \sum_{j=1}^{p} f_{ij}z_j) \\ &+ \sum_{i=1}^{p} s_i(-g_i + \sum_{j=1}^{m} h_{ij}y_j)\}\end{aligned} \tag{9}$$

follows from (4). It is convenient to rewrite (9) as

$$\dot{\rho} = \rho\{\sum_{i=1}^{k} a_i - \sum_{i=1}^{m} d_i r_i - \sum_{i=1}^{p} g_i s_i$$

$$+ \sum_{i=1}^{k} x_i(\sum_{j=1}^{m} e_{ji} r_j - \sum_{j=1}^{k} b_{ji})$$

$$+ \sum_{i=1}^{m} y_i(\sum_{j=1}^{p} h_{ji} s_j - \sum_{j=1}^{k} c_{ji}) \tag{10}$$

$$- \sum_{i=1}^{p} z_i \sum_{j=1}^{m} f_{ji} r_j\}$$

Now let the constants r_i and s_i be nonnegative numbers satisfying the inequalities

$$\sum_{j=1}^{m} e_{ji} r_j - \sum_{j=1}^{k} b_{ji} \geq 0 , \quad 1 \leq i \leq k \tag{11}$$

$$\sum_{j=1}^{p} h_{ji} s_j - \sum_{j=1}^{k} c_{ji} \geq 0 , \quad 1 \leq i \leq m \tag{12}$$

and let

$$\mu = \mu(\{r_i\},\{s_i\}) = \sum_{i=1}^{k} a_i - \sum_{i=1}^{m} d_i r_i - \sum_{i=1}^{p} g_i s_i .$$

Theorem. Top predator persistence holds in (4) if

$$\mu = \mu(\{r_i\},\{s_i\}) > 0 \tag{13}$$

Proof. Suppose $\varphi(t)$ is a solution of (4) with initial value $\varphi(0) \in R_+^{n,o}$ having z_i-component $z_i(t)$ tending to zero as $t \to \infty$, for all i. Taking ρ defined as above with the r_i and s_i satisfying (11) and (12), it follows from (10) that

$$\frac{d\rho(\varphi(t))}{dt} = \dot{\rho}(\varphi(t)) \geq \rho(\varphi(t))\{\mu - \sum_{i=1}^{p} z_i(t) \sum_{j=1}^{m} f_{ji} r_j\} \tag{14}$$

The assumption that $z_i(t) \to 0$ as $t \to \infty$, for all i, $1 \leq i \leq p$ means that the second term in the bracket in (14) becomes arbitrarily small for sufficiently large t. Therefore if λ is any positive constant less than μ,

$$\dot{\rho}(\varphi(t)) \geq \lambda\rho(\varphi(t))$$

for sufficiently large t, and this completes the proof.

The biological interpretation of (13) is straightforward: if the combined intrinsic growth rate of the prey exceeds a linear combination of the intrinsic death rates of the predators where the coefficients of the linear combination are required to satisfy certain relations involving the interaction rates, then top predator persistence is assured. It is clear that the "best" such persistence criterion (13) is obtained by choosing coefficients r_i and s_i which solve the linear programming problem:

$$\text{minimize} \sum_{i=1}^{m} d_i r_i + \sum_{i=1}^{p} g_i s_i$$

subject to the constraints $r_i \geq 0$, $s_i \geq 0$, (11) and (12). The corresponding criterion is "best" in the sense that it places the weakest restriction on the prey growth rates possible. That the result is sharp is indicated by the fact that for μ obtained in the same way for food chains in Gard and Hallam (1979), it was shown that $\mu < 0$ implied top predator extinction.

The main result can be extended to food webs with more than three trophic levels, food webs exhibiting competition for space among predators, and food webs with arbitrary degrees of omnivory. It is not difficult to see the modifications of the persistence criterion and constraints required in each of these situations. In particular, that omnivory enhances top predator persistence can be readily observed from the form of the corresponding constraint relations.

Example. Consider the Lotka-Volterra model of the two intermediate predator food web given by

$$\begin{aligned}\frac{dx}{dt} &= x(a - bx - c_1y_1 - c_2y_2) \\ \frac{dy_i}{dt} &= y_i(-d_i + e_ix - f_iz), \quad i = 1,2 \\ \frac{dz}{dt} &= z(-g + h_1y_1 + h_2y_2) \quad .\end{aligned} \tag{15}$$

For $\rho = xy_1^{r_1} y_2^{r_2} z^s$, (10) has the form

$$\dot{\rho} = \{a - d_1 r_1 - d_2 r_2 - gs$$

$$+ x(e_1 r_1 + e_2 r_2 - b)$$

$$+ y_1(h_1 s - c_1) + y_2(h_2 s - c_2)$$

$$- z(f_1 r_1 + f_2 r_2)\}\rho$$

Then, from the theorem,

$$\mu = a - d_1 r_1 - d_2 r_2 - gs > 0 \qquad (16)$$

implies persistence if the nonnegative constants r_1, r_2, and s are chosen so that

$$e_1 r_1 + e_2 r_2 \geq b \qquad (17)$$

and

$$h_i s \geq c_i \ , \quad i = 1,2 \ . \qquad (18)$$

The sharpest such criterion obtains from a solution of the problem

$$\text{minimize} \quad d_1 r_1 + d_2 r_2 + gs \qquad (19)$$
$$\text{subject to} \quad r_1,\ r_2, \text{ and } s \geq 0, \text{ (17), and (18).}$$

The solution of (19) consists of taking

$$s = \max\{c_1/h_1, c_2/h_2\} \ ,$$

and $r_1 \geq 0$, $r_2 \geq 0$ minimizing $d_1 r_1 + d_2 r_2$ subject to (17), i.e., either

$r_1 = b/e_1$ and $r_2 = 0$ if $d_1 e_2 < d_2 e_1$,

$r_1 = 0$ and $r_2 = b/e_2$ if $d_1 e_2 > d_2 e_1$, or

r_1 and r_2 are the coordinates of any point on the line $e_1 r_1 + e_2 r_2 = b$ in the nonnegative quadrant if $d_1 e_2 = d_2 e_1$.

The corresponding persistence criterion then is

$$a - b \min \{d_1/e_1, d_2/e_2\} - g \max \{c_1/h_1, c_2/h_2\} > 0 .$$

Acknowledgement

This research was supported, in part, by the U.S. Environmental Protection Agency under Cooperative Agreement No. CR 807830.

Literature Cited

Allen L. (1981). Applications of differential inequalities to persistence and extinction problems for reaction-diffusion systems. Ph.D. dissertation. Univ. of Tennessee.

Botkin, D.B. and Sobel, M.J. (1974). The complexity of ecosystem stability. Ecosystem Analysis and Prediction. S. Levin (ed.). SIAM, Philadelphia.

Freedman, H.I. (1980). Deterministic Mathematical Models in Population Ecology, Marcel Dekker, New York.

Freedman, H.I. and Waltman, P. (1977). Mathematical analysis of some three-species food-chain models. Math. Biosci. 33: 257-276.

Gard, T.C. (1980). Strongly flow-invariant sets. Appl. Analysis 10: 285-293.

Gard, T.C. and Hallam, T.G. (1979). Persistence in food webs: I. Lotka-Volterra food chains. Bull. Math. Biol. 41: 877-891.

Harrison, G.W. (1979a). Stability under environmental stress: Resistance, resilience, persistence, and variability. Amer. Natur. 113: 659-669.

Harrison, G.W. (1979b). Persistent sets via Lyapunov functions. Nonlinear Analysis 3: 73-80.

Holling, C.S. (1973). Resilience and stability of ecological systems. Annu. Rev. Ecol. Syst. 4: 1-24.

Innis, G. (1974). Stability, sensitivity, resilience, persistence. What is of interest? Ecosystem Analysis and Prediction. S. Levin (ed.). SIAM, Philadelphia.

Ludwig, D. (1975). Persistence of dynamical systems under random perturbations. SIAM Rev. 17: 605-640.

Maynard Smith, J. (1974). Models in Ecology. Cambridge University Press, Cambridge.

McGehee, R. and Armstrong, R.A. (1977). Some mathematical problems concerning the ecological principle of competitive exclusion. J. Diff. Eq. 23: 30-52.

Paine, R.T. (1966). Food web complexity and species diversity. Amer. Natur. 100: 65-76.

Redheffer, R. and Walter, W. (1975). Flow-invariant sets and differential inequalities in normal spaces. Appl. Analysis 5: 149-161.

Seifert, G. (1976). Positively invariant closed sets for systems of delay differential equations. J. Diff. Eq. 22: 292-304.

Tier, C. and Hanson, F.B. (1981). Persistence in density dependent stochastic populations. Math. Biosci. 53: 89-117.

Wu, L. (1974). On the stability of ecosystems. Ecosystem Analysis and Prediction. S. Levin (ed.). SIAM, Philadelphia.

The Structure of Cycling in the Ythan Estuary

Robert E. Ulanowicz
University of Maryland
Chesapeake Biological Laboratory
Solomons, Maryland 20688-0038 USA

Abstract

A method for identifying all the simple, directed cycles in a network of ecosystems flows is described. Furthermore, the cycles may be divided into distinct groups, or nexuses, distinguished by certain critical links. These groupings may also be used to analytically separate the network into two constituent graphs - one graph containing only cycled flow and the other consisting of only once-through pathways.

The analysis, when applied to the carbon flow in the Ythan estuary, graphically portrays both the importance and the vulnerability of the higher trophic level transfers.

Introduction

In his seminal article, "The Strategy of Ecosystem Development," E. P. Odum (1969) suggests the amount of material cycling as a prime indicator of mature, developed ecosystems. Later, he becomes even more specific by comparing the importance of detritus in slowly-varying benthic, marsh and terrestrial ecosystems with the predominantly once-through foodwebs of transitory pelagic food chains. Mathematical ecologists have been somewhat slow to give quantitative form to Odum's notions about cycling so that his hypothesis may be formally tested. It was not until seven years later that Finn (1976) was able to estimate the aggregate amount of material or energy which was being cycled in a given flow network. If Finn's index of cycled flow properply characterizes Odum's hypothesis, then presumably the index would possess a higher value in more developed ecosystem networks.

Finn's results were an outgrowth of economic input-output analysis. In the notation common to input-output analysis if one calls P_{ij} the flow of any given medium from i to j and T_i the total output from i, then $f_{ij} = P_{ij}/T_i$ represents the fraction of the total output of i which is contributed directly to j. If F is a matrix with components f_{ij}, then it is not too difficult to discern that the matrix

F^2 (F multiplied by itself) will consist of components representing the fraction of total output from i which flows to j over all pathways of length 2. Similarly, the components of F^3 will be the fractions of T_i which flow to j over all pathways of length 3, etc. By summing all powers of F, one obtains the output structure matrix, S, wherein S_{ij} represents the fraction of T_i which flows to j over all possible pathways:

$$S = F^o + F^1 + F^2 + F^3 + \dots$$

Because F was conveniently normalized, this infinite series converges (Yan, 1969) to the value

$$S = (I - F)^{-1},$$

where I is the identity matrix, and the minus one exponent represents matrix inversion.

Of particular interest to those concerned with cycling are the diagonal elements of S. In a closed system each S_{ii} has a minimum value of unity, and any excess over one represents the fraction of T_i which returns to i over all cycles of all lengths. Because this fraction of flow is inherent in all members having communication with i, one may imagine a matrix with entries of $(1 - S_{ii})$ in each of the ith rows for which S_{ij} is not zero. If this cycling matrix is multiplied by the vectors of total outputs (that is the T_i), and the elements of the consequent vector are then summed, the total amount of cycled flow will result. The ratio of this amount of cycled flow to the total amount of flow in the system has come to be known as Finn's cycling index. One infers from Odum's remarks that the cycling index might be greater in more mature communities, or conversely less in networks subject to perturbations.

As intriguing as this analysis might be, the results when applied to real networks gave ambiguous results (Richey et al., 1978). Perturbed and eutrophic systems sometimes possess larger cycling indices. Clearly, events are occurring which are not readily discernible in terms of the aggregate index. It appears that a more detailed knowledge about the structure of the flow cycles might be helpful.

Cycle Analysis

Given a flow network or graph, it is a rather easy computational task to identify a set of fundamental cycles from which all possible cycles may be generated by linear combinations (Knuth, 1973). Unfortunately, there are in general many possible combinations of fundamental cycles, and it is by no means clear which of the combinations would be (biologically speaking) most pertinent. The next step up in detail from this rudimentary description would be to enumerate all the simple cycles in the network. A simple cycle is one in which no compartment appears more than once (see Fig. 1).

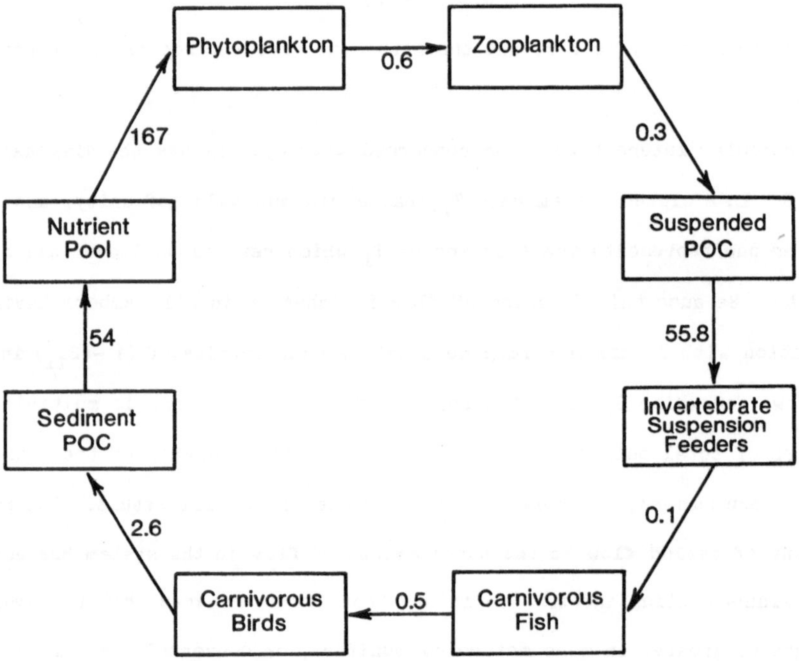

Figure 1 - A simple cycle. No repeated elements. Extracted from Baird and Milne (1981). See legend to Figure 2.

While it might appear that the systematic enumeration of the simple cycles in a graph should be a rather straightforward programming task, the time required for even a modern calculator to identify all cycles can easily get out of bounds. Those familiar with combinatorics will recognize that the number of possible cycles in a graph increases as the factorial of the number of nodes. Such an exponential increase in search time means that relatively small networks have the potential for exceeding the capabilities of even the faster machines available (n.b., 20! = 2.4 x 10^{18}).

Fortunately, ecosystem networks do not appear to be highly connected (May, 1973). Typically only 15-25% of the possible connections are realized, depending upon the strength of the interactions. Even with such simplification, it nonetheless behooves an investigator to choose algorithms which are as efficient as possible. Mateti and Deo (1976) in evaluating various methods for identifying cycles have concluded that backtracking search algorithms with suitable pruning methods (to eliminate many spurious search pathways) are the most efficient programs under the greatest number of circumstances.

In the backtracking algorithm one works with either a matrix of flows or a vector list of arcs. One orders the nodes in some convenient way (see below) and imagines the same order of n nodes to be repeated at n levels (see Figure 2). One begins with the first node and searches the node in the next level until an existing flow connection is found. One progresses to that node in the next level and searches the succeeding level for yet a subsequent connection. Only those nodes in the next level are considered which have not already appeared in the current pathway. One proceeds to increasing levels until one of two things happen. If an arc to the next level brings one back to the starting node, then a simple cycle has been identified and is reported and/or stored. If the search for a link from the th j^{th} node at the mth level for a connection to the $(m+1)^{th}$ level has been exhausted, one "backtracks" to the ancestor node at the $(m-1)^{th}$ level and continues searching the mth level beginning at the $(j+1)^{th}$ position. When further backtracking is impossible, one has identified all simple cycles containing the starting node. The starting node may be dropped from all future consideration and the dimension of the subsequent search may be decreased by one.

Read and Tarjan (1975) and Johnson (1975) give examples of constraints on the backtracking procedure which result in efficient searching. Ulanowicz (in press, Math. Biosciences) has found that for most ecological applications it suffices simply to order the nodes judiciously. In particular, one wishes to first search those nodes for which the probability of completing a cycle at any step is the greatest. This probability varies directly with the number of cycle arcs terminating in the node under consideration. (A cycle arc is a connection from a given arc to one of its ancestors.) As all the descendents of a given arc may readily be determined (Knuth, 1973), one may total the number of cycle arcs back into each node and order the nodes accordingly. Those nodes with no incoming cycle arcs need not be considered in the backtracking subroutine.

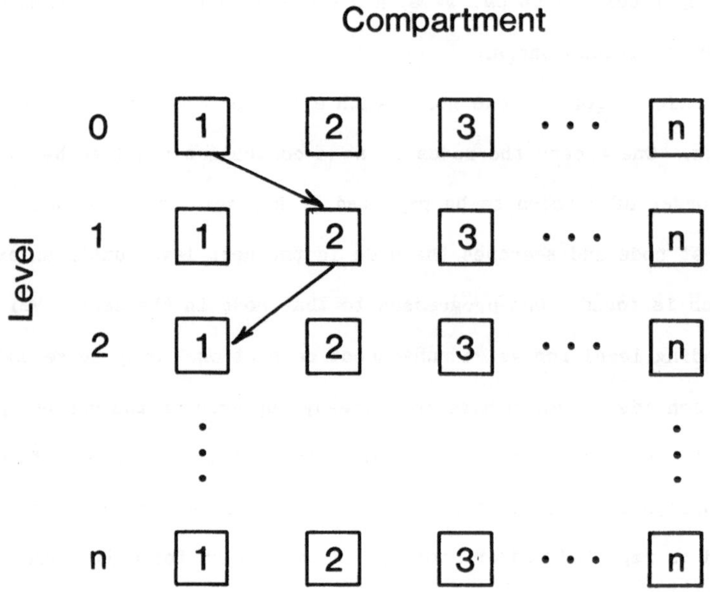

Figure 2 - Mnemonic diagram useful in description of the backtracking algorithm. Here the first cycle to be identified is 1-2-1.

Qualitatively knowing what the cycles in a network are can be very useful information in assessing autonomous behavior in an ecosystem. As cause and effect often follow material or energetic pathways in a system, a loop of flow back upon itself is a likely indicator of a cybernetic effect at operation in the system. Such causal loops (Hutchinson, 1948) frequently behave in ways which appear autonomous of the exchanges with the external world. The constellation of cycles in which a given compartment participates then defines the domain of self-regulation of the compartment in question. If one is interested in the fate of a particular species, it is certainly helpful to know its constellation of cycles. Similarly, if one is studying a given flow pathway, say of the predation of a particular species on another, it is useful qualitative information to know the constellation of cycles in which that particular transformation appears.

While a description of the various constellations is likely to be quite helpful, quantitative information about the relative strengths of the flows still needs to be exploited. It is said that a chain is only as strong as its weakest link, and this notion may apply to cycles as well. Tracing around the arcs of a cycle, one should in principle be able to identify a most vulnerable link according to some criterion. If no information is available other than the weighted network, one may make the assumption that the smallest (or slowest) link in the cycle dominates the behavior of that loop. (Control by the smallest link is analogous to the idea of the rate limiting step in chemical kinetics.) This assumption is only a pedagogical expedient, and if other information about the arcs of a cycle is available, those clues should be investigated to see if another criterion for the most vulnerable arc might be more appropriate.

All of the cycles in a graph may be individually examined to identify a set of most vulnerable arcs. Because some cycles share the same vulnerable arc, the number of most vulnerable arcs is smaller than the number of simple cycles. This homomorphic mapping of cycles into vulnerable arcs serves to define subsets of cycles called nexuses. A nexus is a collection of cycles all sharing the same most vulnerable arc. The magnitude of the most vulnerable arc quantifies the nexus.

In identifying the nexuses one has simplified the description of the structure of cycling and thereby made it more meaningful. The description is simplified because there are fewer nexuses than constellations (the set of most vulnerable arcs being a subset of all arcs), and the nexuses are somewhat smaller in the number of constituent cycles (the cycles for which an arc is most vulnerable being a subset of all the cycles in which it participates). If no two of the most vulnerable arcs are exactly equal in magnitude, the nexuses are disjoint. The nexuses are more meaningful than the constellations in that they focus attention upon a subset of critical interactions and better delimit their domains of influence.

Another advantage of the concept of a nexus is a convenient tool for separating the cycled flow from that flow which only passes straight through the system. If the vulnerable arc is defined as the smallest arc in a cycle, then the following algorithm will effect such a separation:

1. Zero the elements of the matrix of cycled flow.
2. Find the smallest non-zero vulnerable arc and call its magnitude V.
3. If no vulnerable arcs remain, go to 7.
4. For each of the m cycles in the nexus defined by the smallest vulnerable arc calculate the probability P_i ($i = 1,2, \ldots, m$) that a bit of medium starting at any node in the cycle will exactly follow the cycle to return to its starting point. This probability is simply the composite product of the f_{ij}'s associated with each arc in the cycle.
5. Go around each of the m cycles of the nexus defined by the smallest vulnerable arc, subtracting $V P_i / \sum_{j=1}^{m} P_j$ from each link of the starting graph and adding the same quantity to the corresponding entry of the cycles matrix.
6. Go to 2.
7. STOP.

As a result of step 5 the smallest vulnerable arc will be eliminated (set=0), thereby breaking all the cycles of the accompanying nexus. All of the other arcs associated with the nexus will remain positive.

In steps 4 and 5 the magnitude of the vulnerable arc is divided among the constituent cycles of the associated nexus in proportion to the probability that a particle in the vulnerable flow would actually complete a given cycle (suggested to the author by W. Silvert, personal communication). While this scheme is a very plausible one; it is, nonetheless, somewhat arbitrary. The reader wishing to implement a cycle analysis might wish to apportion the vulnerable flow according to some other scheme, such as the principle of maximum entropy (Jaynes, 1958) or the hypothesis of maximum ascendency (Ulanowicz, 1980).

<u>An Example: Carbon Flow in the Ythan Estuary</u>

An appropriate network with which to demonstrate the analysis is the graph of carbon flows among the compartments of the Ythan estuary ecosystem (Baird and Milne, 1981). The schematic diagram which Baird presented did not unambiguously define all the individual intercompartmental flows. Fortunately, Dr. Baird graciously assisted the author in estimating how the lumped flows might be separated into strictly pairwise transfers, and the resulting network is presented in Figure 3.

Three abiotic and ten living compartments of the Ythan ecosystem are identified. There are 2 exogenous inputs of carbon, ten exports of useable carbon and 39 internal exchanges (23% connectance). The degree of aggregation is about uniform over all trophic levels. All flows were reported in g carbon $m^{-2}y^{-1}$. The nutrient pool was originally measured in terms of nitrogen, phosphorous and silicon, but has been converted into equivalent dissolved organic carbon for the purposes of this analysis.

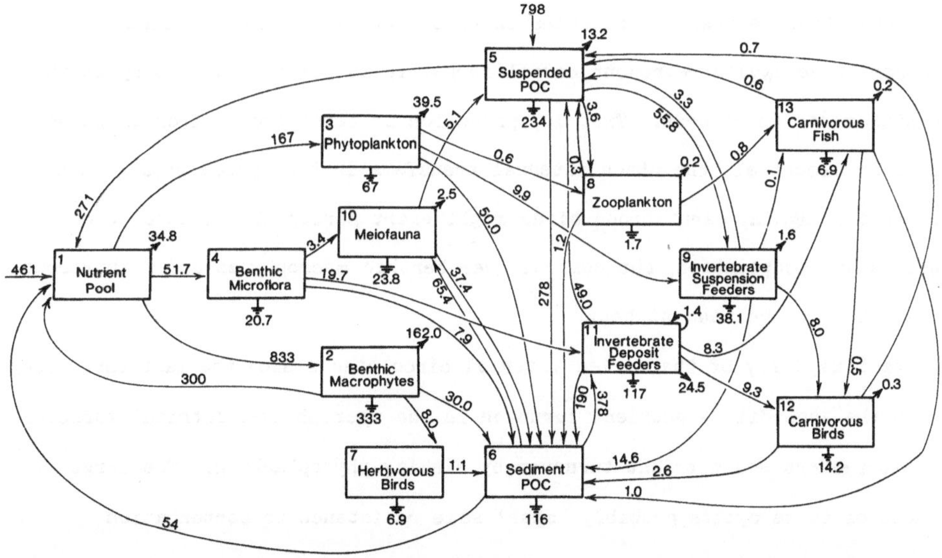

Figure 3 - Schematic diagram of carbon flow in the Ythan estuary, Scotland. Flows measured in $gCm^{-2}y^{-1}$. Ground symbols represent respiration; arrows with no origin, exogenous inputs; arrows not terminating in a box, exports a from the system. After Baird and Milne (1981).

A total of 170 simple cycles may be identified in the graph. These cycles may be grouped into 25 separate nexuses as seen in Appendix 1. The order in which the nexuses appear is interesting. Not unexpectedly, the smallest vulnerable arcs are associated with the higher trophic level compartments. These critical arcs define nexuses with many cycles, and the individual cycles tend to be long. Conversely, the vulnerable arcs of greater magnitude are associated with lower trophic components and have fewer (often only one) cycles. Also, those cycles are very short.

As an example of the complex nexuses, consider the third nexus in Appendix 1. The vulnerable arc represents the predation on carnivorous fish by carnivorous birds. The associated nexus, depicted in Figure 4, involves all of the compartments of the ecosystem in 29 simple cycles. The ecological significance of this nexus can be interpreted in either of two ways. First, a perturbation anywhere in the system is capable of propagating to affect the vulnerable arc, although it is obvious from Figure 4 that perturbations to certain pathways are likely to have more impact than some other disturbances. For example, grazing by herbivorous birds is practically inconsequential to the feeding of their carnivorous counterparts (niche separation). However, disturbances to the invertebrate deposit feeders could have serious consequences on predation by carnivorous birds.

Because these cycles represent causal loops, it is possible to interpret the latter statements in reverse. That is, predation by carnivorous birds is capable of affecting almost all the other internal transfers in the system to a greater or lesser degree. Again, carnivorous birds might exert perceptible control over deposit feeders and zooplankton, but control over benthic macrophytes and herbivorous birds is likely to be weak at best.

The vast majority of the cycled material circulates inside the last three loops listed in the Appendix -- nutrient turnover in the macrophytes, detrital turnover by deposit feeders and nutrient turnover by sinking phytoplankton. The large magnitudes of these cycles probably impart some resistance to perturbation (Ulanowicz, in press). Their relative isolation from other compartments helps to insulate them from disturbances elsewhere in the ecosystem.

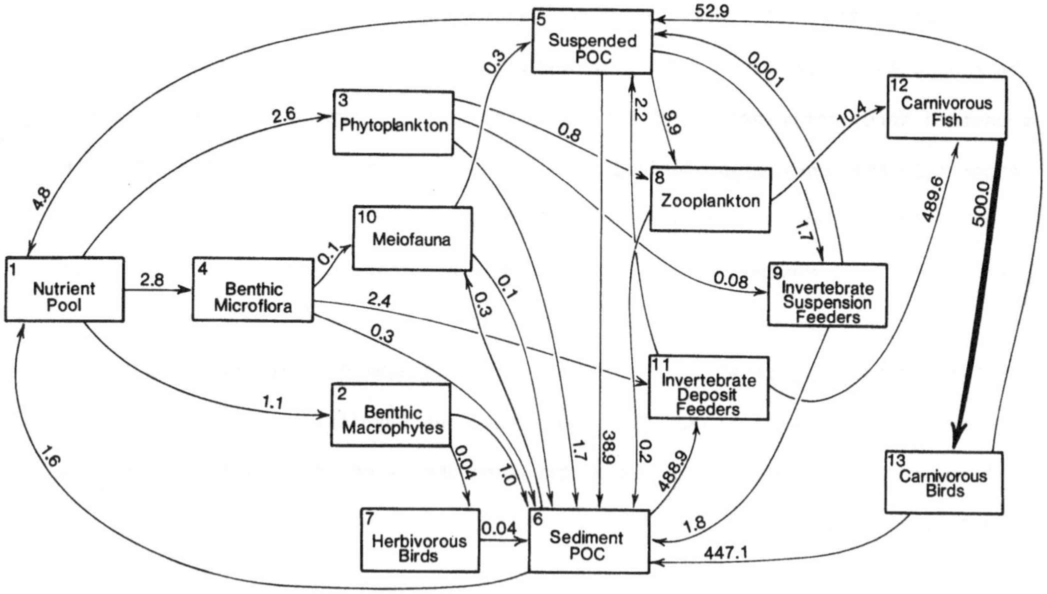

Figure 4 - The nexus of cycles associated with the predation on carnivorous fish by carnivorous birds (indicated by heavy black arrow). Extracted from graph in Figure 3. Flows are in mg Carbon m^{-2} y^{-1}.

It should be kept in mind that, like all other forms of flow analysis, this description of cycling is based upon a steady-state network. One may not infer that the dynamics of the system have thereby been accurately defined. In fact the system can play by very different rules after perturbation and conclusions drawn from the steady-state could be misleading.

Cycle analysis, however, does provide a graphic and semi-quantitative way of portraying what has long been known from experience -- namely, that transfers among higher trophic level components may be both more vulnerable and, obversely, exert more control over a wider domain of an ecosystem. One expects these critical transfers and their accompanying nexuses to be early victims of any disturbances. By contrast, the trophically lower, faster cycles appear less vulnerable to perturbations and might even benefit from the disappearance of higher structure. While these statements are not immediately obvious from the analysis of a single network presented here, comparative studies show that this appears to be precisely what happened in the ecosystems of tidal marsh creeks near the thermal effluent of the Crystal River power generating station (Ulanowicz, 1982) and what was predicted to happen if Gulf of Mexico benthic communities were exposed to brine releases from salt domes (B.C. Patten, personal communication).

Cycling index by itself, is thus seen to be an equivocal indicator of the response to ecosystem stress. It is only when the magnitudes of cycled flows are combined with a description of their structure that a coherent picture of ecological impact emerges.

Acknowledgements

This work, as well as the author's travel to the symposium, was supported by a grant from the U.S. National Science Foundation (ECS-8110035). The Computer Science Center of the University of Maryland also contributed some free computer time. The author extends special thanks to Professor Dan Baird, who spent an entire day with the author estimating how his measured flows might be partitioned so as to make the cycle analysis possible.

A Fortran subroutine for the identification and removal of cycles from a network is available from the author upon request.

Literature Cited

Baird, D. and Milne, H. (1981). Energy flow in the Ythan estuary, Aberdeenshire, Scotland. Estuarine, Coastal and Shelf Science 13:455-472.

Finn, J. T. (1976). Measures of ecosystem structure and function derived from analysis of flows. J. theor. Biol. 56:363-380.

Hutchinson, G.E. (1948). Circular causal systems in ecology. Ann. N.Y. Acad. Sci. 50:221-246.

Jaynes, E.T. (1958). Probability Theory in Science and Engineering. Colloquium Lectures in Pure and Applied Science No. 4. Socony Mobil Oil Company, Dallas, TX. 189 pp.

Johnson, D.B. (1975). Finding all the the elementary circuits of a directed graph. SIAM J. Comput. 4:77-84.

Knuth, D.E. (1973). Fundamental Algorithms, Vol. 1. Addison-Wesley, Reading, MA. 634 pp. Mateti, P. and Deo, N. (1976). On algorithms for enumerating all circuits of a graph. SIAM J. Comput. 5:90-99.

May, R.M. (1973). Stability and Complexity in Model Ecosystems. Princeton Univ. Press, Princeton, NJ. 235 pp.

Odum, E.P. (1969). The strategy of ecosystem development. Science 164:262-270.

Read, R.C. and Tarjan, R.E. (1975). Bounds on backtrack algorithms for listing cycles, paths, and spanning trees. Networks 5:237-252.

Richey, J.E., Wissmar, R.C., Devol, A.H., Likens, G.E., Eaton, J.S., Wetzel, R.G., Odum, W.E., Johnson, N.M., Loucks, O.L., Prentki, R.T., and Rich, P.H. (1978). Carbon flow in four lake ecosystems: a structural approach. Science 202:1183-1186.

Ulanowicz, R.E. (1980). An hypothesis on the development of natural communities. J. theor. Biol. 85:223-245.

Ulanowicz, R.E. (1982). Community measures of marine food networks and their possible applications. In M.J.R. Fasham (ed.) Measurement of Fluxes in Marine Ecosystems. UNESCO Press, Paris. (in press)

Yan, C.S. (1969). Introduction to Input-Output Economics. Holt, Rinehart and Winston, NY.

APPENDIX

*** CYCLE ANALYSES ***

27-CYCLE NEXUS WITH WEAK ARC (9,13) = .100

```
 1.  6- 1- 3- 8- 5- 9-13- 6-
 2.  6- 1- 3- 8- 5- 9-13-12- 6-
 3.  6- 1- 3- 9-13- 6-
 4.  6- 1- 3- 9-13- 5- 6-
 5.  6- 1- 3- 9-13- 5- 8- 6-
 6.  6- 1- 3- 9-13-12- 6-
 7.  6- 1- 3- 9-13-12- 5- 6-
 8.  6- 1- 3- 9-13-12- 5- 8- 6-
 9.  6- 1- 4-11- 5- 9-13- 6-
10.  6- 1- 4-11- 5- 9-13-12- 6-
11.  6- 1- 4-11-12- 5- 9-13- 6-
12.  6- 1- 4-10- 5- 9-13- 6-
13.  6- 1- 4-10- 5- 9-13-12- 6-
14.  6-11- 5- 1- 3- 9-13- 6-
15.  6-11- 5- 1- 3- 9-13-12- 6-
16.  6-11- 5- 9-13- 6-
17.  6-11- 5- 9-13-12- 6-
18.  6-11-12- 5- 1- 3- 9-13- 6-
19.  6-11-12- 5- 9-13- 6-
20.  6-10- 5- 1- 3- 9-13- 6-
21.  6-10- 5- 1- 3- 9-13-12- 6-
22.  6-10- 5- 9-13- 6-
23.  6-10- 5- 9-13-12- 6-
24.  5- 1- 3- 9-13- 5-
25.  5- 1- 3- 9-13-12- 5-
26.  5- 9-13- 5-
27.  5- 9-13-12- 5-
```

5-CYCLE NEXUS WITH WEAK ARC (8, 5) = .300

```
28.  6- 1- 3- 8- 5- 6-
29.  6- 1- 3- 8- 5- 9- 6-
30.  6- 1- 3- 8- 5- 9-12- 6-
31.  5- 1- 3- 8- 5-
32.  5- 8- 5-
```

29-CYCLE NEXUS WITH WEAK ARC (13,12) = .500

```
33.  6- 1- 3- 8-13-12- 6-
34.  6- 1- 3- 8-13-12- 5- 6-
35.  6- 1- 3- 8-13-12- 5- 9- 6-
36.  6- 1- 3- 9- 5- 8-13-12- 6-
37.  6- 1- 4-11- 5- 8-13-12- 6-
38.  6- 1- 4-11-13-12- 6-
39.  6- 1- 4-11-13-12- 5- 6-
40.  6- 1- 4-11-13-12- 5- 8- 6-
41.  6- 1- 4-11-13-12- 5- 9- 6-
42.  6- 1- 4-10- 5- 8-13-12- 6-
43.  6-11- 5- 1- 3- 8-13-12- 6-
44.  6-11- 5- 8-13-12- 6-
45.  6-11-13-12- 6-
46.  6-11-13-12- 5- 6-
47.  6-11-13-12- 5- 1- 2- 6-
48.  6-11-13-12- 5- 1- 2- 7- 6-
49.  6-11-13-12- 5- 1- 3- 6-
50.  6-11-13-12- 5- 1- 3- 8- 6-
51.  6-11-13-12- 5- 1- 3- 9- 6-
52.  6-11-13-12- 5- 1- 4- 6-
```

```
53.  6-11-13-12- 5- 1- 4-10- 6-
54.  6-11-13-12- 5- 8- 6-
55.  6-11-13-12- 5- 9- 6-
56.  6-10- 5- 1- 3- 8-13-12- 6-
57.  6-10- 5- 1- 4-11-13-12- 6-
58.  6-10- 5- 8-13-12- 6-
59.  5- 1- 3- 8-13-12- 5-
60.  5- 1- 4-11-13-12- 5-
61.  5- 8-13-12- 5-
```

12-CYCLE NEXUS WITH WEAK ARC (3, 8) = .600

```
62.  6- 1- 3- 8- 6-
63.  6- 1- 3- 8-13- 6-
64.  6- 1- 3- 8-13- 5- 6-
65.  6- 1- 3- 8-13- 5- 9- 6-
66.  6- 1- 3- 8-13- 5- 9-12- 6-
67.  6-11- 5- 1- 3- 8- 6-
68.  6-11- 5- 1- 3- 8-13- 6-
69.  6-11-12- 5- 1- 3- 8- 6-
70.  6-11-12- 5- 1- 3- 8-13- 6-
71.  6-10- 5- 1- 3- 8- 6-
72.  6-10- 5- 1- 3- 8-13- 6-
73.  5- 1- 3- 8-13- 5-
```

18-CYCLE NEXUS WITH WEAK ARC (13, 5) = .600

```
74.  6- 1- 4-11-13- 5- 6-
75.  6- 1- 4-11-13- 5- 8- 6-
76.  6- 1- 4-11-13- 5- 9- 6-
77.  6- 1- 4-11-13- 5- 9-12- 6-
78.  6-11-13- 5- 6-
79.  6-11-13- 5- 1- 2- 6-
80.  6-11-13- 5- 1- 2- 7- 6-
81.  6-11-13- 5- 1- 3- 6-
82.  6-11-13- 5- 1- 3- 8- 6-
83.  6-11-13- 5- 1- 3- 9- 6-
84.  6-11-13- 5- 1- 3- 9-12- 6-
85.  6-11-13- 5- 1- 4- 6-
86.  6-11-13- 5- 1- 4-10- 6-
87.  6-11-13- 5- 8- 6-
88.  6-11-13- 5- 9- 6-
89.  6-11-13- 5- 9-12- 6-
90.  5- 1- 4-11-13- 5-
91.  5- 8-13- 5-
```

20-CYCLE NEXUS WITH WEAK ARC (12, 5) = .700

```
 92.  6- 1- 3- 9-12- 5- 6-
 93.  6- 1- 3- 9-12- 5- 8- 6-
 94.  6- 1- 3- 9-12- 5- 8-13- 6-
 95.  6- 1- 4-11-12- 5- 6-
 96.  6- 1- 4-11-12- 5- 8- 6-
 97.  6- 1- 4-11-12- 5- 8-13- 6-
 98.  6- 1- 4-11-12- 5- 9- 6-
 99.  6-11-12- 5- 6-
100.  6-11-12- 5- 1- 2- 6-
101.  6-11-12- 5- 1- 2- 7- 6-
102.  6-11-12- 5- 1- 3- 6-
103.  6-11-12- 5- 1- 3- 9- 6-
104.  6-11-12- 5- 1- 4- 6-
105.  6-11-12- 5- 1- 4-10- 6-
106.  6-11-12- 5- 8- 6-
107.  6-11-12- 5- 8-13- 6-
108.  6-11-12- 5- 9- 6-
```

```
109.   5- 1- 3- 9-12- 5-
110.   5- 1- 4-11-12- 5-
111.   5- 9-12- 5-
```

5-CYCLE NEXUS WITH WEAK ARC (8,13) = .800
```
112.   6- 1- 3- 9- 5- 8-13- 6-
113.   6- 1- 4-11- 5- 8-13- 6-
114.   6- 1- 4-10- 5- 8-13- 6-
115.   6-11- 5- 8-13- 6-
116.   6-10- 5- 8-13- 6-
```

3-CYCLE NEXUS WITH WEAK ARC (13, 6) = 1.000
```
117.   6- 1- 4-11-13- 6-
118.   6-11-13- 6-
119.   6-10- 5- 1- 4-11-13- 6-
```

3-CYCLE NEXUS WITH WEAK ARC (7, 6) = 1.100
```
120.   6- 1- 2- 7- 6-
121.   6-11- 5- 1- 2- 7- 6-
122.   6-10- 5- 1- 2- 7- 6-
```

5-CYCLE NEXUS WITH WEAK ARC (8, 6) = 1.200
```
123.   6- 1- 3- 9- 5- 8- 6-
124.   6- 1- 4-11- 5- 8- 6-
125.   6- 1- 4-10- 5- 8- 6-
126.   6-11- 5- 8- 6-
127.   6-10- 5- 8- 6-
```

1-CYCLE NEXUS WITH WEAK ARC (11,11) = 1.400
```
128.   11-11-
```

10-CYCLE NEXUS WITH WEAK ARC (12, 6) = 2.600
```
129.   6- 1- 3- 9-12- 6-
130.   6- 1- 4-11- 5- 9-12- 6-
131.   6- 1- 4-11-12- 6-
132.   6- 1- 4-10- 5- 9-12- 6-
133.   6-11- 5- 1- 3- 9-12- 6-
134.   6-11- 5- 9-12- 6-
135.   6-11-12- 6-
136.   6-10- 5- 1- 3- 9-12- 6-
137.   6-10- 5- 1- 4-11-12- 6-
138.   6-10- 5- 9-12- 6-
```

3-CYCLE NEXUS WITH WEAK ARC (9, 5) = 3.300
```
139.   6- 1- 3- 9- 5- 6-
140.   5- 1- 3- 9- 5-
141.   5- 9- 5-
```

5-CYCLE NEXUS WITH WEAK ARC (4,10) = 3.400
```
142.   6- 1- 4-10- 6-
143.   6- 1- 4-10- 5- 6-
144.   6- 1- 4-10- 5- 9- 6-
145.   6-11- 5- 1- 4-10- 6-
146.   5- 1- 4-10- 5-
```

7-CYCLE NEXUS WITH WEAK ARC (10, 5) = 5.100
```
147.   6-10- 5- 6-
148.   6-10- 5- 1- 2- 6-
149.   6-10- 5- 1- 3- 6-
150.   6-10- 5- 1- 3- 9- 6-
151.   6-10- 5- 1- 4- 6-
```

152. 6-10- 5- 1- 4-11- 6-
153. 6-10- 5- 9- 6-

2-CYCLE NEXUS WITH WEAK ARC (4, 6) = 7.900
154. 6- 1- 4- 6-
155. 6-11- 5- 1- 4- 6-

2-CYCLE NEXUS WITH WEAK ARC (3, 9) = 9.900
156. 6- 1- 3- 9- 6-
157. 6-11- 5- 1- 3- 9- 6-

2-CYCLE NEXUS WITH WEAK ARC (9, 6) = 4.600
158. 6- 1- 4-11- 5- 9- 6-
159. 6-11- 5- 9- 6-

3-CYCLE NEXUS WITH WEAK ARC (4,11) = 19.700
160. 6- 1- 4-11- 6-
161. 6- 1- 4-11- 5- 6-
162. 5- 1- 4-11- 5-

2-CYCLE NEXUS WITH WEAK ARC (2, 6) = 30.000
163. 6- 1- 2- 6-
164. 6-11- 5- 1- 2- 6-

1-CYCLE NEXUS WITH WEAK ARC (10, 6) = 37.400
165. 6-10- 6-

2-CYCLE NEXUS WITH WEAK ARC (11, 5) = 49.000
166. 6-11- 5- 6-
167. 6-11- 5- 1- 3- 6-

1-CYCLE NEXUS WITH WEAK ARC (3, 6) = 50.000
168. 6- 1- 3- 6-

1-CYCLE NEXUS WITH WEAK ARC (11, 6) = 190.100
169. 6-11- 6-

1-CYCLE NEXUS WITH WEAK ARC (2, 1) = 300.000
170. 1- 2- 1-

PART IV
APPLICATIONS: FISHERIES

Constant Yield Harvesting of Population Systems

Fred Brauer
University of Wisconsin

1. The roots of modern mathematical ecology lie in attempts to model population systems whose sizes were estimated from catch data, notably the hare and lynx data compiled by the Hudson's Bay Company and the oscillation of fish populations in the Adriatic Sea [Volterra (1931)]. For more than 25 years the study of systems from which members are harvested has been of importance in the management of fisheries, see [Schaefer (1954)] or [Beverton and Holt (1957)] for the origins of such studies. Here, information about population size is used to estimate feasible catch sizes.

In fishery management it is usually assumed that the number of fish caught is proportional to the effort expended. Therefore, much attention has been paid to constant-effort harvesting, in which members are removed from the population at a time rate which is proportional to population size. Questions studied include not only the maximization of the long-term (equilibrium) yield, but also the maximization of economic yield [Clark (1976)].

In some situations it may be more appropriate to set a quota for the number of members of a population harvested in unit time. This provides a motivation for the study of constant-yield harvesting. As constant-yield harvesting generally has a more pronounced effect on the qualitative behavior of a population system than does constant-effort harvesting, constant-yield harvesting poses more serious problems to a resource manager. While the effect of applying a slightly excessive effort in constant-effort harvesting is likely to be little more serious than a reduced yield, the effect of a slightly excessive constant-yield harvest is to drive the system to collapse.

In addition to the obvious applications to resource management, there are other reasons for studying the effect of harvesting on a population system. Terms such as stability, resilience, and sensitivity have been used to describe the response of a population system to changes in initial state, perturbations, or parameter changes. In describing such responses, one may be concerned with the size of the change which can be absorbed without drastic changes in behavior, with the magnitude of the long-term change, and with the rapidity of return to equilibrium. A constant-yield harvest may be regarded as a persistent disturbance and the response of a population system to such a disturbance may provide a useful measure of its "stability".

In this paper we shall describe some results on constant-yield harvesting and suggestions for further exploration which are connected with these views of harvesting. As the results have appeared or will appear elsewhere, we shall be descriptive rather than detailed. Our description is rather distant from real life in many respects. For example, we have made no mention of the major difficulties in constructing models from incomplete and inaccurate data, and we have taken no account of stochastic effects in our models. The use of purely deterministic models amounts to an assumption that stochastic effects are small enough to be subsumed in higher order terms having negligible effect on the qualitative behaviour. Nevertheless, an examination of constant-yield harvesting may be useful for suggesting results which may also hold in more realistic situations.

2. Predator-prey systems are often used as refinements of single-species models, considering the prey as a food supply and the predator as the species of interest. If the growth rates of the two species at any time are functions of the two population sizes at that time only, a predator-prey system may be modelled by an autonomous system of two ordinary differential equations

$$x' = xf(x,y)$$

$$y' = yg(x,y).$$

Here x denotes the size of the prey population, y denotes the size of the predator population, and primes indicate derivatives with respect to the time t. We make the following assumptions to describe a predator-prey system in which coexistence of the two species is possible:

(i) $f_y(x,y) < 0$, $g_x(x,y) > 0$, $g_y(x,y) \leq 0$ for $x > 0$, $y > 0$. (Subscripts denote partial derivatives.)

(ii) There is a number $K > 0$ such that $f(K,0) = 0$ and $f(x,y) < 0$ for $x > K$, $y \geq 0$.

(iii) There is a number $J, 0 \leq J < K$ such that $g(J,0) = 0$.

Under these assumptions, the system has a unique equilibrium $P_\infty(x_\infty, y_\infty)$ with $x_\infty > 0$, $y_\infty > 0$ which can not be a saddle point. In addition there is an unstable equilibrium or saddle point at the origin and a saddle point at $S(K,0)$. Every solution of the system with initial values in the interior of the first quadrant of the phase plane tends as $t\to\infty$ either to the equilibrium P_∞ or to a limit cycle around P_∞ [Kolmogorov (1936)].

The original Lotka-Volterra model which predicted population oscillations was not satisfactory because of its lack of structural stability. The Kolmogorov theorem provided a reasonable explanation of the lynx-hare population oscillations noted in the Hudson's Bay Company data, and this represented one of the early triumphs of mathematical ecology. Closer examination of the data discloses discrepancies and suggests that the validity of the data is so questionable that no stock should be put in it [Gilpin (1973)]. In fact, if the data were valid no model of the type we are considering could explain the data because the model predicts counter-clockwise oscillation in the phase plane while the data seems to suggest a "figure-eight" type oscillation whose larger loop is clockwise.

The harvesting of predators at a constant time rate H is modelled by the system

$$x' = xf(x,y)$$
$$y' = yg(x,y) - H.$$

The following description of the behavior of solutions of the harvested system may be given [Brauer and Soudack (1979)]. For $0 \leq H < H_c$, where

$$H_c = \max_{f(x,y)=0} yg(x,y),$$

there are two equilibria of the harvested system in the interior of the first quadrant, $P_\infty(H)$ which can not be a saddle point and a saddle point $S(H)$. These equilibria lie on the curve $f(x,y) = 0$ and depend continuously on H

with $P_\infty(0) = P_\infty$, $S(0) = S$. For $H > H_c$ there is no equilibrium in the interior of the first quadrant and every orbit reaches the x-axis in finite time (predator extinction). For $H < H_c$, orbits may tend to $P_\infty(H)$ or to a limit cycle around $P_\infty(H)$ as $t \to \infty$, or may reach the x-axis in finite time. We define the region of coexistence to be the set of initial values for which orbits tend to $P_\infty(H)$ or to a limit cycle around $P_\infty(H)$ as $t \to \infty$. The determination of the region of coexistence depends on the stable and unstable orbits at the saddle point $S(H)$. If there is an orbit from $S(H)$ to $P_\infty(H)$ or to a limit cycle around $P_\infty(H)$ as t runs from $-\infty$ to $+\infty$, then there is a region of coexistence bounded by the two stable orbits which tend to $S(H)$ as $t \to +\infty$. If, on the other hand, there is an orbit from $P_\infty(H)$ or a limit cycle around $P_\infty(H)$ to $S(H)$ as t runs from $-\infty$ to $+\infty$, then the region of coexistence is either empty if $P_\infty(H)$ is unstable or the region inside an unstable periodic orbit if $P_\infty(H)$ is asymptotically stable. Although the equilibrium $P_\infty(H)$ exists for $0 \leq H < H_c$, the region of coexistence may be empty. For H sufficiently small there is a region of coexistence bounded by orbits tending to $S(H)$. If $P_\infty(H_c)$ is unstable then the region of coexistence is empty for H sufficiently close to H_c. Thus there may be a value $H^* < H_c$ which represents the maximum harvest rate for which coexistence is possible. To determine the existence and estimate the size of the region of coexistence, it is necessary to approximate the orbits tending to $S(H)$ numerically; equilibrium analysis is inadequate to determine whether a harvested population system will persist.

The fact that the region of coexistence is not the entire first quadrant of the phase plane if a system is subjected to constant-yield harvesting points to a need for caution in setting yield quotas for real-life systems. A quota set by assuming a particular type of model, fitting parameters to experimental data, and then calculating H_c may lead to accidental extinction of the predator species. A method of avoiding this danger is to set an escapement level and specify that if the population size decreases to this level harvesting must stop until the population size builds up to another preset level. This policy would afford safety, but might reduce the yield. Preliminary computations suggest, however, that the reduction in yield can be made small by suitable choice of escapement level and harvest rates, and in fact that in the case in which $P_\infty(H_c)$ is unstable larger yields can be obtained with an escapement level than without one. A study of such questions is currently in progress, but only fragmentary results have been obtained. It would be useful to have a procedure to set the escapement level and the level at which harvesting may resume in order to maximize the yield. This question has received little attention; presumably numerical simulations would be a first step in resolving it.

Many real-life systems are more complicated than predator-prey systems. For example, the study of management of multispecies fisheries has been begun in [May, Beddington, Clark, Holt, and Laws (1979)] and some work has been done on systems subject to harvesting at more than one trophic level [Beddington and May (1980), Brauer and Soudack (1981)]. However many questions of mathematical interest and biological importance remain for constant-yield harvesting.

Both constant-effort and constant-yield harvesting are idealizations. As effort increases, harvests are likely to grow more slowly than linearly in a constant-effort policy. Similarly, a constant-yield policy is impossible to maintain when population levels fall too low. Actual harvest curves are more likely to display the same sort of saturation effects assumed in the modelling of the functional response of predators. Indeed, harvesting may be regarded as human predation. The study of realistic harvest curves is almost completely untouched.

3. A reasonable question from a biological point of view is "how stable is a population system?" The answer to this question obviously depends on how one measures stability. The resilience of a system has been defined [Holling (1973)] to mean the ability of the system to absorb changes and still persist. Essentially, this measures resilience by the size of the region of asymptotic stability. As this is a non-local concept, it is usually very difficult to estimate. In some cases it may be possible to construct a Lyapunov function which can be used to describe the region of asymptotic stability [Goh (1977,1979)]. For two-species predator-prey systems we have seen in Section 2

that a characterization in terms of separatrices at saddle points is possible.

Another measure of stability of a system is the characteristic return time [Murdoch (1970); Hurd and Wolf (1974); May, Conway, Hassell and Southwood (1974)]. The characteristic return time of a system at an asymptotically stable equilibrium is defined as the negative reciprocal of the largest real part of an eigenvalue of the community matrix at the equilibrium; if the characteristic return time is t_R, then solutions near the equilibrium tend to the equilibrium at least as rapidly as e^{-t/t_R}. It is possible to define the characteristic return time of an asymptotically stable periodic orbit in terms of the characteristic exponents of the variational system with respect to this periodic orbit, but calculations are inevitably difficult.

Another possible measure of the stability of a system is the sensitivity matrix [Astor, Patten, and Estberg (1975)], defined as a fundamental matrix of the variational system with respect to the solution of the system. If the solution of the system is asymptotically stable, then the sensitivity of the system to perturbations, parameter changes, and changes of initial values can be described in terms of the sensitivity matrix.

Let us suggest another measure of the stability of a system, namely its response to constant-yield harvesting. Consider, for example, a predator-prey system which has an unstable equilibrium and an asymptotically stable periodic orbit for $H = 0$. One possibility as H increases is that the equilibrium may become asymptotically stable and the periodic orbit may disappear. Another possibility is that the region of coexistence may become empty for relatively small values of H. Obviously the first situation describes a more stable system than the second.

Consider a single-species population governed by a differential equation

$$x' = g(x)$$

which has an asymptotically stable equilibrium x_∞, with $g(x_\infty) = 0$, $g'(x_\infty) < 0$. The corresponding model with harvesting,

$$x' = g(x) - H$$

has an equilibrium $x_\infty(H)$ with $g(x_\infty(H)) = H$ for sufficiently small $H > 0$.

It is natural to define the response of the system to harvesting as

$$-\frac{d}{dH}x_\infty(H)\Big|_{H=0},$$

the negative of the rate of change of the equilibrium population size with respect to harvest rate at equilibrium level. By implicit differentiation of $g(x_\infty(H)) = H$ it is easy to see that this response is $-1/g'(x_\infty)$, which is identical with the characteristic return time.

For a system of interacting populations governed by a system of differential equations

$$x_i' = g_i(x_1,\cdots,x_n) \qquad [i = 1,2,\cdots,n]$$

with an asymptotically stable equilibrium $(\xi_1,\cdots,\xi_n)$, we may define a harvest response matrix whose $(i,k)^{th}$ element is the negative of the rate of change of equilibrium population size of the ith species with respect to harvest rate of the kth species at equilibrium level. Thus we consider the system

$$x_i' = g_i(x_1,\cdots,x_n) - H_i \qquad [i = 1,2,\cdots,n]$$

with equilibrium $(\xi_1(H),\cdots,\xi_n(H))$ given by

$$g_i[\xi_1(H),\cdots,\xi_n(H)] = H_i \qquad [i = 1,2,\cdots,n].$$

Implicit differentiation gives

$$\sum_{j=1}^{n} \frac{\partial g_i}{\partial x_j}[\xi_1(H),\cdots,\xi_n(H)]\frac{\partial \xi_j(H)}{\partial H_k} = \delta_{ik}[i,k = 1,2,\cdots,n]$$

and from this it is easy to deduce that the harvest response matrix

$$\left[-\frac{\partial \xi_i(H)}{\partial H_k}\Big|_{H=0}\right]$$

is the negative of the inverse of the community matrix $[\partial g_i/\partial x_j(\xi_1,\cdots,\xi_n)]$ at the (unharvested) equilibrium. The smallest eigenvalue of the harvest response matrix is precisely the characteristic return time. Thus the information contained in the harvest response matrix extends that contained in the characteristic return time.

For an unharvested system with an unstable equilibrium and an asymptotically stable periodic orbit, the effect of harvesting with sufficiently small harvest rates will be to move this periodic orbit. It should be possible to define a harvest response describing the rate of change of the periodic orbit with respect to the harvest rates. This will, however, undoubtedly be extremely difficult to calculate.

The information which is desired is qualitative in nature, and may not be completely describable in quantitative terms. In describing the response of a system to harvesting, we should include the effect of harvesting each species, the qualitative changes in behavior which may occur, the sizes of harvest rates which can be tolerated, and the changes in the region of asymptotic stability. Most of this information can be obtained only approximately by numerical simulation. This is quite feasible for two-species models [Brauer and Soudack (1981)] but may be very difficult for systems involving three or more species. Also, different types of models may produce different responses. For example, differential-difference equation models and difference equation models fitting the same data may predict radically different responses to harvesting. The effect of harvesting on age-structured models with age-dependent birth moduli is to destroy equilibrium age distributions.

For systems involving three or more species, surprising results may be obtained by examining response to harvesting. For example, it is possible to construct a system consisting of a predator species which preys on each of two competing species which would have an unstable equilibrium in the absence of predators for which the three-species system has an asymptotically stable equilibrium such that the effect of harvesting predators is to increase the predator equilibrium population size. An example is given by the system

$$x' = x(3 - x - 2y - u)$$
$$y' = y(2 - x - y - \frac{1}{3}u)$$
$$u' = u(3x + y - \frac{25}{6})$$

which has an asymptotically stable equilibrium at (7/6, 2/3, 1/2) and for which harvesting of the u-species increases the equilibrium population size of the u-species. It is not known whether such behavior represents a failure of the model to depict reality or whether such behavior can actually occur in nature. Nevertheless, there seems to be ample justification for studying the response of a system to harvesting in order to understand the stability of the system.

Literature Cited

Astor, P.H., Patten, B.C., and Estberg, G.N. (1975). The sensitivity substructure of ecosystems, in Systems Analysis and Simulation in Ecology, Vol. 4, B.C. Patten, ed., Academic Press, New York.

Beddington, J.R. and May, R.M. (1980). Maximum sustainable yields in systems subject to harvesting at more than one trophic level, Math. Biosciences 51:261-281.

Beverton, R.J. and Holt S.J. (1957). On the Dynamics of Exploited Fish Populations, Ministry of Agriculture, Fisheries and Food (London) Fisheries Investigators Series 2(19).

Brauer, F. and Soudack, A.C. (1979). Stability regions and transition phenomena for harvested predator-prey systems, J. Math. Biol. 7:319-337.

Brauer, F. and Soudack, A.C. (1981). Coexistence properties of some predator-prey systems under constant rate harvesting and stocking, J. Math. Biol. 12: 101-114.

Clark, C.W. (1976). Mathematical Bioeconomics, Wiley, New York.

Goh, B.S. (1977). Global stability in many-species systems, Am. Naturalist 111:135-143.

Gilpin, M.E. (1973). Do hares eat lynx?, Amer. Naturalist 107:727-730.

Goh, B.S. (1979). Robust stability concepts for ecosystem models, in Theoretical Systems Ecology, E. Halfon ed., Academic Press, New York, 467-487.

Holling, C.S. (1973). Resilience and stability of ecological systems, Ann. Rev. Ecol. Systematics 4:1-23.

Hurd, L.E. and Wolf, L.L. (1974). Stability in relation to nutrient enrichment in arthopod consumers of old-field successional ecosystems, Ecol. Monographs 44: 465-482.

Kolmogorov, A.N. (1936). Sulla teoria di Volterra della lotta per l'esisttenza;, Giorn. Inst. Ital. Attuari 7:74-80.

May, R.M.; Beddington, J.R.; Clark, C.W.; Holt, S.J. and Laws, R.M. (1979). Management of multispecies fisheries, Science 205:267-277.

May, R.M.; Conway, G.R.; Hassell, M.P. and Southwood, T.R.E. (1974). Time delays, density dependence, and single-species oscillations, J. Animal Ecol. 43: 747-770.

Murdoch, W.W. (1970). Population regulation and population inertia, Ecology 51:497-502.

Schaefer, M.B. (1957). Some considerations of population dynamics and economics in relation to the management of marine fisheries, Journal of the Fisheries Research Board of Canada 14:669-681.

Volterra, V. (1931). Lecons sur la Théorie Mathématique de la Lutte pour la Vie, Gauthier-Villars, Paris.

Estimating the response of population to exploitation from catch and effort data

J.R. Beddington & J.G. Cooke

International Institute for Environment and Development
10 Percy Street
London W1P 0DR

Introduction

It is implicit in harvesting theory that populations are capable of providing a sustainable yield. It is further supposed that this is produced when exploitation increases the per capita resources available and thus facilitates higher productivity. One of the central problems of fisheries science is therefore to detect the response of populations to exploitation and hence to deduce this potential yield. In this paper we will concentrate on the problem of detecting changes in abundance of the exploited resource.

There are two basic types of data used in assessing changes in abundance of a resource. The first type consists of information on the catch and effort used by the fishery. The second consists of information on the size or age composition of the catch. These data are used in rather different ways to assess historical changes in abundance and we will consider them separately, before considering ways in which they can be combined.

There are of course a number of other sources of data that can be used. In some fisheries direct estimation of abundance is possible using egg or larval surveys or acoustic techniques. In addition in some cases research vessels can obtain estimates of relative abundance using standard survey techniques.

Changes in Catch Rate

The essential idea of analyses of this type is that there is a relationship between the abundance of the stock $\underline{N}$ and the catch per unit effort (C.P.U.E.) defined as the ratio of the catch $\underline{C}$ to effort $\underline{E}$

$$C = F(N,E) \qquad (1)$$

In traditional analysis this relationship has been assumed to be linear:

$$C = qEN \qquad (2)$$

A major use of this information is in the calculation of the parameters of a production model (see Adu-Asamoah and Conrad this volume). Here a population model for the renewal of the resource is specified as

$$\frac{dN}{dt} = rN(1 - N/K) \qquad (3)$$

and this coupled with equation (2) gives an equilibrium catch rate for constant effort E:

$$C = aE - bE^2 \qquad (4)$$

Schnute (1977) gives an alternative formulation which does not require the assumption of equilibrium. The estimation of the parameters of this model from data on catch and effort is then mathematically straightforward(e.g. Fox (1975)) but of questionable validity, as random error in the measurement of effort can bias the estimates and even generate correlations when none exist (e.g. Sissenwine (1978)).

In analysing populations with very slow rates of change,such as whales,a particularly simple population model has been employed. This method is due to de Lury (1947) and has been employed by the Scientific Committee of the International Whaling Commission in estimating changes in abundance of whale populations.

The analysis simply consists of a regression of the CPUE against the accumulated catch. It is based on the assumption that the recruitment rate to the stock is equal to the natural mortality rate; itself independent of harvesting mortality. It is therefore implicitly assumed that there are no density dependent changes in mortality or recruitment.

It is useful to explore the developments and problems of this analysis in some detail as they will illuminate some of the complexities of later work.

<u>The relationship between catch rate and abundance</u>

The linear form of equation (2) depends critically for its validity on the choice of a measure of effort. A salutary example is provided by the use of such a measure in the history of the International Whaling Commission. They used as a measure of effort, the catcher boat day: typically it takes approximately an hour to catch a whale after it has been sighted, hence the relationship between the catch per boat day and the abundance of whales is non-linear and of the approximate form

$$C = \frac{aNE}{1 + bN} \qquad (5)$$

The adjustment of the measure of effort so that the handling time is removed and searching hours remain is unfortunately still insufficient to produce a linear relationship between CPUE and abundance. In fact the relationship is more complex and involves both the distribution of the resource and the way in which the harvester searches and reacts to that distribution over space and time.

Where the encounters between fish and fishermen or whales and whalers are non-random, then the relationship will be non-linear and usually of a form to conceal declines in the stock. The details of such non-linear forms vary greatly, depending on the assumptions concerning the distribution of the resource, and the type of searching by the fishermen (Palaheimo and Dickie (1964) Clark and Mangel 1982, Cooke in prep.) Whatever the form, there appears to be a good empirical justification for considering non-linear forms. e.g. Ulltang (1975), Saville (eds. 1980)

A further problem of this sort of analysis is that catch and effort data are often highly variable. This implies that it will often be difficult to detect even marked changes in abundance of a resource. As an illustration to this point Table 1 gives the probability of detecting a decline in abundance of a whale stock under a typical value for the coefficient of variation of the catch per unit effort.

Table 1

The probability of detecting a decline in population at 5% significance level with a C.P.U.E. coefficient of variation of 50%

Reduction in stock size as a percentage of initial	10 years data	20 years data
50%	.17	.35
25%	.52	.83

Although effort data for other species may have lower levels of variability,this illustration is indicative of the difficulty in detecting change using catch and effort data alone.

The problem of population dynamics

As indicated earlier, there is an implicit assumption in the de Lury procedure that mortality and recruitment balance. It is possible to relax this assumption and to permit a model of the dynamics of the population to be used. Such models differ markedly depending on the type of species concerned. In marine mammals recruitment is closely linked to adult stock size, but in many fish species high variability in recruitment occurs for similar stock sizes. (Hennemuth, Brown, & Palmer, 1980).

Accordingly for fish, each yearly recruitment must be considered as a free parameter or as a random variable. By contrast,for marine mammals, a population model may be used, though such models do not predict behaviour of stock well, in the few cases where verification is possible.

Such a model is typified by those used in the assessments by the IWC (Allen, 1980).

Here the dynamics of the stock are defined as

$$N_{t+1} = (N_t - C_t)\,e^{-M} + e^{-M}F(N_{t-k}) \qquad (7)$$

or more complex variations when the age at sexual maturity is varying, see Grenfell & Beddington (1980). In this formulation M is the coefficient of natural mortality, (assumed constant for all age classes) F(t) the fecundity of the stock, N_t the abundance in year t, C_t the catch, and k the age at sexual maturity and recruitment (assumed for simplicity to be equal). The catch is assumed to occur instantaneously before other sources of mortality. This model is entirely deterministic, so that once the stock size N_o in some reference year is specified, the stock sizes in all other years are uniquely determined provided the catches and biological parameters are known.

The estimation procedure is then a matter of linking the equations (2) and (7) such that q and N_o are estimated by non-linear regression from iterating equation (7) with known catches to minimise the required residuals. Kirkwood (1981) for this type of analysis uses as a criterion function

$$S = \sum_t \left(C_t^{1/2} - (qE_tN_t)^{1/2}\right)^2 \qquad (8)$$

to produce approximate homoscedasticity of residuals. Confidence regions for the parameters are poorly approximated by the standard linearisation, but for this problem, an approximate α -level confidence region in (q, N_o) space is given by the set

$$\{ (q, N_o) : \frac{S(q,N_o - S(q,N_o)}{\bar{S}(q,N_o)} < F_{2, n-2,\alpha} \} \tag{9}$$

Kirkwood notes that the resulting confidence regions in (q, N_o) space are markedly asymetric. As q is not usually of direct interest, it is helpful to eliminate q from (8) and (9) by differentiation, leaving only one parameter to be estimated, N_o.

Defining the recruited population

Once the analysis is expanded to take into account the dynamics of the population it becomes critically important to define which age or size classes are recruited to the fishery. This may be done approximately by defining an average age at recruitment in equation (7) or in a more flexible model by defining the partial recruitment parameters, which are assumed to define the time invariant proportion of an age or size class that is recruited to the fishery. Further analysis depends on estimating these parameters and hence an analysis of the catch at age data.

Catch at age data

The basic data set consists of estimates of the number of fish caught at each age over time C(a,t). If fishing mortality is sufficiently high it is possible to consider the numbers at age in the current year and those of the oldest age classes in earlier years as effectively zero. It is then possible to reconstruct the population by adding up the catches at age and estimates of the natural deaths. Pope (1974) shows that estimates of final age class strength, current age structure and natural mortality are relatively unimportant in determining the estimates of earlier population size, as long as the fishing mortality is high.

Where fishing mortality is not particularly high, the natural mortality has an important effect on the estimates of population abundance.

A common method of estimation of natural mortality, is based on a catch at age sample of the virgin stock (Chapman and Robson 1960). Even where such data are available, the properties of the estimator are poor if either recruitment varies markedly or there is a selection pattern by age of the fishery or the age samples. Often such data do not exist and estimates are obtained in a variety of ad hoc ways,e.g. (Pauly 1979).

In many situations it is unreasonable to assume that the oldest age class in each year and all age classes in the final year are sufficiently small to be ignored. In these situations it is necessary to have some method of estimating the abundance of these age classes. Traditionally they have been estimated using effort data. However for the reasons outlined above the precision of such estimates is often doubtful.

V.P.A. per se is of no use in predicting into the future. It is essentially a method for reconstructing historical abundance.

<u>Fitting catch at age data</u>

Traditional methods of V.P.A. are simply accounting exercises, but given the basic matrix of catch at age it is possible to consider the problem as one of fitting the catch at age data using the following equation:

$$C(a,t) = N(a,t)\,(1 - \exp(-F(t)S(a))) \tag{10}$$

where F(t) is the year effect known as fishing mortality and S(a) the selection at age assumed constant over all years. Thus the mortality imposed on age class a in year t is F(t) S(a). In this formulation natural mortality during the catching season is assumed to be negligible.

If the S(a) = 1 for all a it can simply be shown that there is no unique solution for the F(t) and N(a,t), (Pope & Shepherd 1982). Furthermore Pope & Shepherd show that even if the pattern of selectivity is markedly different from unity, then although a solution exists, it is sufficiently ill-defined when even small amounts of random variation are present in the age data, that no discrimination is possible. Accordingly it may be concluded that using catch at age data alone, it is not possible to obtain a least squares solution. Pope & Shepherd have found that with an estimate of M, the fishing mortality on some reference age class in the final year, and the proportion of that fishing mortality that applies to the last age class it is possible to determine the N(a,t). In essence their analysis is a more sophisticated form of V.P.A. What is different is that traditional V.P.A. makes no assumption that fishing mortality is separable (constant selectivity). Pope & Shepherd make this additional assumption, and so reduce the number of other assumptions to be made.

One practical problem with the analysis of Pope & Shepherd is that they have not analysed the problem for more than one fishing operation. Hence where a stock is subject to fishing from different types of gear, in practice the analysis cannot be applied at its present level of development.

Recruitment variation

As indicated earlier, for most fish stocks, recruitment is only loosely linked to the stock size. Hence it is not possible to utilise any form of population model in formulating estimates, as effectively recruitment is a free parameter. In the case of marine mammals, it is possible to utilise an underlying population model to mimic the dynamics. Essentially this imposes a structural constraint on the possible values of recruitment in the numbers at age matrix. The importance of being able to make this assumption is the fact that, if the history of exploitation is available one can see how the equilibrium or stationary age structure is distorted by the catches. The greater the distortion for a fixed harvest the smaller the underlying population. As an illustration of this we describe how we have used this idea in estimating the response of sperm whale populations to exploitation.

Estimating sperm whale abundance from length data

The dynamics of sperm whales are modelled by a set of sex and age structured equations; the reproduction factor in these equations is modified in practice when the ratio of mature males to mature females drops below a specified amount required for complete fertilisation of the females. The analysis of sperm whale population has utilised the information on the changes in length structure of the population. Unlike fish species where samples are taken of age/length structure the length of each sperm whale caught is recorded. It is thus possible to use such equations with a growth equation to deduce the equilibrium length structure of the population and to estimate how that length structure has been distorted by a history of catches. As in the case of age structure, one seeks to track the distortions away from the equilibrium length structure caused by the catch.

The population dynamics of the males is described by the equations

$$N_{a+1,t+1} = e^{-M_m}(N_{at} - C_{at}) \qquad (a > 1) \tag{11}$$

and

$$N_{1,t+1} = {}^{1}/_{2}e^{-M_j} \cdot B_t \tag{12}$$

where M_m is the male mortality rate, M_j the juvenile mortality (both assumed constant), N_{at} the number of males aged a alive in year t, C_{at} the catch of males aged a in year t, and B_t, the birth rate at time t, supposed to be:

$$B_t = F_t(P + Q(1 - (F_t/F_o)^{1+Z})) \tag{13}$$

where F_t is the number of mature females alive in year t, P the initial pregnancy rate in the unexploited stock, and Q and Z the range and exponent of the density-dependent response respectively, provided that the number of mature males is sufficient to allow maximum fecundity. The sex-ratio at birth is assumed to be 50:50, and the juvenile mortality is assumed equal in both sexes.

The mature female population is described by the model:

$$F_{t+1} = e^{-M_f}(F_t - CF_t) + {}^{1}/_{2}e^{-(T_f-1)M_f}.e^{-M_j}.B_{t+1-T_f} \tag{14}$$

where F_t is the number of mature females alive in year t, CF_t is the catch of females in year t, M_f is the female natural mortality rate (which is not necessarily the same as that for males), and T_f is the female age at maturity, assumed constant. It is assumed that all females caught are mature.

The analysis is based on the central assumption that the catch of each length class is determined solely by the abundance of whales in that length class and a selection factor by length, which is considered to be constant over time. This can be written as

$$E\ (C(j,t)) = N(j,t)\ (1 - \exp(-S(j)F(t))) \tag{15}$$

where $F(t)$ is the rate of exploitation which will vary yearly, $S(j)$ the selectivity factor at length j and $N(j,t)$ the number in each length class in year t.

This is simplified to

$$E\ (C(j,t)) = S(j)\ F(t)\ (N(j,t) - C(j,t)/2) \tag{16}$$

The least squares problem then involves minimising the sum of squared residuals

$$H = \sum_j \sum_t (C(j,t) - S(j)\ F(t)\ N(j,t))^2 \tag{17}$$

A critical aid to this is that when the catches at length are known, the factors $S(j)$ and $F(t)$ do not enter into the simulation of the population. Hence for any set of population parameters they may be estimated by minimising the criterion function, defined in equation (17) by differentiation. (Cooke & Beddington (1981) describe an algorithm that solves these equations).

There are three main aspects of the performance of this estimation method which need highlighting. These were investigated by Cooke and de la Mare (1982) using simulation trials and are:

(1) The level of exploitation: since the method involves quantifying the change in length composition of the catch due to exploitation, one would only expect reliable estimates to be obtainable when the exploitation rate is reasonably high. Simulation studies show that if a stock is reduced in abundance by 50% over a 25-year period, then the estimates of initial stock size would be expected to be accurate to within about 10% depending on the variability of the basic catch-at-length data. If a stock is reduced only by 25% in abundance over the same length of time, the range of error in the initial population estimates extends to about 25%. For lower rates of exploitation, the estimates become very unreliable.

(2) The variability of recruitment: recruitment may vary haphazardly in ways not predicted by the population model. A coefficient of random variation in recruitment of 10% was found to substantially increase the range of error in the population estimates, while a coefficient of variation of 20% made the estimates very unreliable. This makes the estimation procedure inapplicable to species where recruitment is not closely linked to the parent population size.

(3) The natural mortality rate: the estimates of the population are rather insensitive to the assumed value of the natural mortality rate within quite wide limits, which is a useful property given the difficulties of estimating this parameter.

While in principle this procedure could yield estimates of stock size, the main assumptions such as that of constant selectivity by length are difficult to verify and will not necessarily be of any practical value.

Concluding Remarks

The results of the simulation trials on the sperm whale estimation problem indicate that where the coefficient of variation of recruitment is greater than 10%, the estimation of population abundance and trend is poor. Accordingly for fish population which have highly variable recruitment (Hennemuth, Brown & Palmer, 1980) the possibility of estimating the current abundance from catch at age or length data is small.

One possibility that is currently under investigation is to combine catch per unit effort data with catch at age data in an overall estimation procedure. (Cooke in prep.) The essence of this method is to modify the De Lury procedure to regress C.P.U.E. against effective accumulated catch for each cohort. A separate intercept for each cohort is calculated which corresponds to the strength of that cohort. A common slope is estimated and corresponds to the catchability coefficient. There have been other attempts at simultaneous analysis of catch at age and relative abundance data (e.g. Doubleday, 1981) but so far these have not incorporated any simulation studies to test the reliability of the resulting estimators.

This is based on the assumption that selectivity is uniform within the age range in question. If this assumption is not warranted then a slightly different procedure, involving estimation of the selection pattern, must be followed, but the estimates of stock size in this case are more variable. Collie and Sissenwine (1982) have effectively used such a technique in analysing catch at age data from bottom trawl surveys. Their method appears to give good results in certain situations, but has not been the subject of Monte Carlo investigation as yet. Its general performance is therefore uncertain.

The implication of this brief review of techniques used to estimate the response of populations to exploitation is somewhat bleak. Current data and assessment techniques are by and large inadequate for determining small changes in abundance of the harvested stock. A corollary of this is that harvesting strategies, dependent on fine tuning of the harvesting level to the abundance of the resource are unlikely to be achievable.

LITERATURE CITED

Adu Asamoah, A.and Conrad, J.M. (1983). Fishery management: the case of tuna in the Eastern Tropical Atlantic. This volume

Allen, K.R. and Kirkwood, G.P. (1976). Further developments of sperm whale population models Rep. Int. Whal. Commn 26: 106-12

Allen, K.R. (1980). Conservation and management of Whales. Univ. of Washington Press.

Beddington, J.R. and Cooke, J.G. (1980). Development of an assessment technique for male sperm whales based on the use of length data from the catches. Rep. Int. Whal. Commn 31: 747-60.

Chapman, D.G. and Robson, D.S. (1960). The analysis of a catch curve Biometrics 16: 354-368.

Clark, C.W. and Mangel, M. (1978). Aggregation and fishing dynamics: a theoretical study of schooling and purse seine tuna fisheries. Fishery Bulletin 77 (2) 317-337.

Collie, J.S. and Sissenwine, M.P. (1983). Estimating population size from relative abundance data measured with error. Can. J. Fish. Aquat. Sci. In Press.

Cooke, J.G. and Beddington, J.R. (1981). Further developments of an assessment technique for male sperm whales based on the length date from catches. Rep. Int. Whal. Commn. 32: 239-42

Cooke, J.G. and de la Mare, W.K. (1982). Description of and simulation studies on the length specific sperm whale estimation techniques. Rep. Int. Whal. Commn. 33: (In Press)

De Lury, D.B. (1947). On the estimation of biological populations. Biometrics 3: 145-167

Doubleday, W.G. (1981). A method of estimating the abundance of survivors in an exploited fish population using commercial fishing catch at age and research vessel abundance indeces. Can. Spec. Publ. Fish. Aquat. Sci. 58:164-78

Fox, W.W. (1975). Fitting the generalised stock and production model by least squares and equilibrium approximation. Fishery Bulletin 73(1): 23-37.

Grenfell, B.T. & Beddington, J.R. (1981). A new population model and assessment technique based on proportional maturity and recruitment with application to the Minke Whales of the Southern Hemisphere. Rep. Int. Whal. Commn. 31: 233-240.

Hennemuth, R.C., Palmer, J.E. and Brown, B.E., (1980). A statistical description of recruitment in eighteen selected stocks. J. Northw. Atl. Fish. Sci. 1:101-111.

Kirkwood, G.P. (1981). Estimation of stock size using relative abundance data, a simulation study. Rep. Int. Whal. Commn. 31:729-735

Palaheimo, J.E. and Dickie, L.M. (1964). Abundance and fishing success. Cons. Perm. Int. Explor. Mer. Rapp. P.V. 155: 152-163

Pauly, D. (1979). On the interelationships between natural mortality, growth parameters and mean environmental temperature in 175 Fish Stocks. J. Cons. Int. Explor. Mer. 39: 175-192

Pope, J.G. (1972). An investigation of the accuracy of virtual population analysis using cohort analysis. Int. Commn. North W. Atlant. Fish. Res. B.M. 9: 65-74.

Pope, J.G. and Shepherd, J.G. A simple method for the consistent interpretation of catch at age data. J. Cons. Int. Explor. Mer, 40:176-184.

Saville, A. (Ed.) (1980). The assessment and management of pelagic fish stocks. Cons. Intern. Expl. Mer., Rapp. Proc. - Verbaux des Reunions 177.

Schnute, J. (1977). Improved estimates from the Schaefer Production Model: Theoretical Considerations. J. Fish. Res. Bd Can. 34: 583-603.

Sissenwine, M. (1978). Is MSY an adequate foundation for optimum yield? Fisheries, 3(6): 22-42.

Ulltang, (1975). Catch per unit of effort in the Norwegian purse seine fishery for Atlanto-Scandian (Norwegian Spring-Spawning) Herring. FAO Fish. Tech. Pap., 155:91-101.

BIOECONOMICS AND THE MANAGEMENT OF TUNA STOCKS IN THE EASTERN TROPICAL ATLANTIC

by

Richard Adu-Asamoah and Jon M. Conrad*

ABSTRACT

A simple bioeconomic model (the Gordon-Schaefer model) is estimated for three species of tuna in the Eastern Tropical Atlantic. The model conformed well to the data and afforded estimates of maximum sustainable yield, bioeconomic and open access equilibria. Strong marginal stock effects were identified in all three fisheries, resulting in bioeconomic optima with stocks in excess of X_{MSY} While all three fisheries appear to have been economically overfished (fishery rents being driven toward zero), Yellowfin and Skipjack stocks do not appear to be biologically overfished (stocks appear to be at or slightly above $X = K/2$). For Bigeye Tuna there was strong indication of both economic and biological overfishing. A management policy employing transferable quotas and landings taxes is examined. Such a policy has three advantages: (a) optimal yield will be harvested at least cost; (b) potential fishery rents may be distributed in a flexible fashion between West African and foreign flag vessels; and (c) a portion of the potential fishery rents may be captured by the management agency to defray the costs of administration, enforcement and research.

*Richard Adu-Asamoah is a former graduate student and Jon Conrad is an Associate Professor of Resource Economics at Cornell University.

BIOECONOMICS AND THE MANAGEMENT OF TUNA STOCKS IN THE EASTERN TROPICAL ATLANTIC

I. Introduction and Overview

This paper develops bioeconomic models for three species of tuna in the Eastern Tropical Atlantic (ETA). From a theoretical perspective it draws from the seminal work by Gordon (1954), Scott (1955) and Schaefer (1957), as well as the more recent capital-theoretic approach summarized by Clark and Munro (1975). Parameters of a Gordon-Schaefer model are estimated for each species, allowing one to identify maximum sustainable yields, open access equilibria, and bioeconomic optima. These equilibria are useful in evaluating the magnitude of recent landings and in suggesting policies which would establish and maintain the fishery in a more profitable state.

The next section describes the ETA tuna fishery, focusing on the countries participating in this fishery, effort, and catch. This is followed by a brief review of the basic bioeconomic model and the Gordon-Schaefer specification. The fourth section presents estimates for the various bioeconomic parameters and compares three steady state equilibria: maximum sustainable yield, open access, and the bioeconomic optimum. The fifth section examines landing taxes and transferable quotas as policies for managing the fishery and distributing rents among coastal and distant water fleets. The final section collects and qualifies the principal conclusions in light of the limitations inherent in lumped parameter (biomass) models.

II. The Tuna Fishery in the Eastern Tropical Atlantic

Harvests of tuna in the ETA are dominated by three species: Yellowfin Tuna (*Thunnus albacares*), Skipjack Tuna (*Katsuwonus pelamis*) and Bigeye Tuna (*Thunnus obesus*). These species are caught by vessels operating in an area roughly bounded by latitudes 30° North to 30° South and by longitude 30° West to the west coast of Africa (Figure 1).

Modern commercial exploitation of Yellowfin and Skipjack Tuna within the ETA began in the mid-1950s. During that period pole-and-line boats from France and

FIGURE I. LOCATION MAP FOR THE EASTERN TROPICAL ATLANTIC TUNA FISHERY*

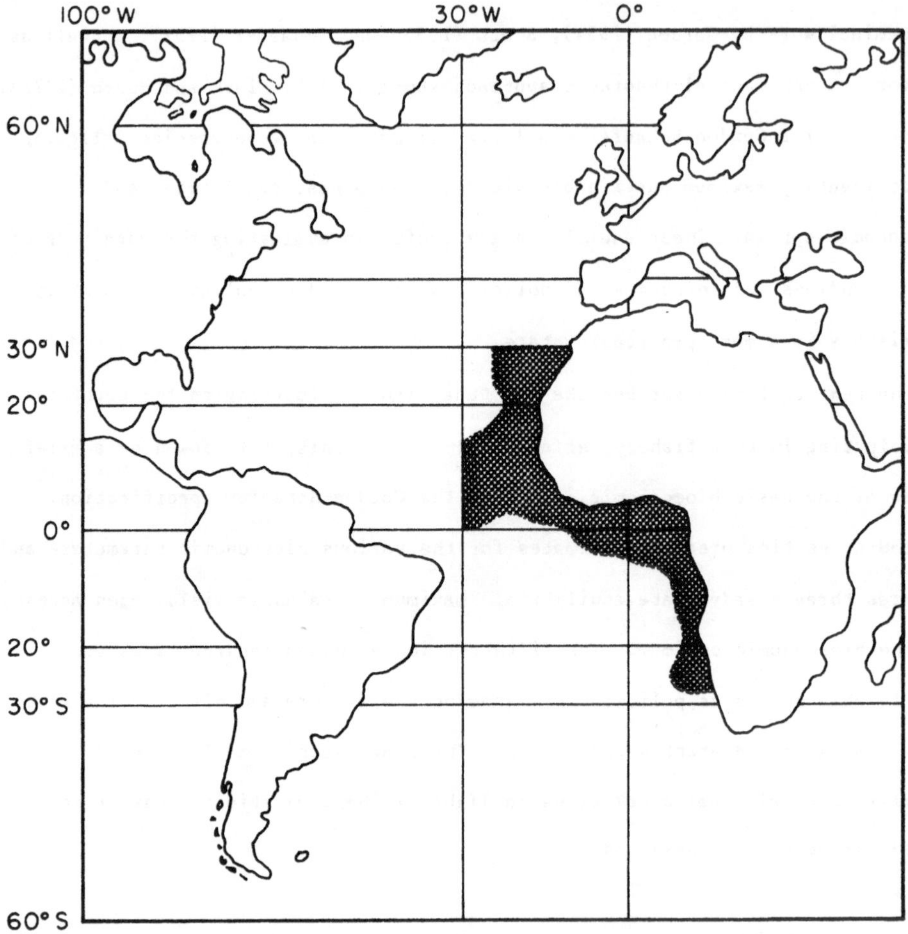

*Cross hatching indicates area of greatest fishing intensity

Spain moved into the waters off present-day Senegal. Their operations expanded, and by the early 1960s, they were fishing throughout the year, ranging from the Canary Islands south to Point Noire in present-day Zaire. During the 1960s, there was an influx of purse seiners registered in Japan, Korea, Taiwan, Panama, the United States and Norway. The bulk of their catch was sold to a single company and landed in Tema (Ghana), which developed into a major transshipment point.

Landings, effort and catch per unit effort (CPUE) for Yellowfin, Skipjack and Bigeye Tuna are shown in Tables 1-3 for 1967 through 1980. The early 1970s saw the formation of several joint ventures and a multinational alliance between France, Ivory Coast, Senegal and Morocco (FISM).1/ Vessels from Korea and Panama joined to harvest all three species, and although their landings are individually reported, Ghana and Japan formed two jointly owned companies in 1972 and 1974.

Total Yellowfin landings range from a low of 53,000 metric tons (MT) in 1967 to a record high of 118,500 MT in 1978. Spain has displaced Japan as a major harvester of Yellowfin. This shift reflects the decision on the part of Japan to develop its longline fleet and reduce the number of purse seiners fishing surface stocks (primarily Yellowfin). The FISM alliance has also been a dominant harvester of Yellowfin; its share ranging from a low of 35.9 percent in 1969 to a high of 48.3 percent in 1980.

Landings of Skipjack ranged from 19,000 MT in 1967 to 113,700 MT in 1977. Japan, the FISM alliance and Spain have accounted for 60 to 80 percent of the total Skipjack landings during the 1967-80 period. There has been a shift in the share of landings from Japan to Spain, although the degree of displacement is less than that which occurred for Yellowfin.

Japan is the dominant harvester of Bigeye Tuna. Its share of total harvest has ranged from 14.1 to 59 percent, with an average of 37.3 percent during

---

1/ During certain years this alliance also included Portugal. Portugese landings were never more than a small fraction of the total landings by the alliance and limited to Yellowfin and Bigeye Tuna. Thus the landings attributed to the FISM alliance may include minor amounts of Yellowfin and Bigeye caught by Portugal.

the 1967-80 period. Vessels from Korea, Panama and Taiwan have also been significant participants in the Bigeye fishery. Total landings of Bigeye have ranged from 9,600 MT in 1967 to 23,900 MT in 1973. Landings of Yellowfin and Skipjack have been two to six times larger than the landings of Bigeye, and in 1980 the Bigeye harvest of 13,600 MT was only 6.4 percent of the total for all three species.

The landings data in Tables 1-3 were compiled by the International Commission for the Conservation of Atlantic Tunas (ICCAT) based on reports from countries with fleets harvesting tuna in the ETA. ICCAT also keeps track of the number and type of vessels harvesting tuna. The effort levels reported in Tables 1-3 are measured in standard days (SDs) at sea, where the number of days fished by small seiners, baitboats and longliners have been converted to large-seiner-day equivalents.[2/]

In many models of commercial fishing, and in the Gordon-Schaefer model to be discussed shortly, catch per unit effort is assumed proportional to the fish stock. While such models are a plausible first approximation, they may be inappropriate for schooling species such as tuna. Within a model with schooling and stochastic search Clark and Mangel (1979) show that a catastrophic collapse can occur as the level of fishing passes some critical level. CPUE proved to be an unreliable index of stock abundance and, for simulations leading to collapse, did not predict any significant decline in abundance until the fishing was "virtually destroyed."

With Clark and Mangel's caveats in mind, we note from Tables 1-3 that all three species have exhibited a declining trend in CPUE during the 1967-80

2/ It was assumed that the fishing power of a small seiner was 0.48 of a large seiner (Fonteneau and Cayre, 1981). Average daily catch rates for baitboats and longliners were divided by the average daily catch rate for large seiners and the resulting fractions were used to weight (convert) baitboat and longline days into large-seiner-day equivalents. Baitboats and longliners were assumed to spend an average of 231 days at sea per year, while large and small seiners were assumed to spend 219 and 198 days per year at sea, respectively.

TABLE 1: LANDINGS, EFFORT, AND CATCH PER UNIT EFFORT FOR YELLOWFIN TUNA IN THE EASTERN TROPICAL ATLANTIC[a]

YEAR	LANDINGS BY NATION, JOINT VENTURE, OR MULTINATIONAL ALLIANCE[b]								TOTAL LANDINGS	EFFORT (SDs)	CATCH PER UNIT EFFORT (CPUE)
	GHANA	JAPAN	KOREA / PANAMA	FISM	ANGOLA	SPAIN	USA	OTHERS			
1967	0	16,600 (31.2)	0	23,400 (44)	900 (1.7)	3,100 (5.8)	900 (1.7)	8,300 (15.6)	53,200 (100)	12,800	4.15
1968	0	19,500 (26.2)	1,600 (2.2)	31,400 (42.2)	1,100 (1.5)	3,300 (4.4)	5,800 (7.8)	11,700 (15.7)	74,400 (100)	15,500	4.78
1969	0	12,100 (14.7)	4,200 (5.1)	29,500 (35.9)	400 (0.5)	5,800 (7.1)	18,800 (22.9)	11,300 (13.8)	82,100 (100)	21,400	3.83
1970	0	4,400 (7.2)	9,300 (15.2)	24,700 (40.4)	300 (0.5)	7,100 (11.6)	9,000 (14.7)	6,400 (10.4)	61,200 (100)	19,900	3.08
1971	0	5,600 (9.7)	6,900 (11.9)	26,800 (46.2)	500 (0.9)	7,600 (13.1)	3,800 (6.6)	6,800 (11.7)	58,000 (100)	15,100	3.84
1972	0	8,300 (10.5)	8,200 (10.4)	32,100 (40.7)	600 (0.8)	9,300 (11.8)	12,000 (15.2)	8,400 (10.6)	78,900 (100)	18,600	4.24
1973	100 (0.1)	9,000 (10.6)	17,900 (21.1)	32,300 (38)	600 (0.7)	14,000 (16.5)	3,000 (3.5)	8,000 (9.4)	84,900 (100)	31,900	2.66
1974	300 (0.3)	8,600 (8.9)	17,500 (18.2)	39,200 (40.8)	800 (0.8)	15,700 (16.3)	5,600 (5.8)	8,600 (8.9)	96,300 (100)	30,400	3.17
1975	700 (0.6)	2,900 (2.6)	13,700 (12.5)	48,000 (43.9)	100 (0.1)	24,800 (22.7)	14,000 (12.8)	5,300 (4.8)	109,500 (100)	41,600	2.63
1976	800 (0.7)	5,200 (4.5)	12,900 (11.3)	54,200 (47.4)	1,000 (0.9)	33,300 (29.1)	1,700 (1.5)	5,300 (4.6)	114,400 (100)	40,200	2.84
1977	600 (0.5)	2,700 (2.4)	12,700 (11.1)	51,300 (44.9)	1,900 (1.7)	33,500 (29.3)	6,400 (5.6)	5,100 (4.5)	114,200 (100)	42,700	2.68
1978	300 (0.3)	1,700 (1.4)	10,100 (8.5)	56,500 (47.7)	2,000 (1.7)	35,300 (29.8)	8,100 (6.8)	4,500 (3.8)	118,500 (100)	54,900	2.16
1979	300 (0.3)	900 (0.8)	6,000 (5.2)	51,000 (44.3)	800 (0.7)	40,300 (35)	2,900 (2.5)	12,900 (11.2)	115,100 (100)	60,200	1.91
1980	300 (0.3)	1,000 (1)	2,800 (2.8)	48,300 (48.3)	0	35,700 (35.7)	4,800 (4.8)	7,200 (7.1)	100,100 (100)	57,700	1.73

[a] Source: International Commission for the Conservation of Atlantic Tunas (ICCAT).
[b] Landings are in metric tons (MT). Percentage of total catch is given in parentheses.

TABLE 2: LANDINGS, EFFORT, AND CATCH PER UNIT EFFORT FOR SKIPJACK TUNA IN THE EASTERN TROPICAL ATLANTIC[a]

YEAR	LANDINGS BY NATION, JOINT VENTURE OR MULTINATIONAL ALLIANCE[b]								TOTAL LANDINGS	EFFORT (SDs)	CATCH PER UNIT EFFORT (CPUE)
	GHANA	JAPAN	KOREA/PANAMA	FISM	ANGOLA	SPAIN	USA	OTHERS			
1967	0	5,900 (31.1)	0	5,300 (27.9)	2,000 (10.5)	3,800 (20)	500 (2.6)	1,500 (7.9)	19,000 (100)	8,300	2.28
1968	0	13,600 (30.5)	0	12,400 (27.8)	4,200 (9.4)	9,500 (21.3)	3,200 (7.2)	1,700 (3.8)	44,600 (100)	12,300	3.62
1969	0	5,600 (21.3)	0	6,500 (24.7)	1,800 (6.8)	7,200 (27.4)	4,700 (17.9)	500 (1.9)	26,300 (100)	9,400	2.81
1970	0	11,000 (23.5)	0	13,200 (28.1)	900 (1.9)	8,300 (17.7)	11,800 (25.2)	1,700 (3.6)	46,900 (100)	15,900	2.94
1971	0	17,900 (24.7)	0	20,000 (27.6)	1,900 (2.6)	14,900 (20.6)	16,200 (22.4)	1,500 (2.1)	72,400 (100)	23,400	3.09
1972	0	13,500 (19)	700 (1)	18,600 (26.3)	1,500 (2.1)	24,200 (34.1)	12,200 (17.2)	200 (0.3)	70,900 (100)	24,100	2.95
1973	100 (0.1)	14,500 (19.9)	1,100 (1.5)	12,700 (17.5)	1,300 (1.8)	21,300 (29.3)	21,200 (29.1)	600 (0.8)	72,800 (100)	25,000	2.91
1974	700 (0.6)	19,600 (17.3)	3,100 (2.7)	28,500 (25.3)	3,400 (3)	37,000 (32.7)	20,000 (17.7)	800 (0.7)	113,000 (100)	46,200	2.45
1975	1,300 (2.3)	3,800 (6.6)	6,300 (11)	13,300 (23.3)	600 (1)	18,900 (33)	7,400 (12.9)	5,700 (9.9)	57,300 (100)	26,600	2.15
1976	2,100 (2.9)	15,000 (20.5)	4,400 (6)	18,600 (25.4)	1,500 (2)	17,400 (23.8)	1,800 (2.5)	12,400 (16.9)	73,200 (100)	40,100	1.82
1977	3,500 (3.1)	16,800 (14.8)	7,600 (6.7)	33,600 (29.6)	3,800 (3.3)	27,700 (24.3)	5,900 (5.2)	14,800 (13)	113,700 (100)	68,700	1.65
1978	2,600 (2.6)	14,600 (14.6)	11,100 (11.1)	28,300 (28.2)	3,200 (3.2)	25,500 (25.4)	6,800 (6.8)	8,100 (8.1)	100,200 (100)	63,800	1.57
1979	3,900 (4.6)	14,700 (17.5)	13,800 (16.4)	21,100 (25.1)	3,600 (4.3)	19,800 (23.6)	2,100 (2.5)	5,000 (6)	84,000 (100)	68,500	1.23
1980	not available	20,000 (20.6)	not available	30,000 (30.9)	5,100 (5.2)	33,500 (34.5)	3,500 (3.6)	5,100 (5.2)	97,200 (100)	79,600	1.22

[a] Source: International Commission for the Conservation of Atlantic Tunas (ICCAT).
[b] Landings are in metric tons (MT). Percentage of total catch is given in parentheses.

TABLE 3: LANDINGS, EFFORT, AND CATCH PER UNIT EFFORT FOR BIGEYE TUNA IN THE EASTERN TROPICAL ATLANTIC[a]

YEAR	LANDINGS BY NATION, JOINT VENTURE OR MULTINATIONAL ALLIANCE[b]								TOTAL LANDINGS	EFFORT (SDs)	CATCH PER UNIT EFFORT (CPUE)
	GHANA	JAPAN	KOREA/PANAMA	FISM	TAIWAN	SPAIN	USA	OTHER			
1967	0	5,700 (59)	100 (1)	0	1,900 (20)	0	0	1,900 (20)	9,600 (100)	11,200	0.86
1968	0	7,200 (54)	200 (1.5)	0	3,800 (28.5)	0	0	2,100 (16)	13,300 (100)	15,200	0.87
1969	0	9,500 (52)	1,400 (7.7)	400 (2)	4,500 (24.8)	0	100 (0.5)	2,300 (13)	18,200 (100)	16,700	1.09
1970	0	4,800 (33)	3,500 (24)	700 (5)	2,400 (16)	100 (0.7)	200 (1.4)	2,900 (19.9)	14,600 (100)	23,100	0.63
1971	0	8,100 (35.4)	5,500 (24)	800 (3.5)	3,100 (13.5)	200 (0.9)	500 (2.2)	4,700 (20.5)	22,900 (100)	33,300	0.69
1972	0	7,900 (38.2)	4,400 (21.3)	900 (4.2)	4,200 (20.3)	200 (1)	200 (1)	2,900 (14)	20,700 (100)	27,600	0.75
1973	0	10,800 (45.2)	3,000 (12.6)	2,200 (9.2)	2,500 (10.4)	400 (1.7)	100 (0.4)	4,900 (20.5)	23,900 (100)	35,400	0.68
1974	100 (0.5)	5,400 (27.7)	4,000 (20.5)	1,600 (8.2)	2,000 (10.3)	700 (3.6)	900 (4.6)	4,800 (24.6)	19,500 (100)	37,800	0.52
1975	100 (0.6)	5,100 (28.7)	4,000 (22.5)	600 (3.4)	2,500 (14)	200 (1.1)	100 (0.6)	5,200 (29.1)	17,800 (100)	45,200	0.39
1976	100 (0.6)	2,300 (14.1)	4,100 (25.2)	600 (3.7)	2,900 (17.8)	400 (2.5)	0	5,900 (36.1)	16,300 (100)	47,800	0.34
1977	200 (1.1)	4,800 (26.2)	3,000 (16.4)	1,300 (7.1)	2,700 (14.8)	800 (4.4)	300 (1.6)	5,200 (28.4)	18,300 (100)	54,900	0.33
1978	100 (0.5)	4,100 (22.3)	5,600 (30.4)	1,100 (6)	2,000 (10.9)	600 (3.3)	200 (1.1)	4,700 (25.5)	18,400 (100)	70,700	0.26
1979	100 (0.5)	7,400 (36.6)	5,500 (27.2)	700 (3.5)	1,900 (9.4)	600 (3)	200 (1)	3,800 (18.8)	20,200 (100)	79,200	0.25
1980	not available	6,800 (50)	5,100 (37.5)	600 (4.4)	200 (1.5)	600 (4.4)	200 (1.5)	100 (0.7)	13,600 (100)	74,600	0.18

[a] Source: International Commission for the Conservation of Atlantic Tunas (ICCAT)
[b] Landings are in metric tons (MT). Percentage of total catch is given in parentheses.

period. Such trends would be associated with harvests in excess of growth (or recruitment) and may be symptomatic of overfishing. Have the stocks of Yellowfin, Skipjack and Bigeye been reduced below a level which would sustain maximum yield; that is, has biological overfishing occurred? Has fishing effort increased to the point where fishery rents have been dissipated; that is, has economic overfishing occurred? Before we present the econometric and numerical analysis which will address these questions, we will briefly review the basic bioeconomic model.

III. Bioeconomics

With the development of the maximum principle, economists gained a more powerful tool for analyzing dynamic allocation problems. This method saw immediate application to the theory of economic growth and subsequently to renewable and nonrenewable resources; and served to highlight the capital-theoretic aspects inherent in the management of resource stocks (Clark and Munro, 1975, and Clark, 1976).

For a single species fishery it is assumed that the resource can be adequately described by a single state variable X(t) representing biomass. The instananeous rate of change in biomass is given by

$$\dot{X} = \frac{dX(t)}{dt} = F(X(t)) - Y(t) \tag{1}$$

where X(t) is the time derivative of the fish stock (biomass), F(.) is net natural growth, and Y(t) is commercial harvest.

Let

$$\pi(t) = \pi(Y(t), X(t)) \tag{2}$$

represent the net revenues from commercial harvest Y(t). Net revenues would depend on fish stock if the cost of harvest depends on stock abundance. Maximization of the present value of net revenues would entail maximization of

$$\pi = \int_0^\infty \pi(Y(t), X(t))\, e^{-\delta t} dt \tag{3}$$

subject to the equation describing the change in biomass and an initial condition on the fish stock $X(0) = X_0$. The instantaneous discount rate is denoted by δ.

The current value Hamiltonian for this problem is

$$H(t) = \pi(Y(t), X(t)) + \mu(t)[F(X(t)) - Y(t)] \tag{4}$$

where $\mu(t)$ is the current value shadow price associated with an incremental change in the fish stock. The first order conditions for a maximum require

$$\frac{\partial H(t)}{\partial Y(t)} = \frac{\partial \pi(\cdot)}{\partial Y(t)} - \mu(t) = 0 \tag{5}$$

$$\dot{\mu}(t) = \frac{-\partial \pi(\cdot)}{\partial X(t)} + \mu(t)[\delta - F'(\cdot)] \tag{6}$$

$$\dot{X} = F(\cdot) - Y(t) \tag{7}$$

In steady state $\dot{\mu}(t) = \dot{X}(t) = 0$ and (5) and (6) imply

$$F'(\cdot) + \frac{\frac{\partial \pi(\cdot)}{\partial X}}{\frac{\partial \pi(\cdot)}{\partial Y}} = \delta \tag{8}$$

which is a fundamental equation for the basic bioeconomic model. The first term on the left hand side of equation (8) is the rate of change in net growth associated with an increment to the fish stock. The second term is referred to as the marginal stock effect. Together they sum to what has been called the resource's own rate of return. In steady state the optimal stock equates the resource's own rate of return to the market rate obtainable on other assets (Clark and Munro, 1975, p. 96).

The Gordon-Schaefer model presumes that equations (1) and (2) take the following form:

$$\dot{X} = \frac{dX(t)}{dt} = rX(t)[1 - X(t)/K] - Y(t) \tag{9}$$

and

$$\pi(t) = [p - \frac{c}{qX(t)}]\, Y(t) \quad . \tag{10}$$

Equation (9) is the logistic growth curve while (10) is the expression for net revenues which results when: (a) the per unit price for fish and the per unit cost for effort are constant and denoted by p and c respectively, and (b) the production function for the fishery is of the form

$$Y(t) = qE(t)X(t) \tag{11}$$

where E(t) is effort and q is referred to as the catchability coefficient.

For the Gordon-Schaefer model, equation (8) leads to a quadratic equation where the optimal stock X^* is the positive root and depends on the bioeconomic parameters c, p, q, δ, r, and K according to

$$X^* = \frac{K}{4}\left[\left(\frac{c}{qpK} + 1 - \frac{\delta}{r}\right) + \sqrt{\left(\frac{c}{qpK} + 1 - \frac{\delta}{r}\right)^2 + \frac{8c\delta}{qpKr}}\,\right] \quad . \tag{12}$$

Alternatively, one could define steady state in terms of the two equation system:

$$Y = \phi(X) = [\delta - r(1 - 2X/K)][X(qpX/c - 1)] \quad , \tag{13}$$

and

$$Y = rX(1 - X/K) \quad . \tag{14}$$

Equation (13) has been referred to as the "catch locus" (Gould, 1972), while equation (14) is the sustainable yield curve equating harvest to logistic growth. Three catch loci and a sustainable yield curve are drawn in Figure 2. The catch locus denoted as $\phi_1(X)$ might result from a combination of a high discount rate, low (stock insensitive) harvest costs, and high market value.

FIGURE 2. CATCH LOCI AND THE SUSTAINABLE YIELD CURVE IN THE SCHAEFER-GORDON MODEL

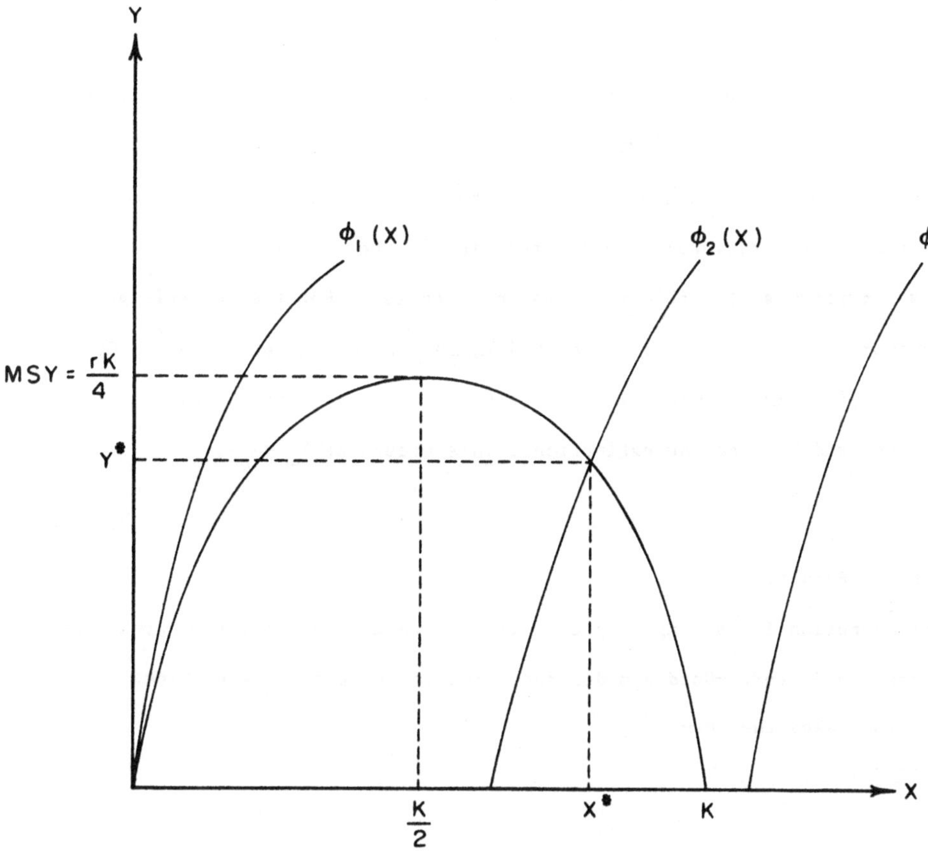

Under such circumstances, it may be optimal to harvest the resource to extinction.[3/] Locus $\phi_2(X)$ shows a situation where the marginal stock effect is greater than the discount rate. It is optimal to maintain a stock in excess of X_{MSY} ($X^* > \frac{K}{2}$ for logistic growth) because of the reduced harvest cost associated with larger stocks. Finally, $\phi_3(X)$ might correspond to a situation of high harvest cost and low market price, making commercial harvest unprofitable ($X^* = K$, its environmental maximum).

In addition to the bioeconomic optimum occurring at the intersection of a catch locus and the sustainable yield curve, it will be useful to note two other equilibria: maximum sustainable yield and open access. For the logistic growth model, maximum sustainable yield is denoted $Y_{MSY} = \frac{rK}{4}$ occurring at $X_{MSY} = \frac{K}{2}$. Open access equilibrium occurs when $\pi(\cdot) = 0$ (fishery rents are dissipated). For positive stock and harvest (no extinction), this occurs at $X_\infty = \frac{c}{pq}$.

IV. Empirical Results

A yield function for a single species fishery relates equilibrium harvest to effort. For the Schaefer-Gordon model specified in the preceding section, the yield function takes the form:

$$Y = qKE(1 - qE/r) \quad . \tag{15}$$

or

$$U = \alpha - \beta E \tag{16}$$

where U is catch per unit effort, $\alpha = qK$, and $\beta = q^2K/r$. Estimates of α and β can be obtained using ordinary least squares and the data contained in Tables 1 - 3, although care must be taken to avoid certain econometric pitfalls (see Uhler, 1979).

3/ Clark (1976, p. 61) shows that extinction is optimal if both $p \geqslant c(0)$ and $\delta > 2F'(0)$, where c(0) denotes the cost of harvesting the last surviving member of the population.

TABLE 4: ESTIMATES OF α, β, q, r, and K FOR TUNA IN THE EASTERN TROPICAL ATLANTIC

SPECIES	$\hat{\alpha}$	$\hat{\beta}$	$\hat{q}$ ($\times 10^{-2}$)	$\hat{r}$	$\hat{K}$ ($\times 10^3$ MT)	R^2	d
YELLOWFIN	4.8188 (21.21)[a]	0.0513 (8.30)	1.372	1.2883	351.2244	0.852	2.07
SKIPJACK	3.2853 (16.15)	0.0260 (5.57)	1.240	1.5686	264.9435	0.721	2.08
BIGEYE	1.0266 (15.86)	0.0114 (8.13)	2.110	1.9018	48.6540	0.846	1.94

[a] Values in parentheses, below estimates of α and β, are t-ratios.

Estimates for α, β, q, r, and K, along with supporting statistics, are given in Table 4 for Yellowfin, Skipjack, and Bigeye Tuna in the ETA. The Gordon-Schaefer specification would seem to conform well to the data. Estimates of α and β are of the expected sign and significant for all three species. The Durban-Watson statistics do not indicate autocorrelation.

Estimates for the cost of a standard day at sea in the ETA were not available. Recall that the effort of small purse seiners, baitboats and longliners had been converted to large seiner equivalents by using fishing power or daily-catch-rate weighting factors. The cost of operating a large purse seiner in the Eastern Tropical Pacific (ETP) has been examined by Flagg (1977), and baitboat costs in the ETP were estimated by the U.S. Bureau of Commercial Fisheries (1970). As with land-based firms, both fixed and variable cost components are present. Fixed costs occur regardless of the level of fishing effort, and include interest charges on vessel, equipment and gear, insurance premiums, depreciation, moorage and certain maintenance. Variable costs are associated with fishing and would include such items as fuel, oil, food and other maintenance. Estimates of cost per day for large seiners in the ETP were adjusted for general inflation and the more rapid escalation in fuel prices. Bioeconomic and open access equilibria were then calculated for \$500.00 cost increments for $c = \$2000/SD$ to $c = \$3500/SD$ in 1980 dollars.

The market prices for Yellowfin, Skipjack and Bigeye Tuna are recorded by the National Marine Fisheries Service (NMFS) Market News Service at Terminal Island, California. Yellowfin and Bigeye Tuna fetch the same price, while Skipjack prices were \$50 to \$100 less per metric ton during the 1967-80 period. The 1980 average monthly prices for Yellowfin/Bigeye and for Skipjack were \$1300 and \$1200 respectively.

Sensitivity of the bioeconomic optimum was also tested with regard to variation in the discount rate δ. It was varied in 0.05 increments from $\delta = 0.00$ to $\delta = 0.20$. Tables 5 and 6 show maximum sustainable yield (MSY), bioeconomic and open access equilibria for Yellowfin and Skipjack Tuna in the ETA. For both

TABLE 5: MSY, BIOECONOMIC, AND OPEN ACCESS EQUILIBRIA FOR YELLOWFIN IN THE ETA

Yellowfin Parameters: $p = \$1300$, $q = 1.372 \times 10^{-2}$, $r = 1.2883$, $K = 351.2244$
Maximum Sustainable: $X_{MSY} = 175.56$, $Y_{MSY} = 113.12$ $E_{MSY} = 46.9495$

	δ	c = \$2,000	c = \$2,500	c = \$3,000	c = \$3,500
BIOECONOMIC EQUILIBRIA	0.00	X* = 231.68	X* = 245.70	X* = 259.71	X* = 273.73
		Y* = 101.59	Y* = 95.10	Y* = 87.18	Y* = 77.81
		E* = 31.96	E* = 28.21	E* = 24.47	E* = 20.72
	0.05	X* = 228.21	X* = 242.81	X* = 257.35	X* = 271.83
		Y* = 102.97	Y* = 96.56	Y* = 88.61	Y* = 79.16
		E* = 32.89	E* = 28.98	E* = 25.10	E* = 21.23
	0.10	X* = 224.85	X* = 240.02	X* = 255.07	X* = 270.00
		Y* = 104.23	Y* = 97.90	Y* = 89.96	Y* = 80.44
		E* = 33.79	E* = 29.73	E* = 25.71	E* = 21.71
	0.15	X* = 221.58	X* = 237.32	X* = 252.87	X* = 268.24
		Y* = 105.37	Y* = 99.15	Y* = 91.23	Y* = 81.65
		E* = 34.66	E* = 30.45	E* = 26.30	E* = 22.19
	0.20	X* = 218.41	X* = 234.71	X* = 250.74	X* = 266.54
		Y* = 106.40	Y* = 100.31	Y* = 92.42	Y* = 82.80
		E* = 35.51	E* = 31.15	E* = 26.87	E* = 22.64
OPEN ACCESS	$\delta \to \infty$	X_∞ = 112.13	X_∞ = 140.17	X_∞ = 168.20	X_∞ = 196.23
		Y_∞ = 98.34	Y_∞ = 108.51	Y_∞ = 112.92	Y_∞ = 111.56
		E_∞ = 63.92	E_∞ = 56.43	E_∞ = 48.93	E_∞ = 41.44

TABLE 6: MSY, BIOECONOMIC, AND OPEN ACCESS EQUILIBRIA FOR SKIPJACK IN THE ETA

Skipjack Parameters: p = \$1200, q = 1.240×10^{-2}, r = 1.5686, K = 264.9435
Maximum Sustainable: X_{MSY} = 132.4718, Y_{MSY} = 103.9005, E_{MSY} = 63.2518

	δ	c = \$2,000	c = \$2,500	c = \$3,000	c = \$3,500
BIOECONOMIC EQUILIBRIA	0.00	X* = 199.68 Y* = 77.16 E* = 31.16	X* = 216.48 Y* = 62.12 E* = 23.14	X* = 233.28 Y* = 43.73 E* = 15.12	X* = 250.08 Y* = 22.01 E* = 7.10
	0.05	X* = 198.32 Y* = 78.23 E* = 31.81	X* = 215.55 Y* = 63.04 E* = 23.59	X* = 232.71 Y* = 44.41 E* = 15.39	X* = 249.83 Y* = 22.35 E* = 7.22
	0.10	X* = 196.99 Y* = 79.25 E* = 32.44	X* = 214.64 Y* = 63.92 E* = 24.02	X* = 232.17 Y* = 45.05 E* = 15.65	X* = 249.59 Y* = 22.68 E* = 7.33
	0.15	X* = 195.71 Y* = 80.22 E* = 33.06	X* = 213.77 Y* = 64.77 E* = 24.44	X* = 231.64 Y* = 45.68 E* = 15.90	X* = 249.36 Y* = 23.01 E* = 7.44
	0.20	X* = 194.46 Y* = 81.15 E* = 33.65	X* = 212.91 Y* = 65.59 E* = 24.84	X* = 231.12 Y* = 46.28 E* = 16.15	X* = 249.13 Y* = 23.32 E* = 7.55
OPEN ACCESS	$\delta \to \infty$	X_∞ = 134.40 Y_∞ = 103.88 E_∞ = 62.33	X_∞ = 168.01 Y_∞ = 96.42 E_∞ = 46.28	X_∞ = 201.61 Y_∞ = 75.59 E_∞ = 30.24	X_∞ = 235.22 Y_∞ = 41.40 E_∞ = 14.19

TABLE 7: MSY, BIOECONOMIC, AND OPEN ACCESS EQUILIBRIA FOR BIGEYE IN THE ETA

Bigeye Parameters: p = \$1300, $q = 2.11 \times 10^{-2}$, r = 1.9018, K = 48.6540
Maximum Sustainable: X_{MSY} = 24.3270, Y_{MSY} = 23.1332, E_{MSY} = 45.0667

	δ	c = \$400	c = \$500	c = \$600	c = \$700
BIOECONOMIC EQUILIBRIA	0.00	X* = 31.62 Y* = 21.05 E* = 31.56	X* = 33.44 Y* = 19.86 E* = 28.18	X* = 35.26 Y* = 18.46 E* = 24.80	X* = 37.09 Y* = 16.77 E* = 21.43
	0.05	X* = 31.28 Y* = 21.24 E* = 32.19	X* = 33.15 Y* = 20.09 E* = 28.72	X* = 35.02 Y* = 18.66 E* = 25.25	X* = 36.89 Y* = 16.96 E* = 21.79
	0.10	X* = 30.94 Y* = 21.42 E* = 32.81	X* = 32.87 Y* = 20.28 E* = 29.24	X* = 34.79 Y* = 18.85 E* = 25.69	X* = 36.70 Y* = 17.15 E* = 22.15
	0.15	X* = 30.61 Y* = 21.59 E* = 33.42	X* = 32.60 Y* = 20.46 E* = 29.75	X* = 34.56 Y* = 19.04 E* = 26.11	X* = 36.51 Y* = 17.33 E* = 22.50
	0.20	X* = 30.29 Y* = 21.74 E* = 34.02	X* = 32.33 Y* = 20.63 E* = 30.25	X* = 34.34 Y* = 19.22 E* = 26.53	X* = 36.33 Y* = 17.51 E* = 22.83
OPEN ACCESS	$\delta \to \infty$	X_∞ = 14.58 Y_∞ = 19.42 E_∞ = 63.12	X_∞ = 18.23 Y_∞ = 21.68 E_∞ = 56.36	X_∞ = 21.87 Y_∞ = 22.90 E_∞ = 49.61	X_∞ = 25.52 Y_∞ = 23.08 E_∞ = 42.86

species we note that the optimal stock decreases with increases in the discount rate (δ) and increases with increases in the unit cost of effort (c). The marginal stock effect is positive and of a greater order of magnitude than the discount rate. Thus, the bioeconomic equilibria occur at stock levels in excess of X_{MSY}. These equilibria are similar to (X*, Y*) at the intersection of $\phi_2(\cdot)$ and the sustainable yield curve in Figure 2.

For Skipjack Tuna both bioeconomic and open access equilibria occur at stocks in excess of X_{MSY}. Thus biological overfishing neither is optimal nor results from open access status. Open access stocks are less than the bioeconomic (X_∞<X*) for all finite values of δ.

For Bigeye Tuna the initial cost vector led to zero fishing (X = K) for both bioeconomic and open access equilibria. The catch locus, similar to $\phi_3(\cdot)$ in Figure 2, did not intersect the sustainable yield curve at a positive yield. Recall that the predominant source of Bigeye harvests was from longliners. Longline vessels are smaller, and the "passive" nature of the fishing technology employed makes them less costly to operate than the larger purse seiners which require considerable power when hauling back after setting the seine. In the conversion to large seiner equivalents the catch-per-day weighting procedure may result in an overestimation of fishing costs for longliners and baitboats within the Bigeye fishery. Revenues from longline trips were examined, and estimates for the unit cost of effort were based on a fraction (about 75 percent) of gross receipts, leaving the remaining portion of gross revenues to cover other fixed costs. This procedure yielded an estimate of $517 as the cost of an equivalent day. Bracketing this cost estimate with the vector c = [$400, $500, $600, $700] leads to the MSY, bioeconomic and open access equilibria shown in Table 7. Again, we see significant marginal stock effects with X*>X_{MSY} for all combinations of δ and c. As with Yellowfin, open access status leads to equilibria with stocks less than MSY for all but the highest cost estimate (c = $700).

In comparing the estimates of MSY to the time series for total landings contained in Tables 1 through 3, one observes landings rates for Yellowfin and

Skipjack which would be associated with stock reductions from $X(t)=K$ toward $X(t) = X_{MSY} = K/2$. While these species would appear to be economically overfished, they would not appear to be biologically overfished.[4/] For Bigeye, however, an examination of landings relative to MSY would indicate a movement from $X(t) = K$ to $X(t) < K/2$, with stocks considerably below all bioeconomic optima. Thus, both biological and economic overfishing would seem to have occurred during the 1967-80 period, and neither biologists nor economists would be sanguine about the current status of Bigeye stocks.

V. Management Policies

In managing fish stocks, particularly transboundary fish stocks such as salmon and tuna, distributional issues often overshadow efficiency issues. This is most certainly the case for tuna in the ETP, where extended jurisdiction by Latin American countries has led to seizure of U.S. registered vessels and confiscation of catch and seine (which may take $300,000 and several months to replace). Distributional issues in the ETA have not reached the level of international controversy and rancor found in the ETP, but questions are being raised by West African nations as to how they might secure a greater share of the wealth generated by tuna stocks migrating through their coastal waters.

Economists have strongly recommended transferable quotas on the grounds that they are: (a) efficient, in the sense that the aggregate quota would be harvested at least cost; and (b) flexible, in the sense that they may be distributed initially to achieve any agreed-upon distribution of prospective fishery rents (Molony and Pearse, 1979). In addition, Clark (1980) has noted that individual quotas can be used in conjunction with a system of landings taxes to allow recapture of some proportion of fishery rents to help defray the costs of resource management and research. In particular, Clark notes that at a

---

4/ Again, it is important to qualify these conclusions in light of the research by Clark and Mangel (1979).

bioeconomic optimum

$$P_* + \tau = \mu \ , \tag{17}$$

where P_* is the equilibrium price emerging from the market for transferable quotas, τ is the landings tax rate, and μ is the current value shadow price defined by

$$\mu = [p - \frac{c}{qX^*}] \ . \tag{18}$$

For Yellowfin Tuna with c = \$3000 and $\delta = 0.10$, we saw that $X^* = 255.07 \times 10^3$MT, $Y^* = 89.96 \times 10^3$MT, and $E^* = 25.71 \times 10^3$SDs. The current value shadow price would be μ = \$442.75/MT. Suppose the aggregate quota of Y^*=89.96x10^3MT was distributed among a group of West African and foreign flag nations according to some formula. Suppose further that ICCAT levied a \$20/MT landings tax (τ = \$20/MT). Then the market quota price would be \$422.75/MT and ICCAT would generate \$1,799,200 to support administration, enforcement and research.

The formula for distributing the transferable quotas and the selection of a landings tax rate would undoubtedly be the subject of considerable debate. The historical share of landings by a country, joint venture or alliance would presumably influence the quota share formula. The formula could be revised periodically to reflect changes in the ability and interest of West African and foreign flag nations to harvest tuna in the ETA. The smaller West African countries with only artisanal fisheries and no previous commercial (offshore) harvesting capacity might be allocated a share of the total quota. Such countries would presumably sell their quotas to countries or companies who wish to harvest more than their initial allocation. The proceeds from sale of their quotas could be used to finance commercial vessels, thereby developing an ability to harvest offshore stocks in the future, or they may be directed toward other resource development or social projects.

VI. Conclusions and Caveats

A simple bioeconomic model (the Gordon-Schaefer model) was estimated with data for three species of tuna in the Eastern Tropical Atlantic. Statistically, the model conformed well to the data and afforded estimates of maximum sustainable yield, bioeconomic, and open access equilibria. Strong marginal stock effects were identified in all three fisheries, resulting in bioeconomic optima with stocks in excess of X_{MSY}. While all three fisheries appear to have been economically overfished, (fishery rents being driven toward zero), Yellowfin and Skipjack stocks do not appear to be biologically overfished (stocks would appear to be at or slightly above $X = K/2$). For Bigeye Tuna there was strong indication of both economic and biological overfishing, and this stock may warrant special management attention from the International Commission for Conservation of Atlantic Tuna (ICCAT).

A management program based on both transferable quotas and landings taxes would promote the least cost harvest of optimal yield and afford a flexible mechanism for distributing potential fishery rents. Rather modest landings taxes seemed capable of generating sufficient revenues to allow for administration, enforcement and research by a management authority.

Biomass models, such as the Schaefer-Gordon model, are not capable of incorporating age or sex-specific characteristics of a fish population. Where these characteristics are important, a multiple cohort model (Conrad 1982) or sex-selective model (Clark and Tait, 1982) would be required.

It is further the case that the three species of tuna, treated independently in this paper, may compete for a common food source. Models with interspecific competition and multi-trophic level predation (including harvesting by man) are complex on a purely biological basis, making bioeconomic analysis a formidable undertaking with only limited progress to date (May et al., 1979).

In light of these and other extenuating factors the empirical results and conlusions presented here should be regarded as preliminary. They are, hopefully, a useful first step; one which places the Eastern Tropical Atlantic tuna fisheries within a bioeconomic perspective and will help to define future management, distributional, and research issues.

LITERATURE CITED

Bureau of Commercial Fisheries, Division of Economic Research. (1970). "Basic Economic Indicators: Tuna." Working Paper No. 61.

Clark, C. W. (1976). Mathematical Bioeconomics: The Optimal Management of Renewable Resources. New York: John Wiley & Sons.

________. (1980). Towards a predictive model for the economic regulation of commercial fisheries. Can. J. Fish. Aqua. Sci. 37:1111-1129.

Clark, C. W. and Munro, G. R. (1975). The economics of fishing and modern capital theory: A simplified approach. J. Environ. Econ. Manag. 2:92-106.

Clark, C. W. and Mangel, M. (1979). Aggregation and fishery dynamics: A theoretical study of schooling and the purse seine tuna fisheries. Fishery Bull. 77:317-337.

Clark, C. W. and Tait, T. E. (1982). Sex-selective harvesting of wildlife populations. Ecological Modelling 14:251-260.

Conrad, J. M. (1982). Management of a multiple cohort fishery: The hard clam in Great South Bay. Amer. J. Agr. Econ. 64(3):463-474.

Flagg, V. G. (1977). "Alternative Management Plans for Yellowfin Tuna in the Eastern Tropical Pacific." Special Report, Center for Marine Studies, San Diego State University.

Fonteneau, A. and Cayre, P. (1981). Analyse de l'Etat des Stocks d'Albacore (*Thunnus albacares*) et de Listao (*Katsuwonus pelamis*) de l'Atlantique au 30 Septembre 1980. ICCAT Collective Volume of Scientific Papers Vol. XV (SCRS-1980), SCRS/80/57:99-107.

Gordon, H. S. (1954). Economic theory of a common-property resource: The fishery. J. Polit. Econ. 62:124-142.

Gould, J. R. (1972). Extinction of a fishery by commercial exploitation: A note. J. Polit. Econ. 80:1031-1038.

May, R. M. et al. (1979). Management of multispecies fisheries. Sci. 205(4403): 267-277.

Moloney, D. G. and Pearse, P. H. (1979). Quantitative rights as an instrument for regulating commercial fisheries. J. Fisheries Research Board of Canada 36:859-866.

Schaefer, M. B. (1957). Some considerations of population dynamics and economics in relation to the management of marine fisheries. J. Fisheries Research Board of Canada 14:669-681.

Scott, A. D. (1955). The fishery: The objectives of sole ownership. J. Polit. Econ. 63:116-124.

Uhler, R. S. (1979). Least squares regression estimates of the Schaeffer production model: Some Monte Carlo simulation results. Can. J. Fish. Aqua. Sci. 37:1284-1294.

THE MULTISPECIES FISHERIES PROBLEM:

A CASE STUDY OF GEORGES BANK[1]

by

M.P. Sissenwine, B.E. Brown

M.D. Grosslein and R.C. Hennemuth

National Marine Fisheries Service

Northeast Fisheries Center

Woods Hole, MA 02543 USA

Abstract

A multispecies data base is the foundation of the Georges Bank fisheries research. Fisheries statistics describe the amount of each species caught, location of capture and fishing effort for most fishing trips to Georges Bank. Samples are taken from the catch in order to estimate its size and age composition.

Standardized research vessel bottom trawl surveys of the Georges Bank region have been conducted, at least annually, since autumn 1963. A stratified random sampling design is applied. The coefficient of variation of the sample mean is about 30% for the primary demersal species (e.g., haddock, cod, yellowtail flounder).

During bottom trawl surveys, fish stomachs are collected. Over the last two decades, the content of tens of thousands of stomachs of about 80 species has been determined quantitatively.

Technological interactions (simultaneous capture of multiple species) prohibit achieving target catches, based on single species conservation criteria, of all of the individual species of Georges Bank. Linear programming was used to determine

[1]The common names of fishes, referred herein, are based on American Fisheries Society nomenclature (Robins 1980).

the "optimum" combination of catches that would not violate catch constraints of any individual species. Cluster analysis was used to define assemblages of species which are treated as a multispecies fishery unit.

Feeding habits data indicate the potential for biological interactions. The estimate of MSY for the aggregate of the finfish and squid of the region is lower than the sum of the estimates of MSY for individual species, possibly as a result of biological interactions. Yet an empirical approach fails to demonstrate, explicitly, biological interactions. Nevertheless, the energy budget of Georges Bank indicates that predation by a few species of fish are a major cause of the mortality of prerecruits.

Introduction

The application of mathematics to fisheries flourished during the era of Beverton and Holt, Ricker, and Schaefer. During the 1950's, surplus production models (Schaefer 1954, 1957), stock recruitment models (Ricker 1954, 1958; Beverton and Holt 1957), and yield-per-recruit analysis (Beverton and Holt 1957, Ricker 1958) were applied. The groundwork was laid for virtual population analysis (VPA) which followed shortly (Gulland 1965, Murphy 1965).

The models developed during the 1950's were applied to individual species. Biological (e.g., competition and predator-prey) and technological (e.g., simultaneous harvesting of multiple species) interactions were ignored. Like elsewhere, Georges Bank (Figure 1) fisheries research of the 1950's and early 1960's was single-species oriented. This oversimplification was partially justified since a relatively few species were exploited and interacting species were usually of little economic interest.

Nevertheless, during the same era the seeds of a multispecies renaissance were germinated. Rounsefell (1957) defined multispecies assemblages of groundfish on Georges Bank. Edwards (1958) studied a multispecies industrial fishery and its effect on single species food fisheries. McHugh (1959) called for "en masse" management, rather than as a series of distinct species. Galtsoff (1962) wrote, "Life in the ocean does not remain stable. Contrarily, it is in a state of unstable equilibrium in which the struggle for existence gives temporary

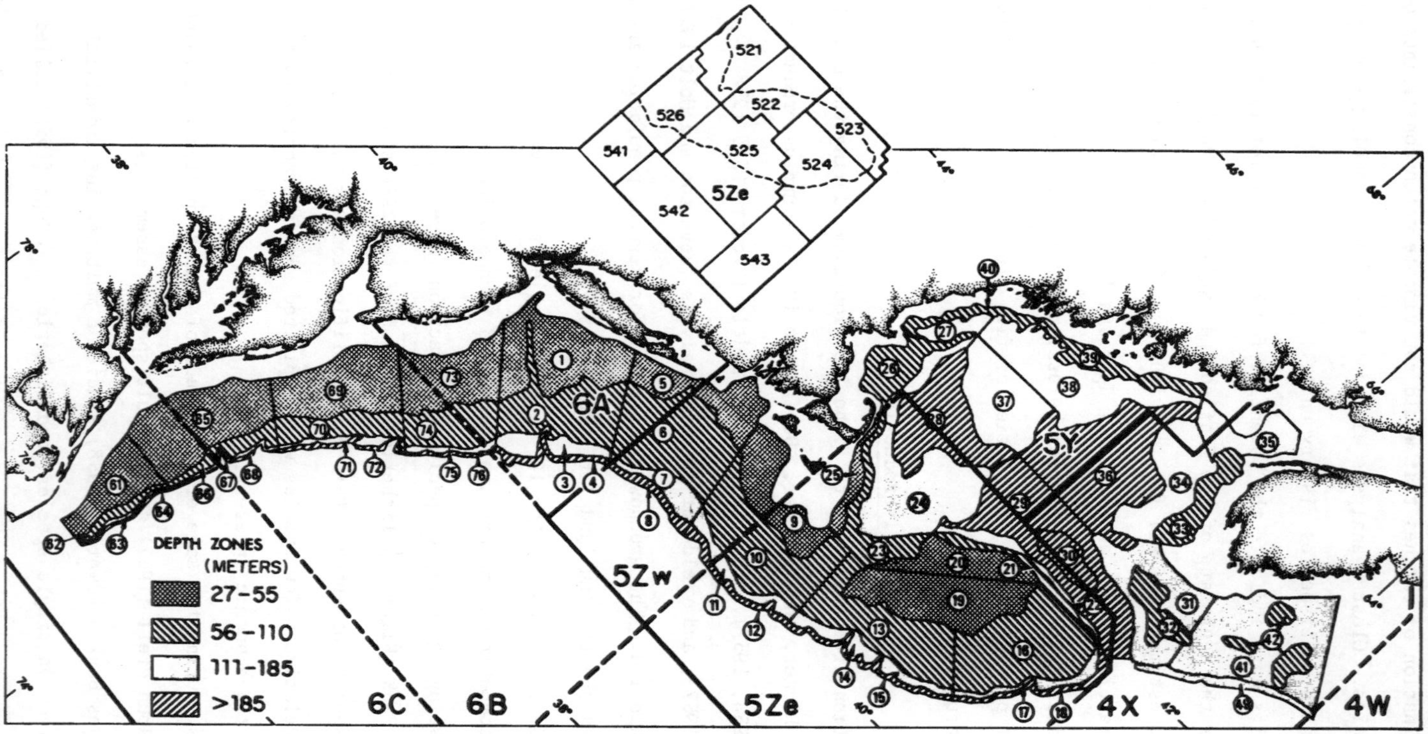

Figure 1. Georges Bank region, from Cape Hatteras to Nova Scotia, with bottom trawl survey strata, NAFO-ICNAF statistical areas and USA statistical areas for Georges Bank (S.A. 5Ze).

predominance to one group which in turn may replace another."

During the 1960's and 1970's, the character of the fishery of the Georges Bank region changed markedly. The total fishing effort and yield of the region increased sharply. Virtually all species were harvested. The biomass of finfish and squid off the northeast coast of the USA (including Georges Bank) was reduced by about 50% (Clark and Brown 1977).

The character of fisheries research changed concurrently with the change in the fishery. Data collection schemes were generalized to describe the entire multispecies fishery. Technological interactions were accounted for by calculating the multispecies yield vector which maximized total yield within individual species constraints. Extensive stomach contents studies were conducted in order to address biological interactions. Multispecies analogs of traditional single species fishery models were developed and/or applied. This paper provides a brief overview of multispecies aspects of the fisheries research of Georges Bank.

Multispecies Fisheries Data Base

"Fisheries statistics" are the foundation of fisheries research. These are the data which describe how much of each species is caught, its age and/or size composition, the location of capture, and the amount of fishing effort (in a form which can be related to fishing mortality rate). Burns et al. (In Press) describe the fisheries statistics collection program for the northeast region of the USA, including Georges Bank.

There is a more or less continuous record of landings for some important species (e.g., haddock, cod) since the beginning of the twentieth century. The fisheries statistics collection program for the Georges Bank region was gradually expanded until by the early 1960's landings data for all species were collected on a fishing trip by fishing trip basis.

The data is collected from the dealers purchasing the fish and from vessel captains. The data base includes (1) date of sale, (2) amount of each species landed, (3) price per pound, (4) fishing effort (e.g., days absent from port, number of trawl hauls, and average duration of trawl hauls), and (5) the location of capture. The catch is divided among statistical areas as defined in Figure 1. For those trips when vessel captains are actually interviewed (approximately 70% of the trips to Georges Bank), harvesting location is further specified to 10-minute squares of latitude and longitude.

Catch statistics of foreign fishing vessels are reported to the Northwest Atlantic Fisheries Organization (NAFO), formerly the International Commission for Northwest Atlantic Fisheries (ICNAF). NAFO fisheries statistics for Georges Bank correspond to statistical area 5Ze (Figure 1).

The size and age composition of the catch is determined by sampling. Each year the length of tens of thousands of fish is measured and scales or otoliths are collected for age determinations. The aim of this data collection is to determine the annual age composition of the catch. The best example of the results is for Georges Bank haddock (Table 1, Clark et al. 1982). This data has been used to estimate annual population size and annual age specific fishing mortality rate via application of VPA. Similar data describing the age composition of the catch of mackerel, sea herring, and silver hake are also available, although these time series are shorter. There is also incomplete data describing the size and age composition of the catch of numerous other species.

Fisheries statistics alone are not enough to solve the multispecies fishery problem. Many species are not landed. Some of these species are involved in biological and technological (e.g., they are unintentionally caught and discarded at sea) interactions. Small, young (prerecruit) fish pass through the mesh of commercial fishing gear. Thus, they are not represented by fisheries statistics. Furthermore, commercial fishing effort data is difficult to interpret because of the large number of harvesting units and because of technological improvements that increase efficiency. Therefore, fisheries statistics are not always adequate to monitor trends in population size.

Because of the limitations of fisheries statistics, a standardized research vessel bottom trawl survey of the Georges Bank Region was initiated in the autumn of 1963 (Grosslein 1969, Azarovitz 1981). In 1968 a spring survey was initiated. There have also been occasional summer and winter surveys. A stratified random sampling design is applied with sample allocation proportional to stratum area.

Strata are defined by depth and latitude (Figure 1).

All of the autumn surveys have been conducted with otter trawls of the same standard design. Two different standard otter trawl designs have been used during spring. Sissenwine and Bowman (1978) estimated the relative fishing power (i.e., capture efficiency) of these two trawl designs.

Table 1. Estimated age composition of haddock in commercial landings (all countries) from Georges Bank (Div. 5Ze), 1931-1979, Clark et al. (1982).

	Number of fish caught (thousands)										Nominal catch	Calculated weight[b]	Ratio Nom./
Year	1	2	3	4	5	6	7	8	9+	Total	(tons)	(tons)	Calc.
1931[a]	1,755	8,801	2,041	5,785	9,100	6,045	3,380	1,794	559	39,260	59,486	59,739	1.00
1932	118	2,084	25,871	2,421	3,676	2,894	1,320	664	391	39,439	54,512	54,552	1.00
1933	244	8,476	6,023	10.046	2,092	1.579	1,210	538	647	30.855	42,215	42,161	1.00
1934	341	4,454	5,414	3,734	3,149	1,051	619	250	168	19,180	25,795	25,590	1.01
1935	1.197	11.872	8,819	3,706	2,944	2,458	499	442	109	32,046	40.944	40,588	1.01
1936	880	12,327	11,486	5,431	2,141	1,377	1,362	259	124	35,387	43,445	43,428	1.00
1937	1,288	11,034	10.910	5,629	4,143	1,875	952	481	222	36,534	49,359	49,402	1.00
1938	1,030	20.199	7,755	3,755	2,113	1,600	945	327	173	37,897	47,773	47,691	1.00
1939	607	13,937	19,617	5,163	2,152	967	837	326	239	43,845	54,054	54,058	1.00
1940	2,040	7,254	12,317	8,253	2,510	1,479	752	222	136	34,963	47,906	47,913	1.00
1941	780	23,464	9,808	8,033	5,764	1,781	941	307	384	51,262	62,944	62,980	1.00
1942	310	14,307	16,348	6,531	3,996	2,331	1,036	227	176	45,262	55,376	55,409	1.00
1943	19	4,191	17,738	8,364	3,102	2,693	790	354	178	37,429	46,323	46,376	1.00
1944	64	761	8,437	14,843	5,689	2,281	497	469	108	33,149	49,637	49,667	1.00
1945	121	8,522	2,029	6,386	5,795	2,315	914	265	205	26,552	40,473	40,443	1.00
1946	209	7,466	15,213	2738	5,785	3,840	1,827	272	23	37,373	53,719	53,683	1.00
1947	90	16,621	10.334	7,181	2,127	2,739	1,501	745	457	41,795	54,431	54,376	1.00
1948	80	11,227	19,237	5,116	2,744	1.157	780	450	369	41,160	48,360	48,303	1.00
1949	328	6,472	12,479	9,608	2,347	1.061	624	409	353	33,681	42,254	42,500	0.99
1950	88	28,971	4,107	4,272	3,315	1,131	520	225	250	42,879	41,273	41,255	1.00
1951	645	8,266	26,472	2,177	2,448	2,138	740	297	215	43,398	47,318	47,422	1.00
1952	—	25,120	8,892	8,485	1,361	944	530	182	107	45,621	43,252	43,349	1.00
1953	1,083	1,807	17,588	5,726	3,757	1,012	542	337	152	32,004	35,926	35,930	1.00
1954	108	31,858	5.107	5,611	2,315	2,131	720	353	98	48,301	46,388	46,401	1.00
1955	90	3,941	19,251	3,316	3,278	1,649	1,068	320	173	33,086	40,851	40,881	1.00
1956	52	11,948	6,698	12,066	3,405	3,378	1,348	563	201	39,659	51,144	52,284	0.98
1957	35	6,594	14,046	4,523	5,822	2,357	1,630	473	366	35,846	48,561	48,847	0.99
1958	125	5,571	7,088	6,665	3,784	2,366	903	442	142	27,086	37,322	37,761	0.99
1959	94	5,716	7,994	5,169	3,934	1,758	1,172	424	334	26,595	36,051	37,994	0.95
1960	258	16,010	6,122	4,562	3,067	1,792	787	406	348	33,352	40,877	42,930	0.95
1961	62	10,689	14,927	4,198	2,917	1,856	1,266	496	674	37,085	46,650	48,522	0.96
1962	74	4,455	16,245	10,440	3,448	2,089	1,566	1,185	898	40,400	54,004	56,430	0.96
1963	2,910	4,047	7,418	11,152	8,198	2,205	1,405	721	1,096	39,152	54,846	57,731	0.95
1964	10,101	15,935	4,554	4,776	8,722	5,794	2,082	1,028	1,332	54,324	64,086	67,823	0.94
1965	9,601	125,818	44,496	5,356	4,391	6,690	3,772	1,094	1,366	202,584	150,362	181,774	0.83
1966	114	6,843	100,810	19,167	2,768	2,591	2,332	1,268	867	136,760	121,274	140,715	0.86
1967	1,150	168	2,891	20,667	10,338	1,209	993	917	698	39,031	51,469	52,065	0.99
1968	8	2,994	709	1,921	14,519	3,499	677	453	842	25,622	40,923	41,018	1.00
1969	2	11	1,698	448	654	5,954	1,574	225	570	11,136	22,252	22,336	1.00
1970	46	158	16	570	186	214	2,308	746	464	4,708	11,300	12,376	0.91
1971	—	1,375	223	40	289	246	285	1,469	928	4,855	10,862	11,998	0.91
1972	156	2	450	81	32	120	78	66	1,236	2,221	5,733	6,464	0.89
1973	2,560	2,057	3	386	53	30	77	15	447	5,628	5,331	6,790	0.79
1974	46	1,820	657	2	70	2	2	53	249	2,901	4,290	4,647	0.92
1975	192	1,034	1,864	375	4	42	4	4	88	3,607	5,420	5,545	0.98
1976	144	473	550	880	216	—	23	4	112	2,402	4,324	4,287	1.01
1977	—	6,130	187	680	515	357	4	39	111	8,023	10,843	11,582	0.94
1978	—	761	11,315	305	567	517	139	14	67	13,685	22,339	22,403	1.00
1979	—	26	1,726	7,169	525	410	315	96	46	10,313	19,461	21,100	0.92

[a] Data for 1931–55 reported in terms of "biological year" (February–January); data for 1956–79 reported by calendar year.

[b] Obtained by multiplying numbers caught at age by corresponding mean weight-at-age data in Table 7.

About 300 to 500 trawl hauls are made during each survey, with approximately 30% of the stations on Georges Bank. Virtually all demersal species and most important pelagic species (e.g., herring, mackerel, squid) are captured. Since a small mesh liner (1.25 cm) is used, prerecruits are captured as well. Since the distribution of most fish populations is highly contagious, the results are subject to considerable variance.

In Figure 2, the natural logarithm of the variance of the mean is plotted against the natural logarithm of the mean for Georges Bank haddock. The least squares fit of the data indicates a slope of about two (actually 1.886). This implies that the coefficient of variation (cv) of the sample mean is independent of the magnitude of the sample mean (or population size). This result appears to be quite general. A slope of about two is obtained for virtually every species captured by the survey (Hennemuth 1976, Sissenwine 1978).

The cv of the sample mean of each species is given by the exponential of one-half of the intercept (if the slope is approximately two). For Georges Bank haddock, the cv is about 30%. A similar cv applies to other important demersal species (e.g., cod, yellowtail flounder; Grosslein 1971). The cv is higher for minor components of the biomass and especially for schooling pelagics like mackerel (cv equals about 700%). Nevertheless, since many of the species of the Georges Bank region have exhibited a one or two order of magnitude range in abundance, the survey is clearly useful to monitor trends. Furthermore, since abundance is a function of several years' recruitment, the information content of a series of surveys is useful for monitoring.

The cv of the sample mean of each survey reflects a population's spatial distribution. Unfortunately, there is another source of error which remains unestimated. Differences in environmental conditions between surveys affect the performance of the sampling gear and the distribution of the fish. Thus, the results of any survey may have been different if conducted under different environmental conditions, even with the same population size. Since random sampling with respect to environmental conditions is impossible, an estimate of the environmental component of variance is not obtainable from the data of a single

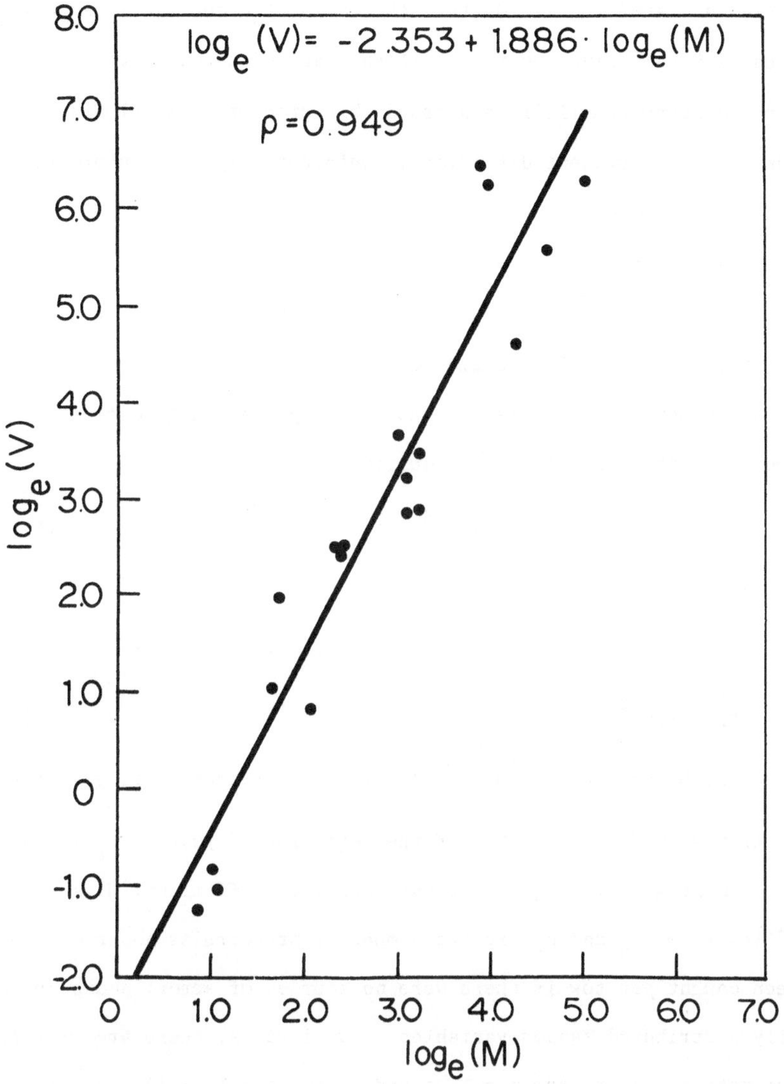

Figure 2. Mean catch per tow and variance of the mean, on $\log_e$ - $\log_e$ scale, for Georges Bank haddock bottom trawl survey data.

survey (Byrne et al. 1981, Sissenwine et al. 1983). More sophisticated statistical methods are required.

Trawl survey data for Southern New England yellowtail flounder demonstrate the importance of environmental variability (Figure 3). The mean catch per tow during 1972 was the highest ever. Yet, the commercial fishery was failing and there was little evidence of recruitment during the previous year. Thus, the 1972 result was viewed as "anomalous." On the other hand, the decline in abundance of 1966 was viewed as "real" since poor recruitment was indicated by the previous survey. Brown and Hennemuth (1971) addressed this problem by developing an abundance index from recruitment data alone. Unfortunately, the information content of the data for older fish is wasted.

Collie and Sissenwine (In Press) introduced a method of filtering survey data so as to distinguish between "anomalous" and "real" results. The catchability coefficient of the survey is also estimated.

Collie and Sissenwine (In Press) fit numerically (by finite difference Lavenberg-Marquardt) three simultaneous equations:

$$n_{t+1} = (n_t - qC_t + r_t)\, e^{-M} + \tilde{\varepsilon} \tag{1}$$

$$n_t' = n_t\, e^{\tilde{\eta}_t} \tag{2}$$

$$r_t' = r_t e^{\tilde{\delta}_t} \tag{3}$$

where r_t' is the catch per tow of young fish which will recruit during year t; n_t' is the catch per tow of fish recruited at the beginning of year t; C_t is the commercial catch during year t; q is the catchability coefficient; M is the known natural mortality rate; r_t and n_t are the number of prerecruits and recruits that would have been caught per tow if there were no sources of error; and $\tilde{\varepsilon}$, $\tilde{\eta}$, and $\tilde{\delta}_t$ are normally distributed random variables. For i years, there are 3i-1 residual errors, 2i parameters, (q, n_t for t = 1, i and r_t for t = 1, i-1); and i-2 degrees of freedom. The results of the analysis are given in Figure 3. The anomalous situation in 1972 is demonstrated.

It is clear that trawl survey data is not only rich in ecological significance, but it is also rich in analytical problems. These problems are considered by Pennington and Grosslein (1978), Doubleday and Rivard (1981), and Sissenwine et al. (1983).

In addition to determining the species and size composition of the survey catch, biological samples are taken for further analysis. Scales and otoliths are collected for age determinations. These data are used to fit growth models and to

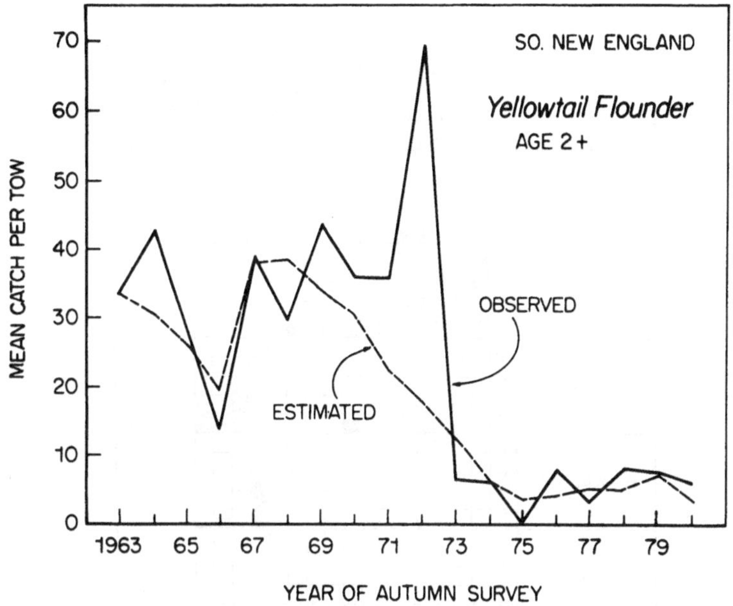

Figure 3. Mean catch per tow in kg observed during research vessel bottom trawl surveys and estimated by Collie and Sissenwine (In Press).

---

monitor changes in growth and mortality, either density dependent or independent. Furthermore, gonad condition is monitored and ovaries of selected species are removed for fecundity studies. Fish are examined for evidence of pathology.

Trawl surveys have been the source of feeding habits data. The stomach content of tens of thousands of specimens collected during these surveys has been determined. An overview of these studies is provided by Cohen et al. (1981). Quantitative estimates of the stomach content of about 80 species have been reported (e.g., Bowman et al. 1976, Grosslein et al. 1980, Figure 4).

While the focus of Georges Bank research is the exploited fishery resource, a much broader ecological research program is in place (Grosslein et al. 1979). Zooplankton, particularly ichthyoplankton, are monitored by surveys about six times per year. There have been mesoscale studies of ichthyoplankton patches involving as many as eight oceanographic vessels from five countries. There are stratified random surveys of the economically valuable macrobenthos, e.g. scallop, surf clams, and ocean quahogs. Primary productivity has been determined by the $_{14}C$ method during a three-year study including all seasons, 515 stations, and thousands of determinations (O'Reilly and Busch, In Press).

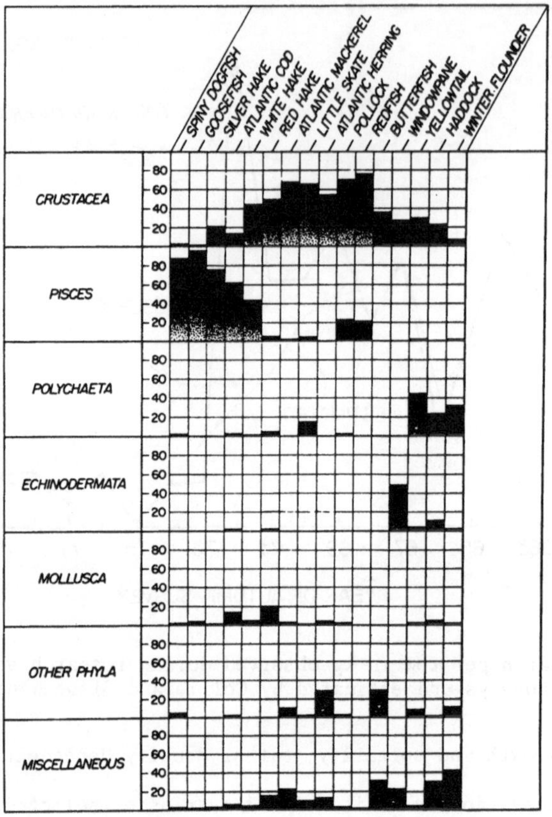

Figure 4. Percent, by weight, of diet of 16 species of fish collected on Georges Bank, 1969-1972 (after Grosslein et al. 1980).

During the year beginning March 1977, a total of 864 research vessel days and 5,309 staff sea days were expended on the research program described above. Nearly half the research vessel time was contributed by foreign nations, most notably the USSR, Poland, and the Federal Republic of Germany. In 1977, the USA extended its fisheries jurisdiction to 200 miles (323 km). As a result the role of foreign vessels has been greatly reduced in recent years. Nevertheless, major components of the research program remain in place. The USA and Canada both claim jurisdiction over the northern third of Georges Bank. Both countries conduct research within this zone.

Technological Interactions

Technological interactions occur when more than one species is captured simultaneously. This situation results from non-selective harvesting gear, such as an otter trawl, the predominant gear of Georges Bank. Thus, the target catches or fishing mortalities of a group of simultaneously harvested species may not all be achievable.

It became apparent, during the early 1970's, that attempts by ICNAF to control fishing mortality were frustrated by technological interactions (Brown et al. 1973). By 1973, ICNAF had established catch quotas for each species or species group and allocated these quotas to each nation. Directed fishing was terminated when national allocations were achieved. The term "directed fishing" refers to the intentional capture of a species. Since intention is difficult to determine, fishing effort is usually considered to be directed at the species which is a simple plurality of the catch. In spite of the closures of directed fisheries, catch continued to accumulate as a result of unintentional catch or by-catch.

Brown et al. (1973, 1979) used linear programming to estimate a multispecies national quota, which maximized the total yield. The objective function was

$$x_{...} = \sum_i \sum_j \sum_k r_{ijk} \cdot x_{ijk} \quad (4)$$

where x_{ijk} is the catch of species i by nation j while directing effort at species k; r_{ijk} is the catch of species i, as a fraction of the catch of species k, when nation j is directing effort at species k. Equation 4 is solved for x_{ijk} (for all i and j) so as to maximize $x_{...}$ within constraints;

$$u_{ij.} > x_{ij.} > 0 \text{ for all } i,j \quad (5)$$

Species quotas ($u_{i..}$) were established by ICNAF based on conservation considerations of each species. They were allocated among participating nations ($u_{ij.}$) based on policy considerations. By-catch ratios (r_{ijk}) were estimated from the most recent available catch data.

An example of the solution of the linear programming problem, specified by Equations 4 and 5, is given in Table 2 (Brown et al. 1979). Only about 60% of the total allowable catch constraint (based on the analysis of individual species) is achievable. About 30% of the yield should be reserved for by-catch.

Of course these results depend on the by-catch ratios used in the analysis. These ratios are a function of the relative abundance of each combination of species caught (i) and species sought (k). A more robust approach would be to recast the by-catch problem in terms of directed and non-directed fishing mortality

Table 2. Sum of individual country's linear programming simulation of 1975 catches, maximizing total catch (1,000 t), and using 1973 by-catch ratios for the ICNAF area. (After Brown et al., 1979.)

Species sought	Total allowable catch restraint	Directed catch	Total catch
Atlantic cod	45.00	16.39	31.48
Haddock	6.00	0.00	5.25
Redfish	25.00	18.24	22.25
Silver hake	175.00	74.69	85.72
Red hake	65.00	11.83	26.51
Pollock	21.30	9.57	20.28
American plaice	2.70	--	1.15
Witch flounder	4.30	--	1.70
Yellowtail flounder	16.00	11.02	15.06
Other flounder	18.00	--	6.54
Other groundfish	65.70	27.38	40.96
Atlantic herring	175.00	107.38	120.01
Atlantic mackerel	285.00	127.51	150.60
Other pelagic	26.90	16.97	26.45
Other fish	56.40	9.33	33.35
Squids	71.00	25.93	40.30
Total	1,058.30	456.24	626.75

instead of catch. Furthermore the objective function could be modified to take account of the relative value of each species.

It is clear that the technological aspect of the multispecies problem cannot be ignored. The definition of directed fishing effort is particularly perplexing. Murawski et al. (In Press) used a different approach. They define, via cluster analysis, multispecies assemblages, with relatively homogeneous species compositions, which co-occurred in time and space. The assemblages persisted from year to year in spite of changes in stock abundance. Once defined, these multispecies assemblages can be treated as a fishery unit. Fishing effort is treated as directed at the entire assemblage, not any particular species within it. Exploitation and management strategies must also be directed at the assemblage.

Murawski et al. (In Press) defined three major fishery units on Georges Bank. Murawski (1982) applied multispecies yield-per-recruit analysis to these fisheries. Technological interactions were taken account of by a multispecies fishing mortality vector,

$$\tilde{F} = \tilde{Q} \cdot \tilde{E} \qquad (6)$$

where $\tilde{E}$ is a column vector (m x 1) of fishing efforts applied to each fishery unit, $\tilde{Q}$ is a (n x m) matrix of fishery and species specific catchability coefficients, and $\tilde{F}$ is the resulting column vector (n x 1) of species specific fishing mortality rates. Murawski (1982) demonstrated tradeoffs between species and total yield depending on $\tilde{E}$.

Biological Interactions

The feeding habits of fish of the Georges Bank region indicate the potential for biological interactions (Figure 4). The diets of several species are similar (i.e., potential for competition) and several others feed primarily on pisces (i.e., potential for predator-prey interactions).

Brown et al. (1976) treated biological interactions implicitly by applying a surplus production model (e.g., Schaefer 1954) to the aggregate catch of all species of finfish and squid, except menhaden (which are captured close to shore in the southern part of the region) and large pelagic species (e.g., swordfish, tuna). Fishing effort of different gear types and nations was calibrated and combined to provide a standardized index of days fished. A multiplicative learning function was applied as a correction to fishing effort in newly developing fisheries.

Brown et al. (1976) demonstrated a sixfold increase in standardized fishing effort, and a 55% decline in abundance during the period 1961-1972. They used Gulland's (1961) method to fit a surplus production model to aggregate catch and standardized effort data. The resulting estimate of maximum sustainable yield

(MSY) of the region (NAFO SA 5-6) was 900,000 tons, much less than the sum of MSY's of individual species or species groups (1,352,000 tons; Table 3). Brown et al. (1976) inferred that the difference reflected biological interactions.

Table 3. Individual stock estimates of MSY for finfish stocks in SA 5 + 6. (After Brown et al., 1976.)

Species	Estimate of MSY (000's tons)
Herring	335
Mackerel	310
Silver hake	200
Squid	80
Red hake	70
Haddock	50
Cod	45
Yellowtail flounder	37
Redfish	30
Pollock	20*
Other flounder	25
Other finfish	150
Sum of species assessments	1,352

*MSY estimated to be 50,000 tons including Div. 4VWX (ICNAF, 1972b), 20,000 tons based on catch ratios assigned to SA5.

The analysis of aggregate MSY provided the impetus for futher study of the multispecies problem. As a result of the work of Brown et al. (1973, 1976, 1979), ICNAF established, in 1974, a "second tier" quota which limited aggregate catch to less than the total of the individual species quotas. This action was an explicit response to the multispecies problem.

Notwithstanding the significance of the analysis of aggregate catch, there are several limitations to this approach. Mathematically, it is identical to traditional single species surplus production modeling and suffers from the same weaknesses (Sissenwine 1978). Furthermore, a variety of methods were used to calculate the individual species' MSY's, therefore, their sum is not strictly comparable to the aggregate MSY. In addition, the specific interactions between sets of two or more species, included within the aggregate, are not identifiable.

The surplus production modeling approach may be readily extended to include explicit consideration of biological interactions (Pope 1976, Horwood 1976, May et al. 1979). Sissenwine et al. (1982) applied such an approach to fisheries of the Georges Bank region. They extended Schaefer's (1954) model to two species,

$$\frac{1}{B_i}\frac{dB_i}{dt} = a_i - b_i B_i - q_i \frac{C_i}{B_i} \pm c_{ij} B_j \tag{7}$$

where B_i is proportional to population biomass of species i; C_i is the rate of catch; q_i is the coefficient of proportionality between B_i and actual biomass; a_i, b_i, and c_{ij} are parameters. Fishing mortality was replaced in Equation 7 by $q_i C_i/B_i$. A biological interaction between species i and j is indicated by a non-zero value of c_{ij}.

A discrete-time version of Equation 7 was applied pairwise to populations of the Georges Bank region. Twenty-nine different measures of biomass (B_i) were considered for populations of Georges Bank (Figure 1, strata 13-25), Southern New England (strata 1-12), and both combined. Twenty-four of the biomass measures were from standardized autumn bottom trawl surveys. The remaining five measures of biomass were from VPA. Equation 7 was fit for all combinations of geographically co-occurring species.

The results are summarized in Table 4. The partial correlation coefficient (PCC) of the interaction term (c_{ij}) of 724 combinations are given. Only 26 PCC's are significant at the 5% level. Since less than 5% are significant, it must be concluded that many of the "significant" PCC's are due to Type I errors (i.e., due to chance alone).

Sissenwine et al. (1982) also considered a time lag for those species which were fish predators. Another 198 combinations were fit. The results were essentially the same, about 5% significant at the 5% level.

Table 4. Partial correlation coefficients (X 100) for the interactions between combinations of 29 measures of population size based on the linearized version of equation (7). An asterisk below any coefficient indicates statistical significance at the 5% level. (After Sissenwine et al. 1982).

Populations (dependent variables)	Populations (independent variables) 1	2	3	4	5	6	7	8	9	10	11	12	13	14	15	16	17	18	19	20	21	22	23	24	25	26	27	28	29
1. Silver hake G.B.	—			26		18	22	−29	−03	−29		−19			−22	16	−05	22	16	23	−13	18	47	−15	−21	−14		−14	−26
2. Silver hake S.N.E.		—			−27	03	−11		−41		−19		04	28	−30	−14	02	−19	−35	−30	−18	22	32	−21	−07	−23			
3. Silver hake G.B. and S.N.E.			—	21	−09	15	05	−03	−26	−21	−11	−05	06	11	−18	−05	−40	−02	−20	−13	−13	13	36	−18	−20	−15		−06	−03
4. Red hake G.B.	−46		−38	—			03	−41	−15	−37		−21			−34	14	−24	04	32	21	−40	39	45	−46	−37	09	−40	−40	−43
5. Red hake S.N.E.		05	−29		—		−53*		−43		25		02	−09	00	−27	19	−48	−06	−38	−13	00	−01	−34	19	22			
6. Red hake G.B. and S.N.E.	−58*	00	−27			—	−32	−21	−20	−11	00	−36	−09	07	−09	00	00	−17	05	−10	−21	14	25	−35	−09	14	−26	−16	−20
7. Cod G.B. and S.N.E.	−18	−02	−12	−06	03	−02	—	−18	−01	−26	04	04	09	−17	06	−03	−01	05	−04	00	−18	07	19	−06	−10	06	−28		−17
8. Haddock G.B.	18		08	17		52*	−27	—	13	09		−17			11	−20	−04	−10	10	−03	40	−34	31	36	34	09	03	09	
9. Pollock G.M., G.B., and S.N.E.	19	04	08	21	02	14	−10	−20	—	−17	−18	−41	−28	−11	−23	−05	−10	12	11	12	−36	03	13	−34	−08	11	−12	−15	−16
10. Yellowtail flounder G.B.	22		15	07		08	09	00	38	—		−04			31	15	00	33	−17	13	−18	−20	−14	−05	−01	16	01	07	03
11. Yellowtail flounder S.N.E.		11	20		01	12	−19		21		—		−15	−19	−10	08	03	25	−10	09	13	−29	04	09	06	04		14	
12. Winter flounder G.B.	33		17	22		37	07	07	35	02		—			−05	21	−11	21	31	33	34	−06	33	35	−04	−15	03	05	07
13. Winter flounder S.N.E.		−09	02		22	29	−36		30		07		—	27	−05	33	−39	11	14	16	26	03	28	20	−16	−09		09	
14. Summer flounder S.N.E.		−27	−21		−09	−14	30		05		11		−28	—	18	−08	−22	−02	20	11	−24	31	−11	−23	−08	27		−25	−25
15. Total flounder G.B. and S.N.E.	27	08	20	20	22	21	−41	02	51*	−40	−04	29	00	−22	—	25	00	48	−18	19	−03	−26	04	01	07	01	−16	10	05
16. Butterfish G.B. and S.N.E.	26	00	12	26	−01	14	08	−02	06	07	−01	−02	−24	04	00	—	−13	09	−14	−06	−24	−03	00	−18	−06	10	19	−04	01
17. River herring G.B. and S.N.E.	42	21	36	40	24	43	−03	47	58*	−05	06	36	27	18	08	34	—	40	−37	13	20	−26	34	41	−01	−32	26	46	47
18. Shortfinned squid G.B. and S.N.E.	15	−03	01	31	14	28	13	−04	−23	−24	−01	03	10	01	−05	−31	−11	—	−11	−14	−08	02	46	−06	−13	−10	01	−06	−04
19. Longfinned squid G.B. and S.N.E.	−34	50*	08	−32	−01	−27	03	−21	−29	17	13	−60*	−13	−24	10	−16	64*	−18	—	−19	−45	−07	−26	−41	38	27	−21	−04	−21
20. Total squid G.B. and S.N.E.	−11	43	24	−07	05	−02	16	−08	−18	01	01	−53*	−14	−20	−06	−07	51*	05	−01	—	−32	−14	06	−22	41	08	−17	10	−04
21. Dogfish G.B. and S.N.E.	31	09	24	37	28	44	30	04	13	13	00	−43	−06	18	04	16	−10	12	09	12	—	19	37	26	00	−23	20	14	10
22. Bluefish G.B. and S.N.E.	13	−02	07	00	21	11	−13	−08	06	20	07	−06	−08	−09	18	06	−25	−06	13	03	−24	—	−31	−16	−11	16	11	−12	−09
23. Scup G.B. and S.N.E.	−06	00	−02	22	02	11	−12	−12	−13	−01	02	−37	−10	10	09	00	−15	−16	−06	−16	−34	25	—	−28	−05	00	09	−08	−09
24. Total finfish and squid G.B. and S.N.E.	42	02	21	54*	44	60*	13	−18	27	−07	01	−57*	−22	20	01	31	−33	18	23	30	−14	27	59*	—	−25	−14	15	−10	−10
25. Herring VPA G.B. and S.N.E.	−11	41	26	18	32	31	−16	12	−12	04	31	−41	31	−02	22	−01	05	−36	−34	−45	−30	26	25	−21	—	−13	21	12	13
26. Mackerel VPA NW Atl.	−19	−17	−29	−31	−50*	−54*	−05	−30	−05	−53*	−29	31	−20	−28	−30	−30	38	15	−26	−03	−21	−18	−28	−32	19	—	−50*	00	−23
27. Silver hake VPA G.B.				05		14	11	−60*	−06	−22		02			18	−08	−55*	−05	−07	−09	−58*	27	−06	−64*	−49	28	—	−59*	−62*
28. Cod VPA G.B. and S.N.E.	08	09	10	46	−30	17		−34	−12	−43	−41	−36	−38	28	−50*	06	06	15	07	11	−31	29	62*	−38	−14	−38	−21	—	−16
29. Haddock VPA G.B.	25		34	15		44	07		−15	08		−17			15	−47	01	−17	−12	−22	02	−27	14	03	35	03	18	−02	—

Thus, empirical examination failed to demonstrate biological interactions. This result does not imply that biological interactions are unimportant. The result probably implies that an empirical approach is inadequate.

Sissenwine et al. (1982) discussed the potential reasons for the failure of the empirical approach (e.g., measurement error, colinearity of variables). The most important factor that makes empirical examination of biological interactions impractical is probably unexplained variability in recruitment that is induced by a fluctuating environment. For the same reason, models of the relationship between stock and recruitment (S-R) are generally inadequate (e.g., Georges Bank haddock, Figure 5). Hennemuth et al. (1980) examined recruitment for 18 species and concluded that it was effectively stochastic.

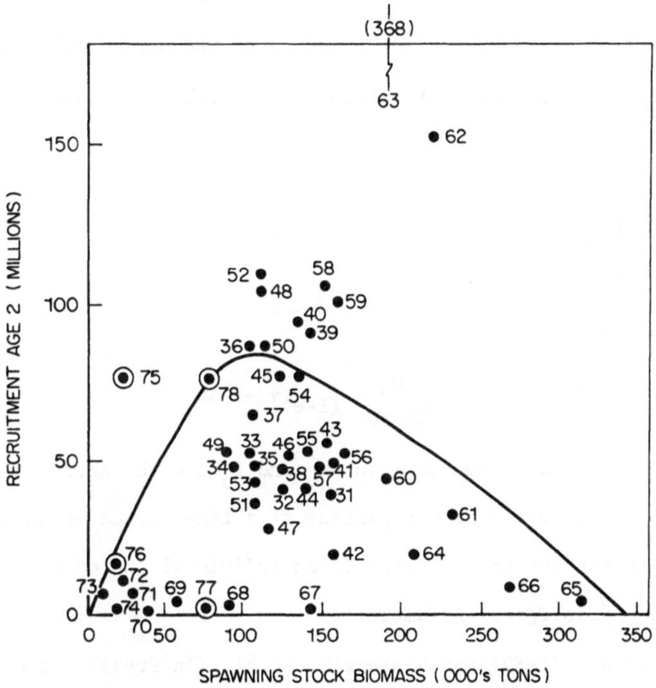

Figure 5. Stock-recruitment of Georges Bank haddock (after Clark et al. 1982). Values for 1975-1977 are preliminary.

Since most surplus production results from recruitment, it should not be surprising that noise induced by fluctuating environment obscures biological interactions. Thus, a more inductive approach is necessary.

Cohen et al. (1982) and Sissenwine et al. (In Press) describe the energy budget of Georges Bank. The latter study is more complete. It includes estimates of production and consumption by young (pre-exploitable) fish.

The most reliable information contained in the energy budget is for the lowest (primary producers) and highest (pelagic and demersal fish) trophic levels. There are also independent estimates of macrozooplankton and macrobenthos biomass. These are the components upon which the fish are most dependent. Biomass and production of other components are derived by interpolation.

The estimation of young fish consumption and production is particularly important. Little is known, quantitatively, about the life history of fish during their first year. Nevertheless, it is still feasible to estimate production during this period since the energy content of individual fish and of the entire cohort of young fish is known at the beginning of the period (when they are eggs) and the end of the period (when they become exploitable). Assuming exponential growth of individuals and of the cohort,

$$P = \begin{cases} \dfrac{GB_o}{Z-G}\left(1-e^{(G-Z)\Delta t}\right) & \text{when } G \neq Z \\ GB_o\Delta t & \text{when } G = Z \end{cases} \tag{8}$$

$$C = \frac{1}{\alpha}\dot{P} + \frac{\beta}{\alpha} \cdot \frac{B_oW_o^{\gamma-1}}{Z-\gamma G}\left(1-e^{(\gamma G-Z)\Delta t}\right) \tag{9}$$

where P is production; C is consumption; W_o and B_o are the energy content of an individual and of a cohort at the beginning of a time period of duration Δt; G and G-Z are the instantaneous growth rates of an individual and of the cohort; and α, β, and γ are physiological parameters.

Grosslein et al. (1980) and Sissenwine et al. (In Press) noted that G-Z is relatively small. When this is the case, and if a cohort grows monotonically (except during the egg period), then the assumption of exponential growth is robust. Furthermore, estimates of young fish production and consumption reported in Figure 6 cannot be too far in error because they are bounded by production of suitable prey and by consumption of their predators. Demersal fish are their primary predator.

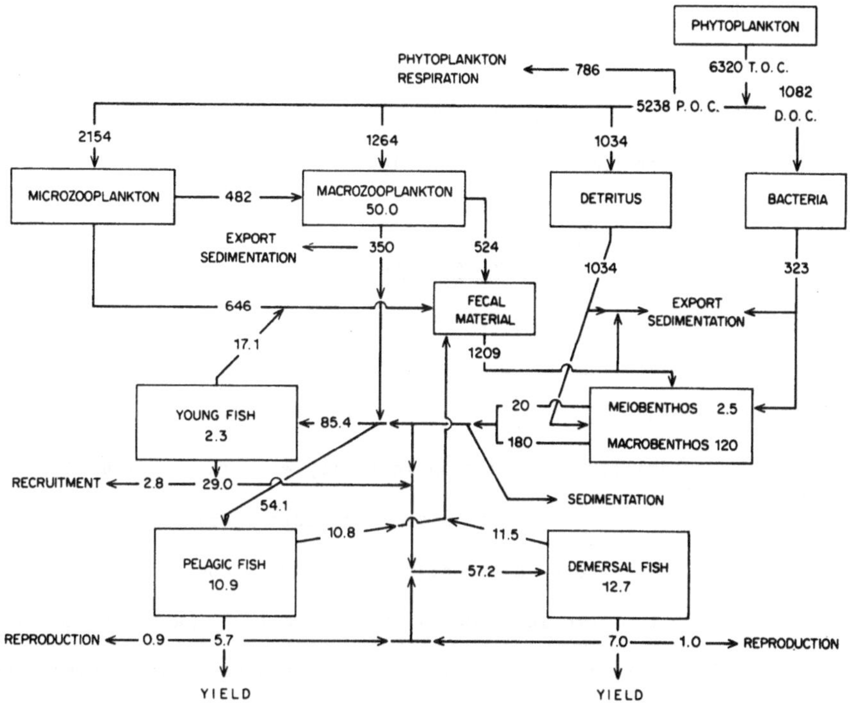

Figure 6. Georges Bank energy budget in Kcal/m^2 or Kcal/m^2 yr. (After Sissenwine et al., In press).

Demersal fish consumption was 57.2 Kcal/m^2yr during 1973-75. Figure 4 indicates that some demersals, which are major components of the biomass (e.g., silver hake and cod), consume a significant amount of fish. In fact, about half the total consumption by demersal fish consists of fish. Total production of fish (including young fish) was only 41.7 Kcal/m^2yr, therefore, predation mortality must be high. Since a relatively small species (silver hake) is the primary predator, most of this mortality must be inflicted on young fish.

Production by fish and their consumption of fish are so nearly equal that one or the other is probably in error. It seems most likely that young fish production is underestimated, but this underestimation cannot be severe since there is a limited amount of potential prey. Therefore, there is strong evidence of biological interactions. Specifically, it appears that predation by a few fish species affects recruitment of others, even if biological interactions cannot be demonstrated empirically (i.e., Table 4).

Another implication of the energy budget is that the mortality rate of young fish should remain high beyond the larval stage since predation by silver hake on larvae has not been observed. Furthermore, year class strength (recruitment success may be established during the postlarval period. Sissenwine et al.'s (In Press) preliminary analysis of larval abundance data is consistent with this inference.

Future Research

The static energy budget of Georges Bank demonstrates the potential importance of biological interactions. Some of the variability in recruitment might be explained by estimating the annual predation mortality suffered by the prerecruits of each species. Variability in predation mortality will result from fluctuations in (1) predator abundance, (2) predator size composition, (3) predator diet composition, and (4) abundance of alternative prey. Multispecies VPA is an analytical approach to this problem (Pope 1979, Sparre 1980, Majkowski 1981).

Much variability in recruitment is likely to remain unexplained even after the effect of predation is taken into account. In order to explain environmentally induced variability, it will probably be necessary to initiate new surveys to sample young postlarval fish before they become vulnerable to traditional samplers. Such data will help to resolve when year class strength is established.

In order to understand the role of environmental variability, it will be necessary to monitor mesoscale oceanographic processes. Remote sensing from satellites may help to make such monitoring feasible. Numerical hydrodynamic models might be used to simulate historical circulation, based on storms and atmospheric pressure records.

While much progress has been made, it is clear that uncertainty will remain a component of the multispecies fisheries problem. Therefore, more robust strategies for exploitation and management are needed.

Galtsoff's (1962) plea of two decades ago for "long continued and well planned observations" is still valid. Such information is extremely expensive. Thus, the proper mathematical and statistical tools to guide collection and interpretation are invaluable.

Literature Cited

Azarovitz, T. R. (1981). A brief historical review of the Woods Hole Laboratory trawl survey time series. Can. Spec. Publ. Fish. Aquat. Sci. 58:62-67.

Beverton, R. J. H., and Holt, S. J. (1957). On the dynamics of exploited fish populations. U.K. Min. Agric. Fish., Fish. Invest. (Ser. 2) 19, 533 pp.

Bowman, R. E., Maurer, R. O., and Murphy, J. B. (1976). Stomach contents of 29 fish species from five regions in the Northwest Atlantic data report. U.S. Dept. of Comm., NOAA, NMFS, NEFC, Woods Hole Lab. Ref. Doc. No. 76-10. 37 pp.

Brown, B. E., and Hennemuth, R. C. (1971). Prediction of yellowtail flounder population size from pre-recruit catches. Int. Comm. Northw. Atl. Fish. Redbook, Part III:221-228.

Brown, B. E., Brennan, J. A., Heyerdahl, E. G., and Hennemuth, R. C. (1973). Effects of by-catch on the management of mixed species fisheries in Subarea 5 and Statistical Area 6. Int. Comm. Northw. Atl. Fish.Redbook, Part II, 217-231.

Brown, B. E., Brennan, J. A., Heyerdahl, E. G., Grosslein, M. D., and Hennemuth, R. C. (1976). The effect of fishing on the marine finfish biomass in the Northwest Atlantic from the eastern edge of the Gulf of Maine to Cape Hatteras. Int. Comm. Northw. Atl. Fish. Res. Bull., 12:49-68.

Brown, B. E., Brennan, J. A., and Palmer, J. E. (1979). Linear programming simulations of the effects of by-catch on the management of mixed species fisheries off the northeastern coast of the United States. Fish. Bull., U.S. 76:351-860.

Burns, T. S., Schultz, R., and Brown, B. E. (In Press). The commercial catch sampling program on the northeastern United States. Can. Spec. Publ. Fish. Aquat. Sci.

Byrne, C. J., Azarovitz, T. R., and Sissenwine, M. P. (1981). Factors affecting variability of research vessel trawl surveys. Can. Spec. Publ. Fish. Aquat. Sci. No. 58, pp. 258-273.

Clark, S. H., and Brown, B. E. (1977). Changes in biomass of finfishes and squids from the Gulf of Maine to Cape Hatteras, 1963-1974, as determined from research vessel survey data. Fish. Bull., U.S. 75:1-21.

Clark, S. H., Overholtz, W. J., and Hennemuth, R. C. (1982). Review and assessment of the Georges Bank and Gulf of Maine haddock fishery. J. Northw. Atl. Fish. Sci. 3:1-27.

Cohen, E. B., Grosslein, M. D., Sissenwine, M. P., Serchuk, F., and Bowman, R. (1981). Stomach contents studies in relation to multispecies fisheries analysis and modeling for the Northwest Atlantic. Int. Counc. Explor. Sea C.M. 1981/G:66. 18 p.

Cohen, E. B., Grosslein, M. D., Sissenwine, M. P., Steimle, F., and Wright, W. R. (1982). An energy budget for Georges Bank. Can. Spec. Publ. Fish. Aquat. Sci. 59:95-107.

Collie, J. S., and Sissenwine, M. P. (In Press). Estimating population size from relative abundance measured with error. Can. J. Fish. Aquat. Sci.

Doubleday, W. G., and Rivard, D. (ed.). (1981). Bottom trawl surveys. Can. Spec. Publ. Fish. Aquat. Sci. 58:1-273.

Edwards, Robert L. (1958). Species composition of the industrial trawl landings in New England, 1957. United States Department of the Interior, U.S. Fish and Wildlife Service, Special Scientific Report--Fisheries No. 266.

Galtsoff, P. S. (1962). Story of the Bureau of Commercial Fisheries Biological Laboratory, Woods Hole, Massachusetts. U.S. Dept. Interior, Cir. 145, 121 p.

Grosslein, M. D. (1969). Groundfish survey program of BCF Woods Hole. Commer. Fish. Rev. 31(8-9):22-35.

Grosslein, M. D. (1971). Some observations on accuracy of abundance indices derived from research vessel surveys. Int. Comm. Northw. Atl. Fish. Redbook 1971(3):249-266.

Grosslein, M. D., Brown, B. E., and Hennemuth, R. C. (1979). Research, assessment and management of a marine ecosystem in the Northwest Atlantic: a case study. In environmental biomonitoring, assessment, prediction, and management - certain case studies and related quantitative issues. J. Cairns, Jr., G. P. Patil, and W. E. Waters (eds.) International Co-operative Publishing House, Fairland, Maryland, pp. 289-357.

Grosslein, M. D., Langton, R. W., and Sissenwine, M. P. (1980). Recent fluctuations in pelagic fish stocks of the Northwest Atlantic, Georges Bank, in relationship to species interactions. Rapp. P.-v. Reun. Cons. Int. Explor. Mer 177:374-404.

Gulland, J. A. (1961). Fishing and the stocks of fish at Iceland. Fish. Invest. Minist. Agric. Fish. Food (G.B.), Ser. II, 23(4), 52 pp.

Gulland, J. A. (1965). Estimation of mortality rates. Annex to Rep. Arctic Fish. Working Group. Int. Counc. Explor. Sea C.M. 1965(3), 9 pp.

Hennemuth, R. C. (1976). Variation in survey abundance indices. Int. Comm. Northw. Atl. Fish. Res. Doc. 74/104.

Hennemuth, R. C., Palmer, J. E., and Brown, B. E. (1980). A statistical description of recruitment in eighteen selected fish stocks. J. Northw. Atl. Fish. Sci. 1:101-111.

Horwood, J. W. (1976). Interactive fisheries. A two species Schaefer model. Int. Comm. Northw. Atl. Fish., Sel. Pap. 1:151-155.

Majkowski, J. (1981). Application of a multispecies approach for assessing the population abundance and the age-structure of fish stocks. Can. J. Fish. Aquat. Sci. 38:424-431.

May, R. M., Beddington, J. R., Clark, C. W., Holt, S. J., and Laws, R. M. (1979). Management of multispecies fisheries. Science 205:267-277.

McHugh, J. L. (1959). Can we manage our Atlantic coastal fishery resource? Trans. Am. Fish. Soc. 88:105-110.

Murawski, S. A. (1982). Deterministic yield per recruitment simulations of mixed-species fisheries. Int. Counc. Explor. Sea C.M. 1982/G:35, 46 p.

Murawski, S. A., Lange, A. M., Sissenwine, M. P., and Mayo, R. K. (In Press). Definition and analysis of multispecies otter trawl fisheries off the northeast coast of the United States. J. Cons. Perm. Int. Explor. Mer.

Murphy, G. I. (1965). A solution of the catch equation. J. Fish. Res. Board Can. 22:191-202.

O'Reilly, J. E., and Busch, D. A. (In Press). The annual cycle of phytoplankton primary production (net-plankton, nannoplankton and release of dissolved organic matter) for the Northwestern Atlantic Shelf (Middle Atlantic Bight, Georges Bank, and Gulf of Maine). Rapp. P.-v. Reun. Cons. Int. Explor. Mer.

Pennington, M. R., and Grosslein, M. D. (1978). Accuracy of abundance indices based on stratified random trawl surveys. Int. Comm. Northw. Atl. Fish. Res. Doc. 78/VI/77, 42 p.

Pope, J. G. (1976). The effect of biological interaction on the theory of mixed fisheries. Int. Comm. Northw. Atl. Fish., Sel. Pap. 1:157-162.

Pope, J. G. (1979). A modified cohort analysis in which constant natural mortality is replaced by estimates of predation levels. Int. Counc. Explor. Sea, Pelagic Fish Committee, C.M. 1979/H:16, 7 p.

Ricker, W. E. (1954). Stock and recruitment. J. Fish. Res. Board Can. 11:599-623.

Ricker, W. E. (1958). Handbook of computations for biological statistics of fish populations. Bull. Fish. Res. Bd. Canada, 119, 300 pp.

Robins, C. R. (1980). A list of common and scientific names of fishes of the United States and Canada. Am. Fish. Soc. Spec. Publ. No. 12, 174 p.

Rounsefell, G. A. (1957). A method of estimating abundance of groundfish on Georges Bank. Fish. Bull., U.S. 113(57), 265-278.

Schaefer, M. B. (1954). Some aspects of dynamics of populations important to the management of commercial marine fisheries. Bull. Inter-Am. Trop. Tuna Comm. 1(2):27-56.

Schaefer, M. B. (1957). A study of the dynamics of the fishery for yellowfin tuna in the eastern tropical Pacific Ocean. Bull. Inter-Am. Trop. Tuna Comm. 2: 247-268.

Sissenwine, M. P. (1978). Is MSY an adequate foundation for optimum yield? Fisheries 3(6):22-42.

Sissenwine, M. P. (1978). Using the USA research vessel spring bottom trawl survey as an index of Atlantic mackerel abundance. Int. Comm. Northw. Atl. Fish., Selected Papers No. 3, 49-55.

Sissenwine, M. P., and Bowman, E. W. (1978). An analysis of some factors affecting the catchability of fish by bottom trawls. Int. Comm. Northw. Atl. Fish., Res. Bull. 12:81-87.

Sissenwine, M. P., Brown, B. E., Palmer, J. E., Essig, R. J., and Smith, W. (1982). Empirical examination of population interactions for the fishery resources off the northeastern USA. Can. Spec. Publ. Fish. Aquat. Sci. 59:82-94.

Sissenwine, M. P., Azarovitz, T., and Suomala, J. B. (1983). Determination of fish abundance. In "Experimental Biology at Sea". Academic Press, pp. 51-101.

Sissenwine, M. P., Cohen, E. B., and Grosslein, M. D. (In Press). Structure of the Georges Bank ecosystem. Rapp. P.-v. Reun. Cons. Int. Explor. Mer.

Sparre, P. (1980). A goal function of fisheries (Legion Analysis). Int. Counc. Explor. Sea, Demersal Fish Committee, C.M. 1980/G:40, 81 p.

THE LEGACY OF BEVERTON AND HOLT

George N. White, 3rd.
Department of Fisheries and Oceans, Canada

Abstract

Beverton and Holt's yield per recruit models have exerted a profound influence on the management of many of the world's fisheries. Despite successes in rebuilding depleted stocks, profitability of fishing enterprises remains chronically low. Efforts are now underway to increase the economic efficiency of the fisheries.

Current policy models allow biological processes to be analyzed separately from economic processes. It is argued that this separation cannot be made at the operating level. New economic policies will need the support of operational models which integrate economic and biological views of the system.

Introduction

There are few places where mathematical models have a more direct influence on the lives of ordinary men and women than in the fishing communities of Nova Scotia and Newfoundland. These people live by what they can take from the sea; and what they take is, in law if not always in practice, regulated. In Atlantic Canada, as in many jurisdictions, catches have been controlled since the early 1970's by a quota system which relies on the yield per recruit models of Beverton and Holt.

The acceptance of these models as a basis for management is due, in large part, to a remarkable book entitled "On the Dynamics of Exploited Fish Populations" by Beverton and Holt (1957). Their legacy is an evolving collection of mathematical models and (sometimes heuristic) solution procedures. It also includes the data collection and processing system which has been created to support these models. Over the years many of the specifics supplied by Beverton and Holt have been modified or replaced, but always in concert with the authors' original aims.

The depressed state of many of the world's great fisheries was the dominant concern of fisheries managers in 1957. Rebuilding fish stocks through conservation measures was seen as a prerequisite to improving the economic status of the fishing industry. Sadly, the experience of the last decade has been that recovery of depleted stocks is not enough to ensure that the needs of fishermen are met. Despite dramatic increases in many stocks, the average household income for full-time fishermen in Atlantic Canada was $21,900 in 1981. This figure includes seasonal unemployment benefits, which average $2,466 for each full-time fisherman. Nearly a third of these households were below Canada's official rural poverty level of $12,035 for a four person household (Task Force on Atlantic Fisheries 1982).

Economic efficiency has been a Canadian policy objective since 1976 (Fisheries and Marine Service 1976). Towards this end, major reviews of fisheries policy have been initiated on the Pacific and Atlantic coasts. The final report of the Commission on Pacific Fisheries Policy (Pearse 1982) recommends sweeping changes in policies affecting the allocation of catches, but few alterations to the way harvesting levels are determined. The Task Force on Atlantic Fisheries (1982) has likewise focused on the allocation of harvests.

New policies intended to improve the economic performance of fisheries will place new demands on fisheries management organizations. Fisheries managers in Atlantic Canada rely heavily on scientific advice provided by the Canadian Atlantic Fisheries Scientific Advisory Committee (CAFSAC). This paper will try to anticipate some of the changes which will be required to meet the needs of fisheries managers in the 1980's. The next section briefly recalls the historical development of theoretical models which have governed policies for fisheries management. This is followed by a brief description of the system used to generate quota advice in Atlantic Canada. Then a specific proposal for economic rationalization will be examined using a simple microeconomic model. This analysis suggests two fundamental principles: 1) policy models do not reflect the structure of operational models, and 2) the scales appropriate for the study of different classes of phenomena do not, in general, coincide. In terms of the fisheries, this translates into an urgent need to refine the scales of spatial and temporal aggregation used by current operational models.

Policy Models

Fisheries policy decisions are, ultimately, political. While mathematical models seldom play a direct role in such decisions, they are useful in describing the logic which motivates a particular policy.

The key variables which enter into discussions of fisheries policy are the resource biomass, x, a biological measure of effort, E, and a measure of the efficiency of harvesting, q. In the early years, effort was measured in units of time. To discuss economic performance, however, an economic definition of effort is required. When measured in monetary units the symbol for effort will be E', and for the corresponding efficiency, q'.

The basic models employed in discussions of policy obtained from the following system:

$$\frac{dx}{dt} = x f(x) - H,$$

$$\frac{dE}{dt} = g(R - C) \quad \text{or} \quad \frac{dE'}{dt} = g'(R' - C),$$

$$\frac{dq}{dt} = k \quad \text{or} \quad \frac{dq'}{dt} = k',$$

where

H = rate of harvesting

= qEx = q'E'x

R = total revenues = price x H

C = total costs = cE = E'

k or k' = a constant (often 0).

It is further assumed that

$$f(x) \begin{cases} < 0 & 0 < x < x_0 \\ > 0 & x_0 < x, \end{cases}$$

$$g(0) = 0, \text{ and}$$

$$g'(0) > 0.$$

There have been three fundamentally different periods in the development of fisheries policy models (Table 1). The concerns of each succeeding period were anticipated in some quarters long before they became topics for public policy discussions. Because precise dates would have little meaning, those given in Table 1 are rounded to the nearest decade.

Table 1. Historical periods in the evolution of fisheries policy.

Open access (1930-1950)	characterized by efforts to encourage development, research centers on vessels and catching methods.
Conservation (1950-1980)	characterized by efforts to rebuild stocks, research centers on sampling and parameter estimation.
Economic rationalization (1980-?)	characterized by efforts to increase economic efficiency through rationalization of the industry and introduction of usufructuary rights.

During the open access era, fisheries policy modelling was concerned primarily with the decline of stocks as effort expanded. Changes in q and q' were generally ignored. In fact, both q and q' probably increased (k and k' > 0) during this period, thus accelerating the decline of the resource (Figure 1).

Recovery of North Sea stocks which had not been fished during the war provided convincing evidence that controls on fishing could reverse the decline. This marked the beginning of the conservation era. It must be emphasized that rebuilding depleted stocks was seen not as an end in itself, but as a vital first step towards reversing the decline of the fishing industry that resulted from the collapse of the stocks.

Many strategies, including mesh regulation, effort limitation, and finally catch quotas have been tested. In Atlantic Canada a system of catch quotas, supported by restrictions on mesh size and a licensing system, has proven successful in rebuilding groundfish stocks and, to a lesser degree, pelagic stocks. In terms of the general policy models being considered here, quota management simplifies the model by making effort a function of the harvesting rate, Q, determined by the quota:

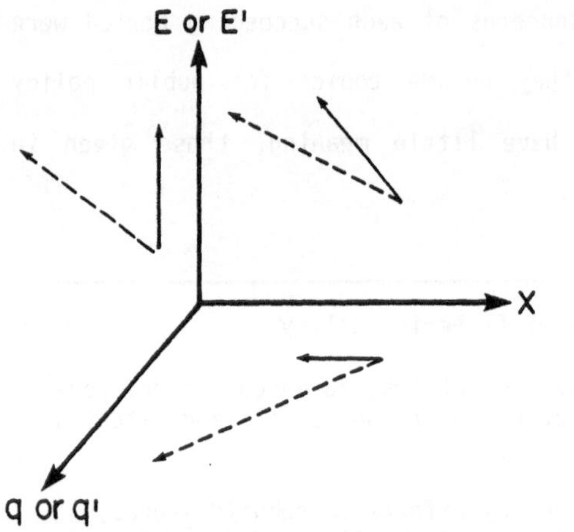

Figure 1. Open access policy models (1930-1950). Expected (solid line) trajectory underestimated the extent to which changes in harvesting efficiency (both q and q') would allow harvests to increase in relation to effort, as shown by the systems observed behaviour (dashed line).

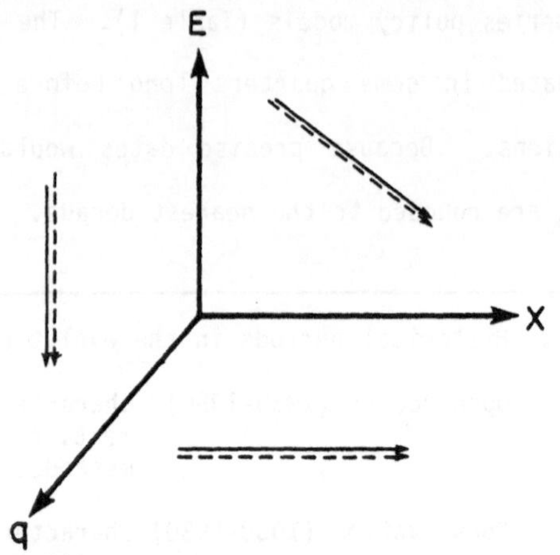

Figure 2. Quota management (1950-1980) as viewed through a biological measure of effort. Observed (dashed line) behaviour matched that predicted by policy models (solid line).

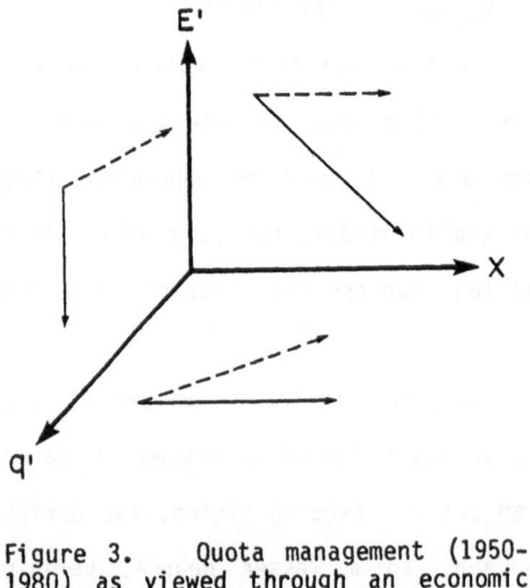

Figure 3. Quota management (1950-1980) as viewed through an economic measure of effort. The expected (solid line) decrease in effort E' did not materialize (dashed line).

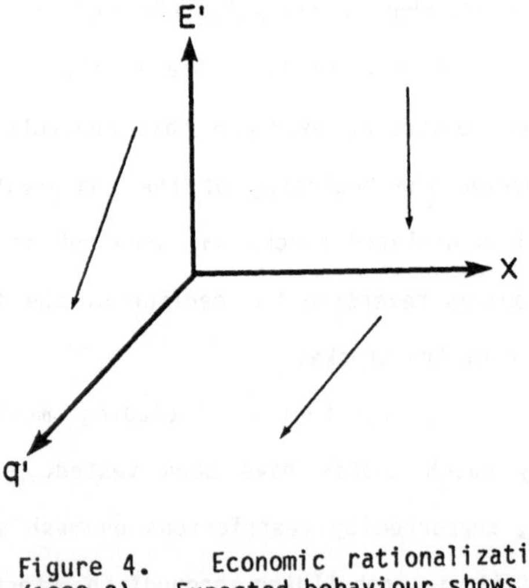

Figure 4. Economic rationalization (1980-?). Expected behaviour shows an increase in the efficiency q' of harvesting while resource levels remain high.

$$E = \frac{Q}{xq} \quad \text{or} \quad E' = \frac{Q}{xq'}$$

Again, it was assumed initially that q and q' would not change. It is clear that the expected increase in the resource could be harvested with a much reduced biological effort, E. In retrospect, however, it was politically impossible to reduce participation in the fishery while stocks were increasing. A more realistic model for quota management would allow E' to change, while assuming that q' satisfies $q' = \frac{Q}{xE'}$ (Figure 2). Then

$$(1) \qquad \frac{dq'}{dt} = -\frac{Q}{(xE')^2}\left(E'\frac{dx}{dt} + \frac{dE'}{dt}\right)$$

Suppose the system was initially at the bionomic equilibrium (x, E'), where

$$\bar{x} = \frac{1}{pq'}, \quad \text{and} \quad \bar{E}' = \frac{f(\bar{x})}{q'},$$

and that Q is chosen somewhat smaller than $\bar{H} = \bar{q}'\ \bar{x}\ \bar{E}'$. Then the right side of (1) involves the sum of a positive term representing the growth of the stock and a negative term representing the decline of effort due to reduced revenue. Events have shown that introduction of restrictive quotas resulted in a decline in revenue to individuals and organizations involved in fishing. This created political pressure for compensatory measures, and even included subsidies aimed at stimulating the industry. The result was that E' tended to remain constant, and q' declined, as the stocks rebuilt (Figure 3).

Concern with rebuilding stocks changed to concern with economic problems in Canada when it became apparent that fishing enterprises remain unprofitable despite increases in stock abundance. This is reflected in the dramatic difference between the dynamics portrayed in Figure 2 and that in Figure 3. Thus a major concern for fisheries management today is to reverse the decline in q' while maintaining stock abundance (Figure 4).

It is important to note that, under quota management, the coupling between the resource, x, and the variables q' and E' is through the quota harvesting rate, Q. Thus, if Q remains constant, the problem of altering the dynamics of q' can be addressed independently of the biological system. While this observation has been used in the search for new policies, one must not assume that separation of biological and economic analyses can be made in the implementation of a policy. Indeed, the following sections will show that economic analyses are likely to require biological information that is not provided by existing institutions. In order to appreciate the nature of the differences it is necessary to examine both current biological models (the legacy of Beverton and Holt) and a representative economic model in some detail.

Beverton and Holt (1957, pp. 378 ff.) were well aware that management actions and the profitability of fishing enterprises are related. Yet, in contrast to biological investigations, a quantitative basis for analyzing the human side of fisheries is lacking. In the following section it will be seen that yield per recruit models require an ongoing effort to produce estimates for the parameters of the models. Had there been a comparable routine effort to produce estimates for economic parameters during the 1970's, it seems apparent that the situation today would be quite different.

Scientific Advice

Yield per recruit models are used by CAFSAC scientists to provide quantitative recommendations of quota levels for the major groundfish stocks. The primary mandate of CAFSAC, as given in its terms of reference, reads:

> CAFSAC is responsible for providing scientific advice to the Atlantic Directors-General Committee on the management, including the full range of conservation measures taking into account economic objectives, of all stocks of interest or potential interest to Atlantic coast fishermen. Resource management advice will be provided in accordance with specific fisheries management objectives and strategies and will normally be published as a matter of routine. (CAFSAC 1982)

The specific objective of management has, in recent years, been a gradual rebuilding of fish stocks. The strategy involves, among other measures, catch quotas. CAFSAC has been asked to estimate the catch which would produce an $F_{0.1}$ (Gulland and Boerema 1973) level of fishing mortality. By definition, $F_{0.1}$ satisfies

$$\left.\frac{dY}{dF}\right|_{F_{0.1}} = 0.1 \left.\frac{dY}{dF}\right|_{0}$$

where Y(F) is the yield per recruit. For most stocks $F_{0.1}$ is comparable to estimates for M, the natural mortality. When a stock is very low CAFSAC will advise a reduced quota. In practice, this strategy resembles that suggested by Goh (1980). An important advantage of the $F_{0.1}$ criterion is its nearly linear response to parameter variation, in sharp contrast to the nonlinear response of F_{MAX} (White 1983a).

Beverton and Holt were writing for an audience with limited mathematical ability and rudimentary calculating machines. In describing yield per recruit modelling as it is used today it will be useful to formulate the model in terms of a conservation law:

$$\text{(2)} \qquad \frac{\partial N}{\partial t} + \frac{\partial N}{\partial a} + \frac{\partial g(w)N}{\partial w} = -MN - \frac{dC}{dt}$$

where N(t,a,w) is the number of individuals in the stock at time T, age a, and weight w; g is the growth rate; M the per capita rate of natural mortality; and $\frac{dC}{dt}$ the rate of death due to fishing. Recruitment at age a_0 is specified through a boundary condition; $N(t,a_0,w) = N_0(t,w)$. In practice it is assumed that recruitment is concentrated at one year intervals and, after a certain stage in the analysis, that weight is a function of age. An important innovation since 1957 is the use of age-length keys. An age-length key is a double frequency:

$$n(t,a,w(l)) = N(t,a,w(l)) \,/\, \int_0^\infty\!\int_0^\infty N(t,a,w)\, da\, dw$$

where weight, w, is assumed to be a function of length, l. These may be constructed directly from sampling data, and are used at a variety of steps in the development of quota advice.

Solving (2) by the method of characteristics yields:

$$\text{(3)} \qquad \frac{dN}{ds} = -MN - \frac{dC}{ds} \quad \text{and} \quad \frac{dw}{ds} = g(w)$$

From the practical standpoint, the characteristic solutions (3) have the important advantage that processes of growth, mortality, and recruitment can each be analyzed separately.

The system required to support this model can be divided into four main activities. These do not necessarily correspond to organizational boundaries. Each activity involves different academic disciplines, as shown in Table 2. These activities produce a rather complex information flow (Table 3).

The procedures of the synthesis are direct descendants of those used by Beverton and Holt (1957). Some of the most important changes to the original theory include:

1. use of sequential population analysis to estimate historical abundance,
2. increasing use of statistical tools and automatic computing machinery, and
3. the calculation of the catch which would produce a fishing mortality of $F_{0.1}$.

This final step goes part way towards overcoming Beverton and Holt's reliance on steady-state solutions of their models, but may be inappropriate for some stocks (Sinclair et al. 1983).

It should be noted that CAFSAC analyses are directed towards short-term behaviour of the stocks. Interactions between species and effects of stock size on growth and recruitment are too poorly understood to be useful in predicting behaviour over the next few years. Thus CAFSAC's projections have been based on typical levels of recruitment rather than a relationship between stock and recruitment, and on growth rates observed in recent years.

One aspect of the system represented in Table 3 requires further comment. Most information processing systems involve tradeoffs between levels of aggregation and the volume of data to be sorted and analyzed. Although modern automatic computing machinery has made it possible to process greatly enlarged volumes of data, existing systems still reflect the constraints imposed by the computational tools available to Beverton and Holt. In particular, estimates of harvest levels and population size are made on an annual basis for each fish stock.

The models used by CAFSAC bear little relation to the models used in policy analysis. In particular, quota levels can, and do, vary from year to year with changes in stock age composition and abundance. The data requirements of a system which generates quantitative management advice have produced a complex and

Table 2. Activities of the systems defined by the legacy of Beverton and Holt.

Activity	Disciplines	Role
Sampling	Field Biology	Collection of data
Data Management	Information Science	Organize, edit, store and retrieve data
Analysis	Biology Statistics	Estimation of fundamental parameters from sampling data
Synthesis	Biology Mathematics	Calculation of results

Table 3. Information flow for stock assessment. Only major quantitative data objects are shown. Qualitative information is often vitally important to the conduct of assessments.

Sampling Data Sources	Information Processing Data Stores	Analysis Products	Synthesis Products[1]
Fish buyers	Commercial vessel landings	Comprehensive landings and fishing effort by stock	
Vessel Logbooks	Fishing effort and area of capture (stock)		
Biological sampling of commercial landings	Commercial sample age, length data for landings	Age-length composition of landings, growth parameters	Sequential population analysis: estimate of absolute abundance estimate of stock biomass
Observers at-sea sampling	Observer catch, effort, bycatch, discards; observer sample age, length data for catches	Sample catch and effort data, bycatch and discards, age-length composition of catches, growth parameters.	estimate of fishing mortality Yield per recruit analysis: $F_{0.1}$ Catch projections: $F_{0.1}$ catch biomass
Research vessel surveys	Research vessel catch, effort, age, length, weight data	Indices of relative abundance for fishable stock and pre-recruits	

[1]The best published description of the calculations used in synthesis is the documentation of the programs (Rivard 1982). See also White 1983a,b,c, and Beyer and Sparre (1982).

expensive apparatus. Many aspects of the design of this system were dictated by the limited data management and calculating tools available to Beverton and Holt. In particular, aggregation in time and space reduces the volume of data which must be processed. This minimizes computational work and simplifies data transfers between different parts of the organization, but may reduce the value of the results for new kinds of analyses.

Economic Rationalization

Usufructuary rights schemes are seen by many as an answer to the economic problems of the fisheries (Scott and Neher, eds. 1981). Both Pearse (1982) and the Task Force on Atlantic Fisheries (1982) have proposed that a system of quota licenses be adopted for stocks currently managed by quota. Under this system each license entitles the holder to harvest a predetermined quantity (or a fraction of the quota) of a particular species in a specified area. The licenses are to be sold using a system of competitive bids so that their cost will be determined by market forces.

Implementation of a quota license policy requires that the fishing grounds be partitioned into licensing areas. It can be argued that many stock areas should be split into several quota license areas. If this is not done fishing effort will continue to concentrate in the most profitable areas, leading to overcrowding of port and processing facilities while similar facilities in other areas lie idle.

To understand the implications of a quota license policy it is necessary to model the behaviour of an individual fishing enterprise. The analysis presented in this paper is designed to explore the kinds of information that will be needed from biological models. Although the actions of fishermen may not always be rational, a model for rational decision making would be expected to involve more kinds of relevant information than are actually used, and is therefore suitable to the task at hand.

A standard model for rational decision making in the presence of uncertainty is the maximization of expected <u>utility</u> (Luce and Raiffa 1957). In this model a decision, $\underline{\delta}$, produces a net return, $r(\underline{\delta},\omega)$, where ω is a random variable representing the state of nature. In the case at hand, $\underline{\delta}$ will be a vector of fishing efforts, $\underline{\delta} = (E_1,\ldots E_n)$, and $r(\underline{\delta},\omega) = \Sigma E_i \rho_i(\omega)$ where ρ_i is the net rate of return. This depends on the market price of rights in category i as well as many uncertain elements.

The utility of a particular net rate of return will be denoted by u(r). The decision rule stated above implies that a decision $\underline{\delta_1}$ is prefered to $\underline{\delta_2}$ if $E[u(r(\underline{\delta_1},\omega))] > E[u(r(\underline{\delta_2},\omega))]$. It will be assumed that u(r) is monotonic increasing and concave. Then, by Jensen's inequality,

$$E[u(r)] \leq u(E[r]). \tag{4}$$

Equality holds in (4) if the return is certain. This raises the possibility that fishing enterprises may adopt strategic behaviour involving tradeoffs between uncertainty and expected returns.

One possibility for strategic behaviour under a system of quota licenses is purchase of a portfolio including rights to a variety of species and areas. Since profitability is likely to vary with species and area, this requires a sacrifice of expected returns for a gain in expected utility. In particular, if the licensing units correspond to different geographic areas occupied by a population with unpredictable distribution, a decline in abundance in one area must be accompanied by an increase in some other area. By fishing in all areas that may be occupied, a fisherman is assured of finding fish. This idea is illustrated in the following example.

Example

It is assumed that a stock has been divided into two quota license areas, but in any given year is abundant in only one of these areas. Let r_i be the return from harvesting operations (price received less cost of capture) and h_i the number of quota units purchased by a particular enterprise for the ith license area (i = 1,2). $r_1 = 1-r_2$ is assumed to be a coin-tossing game with values 0 or 1, and the enterprise is assumed to have a maximum capacity of $H \leq 2$ units. The net return is $r = r_1h_1 + r_2h_2 - p(h_1 + h_2) - cH$, where p is the price established by the market for a quota unit and c the cost of harvesting capacity. Since $r_1 + r_2 = 1$, $E[u(r(1,1,2))] = u(1-2p-2c)$. Suppose that this zero risk decision is optimal. Since the fishery cannot exist with negative net return, the market price p must satisfy

$$0 \le 1-2p-2c$$

Recalling (4), the other decisions yield

$$E[u(r(0,1,1))] \le u(\tfrac{1}{2}-p-c)$$

$$E[u(r(0,2,2))] \le u(1-2p-2c)$$

The decision (1,1,2) is an expected utility maximum. Actually, it is possible to improve profitability if harvesting capacity can be reduced to the single unit which yields a positive return. This analysis appears to contradict a statement of Pearse (1982, p. 91) to the effect that investment in the purchase of quota rights is an incentive to exercise these rights, and points out a serious weakness in the proposal for quota licensing. It should, however, be noted that Pearse also recommended consideration of an insurance system. By reducing uncertainties this could encourage fulfillment of quotas.

The preceding analysis suggests one way in which a quota license market may have undesirable consequences for a fishery. Despite the simplicity of the model, it is apparent that the determination of quota license areas is crucial to the success of a quota license policy. Furthermore, if the management apparatus is unable to monitor the relevant variables, problems which arise may be recognized late and their causes misunderstood. These management functions will require data which describe the distribution of a resource in space and time. This requires a reduced level of aggregation from that currently used in providing quota advice.

Discussion

Beverton and Holt have exerted a profound influence in the development of fisheries management. Their yield per recruit models provided a practical basis for rational management towards conservation goals. The strength of their contribution lies in its sound theoretical foundations. Beverton and Holt were, however, keenly aware that the limitations of the information processing machinery available in 1957 placed severe constraints on their analyses. The legacy of Beverton and Holt suffers, consequently, from limited use of statistical procedures and high levels of aggregation. In particular, information regarding the seasonal distribution and growth patterns in a stock is not produced by this system on a routine basis.

Beverton and Holt (1957) discussed many extensions to their simple theory of fishing, including models for the dependence of growth and mortality on population density, but also for spatial variation. Their models were formulated as differential equations. They recognized that a fishery has both a human and a biological side. While Beverton and Holt have clearly influenced the methods used by CAFSAC, it would be unfair to blame them for any shortcomings that exist today.

One must also remember that the purpose of the CAFSAC system is to supply advice to fisheries managers. Many desirable refinements would be costly to implement. It can be argued with some justification that improved sampling should have priority over refinement of the analyses and synthesis. In particular, many minor stocks have not been adequately sampled, making the analyses described in Table 3 impossible. In such cases biologists have resorted to the use of inappropriate models, and have, justifiably, been criticized for it (Beddington et al. 1980).

Quantitative models are useless without adequate data. Needs of particular models are unlikely to be satisfied unless anticipated during the design of information systems. The Task Force on Atlantic Fisheries was hampered in its work by limitations in the available data. While specific data relating to costs and revenues of fishing enterprises were not available, the Task Force also encountered problems with the way biological data are processed and analysed.

Reductions in levels of aggregation for biological data are required not only to understand interactions between man and the fishery, but also between other components of marine ecosystems. The fundamental factor preventing reduced levels of aggregation is the need to transfer data between different parts of the management organization.

Today's technology makes possible a data processing system which is dedicated not to the requirements of a particular model, but to those of a generalized model. Such a system will not discard information, but be able to provide it at many levels of aggregation. It will include new kinds of information and will have the flexibility to accept change.

The impetus for the creation of such a system can come only from theory, for theory will provide the foundation for future models. Theoreticians must work to establish generality and flexibility of information as goals ranking in importance with the current requirements of operational models. This will require new operational models in which biological and economic views of the systems are inseparably intertwined.

Acknowledgements

The author is indebted to R.J.H. Beverton and S.J. Holt for their comments on an earlier draft. Critical readings by R. Halliday, R. Mahon and S.J. Smith were also greatly appreciated. M. Mingo's able technical assistance freed the author from many onerous chores so he could concentrate on the writing.

Bibliography

Beddington, J., D. Botkin, and S.A. Levin. (1980). Mathematical models and resource management, p. 1-5. IN T.L. Vincent and J.M. Skowronski (eds.). Renewable Resource Management. Springer-Verlag L.N. Biomath. 40, New York.

Beverton, R.J.H. and S.J. Holt. (1957). On the dynamics of exploited fish populations. Fish. Invest. Ser. II. Mar. Fish. G.B. Minist. Agric. Food Fish 29: 1-533.

Beyer, J. and P. Sparre. (1982). Modelling exploited fish stocks. IN S.E. Jorgensen (ed.) Application of ecological modelling in environmental management. Elsevier Scientific Publ. Co., Amsterdam.

CAFSAC. (1982). Annual report. Vol. 3. Canadian Atlantic Fisheries Scientific Advisory Committee, Dartmouth, Nova Scotia.

Fisheries and Marine Service. (1976). Policy for Canada's Commercial Fisheries.

Department of Environment, Ottawa. 70 p.

Goh, B.S. (1980). Modelling and management of fish populations with high and low fecundities, p. 54-63. IN T.L. Vincent and J.M. Skowronski (ed.). Renewable Resource Management. Springer Verlag L.N. Biomath 40, New York.

Gulland, J.A. and L.K. Boerema. (1973). Scientific advice on catch levels. U.S. Fish Wildl. Serv. Bull. 71: 325-335.

Luce, R.D. and H. Raiffa. (1957). Games and decisions. Wiley, New York, N.Y.

Pearse, P.H. (1982). Turning the tide; a new policy for Canada's Pacific Fisheries Commission on Pacific Fisheries Policy; Final Report. Vancouver.

Rivard, D. (1982). APL programs for stock assessment (revised). Can. Tech. Rep. Fish. Aquat. Sci. 1091: 146 p.

Scott, A. and P.A. Neher (ed.). (1981). The public regulation of commercial fisheries in Canada. Economic Council of Canada, Ottawa. 76 p.

Sinclair, M., R.N. O'Boyle, and T.D. Iles. (1983). Consideration of the stable age distribution assumption in "analytical" yield models. Can. J. Fish. Aquat. Sci. 40: 95-101.

Task Force on Atlantic Fisheries. (1982). Navigating troubled waters. A new policy for the Atlantic fisheries. Minister of Supply and Services Canada, Ottawa. 379 p.

White, G.N. 3rd. (1983a). Identification of influential variables in yield per recruit analyses. Can. Spec. Publ. Fish. Aquat. Sci. In Press.

White, G.N. 3rd. (1983b). Software quality in open systems: The fisheries management experience. Manuscript.

White, G.N. 3rd. (1983c). Uncertainty in implementions of a model for mortality in exploited stocks. Manuscript.

Department of Environment, Ottawa. 70 p.

Goh, B.S. (1980). Modelling and management of fish populations with high and low fecundities. p. 54-55. IN T.L. Vincent and J.M. Skowronski (ed.). Renewable Resource Management. Springer Verlag Lect. Notes Biomath 40, New York.

Gulland, J.A. and L.K. Boerema. (1973). Scientific advice on catch levels. U.S. Fish Wildl. Serv. Bull. 71: 325-335.

Luce, R.D. and H. Raiffa. (1957). Games and Decisions. Wiley, New York, N.Y.

Pearse, P.H. (1982). Turning the tide: a new policy for Canada's Pacific Fisheries. Commission on Pacific Fisheries Policy, Final Report. Vancouver.

Rivard, D. (1982). APL programs for stock assessment (revised). Can. Tech. Rep. Fish. Aquat. Sci. 1091: 146 p.

Scott, A. and P.A. Neher (ed.). (1981). The public regulation of commercial fisheries in Canada. Economic Council of Canada, Ottawa. 75 p.

Sinclair, M., R.K. O'Boyle, and T.D. Iles. (1983). Consideration of the stable age distribution assumption in "analytical" yield models. Can. J. Fish. Aquat. Sci. 40: 95-101.

Task Force on Atlantic Fisheries. (1982). Navigating troubled waters, a new policy for the Atlantic fisheries. Minister of Supply and Services Canada (Ottawa). 379 p.

White, G.N. 3rd. (1983a). Identification of influential variables in yield per recruit analyses. Can. Spec. Publ. Fish. Aquat. Sci. In Press.

White, G.N. 3rd. (1983b). Software quality in open systems: the fisheries management experience. Manuscript.

White, G.N. 3rd. (1983c). Uncertainty in improvements of a model for mortality in exploited stocks. Manuscript.

PART V

APPLICATIONS: EPIDEMIOLOGY

ERADICATION STRATEGIES FOR VIRUS INFECTIONS

K. Dietz and D. Schenzle
Tübingen University

The most important results in mathematical epidemiology are the threshold theorems which specify critical conditions for the spread of an epidemic and the stability of an endemic state. In the case of virus infections these conditions imply a minimum proportion of the population which must be immunised in order to prevent an epidemic and to eradicate an endemic infection. Thus they have practical consequences for the design and the evaluation of vaccination programs.

The simplest model for an endemic state (Dietz, 1976) assumes an age-independent contact and death rate β and μ of the human host. In the equilibrium state it can be shown that the proportion of susceptibles $\bar{u}$ in the population equals R^{-1}, where R is the basic reproductive rate of the infection, i.e. the number of secondary cases which one case would generate if the total population were susceptible. In order to render the equilibrium state with zero infectives stable, the minimum proportion p to be immunised is $1-\bar{u}$, i.e. $1-R^{-1}$. The basic reproductive rate is the product of the contact rate β and the average duration of one infectious period.

This oversimplified theory has been applied recently to a number of virus diseases, such as measles, whooping cough and rubella (Anderson and May, 1982; Hethcote, 1983). The proportion of susceptibles in the population is estimated from a simple exponential fit to the age-specific prevalence of susceptibles:

$$\bar{u} = \int_0^\infty \ell(a) e^{-\lambda a} da \Big/ \int_0^\infty \ell(a) da, \tag{1}$$

where $\ell(a)$ is the survivor function and λ is the equilibrium incidence of the infection. Here it is assumed that the population is in demographic equilibrium, i.e. that the density of the age distribution is proportional to the survival function. The life expectancy of a newborne is

$$L = \int_0^\infty \ell(a)da. \tag{2}$$

For age independent death rates μ, Eq. (1) reduces to

$$\bar{u} = \mu/(\mu+\lambda) = 1/(1+L/A) = 1/R, \tag{3}$$

if we denote the average age at infection by A. For measles the basic reproductive rate was estimated to be around 16 which implied that the minimum percentage to be immunised for eradication is about $100(1-1/16)\% = 94\%$.

It is unlikely that such a high level of effective protection can be achieved in a population as large as that of the U.S. and yet recent reports from the Centers for Disease Control (1983) suggest that herd immunity is achieved. The total of 1697 reported cases in 1982 resulted from 639 index cases out of which 625, i.e. 98%, did not produce a secondary case. These figures raise the question whether the target of 94% immunisation is really necessary.

The oversimplified model on which this target is based makes assumptions which cannot be maintained if one examines the data in more detail. In reality the contact rates vary considerably by age, place and season.

A detailed analysis by Fine and Clarkson (1982) revealed particularly a strong age-dependence of the risk of infection λ: it rises from a level of 0.1 per year at preschool age to 0.65 per year at the age of seven, but then declines again to 0.1 per year at age ten.

The age-specific risk of infection $\lambda(a)$ depends on the age-specific proportion of infectives $y(a')$ and the contact rates $\beta(a,a')$ between infectives of age a' and susceptibles of age a:

$$\lambda(a) = \int_0^\infty \ell(a')y(a')\beta(a,a')da'/L. \tag{4}$$

The epidemiological records allow estimation only of $\lambda(a)$. The age-specific contact rates $\beta(a,a')$ cannot be determined uniquely for given $\lambda(a)$ without additional *a priori* assumptions.

In Schenzle and Dietz (1983)*we explored the consequences of one particular choice of $\beta(a,a')$ which was compatible with the observed $\lambda(a)$. There we assumed $\beta(a,a')$ to be symmetric in its arguments. Contacts were highest within the same yearly age class for children up to age ten. The remaining contact rates within and between age classes have three different levels specific for children, young adults, and older adults. The survivor function $\ell(a)$ equals one up to age 35 and then declines exponentially at the rate μ = 1/40 per year such that L = 75 years.

For age-dependent $\lambda(a)$, the proportion of susceptibles $\bar{u}$ in the population is given by

(submitted for publication)

$$\bar{u} = \int_0^\infty \ell(a)\exp\{-\int_0^a \lambda(a')da'\}da/L. \tag{5}$$

In the present example we get $\bar{u} \approx 8\%$. If the formulas for the age-independent model would still be valid as Anderson and May (1982) suggest, one would have to vaccinate 92% and the basic reproduction rate would be about $1/0.08 = 12.5$. But a stability analysis of the equilibrium with just susceptibles shows that the decisive eigenvalue is only 4.8, which yields a minimum protection level of 79%.

Our emphasis of the present note is not the numerical determination of the critical measles vaccination level for the U.S.. We only wanted to point out the methodological problem of the nonidentifiability of the crucial contact rates on the basis of age-specific risks of infection. The comparison of observed and expected incidence patterns and response to known interventions may help to narrow down the uncertainty about the minimum effort required to eradicate a virus infection.

LITERATURE CITED

Anderson, R.M. and May, R.M. (1982). Directly transmitted diseases: control by vaccination. *Science* 215: 1053-1060.

Centers for Disease Control (1983). Measles-United States, 1982. *Morbidity and Mortality Weekly Report* 32: 49-51.

Dietz, K. (1976). The incidence of infectious diseases under the influence of seasonal fluctuations. *Lecture Notes in Biomathematics* 11: 1-15.

Fine, P.E.M. and Clarkson, J.A. (1982). Measles in England and Wales-II: The impact of the measles vaccination programme on the distribution of immunity in the population. *Int. J. Epidemiology* 11: 15-25.

Hethcote, H.W. (1983). Measles and rubella: will they be eradicated in the USA? *Amer. J. Epidemiology* 117: 2-13.

MODELS FOR A CLASS OF MAN-ENVIRONMENT EPIDEMIC DISEASES†

V. Capasso*

Pomona College
Claremont, CA 91711

1. Introduction

The mathematical theory of man-environment epidemic diseases is not yet well developed as such [1], even if many attempts have been made to describe particular examples of such diseases (see for example the brilliant work on schistosomiasis and helminthic infections [3, 14, 15, 16]). Our own interest started from the analysis of fecal-oral transmitted diseases (such as cholera, typhoid fever, infectious hepatitis, poliomyelitis, etc.), which can be regarded as a particular class of man-environment diseases, but we would like to propose these models as a possible way to describe other diseases which are spread via an interaction (direct or indirect) with the environment. We start with the analysis of the basic model (Section 2) which had been originally proposed to describe the mechanism of transmission of cholera epidemics in the European Mediterranean regions in 1973 [11]. This model suggests the introduction of a basic threshold parameter θ which has been the guideline for all subsequent work; it is important to mention that, for preventive purposes, we need to suggest, to the health authorities of a country, parameters that must be maintained below a suitable threshold in order to prevent the outbreak of future epidemics. In Section 3 we analyze modifications of the basic threshold parameter to take into account seasonal fluctuations of the parameters of the epidemic model. A behavior of this kind was revealed by the statistical analysis of the experimental data relative to the typhoid fever in the city of Bari, Southern

†Work performed under the auspices of the GNFM-CNR (Italy) in the context of the Program for Preventive Medicine (MPP1-CNR). Partially supported by Pomona College and Harvey Mudd College, Claremont.

*On leave from the Department of Mathematics, University of Bari, Palazzo Ateneo, 70121 Bari, Italy.

Italy [8], where to a mean endemic level, we could see a superimposed periodic fluctuation of a seasonal kind. In Section 4 we consider another important modification of the basic model to take into account the spatial heterogeneity of the environment; in fact usually the human population does not contribute to it in an homogeneous way; human infectives send the multiplied agent of the infection (bacteria, viruses, etc.) to a reservoir (sea, ponds, underground water supply, etc.), from which the agent then spreads to the whole habitat via many different mechanisms which here have been described generally as random diffusion. More typical mechanisms of spread may be considered when more and different information are acknowledged. Other sources of spatial heterogeneity have been studied for this class of epidemic phenomena; we just mention the case in which the force of infection of human individuals $g(z)$ due to the agent (see Section 2) is not always concave but S-shaped; for the sake of mathematical simplicity we will not consider here this case, limiting ourselves to refer to [10] (see also [15]).

We only need to mention that many other modifications may be important, from case to case, to take into account other particular features of the mechanism of infection, such as age dependency, presence of carriers, etc. and in fact some of these modifications are under study.

2. The Basic ODE Model

The basic mathematical model suggested in [11] was based on the following ordinary differential equation system

$$(2.1)\qquad \begin{aligned} \frac{dz_1}{dt} &= -a_{11}z_1(t) + a_{12}z_2(t) \\ \frac{dz_2}{dt} &= -a_{22}z_2(t) + g(z_1(t)) \end{aligned} \qquad , \quad t > 0$$

with the following interpretation:

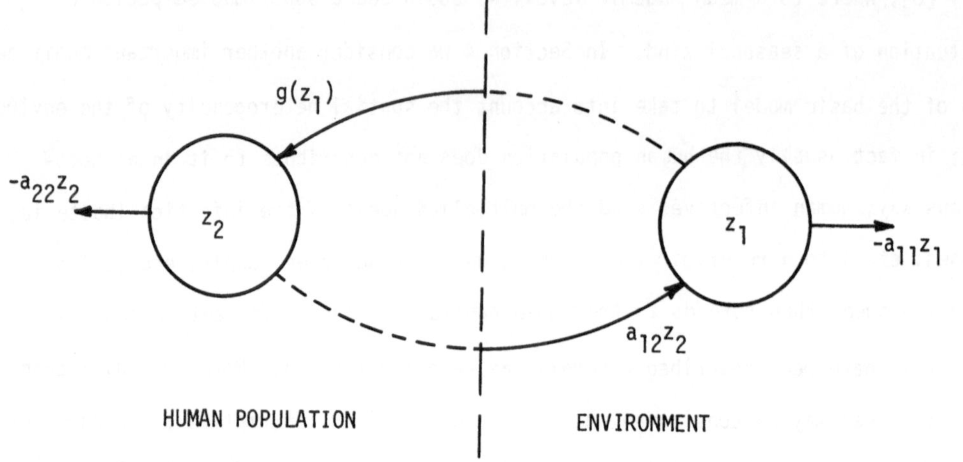

Here

$z_1(t)$ denotes the (average) concentration of the infecting agent in the environment, at time t;

$z_2(t)$ denotes the infective human population at time t;

$1/a_{11}$ is the mean lifetime of the agent in the environment;

$1/a_{22}$ is the mean infectious period of the human infectives;

a_{12} is the growth rate of the agent due to the human population;

$g(z_1)$ is the "force of infection" of the human population due to the agent.

The "force of infection" $g(z_1)$ can be explained, for fecal-oral transmitted diseases, as follows [11]:

$$g(z_1) = N\beta p\ h(z_1) \tag{2.2}$$

where N is the total human population in the community (it can be considered very large in respect to $z_2(t)$ and then constant); β is the fraction of susceptible individuals out of N; p is the fraction exposed to the contaminated environment per unit time (probability per unit time for a person to have a "snack" of contaminated food); $h(z_1)$ is the probability (as a function of the concentration of the contaminating agent), for an exposed susceptible, to get the infection. It is plausible to assume that p and β are z_1 and z_2 independent and that $h(z_1)$ has a saturating behavior.

With this in mind we may assume that in the model (2.1), a_{11}, a_{22}, a_{12} are all positive quantities, and that $g:R_+ \to R_+$ is a twice continuously differentiable function satisfying the following assumptions:

(i) $0 < g(z') < g(z'')$, if $0 < z' < z''$

(ii) $g(0) = 0$

(iii) $g''(z) < 0$, for any $z > 0$

(iv) $0 < g'_+(0) < +\infty$ (right first derivative)

(v) $\lim_{z\to+\infty} \frac{g(z)}{z} < \frac{a_{11}a_{22}}{a_{12}}$

In order to analyze the qualitative behavior of the model (2.1) we consider the two isoclines

$$f_1(z_1,z_2) := -a_{11}z_1 + a_{12}z_2 = 0 \tag{2.3}$$

$$f_2(z_1,z_2) := -a_{22}z_2 + g(z_1) = 0$$

in the positive quadrant $K:=R_+ \times R_+$. Due to the assumptions (i)-(v) on $g(z)$, system (2.3), which always has the trivial solution, may have or may not have the nontrivial solution depending upon the value of the following parameter with respect to one:

$$\theta := \frac{g'_+(0)a_{12}}{a_{11}a_{22}} \tag{2.4}$$

In fact the following "threshold theorem" holds true.

Theorem 2.1. (a) If $0 < \theta < 1$ then system (2.1) admits only the trivial equilibrium solution in the positive quadrant K. It is there globally asymptotically stable.

(b) If $\theta > 1$ then two equilibrium solutions exist for the system (2.1) in the positive quadrant K: they are given by the origin $0 = (0,0)'$ and by the only nontrivial solution $z^* = (z_1^*, z_2^*)'$ of the system (2.3) $(z_1^* > 0, z_2^* > 0)$. In this case, the origin is unstable, while z^* is globally asymptotically stable in $K - \{0\}$.

According to this theorem then θ is a threshold parameter. When $\theta > 1$ any epidemic tends to extinction. On the other hand if $\theta > 1$ then a nontrivial endemic steady state appears to which any epidemic eventually tends. Due to the form of θ (see (2.4) and (2.2)) this result may suggest preventive measures to maintain θ below one.

3. Seasonal Fluctuations

As reported in [8] a time-series analysis of the data relative to the typhoid fever in the city of Bari (Southern Italy) in the years 1955-1978, shows a seasonal behavior (one year periodicity) of the endemic level, and this corresponds to an analogous behavior all over the country. This was conjectured to be due to the seasonally varying eating habits of the human population; if we want to take this into account, we need to introduce in the force of infection $g(z_1)$ (see (2.2)) an explicit time dependence through the parameter p which describes in fact the eating habits of the population. It has to be a periodic function of the time t with a minimal period $\omega > 0$ (1 year) [9]:

$$(3.1) \qquad p(t) = p(t+\omega) > 0\,, \quad t \in R.$$

The function g in the basic model will be substituted by

$$(3.2) \qquad g(t;z) = p(t)h(z), \quad t \in R, \quad z \in R_+$$

where now $h(z)$ has the same mathematical properties (i)-(v) listed for g. In particular if we set $p_{max} := \max_{0 \le t \le \omega} p(t)$, we ask that (v') $$\lim_{z\to +\infty} \frac{h(z)}{z} < \frac{a_{11}a_{22}}{a_{12}\, p_{max}}$$

Under these new assumptions it was shown in [9] that

Theorem 3.1. (a) If

(3.3) $$\theta^{max} := \frac{a_{12}p_{max}h'_+(0)}{a_{11}a_{22}} < 1$$

then the epidemic always tends to extinction.

(b) If

(3.4) $$\theta^{min} := \frac{a_{12}p_{min}h'_+(0)}{a_{11}a_{22}} < 1$$

then a unique nontrivial periodic endemic state exists for system (2.1) (with (3.2)) which is globally asymptotically stable in the whole positive quadrant $K - \{0\}$.

As we can see here the basic threshold parameter θ given in (2.4) is now split in two, even if actually only (3.3) is needed for preventive purposes; it can be shown that θ^{max} is an upper estimate of the actual threshold parameter which cannot be evaluated in practice. In a recent paper [12] methods are given to identify the periodic function $p(t)$ and the multiplication rate a_{12}, based upon the knowledge of the other parameters in model (2.1) and the number of new infections per month, for the twelve months of a year.

4. A Reaction-Diffusion Model.

A close look at the data of the cholera epidemic which spread in Bari in 1973 [7] shows that in a very short period of time the epidemic spread to the whole geographic area of the town, even if the first data were reported from a quite small neighborhood. This induces us to think about a diffusive mechanism which spreads the infecting agent from the infectious individuals to the whole habitat.

This must be related to the fact that usually in this area, as in many other regions of the Mediterranean area, sewage goes untreated into the sea, and from there it is likely that the multiplied infective agent goes back to the habitat due to the particular habit of consumption of raw contaminated food; the incidence of cases shows in fact an important dependence upon the eating habits of the population of the various neighborhoods.

We can conjecture then that for man-environment epidemics the basic mechanism of diffusion is the following: The environment cannot be considered spatially homogeneous since the human population sends the contaminating agent to a small subregion (one or more) of the environment itself, that we may call "reservoir" (see also [13]), from which it is spread somehow to the whole environment, e.g., by random dispersal (through the Laplace operator Δ). With this in mind, if we neglect the dispersal of infectives, system (2.1) is modified as follows:

$$\text{(4.1)}\qquad \begin{aligned} \frac{\partial}{\partial t} u_1(x;t) &= d_1 \Delta u_1(x;t) - a_{11} u_1(x;t) + \int_\Omega k(x,x') u_2(x';t)dx' \\ \frac{\partial}{\partial t} u_2(x;t) &= - a_{22} u_2(x;t) + g(u_1(x;t)) \end{aligned}$$

with $(x,t) \in \Omega \times (0,\infty)$.

Here $u_1(x;t)$ is the local concentration of the contaminating agent (bacteria, viruses, etc.) at point $x \in \Omega$, and time $t \in (0,+\infty)$; $u_2(x;t)$ is the local density of infective population at point $x \in \Omega$, and time $t \in (0,+\infty)$.

The habitat is mathematically represented by Ω, an open bounded subset of R^2, whose boundary $\partial\Omega$ is assumed to be sufficiently smooth. At the boundary we assume conditions of the third type for the first integrodifferential equation:

$$\text{(4.1b)}\qquad \beta(x) \frac{\partial}{\partial \nu} u_1(x;t) + \alpha(x) u_1(x;t) = 0, \quad (x,t) \in \partial\Omega \times (0,+\infty)$$

Here $\partial/\partial\nu$ denotes the outward normal derivative on $\partial\Omega$; $\alpha(x)$ and $\beta(x)$ are sufficiently smooth nonnegative functions of $x \in \partial\Omega$, such that $\alpha(x) + \beta(x) > 0$ on $\partial\Omega$. This condition includes, as particular cases, the homogeneous Dirichlet boundary conditions - $\beta(x) = 0$ in $\partial\Omega$ - (completely hostile boundary), and the homogeneous Neumann boundary conditions - $\alpha(x) = 0$ in $\partial\Omega$ - (isolated habitat).

The kernel $k(x,x')$ is a continuous nonnegative real function of $(x,x') \in \bar{\Omega} \times \bar{\Omega}$; it describes mathematically the presence of possible reservoirs to which the infective agent sends the multiplied contaminating agent. The function $g(z)$ has the same meaning as in the basic model (2.1), but now, instead of (v) we assume that

(v") $$\lim_{z \to +\infty} \frac{g(z)}{z} < \frac{a_{11} a_{22}}{a_{12}}$$

where we have denoted, in analogy with the basic model:

(4.2) $$a_{12} := \max_{x \in \bar{\Omega}} \int_{\Omega} k(x,x')dx'.$$

We would like to mention that mechanisms of interaction via integral operators have been considered in the past in connection with the classical Kermack-McKendrick epidemic model (see e.g. [2, 4, 13] and the related literature).

The mathematical analysis of system (4.1), (4.1b) can be found in [6]. Here we will report the threshold theorem. In order to state this we need to recall the following.

Consider the following linear boundary value problem

(4.3)
$$\Delta\phi + \lambda\phi = 0, \quad \text{in } \Omega$$
$$\beta(x) \frac{\partial\phi}{\partial\nu}(x) + \alpha(x)\phi(x) = 0, \quad \text{on } \partial\Omega.$$

It is well known that its first eigenvalue is a nonnegative real number λ_1; if we denote by $\phi_1(x)$ the corresponding eigenfunction we have that $\alpha(x) \geq 0$, $\beta(x) \geq 0$ in $\partial\Omega$ imply $\lambda_1 \geq 0$, $\phi_1(x) > 0$ in Ω; $\alpha(x) > 0$ in $\partial\Omega$ implies $\lambda_1 > 0$; $\beta(x) > 0$ in $\partial\Omega$ implies $\phi_1(x) > 0$ in $\bar{\Omega}$; and if $\beta(x) = 0$ at some points in $\partial\Omega$, then $\phi_1(x) = 0$ at some points in $\partial\Omega$. Anyhow in this last case it can be shown that another function $\tilde{\phi}_1(x) > 0$ in $\bar{\Omega}$ can be used instead of ϕ_1 such that

$$\Delta\tilde{\phi}_1 + \lambda_1\tilde{\phi}_1 \le 0 \text{ , in } \Omega$$

$$\beta(x)\frac{\partial\tilde{\phi}_1}{\partial\nu}(x) + \alpha(x)\tilde{\phi}_1(x) \ge 0, \text{ on } \partial\Omega.$$

We always normalize ϕ_1 so that $\max_{x\in\bar{\Omega}} \phi_1(x) = 1$. We will see that, from our point of view, this eigenfunction ϕ_1 completely describes the "geography" of the habitat.

Consider the integral operator

(4.5) $$H(v)(x) := \int_\Omega k(x,x')v(x')dx', \quad x \in \bar{\Omega}$$

where v is any real continuous function of $x \in \bar{\Omega}$. We define

(4.6) $$\overline{H(\phi_2)} := \frac{\max_{x\in\bar{\Omega}} H(\phi_1)(x)}{\min_{x\in\bar{\Omega}} \phi_1(x)}$$

(take into account the comments about $\tilde{\phi}_1$), and

(4.7) $$\underline{H(\phi_1)} := \frac{\min_{x\in\bar{\Omega}} H(\phi_1)(x)}{\max_{x\in\bar{\Omega}} \phi_1(x)} = \min_{x\in\bar{\Omega}} H(\phi_1)(x).$$

It is clear then that

(4.8) $$\underline{H(\phi_1)}\, \phi_1(x) \le H(\phi_1)(x) \le \overline{H(\phi_1)}\, \phi_1(x), \quad x \in \bar{\Omega} .$$

The following theorem holds [6].

Theorem 4.1. If

$$\theta_1 := \frac{g_+'(0)\ \overline{H(\phi_1)}}{(a_{11}+d_1\lambda_1)a_{22}} < 1 \tag{4.9}$$

then the trivial solution is the unique steady state of system (4.1), (4.1b), and it is globally asymptotically stable in the class of all continuous real functions which are nonnegative in $\bar{\Omega}$, $X_+ := \{u \in C(\bar{\Omega}) \times C(\bar{\Omega}) |\ u = (u_1,u_2)'$, $u_1(x) \geq 0,\ \ u_2(x) \geq 0$ in $\bar{\Omega}\}$.

If

$$\theta_2 := \frac{g_+'(0)\ \underline{H(\phi_1)}}{(a_{11}+d_1\lambda_1)a_{22}} > 1 \tag{4.10}$$

then the trivial solution is unstable.

In this last situation we want to know if some nontrivial steady state $(\bar{u}_1(x),\bar{u}_2(x))'$ exists for system (4.1), (4.1b). If it exists, it must be a solution of the following nonlinear stationary problem

$$\begin{aligned} &d_1\Delta\bar{u}_1(x) - a_{11}\bar{u}_1(x) + \int_\Omega k(x,x')\bar{u}_2(x')dx' = 0 \\ &-a_{22}\ \bar{u}_2(x) + g(\bar{u}_1(x)) = 0 \end{aligned} \tag{4.11}$$

in Ω, with

$$\beta(x)\frac{\partial}{\partial\nu}\bar{u}_1(x) + \alpha(x)\ \bar{u}_1(x) = 0, \quad \text{in } \partial\Omega. \tag{4.11b}$$

From the second of (4.11) we get

$$\bar{u}_2(x) = \frac{1}{a_{22}}\ g(\bar{u}_1(x)), \qquad x \in \Omega \tag{4.12}$$

and then from the first

$$d_1\Delta\bar{u}_1(x) - a_{11}\bar{u}_1(x) + \frac{1}{a_{22}}\int_\Omega k(x,x')g(\bar{u}_1(x'))dx' = 0, \text{ in } \Omega, \tag{4.13}$$

subject to (4.11b). It is clear then, due to the presence of the integral term, that (4.13) and hence (4.11) admits a spatially homogeneous solution, even with homogeneous Neumann boundary conditions ($\alpha(x) = 0$ in $\partial\Omega$). A complete solution of (4.13), (4.11b) and then of (4.12) can be obtained via numerical simulations. Here we will limit to report conditions for the existence of a nontrivial solution for (4.13), (4.11b) [6], thus completing Theorem 4.1.

Theorem 4.2. If $\theta_2 > 1$ (see (4.10)), then system (4.1), (4.1b) admits a unique nontrivial endemic state which is globally asymptotically stable in $X_+ - \{0\}$.

We wish to comment here that we need to find an upper estimate of the threshold parameter θ_1 that can be measured in concrete cases, in order that preventive control measures can be suggested to the health authorities. As can be seen from (4.9), the basic threshold parameter θ, given in (2.4), may be considered again a good upper estimate if we substitute a_{12} here with an upper estimate of $\overline{H(\phi_1)}$.

Literature Cited

[1] Anderson, R.M. and May, R.M., eds., (1982). Population Dynamics of Infectious Diseases Agents. Dahlem Konferenzen, Springer Verlag.

[2] Aronson, D.G. (1977). The asymptotic speed of propagation of a simple epidemic. In Nonlinear Diffusion (Fitzgibbon, W.E. and Walker, A.F., eds.), Pitman, London.

[3] Barbour, A.D. (1978). MacDonald's model and the transmission of bilharzia. Trans. Roy. Soc. Trop. Med. Hyg. 72: 6-15.

[4] Capasso, V. (1978). Global solution for a diffusive nonlinear deterministic epidemic model. SIAM J. Appl. Math. 35: 274-284.

[5] Capasso, V. (1980). Mathematical Models for Infectious Diseases. Quaderni dell' Istituto di Analisi Matematica 2, Bari (in Italian).

[6] Capasso, V. (1982). Asymptotic stability for an integrodifferential reaction-diffusion system. Submitted.

[7] Capasso, V., Grosso, E. and Serio, G. (1977). I modelli matematici nella indagine epidemiologica. I Applicazione all'epidemia di colera verificatasi in Bari nel 1973. Ann. Sclavo 19: 193-208.

[8] Capasso, V., Grosso, E. and Serio, G. (1980). I modelli matematici nella indagine epidemiologica. II. Il tifo addominale: studio delle serie temporali. Ann. Sclavo 22: 89-206.

[9] Capasso, V. and Maddalena, L. (1982). Periodic solutions for a reaction-diffusion system modeling the spread of a class of epidemics. SIAM J. Appl. Math. To appear.

[10] Capasso, V. and Maddalena, L. (1982). Saddle point behavior for a reaction-diffusion system. Application to a class of epidemic models. Math. Comp. Simulation. To appear.

[11] Capasso, V. and Paveri-Fontana, C.L. (1979). A mathematical model for the 1973 cholera epidemic in the European Mediterranean region. Rev. Epidem. Santé Publ. 27: 121-132.

[12] Di Lena, G. and Serio, G. (1982). The identification of the periodic behavior in an epidemic model for oro-fecal transmitted diseases. Submitted.

[13] Kendall, D.G. (1957), in discussion on Bartlett, M.S. (1957). Measles periodicity and community size. J. Roy. Stat. Soc. Ser. A. 120: 48-70.

[14] MacDonald, G. (1965). The dynamics of helminth infections, with special reference to schistosomes. Trans. Roy. Soc. Trop. Med. Hyg. 59: 489-506.

[15] Näsell, I. and Hirsch, W.M. (1972). A mathematical model of some helminthic infections. Comm. Pure Appl. Math. 25: 459-478.

[16] Näsell, I. and Hirsch, W.M. (1973). The transmission dynamics of schistosomiasis. Comm. Pure Appl. Math. 26: 395-453.

MATHEMATICAL MODELS OF VERTICAL TRANSMISSION OF INFECTION

Kenneth L. Cooke
Pomona College, Claremont, California

1. Vertical transmission

Many diseases that are important from the standpoint of public health, animal husbandry, and agriculture are passed from one individual to another by two distinct mechanisms that may be called horizontal transmission and vertical transmission. In horizontal transmission, susceptible individuals become infected by direct or indirect contact with other individuals who are infectious and who are living at the same time. Vertical transmission refers to direct transmission from infected parents to offspring before or during birth.

Before considering possible models of vertical transmission, we shall mention a few examples that illustrate the various pathways or modes of such transmission. Some infections are passed from mother to offspring transplacentally, during the development of the fetus. The rubella virus can be transmitted in this way. A herpes simplex 2 virus infection of the cervix can cause infection of an infant during passage through the birth canal. For perpetuation of a disease, the most reliable and effective mode of vertical transmission is through the germ line. In vertebrates,the ova or sperm may be infected. For example, cytomegalovirus has been found in human semen. In arthropods, the eggs may carry an infection from one generation to the next. Such transovarial infection is known to occur in the ticks that harbor the agent that causes Rocky Mountain fever, which we shall discuss in some detail. It also occurs for certain arboviruses, in plant viruses carried by aphids and leafhoppers, and so on. In plants, transmission through seeds is fairly common.

Additional information on vertically transmitted diseases may be found in the survey articles by Fine (1975) and Mims (1981)*. Fine has proposed a classification according to

*This reference was kindly pointed out by K. Dietz.

the number of different hosts or species involved and according to whether both vertical and horizontal transfer occur.

Although there is an abundance of mathematical models of infectious diseases, very few incorporate vertical transmission. Fine has used discrete-time or generation-to-generation models in several cases, and we refer the reader to Fine and Le Duc (1978) for an example of the application to an encephalitis virus. Anderson and May (1979, 1981) have introduced some continuous-time (differential equation) models and have discussed factors relating to disease persistence in a population. We shall comment further on their model in Section 3, where we report on a model of Busenberg, Cooke, and Pozio (1982). Another approach, utilizing a combination of discrete and continuous dynamics, has been formulated in Busenberg and Cooke (1980) and is described in Section 4 below. Finally, a preliminary model including age structure has been described in Cooke and Busenberg (1980).

2. Rocky Mountain spotted fever

The models to be described in Sections 3 and 4 extend the classical models of mathematical epidemiology, and have the potential to be applied to various infectious diseases. On the other hand, some of the underlying assumptions were framed with special reference to the particular disease called Rocky Mountain spotted fever. Consequently, we shall briefly discuss the epidemiology of this disease before the formulation of the models.

Rocky Mountain spotted fever (hereinafter denoted RMSF) is a disease that affects several species of mammals, including man. It is endemic in large regions of the United States, including the Rocky Mountain states and also the Atlantic states as far north as Nova Scotia. Over 1500 cases in humans were reported in the period 1970-1974, and mortality in untreated cases is over 20 %. The disease is transmitted to humans by ticks, the most important species in the United States being the arthropods *Dermacentor variabilis* (American dog tick) and *Dermacentor andersoni* (Rocky Mountain wood tick). The infectious agent is *Rickettsia rickettsii* and it is vertically transmitted to the progeny of infected female ticks.

The population dynamics of D. variabilis vary with the climate of the region, and we shall here give a very abbreviated description applicable to the more northerly eastern habitats. For additional information, see Garvie et al. (1978) and Burgdorfer (1975). In this northerly region, the tick has essentially a two-year life cycle which begins with overwintered adult ticks seeking a mammalian host on which to feed in the months of April to June. After finding a host and engorging with a blood meal, the ticks mate and the female drops to the ground to lay eggs which hatch into larvae from June to September. Most of the larvae do not find hosts at this time, but overwinter and do so in the next spring. They then molt into nymphs, which in turn seek hosts and molt into adults who overwinter into the third year to start this cycle over.

The transmission of Rickettsia occurs via vertical transmission from infected females through their eggs, and via horizontal transmission that occurs when infected and uninfected larvae or nymphs feed on the blood of a small mammal. An uninfected tick can become infected if the mammal is undergoing severe Rickettsemia, and an infected tick can transfer Rickettsia to a previously uninfected mammal on which it is feeding. The percentage of infected offspring of infected female parent ticks is very high, exceeding 90°/o in laboratory studies. Viability of ticks does not seem to be affected by a Rickettsia infection, but there is some evidence that fecundity diminishes after the fifth generation in a line of infected females. There is little horizontal transfer of infection between different generations since the adults and the immature ticks that are active in the summer feed on different species of mammals. It appears that infected ticks remain infected for the duration of their lives. The factors that seem to limit the population growth of ticks include predation on the immature ticks and unfavorable growing conditions due to climatic changes.

3. A model with two age groups

In this section, we shall describe some models that have been analyzed by Busenberg, Cooke and Pozio (1982). The assumptions of the first model will now be explained. We consider a single-species population, which is asexual or such that it suffices to consider only one sex. There are two age groups: mature (or adult) and immature. Individuals in each age group are separated into two epidemiological classes, susceptible or infective, and all horizontal transmission is assumed to take place among the mature individuals. We assume that there is a maturation period of length $T \geqslant 0$, during which the newborn are not fertile and do not participate in horizontal transmission of the disease. For example, this would be true for eggs laid by mature females. Thus, our model involves the number $I(t)$ of mature infectives at time t, the number $S(t)$ of mature susceptibles at time t, and these same functions evaluated at time $t-T$.

We also assume that mature infectives and susceptibles are removed at the constant rates r' and r, respectively, and immature individuals at rates $r^{*\prime}$ and r^*. These can represent death rates, or removals to an immune class. However, the model does not allow for members of this removed class to participate in reproduction or in transmission of the infection. We also assume that mature susceptibles have a constant birth rate $\tilde{b}$ and mature infectives have a constant birth rate $\tilde{b}'$. Vertical transmission is put into the model by assuming that a proportion p of the offspring of infective parents are susceptible, while the proportion $q = 1-p$ are infective. Discounting for removals, we see that new births into the susceptible class are given by $bS(t-T) + pb'I(t-T)$ and into the infective class by $qb'I(t-T)$, where $b = \tilde{b}e^{-r^*T}$, $b' = \tilde{b}'e^{-r^{*\prime}T}$. Assuming, for simplicity, a mass action term for horizontal transfer, we see that the structure of the model can be depicted as in Figure 1.

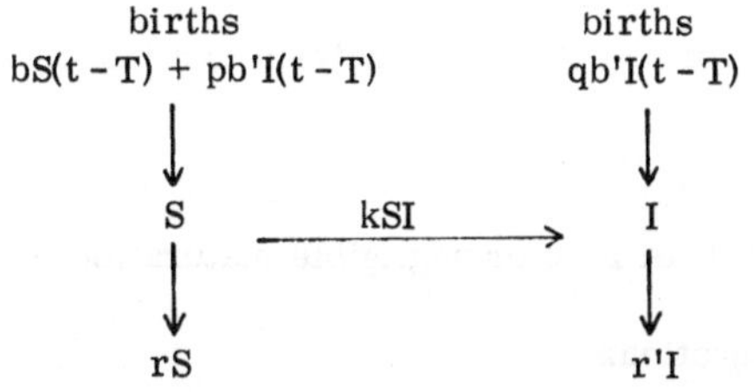

FIGURE 1

The dynamic equations of the model are differential-delay equations

$$dS(t)/dt = -rS(t) + bS(t-T) + pb'I(t-T) - kS(t)I(t)$$

$$(t > 0) \qquad (1)$$

$$dI(t)/dt = -r'I(t) + qb'I(t-T) + kS(t)I(t)$$

to which must be adjoined initial conditions of the form

$$S(t) = S_0(t) , \qquad I(t) = I_0(t) , \qquad -T \leqslant t \leqslant 0 \qquad (2)$$

where S_0 and I_0 are assumed to be known nonnegative functions.

The above model can be regarded as a simplified model for RMSF, in which $S(t)$ and $I(t)$ are numbers of susceptible and infective ticks. The model lumps or averages together all the stages of the tick life cycle, larvae, nymph, and adult, that participate in disease transmission. We point out that we can use an S-I model, rather than an S-I-R model, since ticks seem to remain infectious for life. In the model, we have retained maximum generality by allowing the possibility that $r \neq r'$ and $b \neq b'$. For RMSF, we can apparently assume that $r = r'$. Apparently, infection by Rickettsia does not at first diminish fecundity, but there is some evidence that after the fifth generation in an infective line it does. Therefore, there may be interest in considering both the cases $b = b'$ and $b > b'$.

We note that the only nonlinearity in (1) is due to the transmission term kSI. Without this term, the population would not contain any density dependence. As we shall see, it is possible for the disease alone to regulate the population size for some parameter values. In future work, we shall endeavor to extend the analysis to a model containing density-dependent birth rates.

In the special case $T = 0$ of zero or negligible maturation time, (1) reduces to the pair of ordinary differential equations

$$dS(t)/dt = (b-r)S + pb'I - kSI$$

$$dI(t)/dt = (qb' - r')I + kSI \qquad (3)$$

In this case, we have complete results on the qualitative behavior of the model. Observe first that the isolated equilibrium solutions are

$$(0,0) \text{ and } (S^*, I^*) = \left(\frac{r'-qb'}{k}, \frac{r'-qb'}{k} \cdot \frac{b-r}{r'-b'}\right)$$

when $r' \neq b'$. The solution (S^*, I^*) is of interest only when it is feasible, that is, only when $r'-qb' \geqslant 0$ and $(b-r)/(r'-b') \geqslant 0$. In Busenberg et al. (1982), a complete global analysis of system (3) is given. The results are broken down into 5 cases, depending on the parameter values, and are as follows.

Theorem 1. Assume that $p > 0$, $k > 0$.

Case I. If $r < b$, $r' > b'$, then (S^*, I^*) is feasible and globally stable with respect to all solutions with initial condition $I_0 > 0$.

Case II. If $r \geqslant b$, $r' \geqslant b'$, and one of these is not an equality, then $I(t) \to 0$ as $t \to \infty$ and $S(t)$ tends to a limit that is 0 if $b < r$ and positive if $b = r$.

Case III. If $r = b$, $r' = b'$, and if $P_0 = S_0 + I_0$, then we have $P(t) = P_0$ for all t, and $I(t) \to 0$ if $pb'/k > P_0$ or $I_0 = 0$, whereas $I(t) \to P_0 - pb'/k$ if $pb'/k \leqslant P_0$ and $I_0 > 0$.

Case IV. If $r < b$, $r' \leqslant b'$ or else $r \geqslant b$, $r' \leqslant qb' < b'$, then $S(t) \to pb'/k$ and $I(t) \to \infty$ (unless $S \equiv I \equiv 0$).

Case V. If $r > b$, $qb' < r' < b'$, then the equilibrium (S^*, I^*) is feasible but is a saddle point and there is a separatrix. For initial conditions below the separatrix, $S(t) \to 0$, $I(t) \to 0$, whereas for those above the separatrix, we have $I(t) \to \infty$ and $I(t)/P(t) \to 1$ as $t \to \infty$.

Some observations may be made on the meaning of this theorem. First, the analysis is global, and the theorem rules out any possibility of periodic solutions. Second, there are two cases in which the population is regulated to constant size. One is Case I, in which susceptibles have a lower death rate than birth rate, but the reverse is true for infectives. In this case, there is a stable endemic level of disease, which occurs independent of the initial conditions S_0, I_0 and of k and q (but the values S^*, I^* do depend on k and q). Note that there is no threshold, or minimum community size, required for the disease to become endemic in this case. Also, note that this case may occur when the infectives are at an advantage in birth rate and a disadvantage in death rate ($b' > b$, $r' > r$), or the reverse ($b' < b$, $r' < r$), as well as other possibilities. We do not know whether such a mutualistic situation exists for any real disease. In Case III, the total population remains constant. Here, there is a threshold condition for maintenance of the disease. In Case IV, the population explodes with prevalence tending to 100$^o/_o$, and this also occurs in Case V if a threshold on S_0, I_0 is exceeded. Note that in Case V, as the vertical transmission rate q increases, the separatrix falls, and it becomes more likely that the threshold is exceeded.

As far as RMSF is concerned, at least as a first approximation we can take $b = r$ and $b' = r'$ so Case III applies. The threshold condition is

$$P_0 \geqslant (1-q)b'/k \quad ,$$

where q is the vertical transmission rate, and the endemic prevalence is $1-(1-q)b'/kP_0$. On the other hand, when averaged over several generations, the death rate r' may exceed r. Cases I, II, IV, V are all possible with $r' > r$, and even with $b = b'$ except for Case V. For example, we have Case I if $r < b = b' < r'$, and we see that in this case the disease persists even though there is disease-related mortality and no difference in the birth rates.

Anderson and May (1979, 1981) have formulated an S-I-R-S model that includes vertical transmission, for the case $b = b'$, $r < r'$. Based on local stability analysis of the nontrivial equilibrium, they have given conditions for an endemic state or for the disease to control the population size. It should be useful to provide a general global analysis for their model.

In the paper by Busenberg et al. (1982), equation (1) with positive maturation period T is also considered. It is proved that in the special case $r = r' = b = b'$, the behavior is essentially the same as in Case III of Theorem 1. This result is also extended to a model with spatial diffusion of susceptibles and infectives. The paper also includes another model, in which a delay enters via a different mechanism and in which the existence of periodic solutions is proved. In this model, it is assumed that there is no maturation period, but a proportion p of the offspring of infective parents are immune for a time T_0, after which they become susceptible, while the complementary proportion $q = 1 - p$ become infective after an incubation period T_1.

4. A sequential continuous model

A model combining continuous time dynamics with discrete generation-to-generation dynamics has been proposed by Busenberg and Cooke (1980) for arthropod-carried diseases such as RMSF. In this model, we let $S^{(n)}(t)$ and $I^{(n)}(t)$ denote the proportion of the female population of generation n which is, respectively, susceptible and infected at time t. The time variable is normalized so that each generation or cohort group is involved in the disease dynamics for one unit of time. For example, for Dermacentor variabilis in Nova Scotia a unit of time will be two years, whereas in Virginia it will be one year. It is also assumed, in this model, that the oviposition of generation n is limited by density dependence, where the dependence is on the total female population of generation n. Times $t = 1, 2, \ldots, n, \ldots$ correspond to the emergence of ticks, in the spring. During the active period (summer), the population change is assumed to be governed by equations of the form

$$\begin{aligned} dI^{(n)}(t)/dt &= -c(t)I^{(n)}(t) + k(t)S^{(n)}(t)I^{(n)}(t) \\ & \qquad\qquad\qquad\qquad n < t \leqslant n+1 \qquad (4) \\ dS^{(n)}(t)/dt &= -c(t)S^{(n)}(t) - k(t)S^{(n)}(t)I^{(n)}(t) , \end{aligned}$$

$n = 2, 3, 4, \ldots$. Here $c(t)$ is the removal or death rate and $k(t)$ is the horizontal transmission or contact rate. Births into the susceptible and infective populations from parents of generation $(n-1)$ enter generation n and are assumed to be synchronized so as to arrive at time $t = n$. It is assumed that there is a "window" of fecundity during each season, but for the sake of brevity we here give only the special case in which all reproduction occurs at a fixed time m within each season. The birth law may then be written in the form

$$I^{(n)}(n) = qb'(n-1+m)I^{(n-1)}(n-1+m)\left[1-P^{(n-1)}(n-1+m)\right]$$

$$S^{(n)}(n) = \left[b(n-1+m)S^{(n-1)}(n-1+m) + pb'(n-1+m)I^{(n-1)}(n-1+m)\right]\left[1-P^{(n-1)}(n-1+m)\right] \tag{5}$$

$(n = 2,3,\ldots)$. Here $p + q = 1$ and

$$P^{(n)}(t) = I^{(n)}(t) + S^{(n)}(t) \qquad (n = 1,2,\ldots) \ . \tag{6}$$

For $n = 1$, we have initial conditions

$$I^{(1)}(1) = I_0 \ , \qquad S^{(1)}(1) = S_0 \ . \tag{7}$$

In these equations, the death rate c, the birth rates b for susceptibles and b' for infectives, and the horizontal transmission factor k, are assumed to be known continuous functions that are positive and of period one. The constants m and q satisfy $0 \leq m < 1$, $0 \leq q \leq 1$. The factor $1 - P^{(n-1)}(n-1+m)$ in (5) represents the assumption that fecundity is density-dependent with the limitation depending on the total population $P^{(n-1)}$.

It is not difficult to see that the system (4) - (7) has a unique solution. Indeed, it is a sequence of differential equations, with initial conditions determined by the preceding solution. However, note that (4) can be explicitly integrated to yield $S^{(n)}(t)$ and $I^{(n)}(t)$ in terms of $S^{(n)}(n)$ and $I^{(n)}(n)$. (This would not be true if different death rates were assumed for susceptibles and infectives.)

Due to lack of space, a full statement of the results obtained for the above system cannot be given, and we shall merely sketch some of the results. The simplest case occurs when $m = 0$, that is, when there is no maturation period for the host. In this case, the birth equations (5) are independent of the seasonal dynamics equation (4). We also note that $b(n-1) = b(1)$ and $b'(n-1) = b'(1)$ do not depend on n, because of their periodicity, and for simplicity we write b and b'. We let

$$r_n = \frac{I^{(n)}(n)}{P^{(n)}(n)} \tag{8}$$

be the prevalence of infection at the beginning of the nth generation. Then we have the following result.

(i) If $qb' \leqslant b$ then $r_n \to 0$ as $n \to \infty$.

(ii) If $qb' > b$ then $r_n \to \frac{qb'-b}{b'-b}$ as $n \to \infty$.

This gives a simple threshold condition for the maintenance of the infection through vertical transmission alone ($m = 0$), namely that the rate of reproduction of infectious offspring exceeds the rate of reproduction of susceptible offspring from susceptible parents.

It is noteworthy that although the prevalence r_n has very simple behavior, the populations $P^{(n)}(t)$, $I^{(n)}(t)$, and $S^{(n)}(t)$ may individually exhibit very complicated fluctuations. Indeed, let $a = \max\{b, qb'\} \leqslant 4$ and consider the quadratic map

$$x_n = ax_{n-1}(1-x_{n-1}) . \tag{9}$$

Then for each x_1 in $[0, 1]$ there exists $P^{(1)}(1)$ in $[0, 1]$ such that the total population $P^{(n)}(n)$ obeys $|x_n - P^{(n)}(n)| \to 0$ as $n \to \infty$ and, hence, has the same dynamic behavior as the solutions of (9). Moreover, if $qb' \leqslant b$, then $|x_n - S^{(n)}(n)| \to 0$, while if $qb' > b$ then $S^{(n)}(n)$ and $I^{(n)}(n)$ are asymptotically given by $pb'x_n/(b'-b)$ and $(qb'-b)x_n/(b'-b)$, respectively. The quadratic map (9) has been extensively studied in a number of places and it is known that x_n can have very complicated behavior. As the above remarks show, $P^{(n)}$, $S^{(n)}$, and $I^{(n)}$ can exhibit the same behavior.

We turn now to the more complicated case $m > 0$, but first assume that $b = b'$. As pointed out previously, the assumption of equal birth rates and equal deaths is reasonable for Rickettsia infestations in Dermacentor. Now there is an interplay between horizontal and vertical transmission, and the results depend on the following parameters.

$$a = b(m+n-1) \exp\left[-\int_0^m c(s)ds\right] ,$$

$$d = b(m+n-1)\exp\left[-2\int_0^m c(s)ds\right].$$

Now $P^{(n)}(n) = ax_n/d$ where x_n solves (9) with $x_1 = dP^{(1)}(1)/a$. We see that either an increase in the birth rate $b(t)$ or a decrease in the death rate $c(t)$ has the effect of increasing a and therefore leads to more complicated dynamical behavior. For further information, see the paper by Busenberg and Cooke (1980).

5. Concluding remarks

In Sections 3 and 4, we discussed two of the models that have been proposed for diseases with vertical transmission. This mode of propagation is important for many diseases, but so far very few mathematical treatments have been carried out. We think that this is a promising and exciting field for research. There is need for more theoretical work and for attempts to use models in close association with biological data.

LITERATURE CITED

Anderson, R.M. and May, R.M. (1979). Population biology of infectious diseases. Part I. Nature, Lond. 280: 361-367.

Anderson, R.M. and May, R.M. (1981). The population dynamics of microparasites and their invertebrate hosts. Phil. Trans. Roy. Soc. London B291: 451-524.

Burgdorfer, W. (1975). Rocky Mountain spotted fever, Chap. XXVI in Diseases Transmitted from Animal to Man, 6th Ed., W. Hubbert et al., editors, Charles Thomas Pub., Springfield, Illinois.

Busenberg, S.N. and Cooke, K.L. (1980). Models of vertically transmitted diseases with sequential-continuous dynamics. Nonlinear Phenomena in Mathematical Sciences, V. Lakshmikantham (ed.), Academic Press, New York, 1982, pp. 179-187.

Busenberg, S.N., Cooke, K.L. and Pozio, M.A. (1982). Analysis of a model of a vertically transmitted disease. J. Math. Biol., to appear.

Cooke, K.L. and Busenberg, S.N. (1980). Vertically transmitted diseases. Nonlinear Phenomena in Mathematical Sciences, V. Lakshmikantham (ed.), Academic Press, New York, 1982, pp. 189-197.

Fine, P.M. (1975). Vectors and vertical transmission, an epidemiological perspective. Annals N.Y. Acad. Sci. 266: 173-194.

Fine, P.M. and LeDuc, J.W. (1978). Towards a quantitative understanding of the epidemiology of Keystone virus in the Eastern United States. Amer. J. Trop. Med. Hyg. 27: 322-338.

Garvie, M.B., McKiel, J.A., Sonenshine, D.E., and Campbell, A. (1978). Seasonal dynamics of American dog tick, Dermacentor variabilis (say), population in South Western Nova Scotia, Can. J. Zoology 65: 28-39.

Mims, C.A. (1981). Vertical transmission of viruses, Microbiological Reviews 45: 267-286.

Integral Equations for Infections with Discrete Parasites: Hosts with Lotka Birth Law

K.P.Hadeler
Lehrstuhl für Biomathematik
Universität Tübingen
Auf der Morgenstelle 28
D-7400 Tübingen

Summary: In many (e.g.helminthic) infections the host carries only a small number of adult parasites. Then the state of the population can be described by a generating function which for every age class and every number of parasites gives the corresponding fraction of the host population. The generating function satisfies a partial differential equation with integral terms and integral boundary conditions. It can be reduced to a system of two coupled Volterra integral equations for the parasite acquisition rate and the number of newborns.

In the classical epidemic model of Kermack and McKendrick the population is subdivided into several classes such as susceptibles, infectious, or recovered. The sizes of these classes satisfy certain ordinary differential equations similar to the equations for chemical reactions derived from the mass action law. Such models are called prevalence models since only the prevalence of the disease in an individual is described; the degree of illness has not been taken into account. They are suited to model bacterial or viral infections, where the degree of illness has little to do with the number of germs initially acquired.

On the other hand, in typical helminthic diseases (see [1] for a review on infections, models, and control strategies) such as onchocerciasis (river blindness) the host harbours a small number of adult parasites, and the number (and possibly the age) of these parasites determine the state of the host. In modeling such diseases it makes sense to classify the host population according to age and parasite load.

Let t denote chronological time and a the age of individuals. Let $n(t,a,r)da$ be the number of hosts at time t which have an age between a and $a+da$ and which carry r parasites (n is a continuous, but r is a discrete variable). We make the following assumptions: The hosts age with time, they have a natural death rate $\mu = \mu(a)$. If a host carries r parasites then his death rate is increased to $\mu(a) + \alpha r$, where α is a constant. Parasites within the host die with rate $\delta > 0$ and multiply with rate $\rho \geq 0$. (In helminthic infections one has $\rho = 0$.) Parasites are acquired by hosts with a rate $\varphi(t)$. Multiple events are excluded. Thus parasites within the host are governed by a birth and death process with killing [6].

With these assumptions the functions $n(t,a,r)$ satisfy a system of infinitely many partial differential equations:

for $r = 0,1,2,...$

$$\frac{\partial n(t,a,r)}{\partial t} + \frac{\partial n(t,a,r)}{\partial a}$$

$$= -[\varphi(t) + \mu(a) + (\alpha + \delta + \rho)r]n(t,a,r) \tag{1}$$

$$+ [\varphi(t) + \rho(r-1)]n(t,a,r-1)$$

$$+ \delta(r+1)n(t,a,r+1),$$

where formally $n(t,a,-1) \equiv 0$. On the right hand side of equation (1) the first term counts the hosts who die or acquire or lose a parasite, the second term those who acquire a parasite, the last those which lose a parasite.

The fundamental assumption of our model [2] relates the acquisition rate φ of parasites to the average parasite load. Of course in more sophisticated models one could consider differential exposition of hosts to the risk of infection. We assume

$$\varphi(t) = \beta f(\bar{w}(t)) \tag{2}$$

with

$$\bar{w}(t) = \frac{\int_0^\infty \sum_{r=0}^{\infty} n(t,a,r)\, r\, da}{\int_0^\infty \sum_{r=0}^{\infty} n(t,a,r)\, da}, \tag{3}$$

where f is an influence function and β is a constant. One should think of functions like $f(w) = w/(1+w)$ of $f(w) = \log(1+w)$.
Since it is difficult to treat this system in its present form it is convenient to transform it into a single equation for the generating function

$$u(t,a,z) = \sum_{r=0}^{\infty} n(t,a,r) z^r, \quad 0 \leq z \leq 1 . \tag{4}$$

By comparison of coefficients of equal powers one can show that the function u formally satisfies a first order partial differential equation

$$u_t + u_a + g(z)u_z - [\varphi(t)(z-1) - \mu(a)]\, u = 0 \tag{5}$$

where

$$g(z) = (\alpha + \sigma + \rho)z - \sigma - \rho z^2 . \tag{6}$$

Solutions of (5) which are generating functions give rise to solutions of (1). The acquisition law now assumes the form

$$(t) = f\left(\frac{\int_0^\infty u_z(t,a,1)\, da}{\int_0^\infty u(t,a,1)\, da} \right) . \tag{7}$$

The equation (5)(7) has to be studied together with an initial condition

$$u(0,a,z) = u_0(a,z) \tag{8}$$

which describes the state of the population at time t = 0 and a boundary condition which describes the birth of hosts. In [2,3,4,5] we have used the simple condition

$$u(t,0,z) = N(t,z)$$

or even simpler,

$$u(t,0,z) = N(t). \tag{9}$$

The condition (9) says that new hosts (of age $a = 0$) are produced at a given time-dependent rate independent of the actual population size. Furthermore (9) says that newborns are not infected.

In a more realistic side condition of Lotka type the birth of new hosts is related to the total population size. The individual fertility of the hosts may be related to age and parasite load. A rather general law is the following Lotka condition

$$u(t,0,z) = \int_0^\infty b(a)u(t,a,\omega)\,da \,. \tag{10}$$

Here $b(a)$ is the age dependent birth rate, $\omega\in[0,1]$ is a parameter describing the geometric decrease of the individual fertility with the number of parasites. For $\omega = 1$ parasites have no influence on reproduction, for $\omega = 0$ only noninfected hosts reproduce. The right hand side of (10) is independent of z since newborns are not infected.

The following problems are the most immediate: Does the initial-boundary value problem (5)(7)(8),(9) or (10) have a unique solution? Does the solution exist for all time? Are there stationary solutions and how many: Are these stationary solutions stable?

Concerning the boundary condition (9) partial answers have been given in several papers [2,3,4,5].

Suppose the function $\varphi(t)$ were known. Then (5) is a linear first order equation which can be solved by the method of characteristics. There are two different representations of the solution for $a > t$ (solution depending on the initial data) and for $a < t$ (solution depending on boundary data). For $a > t$

$$u(t,a,z) = u_0(a-t,\, G(-t,z))$$

$$\times \exp\left[\int_0^t [G(s-t,z) - 1]\varphi(s)\,ds - \int_{a-t}^a u(s)\,ds\right]$$

and for $a < t$ (11)

$$u(t,a,z) = N(t-a,\ G(-a,z)) \times \exp\left[\int_0^a [G(s-a,z) - 1]\varphi(t-a+s)ds - \int_0^a \mu(s)ds \right]$$

Here G stands for the solution operator of the Riccati equation

$$\dot{z} = g(z). \tag{12}$$

$G(t,z_o)$ is the solution of equation (12) with initial value $z(0) = z_o$, evaluated at t. An explicit representation of G is

$$G(t,z) = \frac{z_1(z-z_2) - z_2(z-z_1)e^{-\varkappa t}}{(z-z_2) - (z-z_1)e^{-\varkappa t}} \tag{13}$$

where $z_1 \geq 1 \geq z_2$ are the roots of the quadratic equation $g(z) = 0$,

$$z_{1,2} = \frac{1}{2\varrho}\left[\alpha + \sigma + \varrho \pm [(\alpha + \sigma + \varrho)^2 - 4\sigma\varrho]^{1/2}\right] \tag{14}$$

and

$$\varkappa = [(\alpha + \sigma + \varrho)^2 - 4\sigma\varrho]^{1/2}. \tag{15}$$

In the limit case $\varrho = 0$ one has $z_1 = \infty$,

$$G(t,z) = 1 - (1-z)e^{\varkappa t} + \frac{\alpha}{\varkappa}(1 - e^{\varkappa t}) \tag{16}$$

and

$$\varkappa = \alpha + \sigma. \tag{17}$$

Now that the solution is known in terms of φ one can introduce it into equation (7) to obtain an integral equation for the function φ. In order to write down this equation it is convenient to introduce the following operators

$$(A_\omega\varphi)(t,a) = \int_{t-a}^{a} [G(s-t,\omega)-1]\varphi(s)ds ,$$

$$(B_\omega\varphi)(t,a) = \int_{t-a}^{a} G_z(s-t,\omega)\varphi(s)ds ,$$

$$(C_\omega\varphi)(t) = \int_0^t [G(s-t,\omega)-1]\varphi(s)ds, \tag{18}$$

$$(D_\omega\varphi)(t) = \int_0^t G_z(s-t,\omega)\varphi(s)ds \ ,$$

$$M(a) = \int_0^a \mu(s)ds \ .$$

Then the integral equation assumes the form

$$\varphi(t) = (\mathcal{F}\varphi)(t) = \beta f(\bar{w}(t)) \tag{19}$$

where

$$\bar{w}(t) = \frac{\int_0^t e^{A_1\varphi - M(a)}[N_zG_z+NB_1\varphi]da + e^{C_1\varphi}\int_t^\infty e^{-M(a-t)+M(a)}[u_{oz}G_z+u_oD_1\varphi]da}{\int_0^t e^{A_1\varphi-M(a)}N\,da + e^{C_1\varphi}\int_t^\infty e^{-M(a-t)+M(a)}u_o da} \tag{20}$$

where $z = 1$ in the arguments of u_o, u_{oz}, N, N_z as given in (11).

We have proved the following global existence theorem ([3],[5]).

<u>Theorem 1:</u> Let $f:[0,\infty) \to [0,\infty)$ be continuously differentiable, $f(0)=0$, $f'(0) = 1$, and $0 < f(u) \leq f_o u$ for $u > 0$, where $f_o > 0$ is a constant. Let $\beta > 0$. Let $\mu:[0,\infty) \to [0,\infty)$ be continuous, and such that for the function M as given by (18) the integral

$$\int_0^\infty \exp[-M(a)]da \tag{21}$$

is finite.

Let

$$0 < \int_0^\infty u_o(a,1)da < \infty \tag{22}$$

and

$$p = \sup\left\{\frac{u_{oz}(a,1)}{u_o(a,1)} : u_o(a,1) > 0,\ 0 \leq a < \infty\right\}. \tag{23}$$

Then the integral equation (19)(20) has a unique solution for all $t > 0$.

For a stationary solution the acquisition rate φ is a constant which is a solution of an equation

$$\varphi = \beta f \left(\frac{I_1(\varphi)}{I_o(\varphi)} \varphi\right) \tag{24}$$

where

$$I_1(\varphi) = \int_o^\infty e^{-Q_1(a)\varphi - M(a)} q(a)\,da \, ,$$

$$I_o(\varphi) = \int_o^\infty e^{-Q_1(a)\varphi - M(a)} da \, ,$$

$$Q_\omega(a) = -(z_1-1)a + \frac{1}{\varrho} \log \frac{(\omega - z_2) + (z_1 - \omega) e^{\varkappa a}}{z_1 - z_2} \, , \tag{25}$$

$$q(a) = \frac{1}{\varrho} \frac{e^{\varkappa a} - 1}{z_2 + (z_1-1) e^{\varkappa a}} \, .$$

One can prove ([2,5])

Theorem 2: In addition to $\varphi = 0$ there is a branch of nontrivial stationary solutions $\beta = \beta(\varphi)$ with $\varphi > 0$ starting from $(\beta_o, 0)$ where $\beta_o = I_o(0)/I_1(0)$. If the function f satisfies

$$\frac{df(u)}{du} \geq 0 \, , \; \frac{d}{du}\left(\frac{f(u)}{u}\right) \leq 0 \tag{26}$$

then for each $\beta > \beta_o$ there is exactly one $\varphi > 0$.
Usually a stationary solution can only be observed if it is stable. A complete stability analysis for the present problem appears rather difficult. However, it can be shown ([2,4,5])

Theorem 3: Under the hypothesis of Theorem 1 the stationary solution with no parasites is linearly stable for $\beta < \beta_o$.

In the case of the boundary condition (10) the mathematical treatment of the problem becomes more involved. The problem can be reduced to a system of two coupled Volterra integral equations

$$\begin{aligned} \varphi(t) &= \mathfrak{F}(\varphi,N)(t) , \\ N(t) &= \mathcal{G}(\varphi,N)(t) , \end{aligned} \tag{27}$$

for the acquisition rate φ and the number of newborn hosts. Similar to (19)

$$\mathfrak{F}(\varphi,N) = \beta f(\bar{w}) \tag{28}$$

$$\bar{w} = \frac{\int_0^t e^{A_1\varphi-M(a)} NB_1\varphi da + e^{C_1\varphi}\int_t^\infty e^{-M(a-t)+M(a)} [u_{oz}G_z+u_oD_1\varphi] da}{\int_0^t e^{A_1\varphi-M(a)} N da + e^{C_1\varphi}\int_t^\infty e^{-M(a-t)+M(a)} u_o da} \tag{29}$$

$$\mathcal{G}(\varphi,N)(t) = \int_0^t e^{A_\omega\varphi-M(a)} b(a)N da + e^{C_\omega\varphi}\int_t^\infty e^{-M(a-t)+M(a)} b(a)u_o d_z . \tag{30}$$

The problem of existence and uniqueness can be treated with methods similar to those in [2,3,5]. The proofs will be published elsewehere. For the moment we consider only the problem of stationary solutions.

In contrast to the case of the boundary condition (9) we do not expect solutions where the parasites and the host population remain constant. Rather, the host population will show exponential growth or decay with stationary age distribution, and the acquisition rate is constant. The ansatz $\varphi(t) = \bar{\varphi}$, $N(t) = \bar{N}\cdot\exp(\lambda t)$ leads to

$$\varphi = f\left(\frac{\int_0^\infty e^{-Q_1(a)\varphi-M(a)-\lambda a} q(a) da}{\int_0^\infty e^{-Q_1(a)\varphi-M(a)-\lambda a} da}\,\varphi\right) \tag{31}$$

$$\int_0^\infty b(a) e^{-Q_\omega(a)\varphi-M(a)-\lambda a} da = 1. \tag{32}$$

Some properties can be observed immediately. The left hand side of (32) is strictly decreasing in φ and in λ. For given φ , it increases from 0 to $+\infty$, when λ runs from $+\infty$ to some finite negative value.

Thus, for every φ there is exactly one $\lambda = \lambda(\varphi)$ which solves (32). We introduce this λ into equation (31). Then the right hand side of (31) depends only on φ. In equation (31) one can solve for β. Thus, in addition to the trivial solution $\varphi = 0$, there is a branch $\beta = \beta(\varphi)$ of non-trivial stationary solutions. This branch starts at $(\beta_1, 0)$, where

$$\beta_1 = \int_0^\infty e^{-M(a)-\lambda(0)a}da \Big/ \int_0^\infty e^{-M(a)-\lambda(0)a}q(a)da. \tag{33}$$

We discuss a special case in more detail. Let $\mu(t) \equiv \mu$, $b(a) \equiv 1$, and $\omega = 1$. For fixed λ, in view of (25), the numerator in (31) can be evaluated, the denominator is normalized by (32), thus (31) becomes

$$\varphi = \beta f\left(\frac{1-\mu-\lambda}{(z_1-1)(1-z_1)\varrho}\right) \tag{34}$$

On the other hand, from $Q_1(a) = \frac{\alpha}{2}a^2 + \ldots$ and equation (32) follows

$$\int_{\frac{\lambda+\mu}{\sqrt{2\alpha\varphi}}}^\infty e^{-y^2}dy \cdot e^{\frac{(\lambda+\mu)^2}{2\alpha\varphi}}\sqrt{\frac{2}{\alpha\varphi}} \sim 1 \quad \text{for } \varphi \to \infty$$

and thus

$$\lambda + \mu \sim -\sqrt{\alpha\varphi \log\left(\frac{\alpha\varphi}{2\pi}\right)}$$

Finally, for a large φ approximately

$$\varphi \sim \beta f\left(\frac{\sqrt{\alpha}}{(z_2-1)(1-z_2)\varrho} \log\left(\frac{\alpha\varphi}{2\pi}\right) \cdot \sqrt{\varphi}\right) \tag{35}$$

Thus the branch of stationary solutions has approximately the same asymptotic behavior as in the case of boundary condition (9): If $f(u)/u^2 \to 0$ for $u \to \infty$ then for each $\beta > \beta_1$ there is at least one stationary $\varphi > 0$.

Literature cited

[1] Anderson, R.M. and May, R.M. (1982). The transmission dynamics of human helminth infections; Control by chemotherapy. Nature 297:557-563

[2] Hadeler, K.P. and Dietz, K. (1982/83). Nonlinear hyperbolic partial differential equations for the dynamics of parasite populations. Computers and Mathematics with Applications Vol.8 (Special volume, M. Witten Ed.)

[3] Hadeler, K.P. and Dietz, K. (1983) An integral equation for helminthic infections: Global existence of solutions. In: Recent Trends in Mathematics, Conference Proceedings Reinhardsbrunn 1982, Teubner-Verlag Leipzig

[4] Hadeler, K.P. (1983). An integral equation for helminthic infections: Stability of the non-infected population. Proc. Vth Int. Conf. on trends in Theor. and Practice of Nonlin. Diff. Equ. (Ed. V. Lakshmikantham)

[5] Hadeler, K.P. and Dietz, K. A transmission model for multiplying parasites and killing. Manuscript to be submitted.

[6] Karlin, S. and Tavaré, S. (1982). Linear birth and death processes with killing. J. Appl. Prob. 19:477-487

PART VI

THE DYNAMICS OF MOVEMENT: DIFFUSION MODELS

PREDATOR-PREY DYNAMICS IN SPATIALLY STRUCTURED POPULATIONS: MANIPULATING DISPERSAL IN A COCCINELLID-APHID INTERACTION

by

Peter Kareiva

Division of Biology and Medicine
Brown University
Providence, Rhode Island 02912 USA

Naturalists have long recognized that populations change not only as a result of birth and death, but also as a result of individual movement (Elton 1949). Only within the last decade, however, have mathematical ecologists given much attention to dispersal. Nonetheless, because theoretical advances have proceeded so rapidly, mathematical investigations of dispersal's involvement in population dynamics have raced far ahead of empirical studies (see Levin 1978 for a review of the theory). Consequently, whereas there is a plethora of models connecting dispersal and interspecific interactions, none of these models have been tested in the field. The dearth of empirical work addressing theory in this area is especially evident in the literature on predator-prey systems; Huffaker's (1958) classic "orange-and-mites" experiment, which was published a quarter of a century ago, is still the key inspiration for most models of predator-prey interactions in patchy environments (cf Maynard Smith 1974, Hilborn 1975, Gurney and Nisbet 1978, Hassell 1980). Much more exploratory empirical work is clearly needed, both for evaluating existing models and indicating new theoretical avenues.

I report here preliminary results from field studies of dispersal's effect on population dynamics. My experiments were initially designed to analyze the contribution of dispersal to outbreaks of defoliating insects, but they have evolved into an examination of dispersal's role in a single predator-prey interaction, that between coccinellids and aphids; this is what I discuss here. I also use my observations of coccinellids and aphids to illustrate the benefits of models as guides to experimental design and interpretation. Much can be gained by intertwining models with experimental research, so that theory is used iteratively at each step of an empirical inquiry, and vice versa.

OVERVIEW OF EXPERIMENTS

Theory "Targets" Dispersal as a Crucial Variable

In a wide variety of mathematical models, the dynamics of spatially distributed population interactions can be dramatically altered by simply adjusting dispersal rates. For example, theory indicates that species-specific dispersal rates may determine the outcome of interspecific competition (e.g., Levin 1974), the potential for predator-mediated coexistence (e.g., Hastings 1978), and even tendencies toward spatial patterning in uniform environments (e.g., Segel and Jackson 1972). By focusing so much attention on the importance of dispersal, mathematical models prompted me to undertake large-scale field experiments in which I manipulate the opportunities for movement that insects experience in goldenrod (*Solidago* spp.) fields. Dispersal rate is marvelously unambiguous as both a theoretical parameter and a "real-world" population process; it has only two components, the frequencies and distances of individual movement. The concrete nature of dispersal, and the ease with which it can be manipulated, provide a marked contrast to more abstract theoretical terms such as competition coefficients, which are a plague of vagueness to the field biologist. (Is competition simple enough to be adequately represented by a single interaction constant? If so, how does one measure α_{ij}?) Because dispersal is so accessible as an experimental variable, it promises to be a valuable focus for testing general models of spatially distributed populations.

The Biological System

My experiments involve the insects associated with the goldenrod, Solidago canadensis, at Brown University's field station, the Haffenreffer Estate (Bristol Co., RI). To remove vegetation diversity as a complicating factor, I have selected fields where goldenrod grows in large, virtually monospecific stands. Goldenrod is simple enough in growth form that its insects can be accurately censused by visual inspection, without disturbing the fauna. The following species are the dominant insects at my study site:

1. Trirhabda virgata (Coleoptera, Chrysomelidae) is a univoltine beetle that specializes on goldenrods. Typically, it is the dominant herbivore in goldenrod communities, accounting for up to as much as 95% of the total arthropod biomass (Evans 1980). Messina (1982a, b) provides details on its biology. (Hereafter referred to as "leaf beetles".)
2. Corythuca marmorata (Hemiptera, Tingidae) is a lace bug that feeds on Solidago and Aster in Rhode Island. These small sap-feeding bugs can occur in numbers as high as 1,000 per plant, levels at which they are capable of inflicting heavy damage on their host plant. There are 4-5 generations each summer. (Hereafter referred to as "lace bugs".)
3. Dactynotus tissoti (Homoptera, Aphididae) is an aphid that feeds from the stems of several old-field Solidago species (Richards 1972), including S. canadensis. It forms small colonies (averaging < 10 individuals per colony) that are characterized by high extinction rates. Both apterous and alate females move readily among goldenrod shoots, leaving a few offspring on each stem they visit (pers. obs.). (Hereafter referred to as "green aphids".)
4. Dactynotus nigrotuberculatus (Homoptera, Aphididae) is another aphid commonly found on S. canadensis. Unlike D. tissoti, however, D. nigrotuberculatus often forms huge colonies (up to 1000 individuals per stem). Once a female starts a colony, she usually remains on the original plant and continues to produce nymphs (pers. obs.) (Hereafter referred to as "red aphids".)

5. Coccinella septempunctata (Coleoptera, Coccinellidae) is the dominant predator at my goldenrod study sites. Although it will accept a wide variety of arthropods as prey, this ladybird is an especially voracious consumer of aphids (Obrycki et al. 1982). (Hereafter referred to as "7-spotted ladybugs".)
6. Cycloneda sanguinea (Coleoptera, Coccinellidae), about half the size of Coccinella septempunctata, is the second most common predator in my study fields. Because it is less voracious and feeds readily on pollen, Cycloneda sanguinea appears to have much less of an impact on aphids than does Coccinella septempunctata. (Hereafter referred to as the "small ladybugs".)

I began observations on the above species in 1981 and some of those data are presented here. The bulk of my discussion, however, concerns experiments performed in 1982, in which I used manipulations of insect dispersal to develop an understanding of spatially distributed population dynamics.

Experimental Design

To manipulate insect dispersal, I dissected goldenrod stands into archipelagos (by repeated mowing, which produced a grass lawn surrounding parcels of goldenrod) and erected curtain barriers approximately 2 m in height (see Figure 1). There were three treatments: continuous strands, linear arrays of 1 m^2 patches, and linear arrays of patches interrupted by curtain barriers. Although the quantitative details vary (importantly) among insect species, the general qualitative effect of these manipulations was to restrict most movement to within the continuous strands; insects moved much less between patches or across barriers (in prep., and see Figure 4). Clearly, mowing goldenrod into patches and constructing curtain walls may produce secondary alterations in microclimate or in the quality of goldenrod foliage. My data suggest that such complications are insignificant to the insects studied (Kareiva, in prep.). Consequently, experimental results are discussed with the assumption that I have manipulated only dispersal by the scheme portrayed in Figure 1. In the future I plan to measure plant characteristics and perform herbivore feeding and behavior tests to

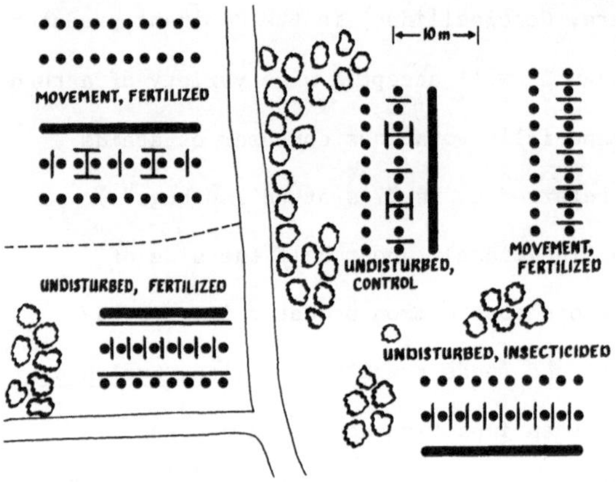

Figure 1. The experimental layout. Each solid circle represents a m^2 patch of goldenrod; the thickened bars represent continuous goldenrod strands; barriers are indicated by narrow solid lines. All the open space is mown field (i.e. grass), with surrounding trees and an access road also pictured on the map. The labels associated with different sets of arrays indicate the use and previous treatment of fields (movement = used for mark-release estimates of dispersal rates; undisturbed = used to weekly census natural populations by visually inspecting 10 stems per square meter of goldenrod). Control, fertilized, and insecticided refer to treatments that entire fields received in 1981. Since all comparisons in this paper are between arrays within the same field (i.e. same treatment) these between-field differences are not relevant to my discussions concerning the effects of manipulating dispersal rates.

further test this assumption that I am manipulating primarily dispersal, not plant quality. I also intend to alter my future experimental design so that dispersal is better isolated as the experimental variable.

PRELIMINARY RESULTS VIEWED THROUGH THEORY

How did Manipulating Dispersal Affect the Insect Community?

Although their movements were affected by my goldenrod manipulations, the distribution and densities of the leaf beetles, lace bugs, green aphids, and small ladybugs did not differ among the "continuous", "patchy" and "patchy + barrier" treatments. (Even when a separate analysis of variance was performed for each census, an approach biased in favor of detecting an effect where none exists, these four species never exhibited a significant response to the goldenrod manipulations.) In contrast, the red aphid and 7-spotted ladybug attained densities in the "patchy" and "patchy + barrier" treatments ten times as high as in the continuous goldenrod strands (see Figures 2 and 3). These results led me to ask why these two species in particular responded so dramatically to my manipulations of dispersal, whereas the remainder of the community was unchanged.

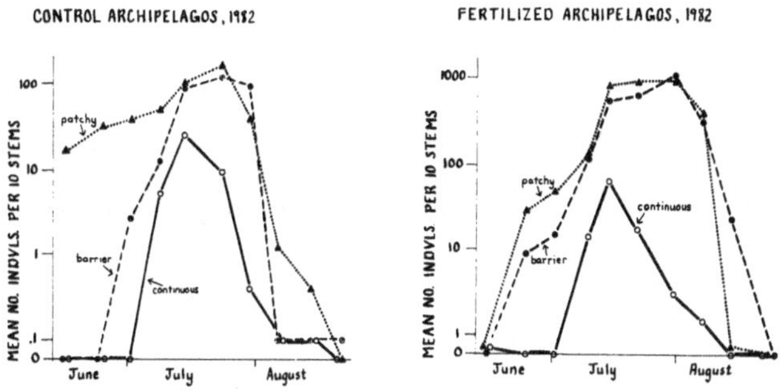

Figure 2. The influence of dispersal manipulations on red aphid density. Each point represents a mean of 8 patches in the fertilized archipelagos and 10 patches in the unfertilized (or control) archipelagos. In the continuous strands, instead of "patches", I censused goldenrod stems from permanently staked 1 m^2 sectors along the goldenrod strip.

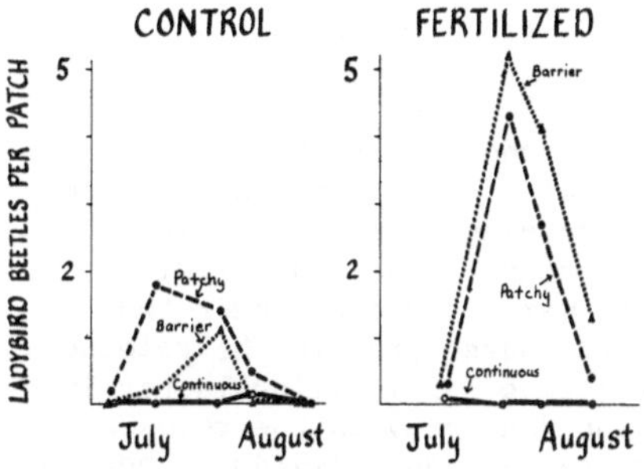

Figure 3. The influence of dispersal manipulations on 7-spotted ladybug density. These data represent means (same sample sizes as in Figure 2) obtained by visually searching each patch in its entirety for ladybugs.

I think the answer is that the 7-spotted ladybugs and red aphids are tightly coupled in a predator-prey interaction, whereas the other species are relatively unaffected by interspecific interactions (within the scale of my experiments). In 1981 I observed that over 90% of the victims consumed by foraging 7-spotted ladybugs were red aphids; in turn, predation by 7-spotted ladybugs was the most important mortality agent in the populations of red aphids I was studying. There are certainly opportunities for interspecific interactions among other species in the goldenrod fauna, but these other interactions are either infrequent (i.e. competition among herbivores) or spread among many different species (e.g., the small ladybug exploiting a great diversity of prey items).

I think the increased densities of both red aphids and 7-spotted ladybugs in "patchy" and "patchy + barrier" treatments are due to altered dispersal rates that in turn change the dynamics of the predator-prey interactions. Dispersal patterns were affected in two notable ways:

(i) both the predator (ladybugs) and prey (aphids) were much less mobile in "patchy" and "patchy + barrier" treatments than in continuous strands (Figure 4).

(ii) the ratio of predator to prey mobility (ladybug mean square displacement ÷ aphid mean square displacement) more than doubled in the "patchy" and "patchy + barrier" treatments relative to the continuous goldenrod strand (calculated from data in Figure 4).

I will now relate the above changes in dispersal patterns to the theory of predator-prey interactions in heterogeneous environments, paying particular attention to ways in which such changes might explain the elevated aphid densities in "patchy" and "patchy + barrier" treatments.

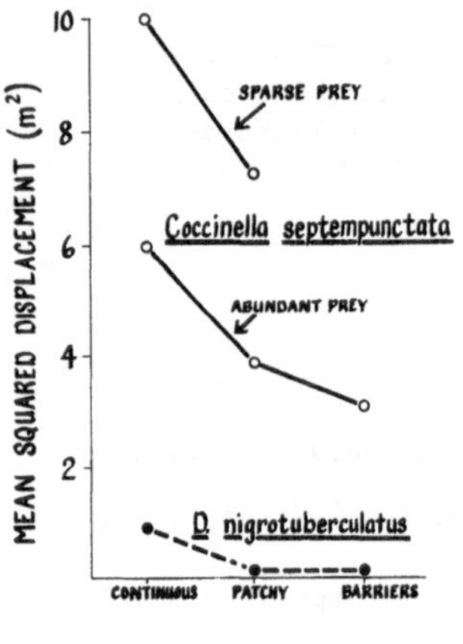

Figure 4. Dispersal one day after a point release of the insects. Open circles represent 7-spotted ladybug movement; closed circles represent aphid movement. The ladybug dispersal experiments were conducted under two circumstances: with sparse prey in the release array (aphids averaging < 5 individuals per 10 stems) and with abundant prey in the release array (aphids averaging > 50 individuals per 10 stems). Sample sizes for aphid movement rates are: 899 in continuous treatment, 1023 in patchy treatment, 1432 in patchy + barrier treatment. Sample sizes for ladybug movement rates where prey were abundant are: 20 in continuous treatment, 22 in patchy treatment, 25 in patchy + barrier treatment. Where prey were sparse ladybug movement sample sizes are: 8 in continuous and 7 in patchy. See Kareiva and Shigesada (1983) for rationale behind squared displacement as a dispersal measure.

Can Theory Help Us Understand These Results?

To understand the results presented in Figures 2 and 3 we must examine the data in more detail. Importantly, red aphid colonies were just as frequent in the continuous goldenrod strips as in the "patchy" or "patchy + barrier" arrays (as measured by the fraction of stems occupied; G-test, $p > .25$). The differences in aphid density shown in Figure 2 largely reflect differences in colony size. In particular, whereas aphids in the patchy or barriered treatments occasionally erupted into local outbreaks of hundreds per stem, these outbreaks were never observed in the continuous strips of goldenrod. I want to focus on these aphid outbreaks because I think they represent the fundamental effect of reducing dispersal, and that the ladybug response (Figure 3) simply mirrors the aphid pattern (as we might expect a consumer to track its resource abundance). The phenomenon that needs explaining is why aphid outbreaks occurred in the "patchy" or "patchy + barrier" manipulations, but not in the continuous goldenrod strands? Theory provides some hints to this puzzle.

Beginning with Nicholson and Bailey (1935), there is a rich tradition of using difference equations to analyze arthropod predator-prey systems. These models have recently been extended to represent the patchy distribution of predators and prey in heterogeneous environments (cf Hassell and May 1974). May (1978) deals with the problem of patchiness by examining the following equations, in which predator aggregation is represented by the zero term of a negative binomial distribution:

$$V_{t+1} = \lambda V_t \left(1 + \frac{\alpha P_t}{K}\right)^{-K} \quad (1a)$$

$$P_{t+1} = V_t - \left(V_{t+1}/\lambda\right) \quad (1b)$$

where V_t is the number of victims at time t, P_t is the number of predators at time t, λ is a reproductive rate for victims, and α reflects the encounter and capture rate between predators and their prey. The degree of predator aggregation is represented by the parameter K. Aggregation is greatest when $K \to 0$ and weakest when $K \to \infty$. This and related models of patchy predator-prey systems (cf Hassell and May 1974) describe predator aggregation as simply a clumping of individuals in space, not a clumping of predators in regions of high prey density. Analyses of these "aggregation models" (see Hassell 1978, pp 61-74) reveal that the asymptotic stability of predator-prey systems generally is enhanced by an increase in the tendency of predators to occur in clumps (e.g. a decrease in K for equations 1a and 1b). These theoretical results suggest one possible connection between aphid outbreaks and altered dispersal patterns--perhaps mowing goldenrods into patches and building curtain walls interferes with the movements and thus the ability of ladybugs to aggregate, thereby destabilizing the interaction. To evaluate this hypothesis, I created clumps of high and low aphid densities in both a continuous goldenrod strip and an array of patches. Then I released ladybugs into these areas and followed their reassortment in space. In Figure 5, we see that, in both patchy and continuous goldenrod, ladybugs tended to aggregate at clumps of high aphid density, but they did so much more effectively in the continuous strand. This suggests that when movement was unhindered, it was indeed the ladybug's aggregation response that prevented aphid outbreaks in the continuous goldenrod strands.

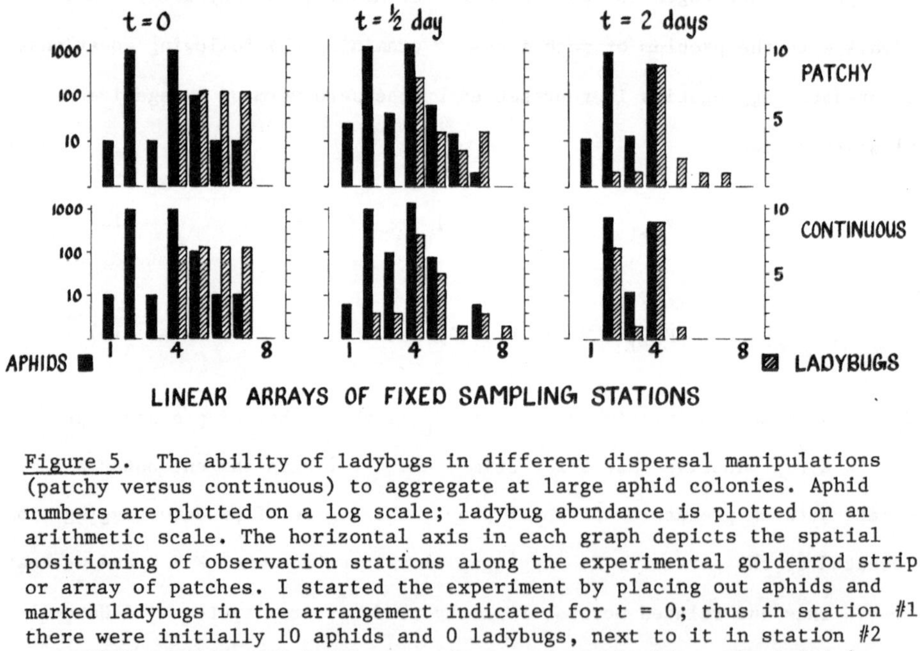

Figure 5. The ability of ladybugs in different dispersal manipulations (patchy versus continuous) to aggregate at large aphid colonies. Aphid numbers are plotted on a log scale; ladybug abundance is plotted on an arithmetic scale. The horizontal axis in each graph depicts the spatial positioning of observation stations along the experimental goldenrod strip or array of patches. I started the experiment by placing out aphids and marked ladybugs in the arrangement indicated for t = 0; thus in station #1 there were initially 10 aphids and 0 ladybugs, next to it in station #2 were 1000 aphids and 0 ladybugs, and so on down the line. The initial arrangements and distance between sampling stations (1 m) were the same for both treatments. Intervening spaces between sampling stations were cleared of aphids and ladybugs. As time proceeds (½ day and 2 days) from when I first placed out the aphids and ladybugs, the above illustration records how the ladybugs move about and reposition themselves.

One problem with these difference equation approaches to patchy predator-prey systems (e.g. Hassell and May 1974, May 1978) is that they describe the distribution of predators in space (through K) but not the process that leads to this distribution. Since I want to connect observed changes in movement patterns (i.e. Figures 4 and 5) to altered predator-prey dynamics (i.e. Figures 2 and 3), I seek predator-prey models that explicitly incorporate movement rather than models that use a phenomenological clumping index to summarize the consequences of movement. Partial differential equations that include both population transport (or a diffusion term) and local population dynamics (or a reaction term) provide a modeling tool that meets this need for integrating dispersal and population dynamics in spatially distributed predator-prey dynamics (see in particular Segel and Jackson 1972, Levin and Segel 1976, and pp. 183-194 of Okubo 1980). A general form of predator-prey models using "reaction-diffusion" equations (along a single dimension) is:

$$\frac{\partial v}{\partial t} = F(v,p) + D_v \frac{\partial^2 v}{\partial x^2} \qquad (2a)$$

$$\frac{\partial p}{\partial t} = G(v,p) + D_p \frac{\partial^2 p}{\partial x^2} \qquad (2b)$$

where v is victim density, p is predator density, D_v is the victim's diffusion or dispersal coefficient, and D_p is the predator's diffusion or dispersal coefficient. The functions F and G represent population growth for victims and predators respectively. This predator-prey system may have a spatially uniform equilibrium that is stable in the absence of dispersal, but which under certain combinations of predator and prey dispersal rates gives rise to spatially nonuniform patterns (cf Segel and Levin 1976). The sorts of spatial nonuniformities that theory predicts superficially resemble the actual spatial patterns of ladybug and aphid outbreaks that I observed in "patchy" and "patchy + barrier" treatments (see Figures 6 and 7). Perhaps the outbreaks of aphids under these particular dispersal manipulations are somehow related to the dispersal-driven instabilities that can be

Figure 6. The spatial patterning of aphid outbreaks which developed in the fertilized, undisturbed field (see Fig. 1). This is typical of patchy and patchy + barrier (but not continuous) treatments in all of the experimental fields. The temporal dynamics can be viewed by reading from top to bottom the graphs in each column.

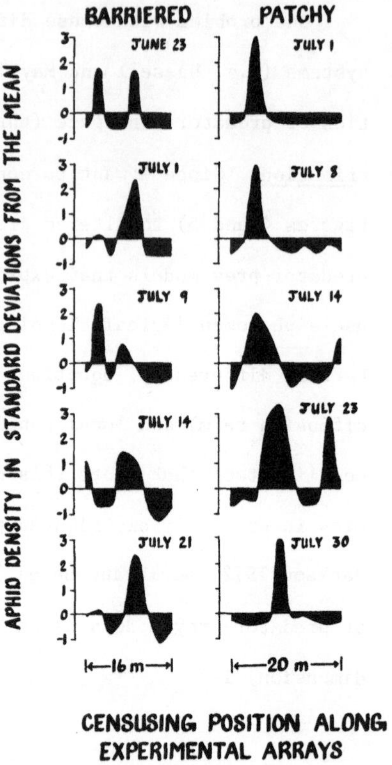

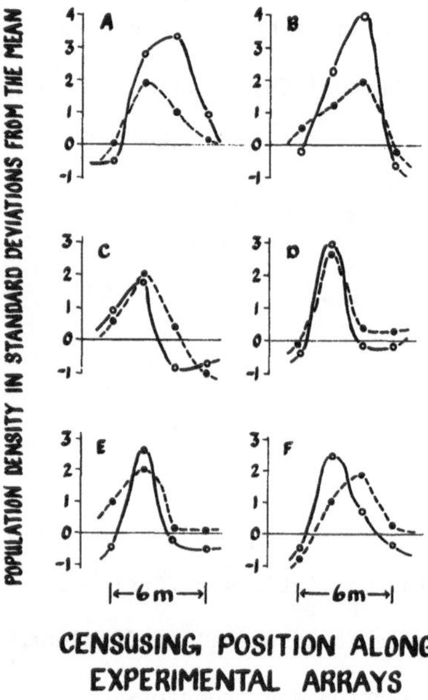

Figure 7. Coincidence of aphids (solid lines) and ladybugs (dashed lines) at outbreak regions. Each of these graphs is simply an enlargement of segments from the larger arrays, such as those pictured in their entirety in Fig. 6. Each point represents a census of a 1 m^2 patch of goldenrod. A is from a patchy array (control, undisturbed) on 7/23; B is from a barrier array (fertilized, undisturbed) on 8/6; C is from a patchy array (fertilized, undisturbed) on 7/30; D is from a patchy array (control, undisturbed) on 7/30; E is from a barrier array (fertilized, undisturbed) on 8/14; F is from a patchy array (fertilized, undisturbed) on 8/6.

generated in models having the form of equations (2a) and (2b). Although there has been much theoretical discussion of such patterning in ecological systems (usually labeled "diffusive instability", see Segel and Jackson 1972, Levin 1974, Segel and Levin 1976, Okubo 1980), field ecologists have never paid much attention to this theory. Nonetheless, because of the striking spatial character of aphid outbreaks (Figures 6 and 7 -- clearly aphid numbers don't simply increase tenfold all along the linear arrays of goldenrod patches or patches + barriers), it is worth asking whether the biology of the 7-spotted ladybug and red aphid interaction is consistent with the mathematical prerequisites for diffusive instability.

These mathematical prerequisites and their biological translations are (see Okubo 1980, pp 204-207):

where v*,p* correspond to the equilibrium densities of victim and predator in system (2a) and (2b) when dispersal is absent

a.) $\left.\frac{\partial F}{\partial v}\right|_{v^*,p^*} = a_{vv} > 0$

An increase in victim density must increase the rate at which the victim population is growing.

b.) $\left.\frac{\partial F}{\partial p}\right|_{v^*,p^*} = a_{vp} < 0$

An increase in predator density must depress victim population growth.

c.) $\left.\frac{\partial G}{\partial v}\right|_{v^*,p^*} = a_{pv} > 0$

An increase in victim density must promote predator population growth.

d.) $\left.\frac{\partial G}{\partial p}\right|_{v^*,p^*} = a_{pp} < 0$

An increase in predator density must depress predator population growth (i.e. standard density-dependent feedback).

e.) $D_p/D_v > \theta_c$ where θ_c is > 1.0 and

$$\theta_c = \frac{\sqrt{a_{vv}\,a_{pp} - a_{vp}\,a_{pv}} \pm \sqrt{|a_{vp}\,a_{pv}|}}{a_{pp}}$$

The predator must be at least θ_c times more mobile than its victim, with the magnitude of θ_c depending on a_{vv}, a_{pp}, a_{vp} and a_{pv}.

Before discussing the conjecture of diffusive instability for the particular ladybug and aphid system I am studying, note that conditions b.), c.) and d.) are likely to hold for any predator-prey system. Condition e.) is a quantitative requirement on the relative mobilities of predator and prey; since diffusion coefficients can be measured (see Kareiva 1983) it represents a condition that can be directly tested. Condition a.) requires that as victim density increases, the rate of victim population growth also increases. This positive feedback, commonly called the Allee effect, is probably the one prerequisite for diffusive instability of which biologists will be most skeptical. Such an Allee effect is not, however, that unlikely for herbivorous insects; experiments indicate that several phytophagous insects appear to use their food more efficiently when they are aggregated on plants (cf. Way and Banks 1967, Way and Cammell 1970, Tsubaki and Shiotsu 1982). An Allee effect may also result because as aphids increase in abundance they saturate their predator's functional response, and this leads to lower predation risks <u>per victim</u>, [cf Kidd's (1982) discussion of grey pine aphid aggregations]. Because equations (2a) and (2b) permit a wide variety of predator-prey dynamics and conditions a.) through e.) are reasonable possibilities, many predator-prey systems may have the potential for spatial patterning as a manifestation of diffusive instability.

The interaction between red aphids and 7-spotted ladybugs could be represented by a variety of coupled partial differential equations having the basic form of (2a) and (2b). One of the simplest models that might apply is

$$\frac{\partial V}{\partial t} = a_0 V + a_1 V^2 - cVP + D_V \frac{\partial^2 V}{\partial x^2} \tag{3a}$$

$$\frac{\partial P}{\partial t} = i - \Phi(V)P + D_P \frac{\partial^2 P}{\partial x^2} \tag{3b}$$

with the aphids as victims or V and the ladybugs as prey or P. Here, $a_0V + a_1V^2$ depicts the Allee effect; $- cVP$ is a simple random encounter predation term, i represents a constant immigration rate for ladybugs; $\Phi(V)$ is a general ladybug emigration function <u>that depends on the density of aphids</u>. Whenever $\partial\Phi/\partial V < 0$, this model admits the possibility of diffusive instability (provided D_P/D_V is $> \theta_c$). Although greatly simplified, equations (3a) and (3b) may be appropriate to the biology of 7-spotted ladybugs and their red aphid victims. To begin with, this model corresponds to local aphid dynamics governed largely by reproduction and predation and local ladybug dynamics governed largely by immigration and emigration.

Within the timecourse of my experiments (in particular, the emergence of the aphid outbreaks portrayed in Figure 6) such a model is consistent with the rapid continuous reproduction of red aphids and the highly vagile 7-spotted ladybugs whose short-term changes in abundance largely reflect movement as opposed to birth or death (these ladybugs are slow to starve and they reproduced in discrete episodes before and after the patterns in Figure 6 emerged). Furthermore, a greater proportion of 7-spotted ladybugs emigrate when their aphid prey are scarce; in mark-release experiments, within one day after their release, 2-3 times as many ladybugs had disappeared when aphids averaged < 5 individuals per goldenrod stem as opposed to > 50 aphid individuals per stem. Since ladybugs do not die of starvation in one day, these density-dependent differences in disappearance rates indicate density-dependent emigration rates such that $\partial\Phi/\partial V < 0$ [as is required to satisfy "diffusive instability condition c.)" of the preceding list]. The Allee effect seems also to be present -- empirically I found a significant positive correlation between the numbers of red aphids per goldenrod stem and their per capita reproductive rate (Kendall's $\tau = .87$, $n = 18$, $p < .05$). Finally, behavioral observations suggest that ladybugs consume aphids as a result of randomly bumping into them, justifying the simple $- cVP$ term of (3a). Thus, it is reasonable to speculate that aphid outbreaks are manifestations of diffusive instability which do not occur in the continuous goldenrod strips because in those circumstance D_p/D_v is not greater than θ_c. Importantly, the ratio of predator to prey mobility is in fact much lower in the continuous goldenrod strips than in the patchy or patchy + barrier treatments (by a factor of ½ as measured by the mean squared displacements shown in Figure 4).

Clearly equations (3a) and (3b) cannot be taken as literal models for the ladybug-aphid system. Ultimately there must be densities at which aphid population growth is self-limited (requiring negative feedback through a term such as $-a_2V^3$). Additionally, the local movement of ladybugs is not simple diffusion (see Figure 5, which necessitates a D_p that is a function of V in equation 3b). Nonetheless, it is intriguing to pursue the ability of equations (3a) and (3b) to explain the localized aphid outbreaks that appear only in patchy or patchy + barrier treatments.

This will require a more thorough experimental and mathematical analysis of the system, as well as numerical identification of a_0, a_1, c, D_p, D_v, i, and Φ.

So far I have discussed the ladybug-aphid interaction as a deterministic process. A stochastic viewpoint provides yet another hypothesis for the occurrence of aphid outbreaks in patchy and patchy + barrier treatments, but not in continuous goldenrod strips. Suppose aphid colonies can guarantee an escape from annihilation by ladybugs only if they first achieve some threshold size at which colony growth exceeds expected predation rates. The opportunity for aphids to escape predation in this manner will depend on colonies surviving undiscovered by ladybugs until they have grown to the threshold size. In turn, the probability that an aphid colony remains undiscovered by ladybugs (per unit time) will be inversely proportional to the rate at which ladybugs search the habitat. Consequently, because ladybug movement is hindered in patchy and patchy + barrier arrays, we may observe local aphid outbreaks in these arrays as a result of a few aphid colonies escaping ladybug detection long enough to grow to a size at which colony reproduction can override the losses due to ladybug predation. In contrast, ladybugs move so freely and rapidly among goldenrod stems in continuous stands that incipient aphid outbreaks may be consistently discovered in these treatments before attaining the escape threshold. This line of argument hinges on the ability of 7-spotted ladybugs to have a dramatic impact on aphid colonies; notably, 7-spotted ladybugs have been reported to average consumption rates greater than 150 aphids per day (Hämäläinen _et al_., 1975) -- a voracity that certainly allows for the potential annihilation of all but the largest aphid colonies by even a single ladybug. A more formal stochastic analysis may offer better insights into this ladybug-aphid interaction, but such stochastic modelling of predator-prey systems has only recently begun to receive attention (cf Chesson 1978).

FUTURE DIRECTIONS FOR THEORY AND EXPERIMENTS

The Question of Density-Dependent Movement in Predator-Prey Systems

Models of predator-prey interactions typically neglect the sort of density-dependent dispersal that is evident in Figure 4 and that is probably characteristic of most predators. It is well-known that predators move less in the presence of abundant prey and more where prey are scarce (cf Hassell 1978). This density-dependent process may be the key to the aggregation of predators at patches of high prey density (i.e. Figure 5) and in turn a prime determinant of predator-prey dynamics. Other forms of density-dependent movement that could play a role in predator-prey systems include: predators moving more when crowded by conspecifics, prey moving more when crowded by conspecifics, prey "running away" from predators (as is reported for pea aphids by Roitberg *et al*. 1979). There are many interesting questions about how to mathematically represent such density-dependent movement, and the consequences of density-dependent movement for phenomena such as diffusive instability are not obvious (Levin, Okubo, and Kareiva *in prep*.). Although a dependence of predator movement on prey density such as that indicated in Figure 4 will oppose the destabilizing feature of passive diffusion, diffusive instability remains a possibility as long as there is still sufficient residual predator movement in regions of high prey density. To better understand spatially distributed predator-prey dynamics, we need population-level models that can represent predator and prey movement behavior such as that evident in Figures 4 and 5.

The Problem of Parameter Identification

To test my hypothesis of diffusive instability, I must estimate each species' diffusion coefficient and population growth function. I do not yet have the techniques for solving this difficult parameter identification problem in its entirety, but much progress has recently been made on the general inverse problem associated with transport equations (Banks *et al*. 1981, Banks and Daniels 1982, Banks and Kareiva 1983). Spline based algorithms have been developed and tested for estimating parameters in equations of the form:

$$\frac{\partial n}{\partial t} = \frac{\partial}{\partial x}\left(D\frac{\partial n}{\partial x}\right) - \frac{\partial}{\partial x}\left(Cn\right) + g\left(n\right) \tag{4}$$

where n is population density, the term involving C represents directed movement (convection, chemotaxis, etc.), D is the coefficient of diffusion and g is a general sink/source function. We are pursuing the application of such transport equations to insect population models because preliminary analyses indicate that insect movement is often well described as a diffusion process, albeit not necessarily simple passive diffusion (cf Dempster 1957; Okubo 1980; Kareiva 1982; Kareiva 1983; Banks and Kareiva 1983; Banks, Daniel and Kareiva 1983). More work is needed on these spline-based identification algorithms so that they can handle coupled reaction-diffusion equations such as (2a) and (2b) with the possibility of density-dependent diffusion coefficients [i.e., with D_V a function of (V,P) and D_P a function of (V,P)]. When attempting to understand specific biological systems as I am with the red aphids and ladybugs, parameter identification is often the unappreciated missing link between mathematical theory and empirical biology.

A Revised Scenario for Experiments

To understand the aphid outbreaks that appeared when I interfered with insect dispersal, I have drawn upon predator-prey theory and interpreted data as though predation were the prime determinant of aphid density and distribution. An alternative, but not mutually exclusive cause of aphid pattern might be variation in the quality of goldenrod as a food resource for aphids (with my dispersal manipulations somehow altering food plant quality). To resolve the relative importance of "predation" versus "plant quality" and to help clarify the mechanisms underlying the effects of my dispersal manipulations, I shall perform the following suite of experiments:

1.) Include only continuous goldenrod stands or continuous stands interrupted by barriers in the experimental design. This will reduce the effects of plant quality and increase the influence of dispersal as a variable.

2.) Repeat the above dispersal manipulations in a nested experimental design so that: all ladybugs are removed from some pairs of barriered and unbarriered strips, while other pairs of barriered and unbarriered strips are left undisturbed (i.e. with ladybugs). If the aphid outbreaks are really due to a predator-prey diffusive instability, removal of the predator should generally elevate aphid densities and remove differences between dispersal treatments.

3.) Artificially establish high aphid densities (outbreaks) and follow their fates in different dispersal treatments and at different "between-outbreak" spacings (to see if there is a wavelength characteristic to the system as should be the case if I really am observing a diffusive instability).

4.) Evaluate plant quality variation by measuring plant characteristics such as nitrogen content, succulence, etc. and by performing reciprocal transplants of aphids between different goldenrod stems.

5.) Experimentally test for an Allee effect. Previously, I had thought that such positive feedback was unlikely. The combination of my preliminary data and the important consequences of an Allee effect in a reaction-diffusion system has alerted me to other reports of such positive feedback (cf Way and Banks 1967; Way and Cammell 1970; Tsubaki and Shiotsu 1982; Kidd 1982). The Allee effect may be a common feature of plant-herbivore interactions and it certainly deserves more attention from field biologists.

6.) Follow the day to day fate of newly started aphid colonies to quantify their opportunities for escaping ladybug annihilation as a function of dispersal regime.

CONCLUDING REMARKS

Manipulations of dispersal produced dramatic changes in a ladybug-aphid interaction associated with goldenrod fields. These experiments raise numerous questions about the theory of spatially distributed predator-prey dynamics and about the importance of dispersal in governing the dynamics of natural populations. For instance, much previous discussion of spatial heterogeneity in predator-prey systems has suggested that "patchiness" may help stabilize predator-prey dynamics (cf Hassell 1980). In contrast, when I increased the "patchiness" of goldenrod fields (by mowing or constructing barriers), the red aphid, Dactynotus nigrotuberculatus, increased in abundance tenfold, presumably because these experimental manipulations somehow interfered with the ability of ladybugs to control aphid populations. Although there are many possible explanations of this experimental result, data are too scarce to choose among the available theoretical explanations. In this paper I have primarily considered theories of spatially distributed predator-prey dynamics for interpretation of data and ideas for future experiments. Clearly, dispersal and spatially distributed dynamics are both generally important in natural systems and poorly understood either empirically or theoretically. The process of dispersal or movement itself is one promising focus for mathematical and empirical research into spatially distributed population interactions. Direct experimental manipulations of dispersal provide a way of testing mathematical models of spatially distributed dynamics. In turn, population models that explicitly incorporate dispersal or movement promise insight into data on spatially distributed dynamics (although there is a need in the models for more attention to density-dependent movement). By intertwining population models that include measurable parameters describing movement with field experiments that vary movement, we may ultimately understand spatially distributed predator-prey dynamics. Progress will be much slower if theory and experiments proceed independently.

ACKNOWLEDGEMENTS

This research was supported by NSF grant DEB 8207117 to P. Kareiva. I am especially grateful to N. Cappuccino for assistance with the fieldwork and to J. Eccleston for data analysis, editing, and the artwork. For generously contributing ideas or commenting on an earlier version of the manuscript I thank P. Bierzychudek, N. Cappuccino, S. Levin, A. Okubo, L. Segel, N. Shigesada and D. Strong. Finally, Ras Schmitt-Tyger provided her usual care and cheer in editing, typing and paginating the final version of the manuscript.

LITERATURE CITED

Banks, H.T. and Daniel, P.L. (1982). Estimation of variable coefficients in parabolic distributed systems. Lefschetz Cent. Dyn. Syst. Report #82-22.

Banks, H.T., Daniel, P.L. and Kareiva, P. (1983). Estimation of temporally and spatially varying coefficients in models for insect dispersal. LCDS Report 83-14, June 1983, Brown University.

Banks, H.T., Crowley, J.M. and Kunisch, K. (1981). Cubic spline approximation techniques for parameter estimation in distributed systems. LCDS Tech. Rep. 81-25, Brown University, Providence, RI.

Banks, H.T. and Kareiva, P. (1983). Parameter estimation techniques for transport equations with application to population dispersal and tissue bulk flow models. J. Math. Biol. in press.

Chesson, P. (1978). Predator-prey theory and variability. Ann. Rev. Ecol. Syst. 9: 323-347.

Dempster, J. (1957). The population dynamics of the Moroccan locust in Cyprus. Anti-Locust Bull. 27:1-59.

Elton, C. (1949). Population interspersion: an essay on animal community patterns. J. Ecol. 37:1-23.

Evans, E.W. (1980). Lifeways of the predatory stinkbugs: feeding and reproductive patterns of a generalist and specialist (Pentatomidae: Podisus maculiventris and Perillus circumcinctus). Ph.D. Diss., Cornell University, Ithaca, NY, USA. 147 pp.

Gurney, W.S.C. and Nisbet, R.M. (1978). Predator-prey fluctuations in patchy environments. J. Anim. Ecol. 47:85-102.

Hämäläinen, M., Markkula, M. and Raij, T. (1975). Fecundity and larval voracity of four ladybird species (Coleoptera, Coccinellidae). Ann. Ent. Fenn. 41:124-127.

Hassell, M.P. (1978). The dynamics of arthropod predator-prey systems. Princeton University Press, Princeton, New Jersey, USA.

Hassell, M.P. (1980). Some consequences of habitat heterogeneity for population dynamics. Oikos 35:150-160.

Hassell, M.P. and May, R.M. (1974). Aggregation of predators and insect parasites and its effect on stability. J. Anim. Ecol. 43:567-594.

Hastings, A. (1978). Spatial heterogeneity and the stability of predator-prey systems: predator-mediated coexistence. Theor. Pop. Biol. 14:380-395.

Hilborn, R. (1975). The effect of spatial heterogeneity on the persistence of predator-prey interactions. Theor. Pop. Biol. 8:346-355.

Huffaker, C.B. (1958). Experimental studies on predation: dispersion factors and predator-prey oscillations. Hilgardia 27:343-383.

Kareiva, P. (1982). Experimental and mathematical analyses of herbivore movement: quantifying the influence of plant spacing and quality on foraging discrimination. Ecol. Monog. 52:261-282.

Kareiva, P. (1983). Local movement in herbivorous insects: applying a passive diffusion model to mark-recapture field experiments. Oecologia 57:322-327.

Kareiva, P. and Shigesada, N. (1983). Analyzing insect movement as a correlated random walk. Oecologia 56:234-238.

Kidd, N.A.C. (1982). Predator avoidance as a result of aggregation in the grey pine aphid, Schizolachnus pineti. J. Anim. Ecol. 51:397-412.

Levin, S.A. (1974). Dispersion and population interactions. Am. Nat. 108:207-228.

Levin, S.A. (1976a). Population dynamic models in heterogeneous environments. Ann. Rev. Ecol. Systematics 7:287-310.

Levin, S.A. (1976b). Spatial patterning and the structure of ecological communities. In: Some Mathematical Questions in Biology, 7. Lectures on Mathematics in the Life Sciences. Levin, S.A. (ed.). Vol. 8, 1-35, Providence, R.I. Amer. Math. Soc.

Levin, S.A. (1978). Population models and community structure in heterogeneous environments. In: Mathematical Association of America Study in Mathematical Biology II: Populations and Communities, Levin, S.A. (ed.), pp. 439-476., Math. Assoc. Amer. Washington.

Levin, S.A. (1981). The role of theoretical ecology in the description and understanding of populations in heterogeneous environments. Amer. Zool. 21:865-875.

Levin, S.A. and Segel, L.A. (1976). Hypothesis for origin of planktonic patchiness. Nature 259:659.

May, R.M. (1978). Host-parasitoid systems in patchy environments: a phenomenological model. J. Anim. Ecol. 47:833-843.

Maynard Smith, J. (1974). Models in Ecology, Cambridge University Press.

Messina, F. (1982a). Comparative biology of the goldenrod leaf beetles, Trirhabda virgata Leconte and T. borealis Blake (Coleoptera: Chrysomelidae). Coleop. Bull. 36:255-269.

Messina, F. (1982b). Food plant choices of two goldenrod beetles: relation to plant quality. Oecologia 55:342-354.

Nicholson, A.J. and Bailey, V.A. (1935). The balance of animal populations. Part I. Proc. Zool. Soc. Lond. 3:551-598.

Obrycki, J., Nechols, J. and Tauber, M. (1982). Establishment of a European lady beetle in New York State. New York Food and Life Science Bull. 94:1-3.

Okubo, A. (1980). Diffusion and ecological problems: mathematical models. Springer-Verlag, New York, NY, USA.

Richards, W.R. (1972). Review of Solidago-inhabiting aphids in Canada with descriptions of three new species (Homoptera: Aphididae). Can. Entom. 104:1-34.

Roitberg, B.D., Nyers, J.H. and Frazer, B.D. (1979). The influence of predators on the movement of apterous pea aphids between plants. J. Anim. Ecol. 48:111-122.

Segel, L.A. and Jackson, J.L. (1972). Dissipative structure: an explanation and an ecological example. J. Theor. Biol. 37:545-549.

Segel, L.A. and Levin, S.A. (1976). Application of nonlinear stability theory to the study of the effects of diffusion on predator-prey interactions. In: Topics in Statistical Mechanics and Biophysics: A Memorial to Julius Jackson. AIP Conf. Proc. No. 27:123-152. Piccirelli, R.A. (ed.). New York: Amer. Inst. Physics.

Tsubaki, Y. and Shiotsu, Y. (1982). Group feeding as a strategy for exploiting food resources in the burnet moth Pryeria sinica. Oecologia 55:12-20.

Way, M.J. and Banks, C.J. (1967). Intra-specific mechanisms in relation to the natural regulation of numbers of Aphis fabae Scop. Ann. Appl. Biol. 59:189-205.

Way, M.J. and Cammell, M. (1970). Aggregation behavior in relation to food utilization by aphids. In: Animal Populations in Relation to their Food Resources, A. Watson (ed.), pp. 229-247. Blackwell Scientific Publications, Oxford.

Oceanic Turbulent Diffusion of Abiotic and Biotic Species*

Akira Okubo

Marine Sciences Research Center
State University of New York
Stony Brook, New York 11794 USA

1. Introduction

When a fluid flows in an orderly fashion, the flow is called "laminar." On the other hand, when the fluid flows in an irregular fashion accompanied with mixing, the flow is called "turbulent." For a neutrally stratified fluid a Reynolds number, which may be interpreted as the ratio of the inertial forces to the viscous forces, determines that a given flow will be laminar or turbulent. Thus in the upper layer of the ocean, the thickness of which ranges from 10 to 100 meters, a mean flow of 10 cm/sec gives Reynolds numbers ranging from 10^6 to 10^7, which are much larger than the critical Reynolds number for transition from a laminar to turbulent flow. This indicates that the flow in the upper layer, where a great many biological activities occur, must be fully turbulent.

Turbulence may be viewed as a type of random motion consisting of many superposed whirls ("eddies") moving in a fashion that is spatially and temporally extremely complicated, so that only statistically distinct average values may be discerned.

The motion of small particles including organisms in the sea is influenced by the turbulence and accordingly is random or stochastic by nature. When the motion possesses randomness, it is accompanied by diffusion. We call the diffusion due to the turbulence of environmental fluids "turbulent diffusion" in order to distinguish it from molecular diffusion due to the random molecular motion.

Some characteristic features of turbulent diffusion in the sea are summarized as follows (Okubo, 1980).

---

*Contribution No. 351 from the Marine Sciences Research Center, State University of New York at Stony Brook.

(i) Motions involved in the oceanic diffusion have scales much larger than the scales of molecular motion. Turbulent diffusion in general is much more effective than molecular diffusion.

(ii) Since oceanic turbulence is composed of a wide range of eddy size from an order of 1 mm to the ocean-wide general circulation (10^8 cm), the scale of the eddies participating in diffusion varies with the scale of phenomenon. As the scale of phenomenon becomes larger, more and larger eddies participate in diffusion, and an effective coefficient of diffusion, i.e., "eddy diffusivity," increases.

(iii) With the exception of very small-scale eddies (typically less than 10 cm), the oceanic turbulence is anisotropic, with horizontal scales that generally exceed vertical scales.

2. The Eulerian model for oceanic diffusion

In dealing with oceanic diffusion it is customary to consider an average value of particle concentration which is viewed in an Eulerian frame. To be specific we first pay attention to an infinitesimal volume around some point $\underset{\sim}{x}$ in a turbulent environment and express the instantaneous particle concentration within it as $S(\underset{\sim}{x},t)$. The turbulence causes the value of S to change randomly. Hence we consider an average value $\bar{S}$ and then query as to what diffusion equation $\bar{S}$ obeys.

The equation of diffusion for instantaneous concentration is given by

$$\frac{\partial S}{\partial t} = - \nabla(\underset{\sim}{u}S) + F(S,\underset{\sim}{x},t) \tag{1}$$

where $\underset{\sim}{u}(\underset{\sim}{x},t)$ is the instantaneous velocity at a given place $\underset{\sim}{x}$, and F represents the instantaneous reaction term or biological term. As usual the molecular diffusion term is ignored in (1).

We now decompose the instantaneous quantities $\underset{\sim}{u}$ and S into their averages $\bar{\underset{\sim}{u}}$, $\bar{S}$ and the turbulent components $\underset{\sim}{u}'$, S'

$$\underset{\sim}{u} = \overline{\underset{\sim}{u}} + \underset{\sim}{u}'$$
$$S = \overline{S} + S' \quad . \tag{2}$$

Substituting (2) into (1) and averaging, we obtain

$$\frac{\partial \overline{S}}{\partial t} = - \nabla(\overline{\underset{\sim}{u}}\ \overline{S}) - \nabla(\overline{\underset{\sim}{u}'S'}) + \overline{F(\overline{S}+S')} \quad . \tag{3}$$

Unfortunately (3) is not of closed form; the covariance function $\overline{\underset{\sim}{u}'S'}$ and $\overline{F}$ are unknowns. Semiempirical theories of turbulent diffusion allow us to make an analogy to the kinetic theory of gases, i.e., molecular diffusion, and relate $\overline{\underset{\sim}{u}'S'}$ to the gradient of the mean concentration by

$$\overline{u_i'S'} = -K_{ij}\ \partial\overline{S}/\partial x_j \tag{4}$$

where the second-order tensor K_{ij} is called the "eddy diffusivity tensor," which may in general be a function of $\underset{\sim}{x}$ and t. It should be emphasized, however, that there is an essential difference between the mechanisms of molecular and turbulent diffusion. The flux of a property by molecular diffusion takes place always from the higher concentration to the lower concentration and is proportional to the concentration gradient, whereas the turbulent flux need not always be in the direction of down-gradient of the mean concentration nor can it always be described as diffusion. The analogy between molecular diffusion and turbulent diffusion is superficial. Accordingly, the notion that turbulent diffusion can be determined in terms of an eddy diffusivity should be regarded as a necessary evil (for details see Corrsin, 1974).

Also note that $\overline{F(\overline{S}+S')}$ is not equal to $F(\overline{S})$ simply because F(S) is in general nonlinear with respect to S. Perhaps at the present stage of biological modeling we may assume the equality

$$\overline{F(S)} = F(\overline{S}) \quad . \tag{5}$$

The above relation is expected to hold approximately when the time scale of biological interactions is much larger than that of physical processes of mixing.

Under these assumptions, (3) can be written as

$$\frac{\partial \bar{S}}{\partial t} = - \frac{\partial}{\partial x_i} (\bar{u}_i \bar{S}) + \frac{\partial}{\partial x_i} (K_{ij} \frac{\partial \bar{S}}{\partial x_j}) + F(\bar{S}) \quad . \tag{6}$$

As was mentioned previously, a very wide spectrum of motions exists in the ocean, in particular in the horizontal direction. As a result, the value of the horizontal eddy diffusivity usually increases with the scale of mixing considered. A mathematical model for horizontal diffusion in the sea must take into account this dependence of the eddy diffusivity. Thus in a model for oceanic relative diffusion, the horizontal diffusivity is assumed to be a function of the distance from the patch centroid (Joseph and Sendner, 1958; Ozmidov, 1958).

The advective and diffusive processes are modeled separately as the advection and diffusion terms in (6). In fact, the vertical and horizontal processes in the sea are strongly coupled in such a manner that the combined action of vertical shear in horizontal currents and transverse vertical mixing can produce an effective dispersion in the horizontal direction (Bowden, 1965). This so-called "shear effect" can be modeled by (6) when we include the velocity shear in the mean velocity $\bar{\underset{\sim}{u}}$ and leave the eddy diffusivity constant (Okubo, 1971).

3. Lagrangian approach to oceanic diffusion

Since the position, not velocity, is the quantity of concern in diffusion, Lagrangian particle displacement should be a natural variable of choice in the theory of turbulent diffusion (Taylor, 1921; Richardson, 1926; Batchelor, 1952; Lin, 1960a, b; Corrsin, 1962; Krasnoff and Peskin, 1971). The Lagrangian form of diffusion equation was first given by Corrsin (1962), but that form has not been studied in detail because of mathematical difficulty, except for estimating the effect of molecular diffusion on the dispersion in turbulent flow (Okubo, 1967; Chevray and Venkataramani, 1979).

Corrsin (1962) presented an instantaneous form for the Lagrangian diffusion equation. It is expressed as

$$\frac{\partial \Gamma}{\partial t} = D\{[[\Gamma,y,z],y,z] + [x,[x,\Gamma,z],z] + [x,y,[x,y,\Gamma]]\} \equiv D\nabla_L^2\Gamma \tag{7}$$

where $\underset{\sim}{x} = (x,y,z)$ denotes the coordinates at a moment t of a fluid particle which is identified by its Lagrangian coordinates $\underset{\sim}{a} = (a,b,c)$ at $t = 0$. $\Gamma(\underset{\sim}{a},t)$ is the concentration in the Lagrangian frame, and D is the molecular diffusivity. [A,B,C,] is the Monin bracket (Monin, 1962) and stands for $\partial(A,B,C)/\partial(a,b,c)$.

The so-called advection terms in the Eulerian diffusion equation do not appear in the Lagrangian equation since fluid particles are followed. A penalty is paid, however, by the fact that complex displacements of fluid particles are intermingled in the diffusion terms of the Lagrangian equation. An exact analytical solution of the Lagrangian diffusion equation has been found only for a uniform Lagrangian deformation field (Okubo, 1966).

Equation (7) is dynamically coupled with the Navier-Stokes equations in the Lagrangian form,

$$\begin{aligned}
\frac{\partial^2 x}{\partial t^2} &= -[P,y,z] + \nu\,\nabla_L^2\left(\frac{\partial x}{\partial t}\right)\\
\frac{\partial^2 y}{\partial t^2} &= -[x,P,z] + \nu\,\nabla_L^2\left(\frac{\partial y}{\partial t}\right)\\
\frac{\partial^2 z}{\partial t^2} &= -[x,y,P] + \nu\,\nabla_L^2\left(\frac{\partial z}{\partial t}\right)
\end{aligned} \tag{8}$$

subject to the equation of continuity for an incompressible fluid

$$[x,y,z] = 1 \tag{9}$$

where $\underset{\sim}{x} = \underset{\sim}{x}(\underset{\sim}{a},t)$, $P = P(\underset{\sim}{a},t)$, P is the modified pressure (Batchelor, 1967) in which the hydrostatic part of the kinematic pressure has been eliminated, and ν is the kinematic viscosity of water.

Since the analytical treatment of the Lagrangian equations is formidable, if not impossible, we will replace them by analogous models. For simplicity let us consider a one-dimensional case. The Langevin-type equation for turbulent diffusion has been used as a model for the Lagrangian equation of motion (Lin, 1960a, b; Krasnoff and Peskin, 1971):

$$\frac{dx}{dt} = v \tag{10}$$

$$\frac{dv}{dt} = f - kv \tag{11}$$

where f is a non-resistive random force per unit mass of fluid, and k is the coefficient of friction. Intuitively we may relate f to the pressure term and -kv to the viscous term in (8). In fact, the pressure forces and viscous forces are dynamically coupled in the Lagrangian equation of motion, and only in the limit of infinite Reynolds number this coupling disappears (Krasnoff and Peskin, 1971).

We first consider the kinematic problem, assuming that the turbulent flow field is statistically stationary with zero mean velocity. The distance x(t) traveled by a particle after time t is obtained by integrating (10)

$$x(t) = \int_0^t v(t')dt' \quad . \tag{12}$$

Since v(t) is a random process with $\bar{v} = 0$, x(t) is also a random process with $\bar{x} = 0$. In diffusion we are interested in the variance of particle displacements $\overline{x^2}$ which is obtained from (12)

$$\overline{x^2}(t) = \int_0^t\int_0^t \overline{v(t')v(t'')}dt'dt'' = 2\int_0^t dt'\int_0^{t'} \overline{v(t')v(t'')}dt'' \quad . \tag{13}$$

In contrast to the random walk of particles, the motion of the particles embedded in turbulent flow is more or less correlated at two instances t' and t". Therefore we need to introduce the Lagrangian velocity autocorrelation coefficient

$$R_L(\tau) \equiv \overline{v(t)v(t+\tau)}/\overline{v^2} \tag{14}$$

where in view of the stationarity of turbulence, $\overline{v^2}$ is constant and $\overline{v(t)v(t+\tau)}$ depends only on the time lag τ. From (13) and (14)

$$\overline{x^2}(t) = 2\,\overline{v^2} \int_0^t dt' \int_0^{t'} d\tau R_L(\tau) \quad . \tag{15}$$

This formula was first given by Taylor (1921) and has been considered to be the cornerstone in the theory of turbulent diffusion. Equation (15) can also be written as

$$\overline{x^2}(t) = 2\,\overline{v^2} \int_0^t (t-\tau) R_L(\tau) d\tau \tag{16}$$

$$= 2\,\overline{v^2} t \int_0^t R_L(\tau) d\tau - 2\,\overline{v^2} \int_0^t \tau R_L(\tau) d\tau \quad . \tag{16'}$$

For very small values of t, $R_L(\tau) \simeq 1$. Hence

$$\overline{x^2}(t) \simeq \overline{v^2}\, t^2 \quad . \tag{17}$$

For ordinary diffusion we anticipate that the particle velocity loses its statistical dependence on past velocities as the dispersion continues. In other words $R_L(\tau)$ decreases to zero at large values of τ such that $\int_0^t R_L(\tau) d\tau$ and $\int_0^t \tau R_L(\tau) d\tau$ converge as $t \to \infty$. We then find from (16) as $t \to \infty$

$$\overline{x^2}(t) = 2Dt \tag{18}$$

where

$$D \equiv \overline{v^2} \int_0^\infty R_L(\tau) d\tau \quad . \tag{19}$$

Clearly (18) is the relation for variance valid for purely random dispersal ("Fickian limit"). In fact a purely random walk is characterized by

$$R_L(\tau) = \lambda\delta(\tau) \tag{20}$$

so that (18) holds for all t.

The assumption of stationarity of the Lagrangian velocity correlation may be valid for "absolute diffusion," i.e., the spread of particles with respect to a fixed frame of reference (Csanady, 1973). In principle the entire spectrum of eddies contributes to absolute diffusion, and hence the velocity autocorrelation of dispersing particles may be assumed to be stationary. On the other hand, the diffusion of a patch of particles released instantaneously is classified as "relative diffusion," i.e., the spread of particles with respect to the moving centroid of the patch. Only eddies whose scales are equal or less than those of the patch contribute appreciably to relative diffusion. As a patch spreads, more and larger eddies participate in the dispersion relative to the centroid, and hence the rate of relative diffusion tends to increase with time. In fact, Richardson (1926) first developed the concept of relative diffusion in the atmosphere and obtained an empirical law that an apparent diffusivity increases as the 4/3 power of the scale of diffusion. The same power law is also applicable to oceanic relative diffusion (Stommel, 1949; Okubo, 1974).

Richardson's law can be derived from the Langevin equation. If we consider the motion of two particles with positions $x_1(t)$ and $x_2(t)$ satisfying the Langevin equation, then their relative separation $x_r(t) = x_2(t) - x_1(t)$ is governed by

$$dx_r/dt = v_r \tag{21}$$

$$dv_r/dt = -kv_r + f_r \tag{22}$$

where $f_r = f_2 - f_1$. Appropriate initial conditions for a pair of particles that start at the same time with sufficiently small separation are

$$x_r(0) = v_r(0) = 0 \quad . \tag{23}$$

In view of the reason previously mentioned, the relative velocity of a pair of particles cannot be assumed to be a stationary random process. However, we may assume that the relative acceleration $f_r(t)$ is a stationary random process so that

$$A(\tau) = \overline{f_r(t)f_r(t+\tau)}/\overline{f_r^{\,2}} \quad . \tag{24}$$

For a turbulent flow with sufficiently large Reynolds numbers such as is seen in the ocean, we expect to have a long duration of intermediate times before the resistive force of viscous origin becomes appreciably large. Mathematically, this imposes the condition k = 0 (Lin, 1960a, b). The solution of (21) and (22) with k = 0 and subject to (23) is

$$x_r(t) = \int_0^t (t-\xi) f_r(\xi) d\xi \quad .$$

Thus the variance of the relative distance is obtained by

$$\sigma^2(t) = \overline{x_r^2(t)} = 2\overline{f_r^2}/3 \int_0^t (t-\tau)^3 A(\tau) d\tau + \overline{f_r^2} \int_0^t (t-\tau)^2 \tau A(\tau) d\tau \quad . \tag{25}$$

If $A(\tau)$ converges very rapidly for the intermediate time such that $\tau > T_c$, we have approximately

$$\sigma^2 = 2/3 \; Bt^3 \qquad \text{for } t > T_c \tag{26}$$

where

$$B = \overline{f_r^2} \int_0^{T_c} A(\tau) d\tau \quad . \tag{27}$$

If we define an apparent diffusivity by

$$K_a = \tfrac{1}{2} \, d\sigma^2/dt$$

we obtain

$$K_a = Bt^2 = (\frac{3}{2})^{2/3} B^{1/3} \sigma_r^{4/3} \qquad (28)$$

which is the relationship discovered by Richardson (1926).

4. Biodiffusion

When the diffusion of ecological systems is considered, another type of diffusion, due to random motion of organisms themselves, appears in addition to turbulent diffusion discussed previously. This is referred to as "biodiffusion." The smaller an organism is, the more subject it is to the effect of the motion of the environmental medium. For instance, free-living bacteria and phytoplankton in the sea diffuse almost passively, while many zooplankters and fishes undergo varying portions of passive and active diffusion.

Theories of turbulent diffusion are expected to hold approximately for organisms that diffuse essentially in a passive manner. On the other hand, the migration and dispersal of animals by and large constitute a ceaseless active effort on the part of the animal to survive. Thus a realistic model of biodiffusion requires a combination of various concepts such as correlated random walks, diffusion incorporating space-time variation of diffusivity and other parameters, effect of nonrandom (behavioral) forces acting on an individual, and interference between individuals.

Among a great variety of problems in biodiffusion we choose two particular topics and apply the methods that have been developed primarily in physical diffusion.

4.1 Statistical theory of animal "swarming"

By the term "swarming" we mean a biological phenomenon such as zooplankton swarming or fish schooling in which a number of animal individuals are involved in more or less random movement as a group. Nevertheless a swarm does not disperse and persists for long times with little change in its dimension.

The distinction in kinematics between diffusion and swarming can be seen in Taylor's formula (16'). To mathematically describe swarming, $\overline{x^2}$ must approach a constant for steady-state swarm maintenance. This is satisfied only when the correlation coefficient $R_L(\tau)$ oscillates about the zero value in such a way that $\int_0^t R_L(\tau)d\tau$ approaches zero asymptotically and the second integral of (16') approaches a negative constant value. In physical terms the individual motions appear to resemble a random pendulum-like motion swinging back and forth about the stationary position of swarm centroid ($x = 0$).

We next consider a simple dynamical model for swarming and calculate $R_L(\tau)$ theoretically. Newton's second law of motion will be applied to swarming animal motion. We assume that (i) the mass of organism is constant with time, (ii) the frictional force is proportional to the velocity of organism, (iii) the nonrandom force is attractive by nature toward the center of the swarm and dependent on the distance from the center, and (iv) the random force is a white noise. Hence the equation of motion in one-dimensional space is given by

$$d^2x/dt^2 = -k\ dx/dt - \omega^2 x - \phi(x) + A(t) \tag{29}$$

where ω is the frequency of harmonic force, $\phi(x)$ is the acceleration due to anharmonic force, and $A(t)$ is the random acceleration of a white noise nature.

Since $\phi(x)$ is nonlinear in x, we use the method of equivalent linearization (Caughey, 1963; Bulsara et al., 1982) to obtain an analytical solution of (29). To this end, (29) is replaced by a linear system with an equivalent linear frequency ω_e,

$$d^2x/dt^2 = -k\ dx/dt - \omega_e^2 x + A(t) \quad . \tag{30}$$

The frequency ω_e is chosen in such a way that the error made by the replacement is minimized.

As a result we obtain

$$\omega_e^2 = \omega^2 + \langle x\phi(x)\rangle / \langle x^2\rangle \tag{31}$$

where $\langle\ \rangle$ denotes a time average over one cycle of oscillation of $x = a \sin\sigma$.

Once ω_e^2 has been determined, the linear equation (30) can be solved, and statistical properties of x can be calculated. In particular the velocity autocorrelation coefficient is found to be

$$R_L(\tau) = e^{-k\tau/2}\{\cos\omega_1\tau - k/2\omega_1\sin\omega_1\tau\} \quad (32)$$

where $\omega_1^2 = \omega_e^2 - k^2/4$. With the use of (32) the variance can be computed from (16'), and is found to approach

$$\overline{x^2} = q^2/k\omega_e^2 \quad (33)$$

as $t \to \infty$, where q^2 is the variance of the random force A(t). That is, this dynamic model enables us to obtain the variance of swarming in terms of parameters that characterize the motion of individuals. The root of (33), $(\overline{x^2})^{\frac{1}{2}}$ scales the spatial extent of the swarm. Applications of the statistical theory to insect swarms and fish schools are found in Okubo (1980).

4.2 Probability density function for organism concentration fluctuations

The advection-diffusion-reaction equation such as (6) serves as a model to discuss the mean concentration of organisms in time and space. In fact, the organism distribution varies in such a stochastic manner that the probability density function is more appropriate to consider rather than the mean concentration.

Let

$$\Pi(x,v,S,t) \quad (34)$$

be the joint probability density function that a parcel of water is found in the phase space of the range (x,x+dx) and (v,v+dv) and contains the organism concentration in the range (S,S+dS) at time t. The stochasticity is due to turbulence in the medium as well as to the randomness of biological activities. We will obtain the basic equation for Π.

The dynamical equations for x, v, and S are modeled by

$$\frac{dx}{dt} = v \tag{35}$$

$$\frac{dv}{dt} = -kv + K(x) + f(t) \tag{36}$$

$$\frac{dS}{dt} = F(x,v,S) + G(x,v,S)\lambda(t) \tag{37}$$

where K(x) is a nonrandom acceleration, F is a nonrandom part of the growth rate of organism, and G is the amplitude of a random part of the growth rate. f(t) and $\lambda(t)$ are random functions of white noise type

$$\begin{aligned} \overline{f(t)f(t+\tau)} &= 2B\delta(\tau) \\ \overline{\lambda(t)\lambda(t+\tau)} &= 2C\delta(\tau) \quad . \end{aligned} \tag{38}$$

In view of (38) we may assume that Π is a Markov process and obeys a Fokker-Planck equation (Arnold, 1974). We thus obtain

$$\frac{\partial\Pi}{\partial t} = -v\,\frac{\partial\Pi}{\partial x} - \frac{\partial}{\partial v}\,\{(-kv + K(x))\Pi\} + B\,\frac{\partial^2\Pi}{\partial v^2} - \frac{\partial}{\partial S}\,(F\Pi) + C\,\frac{\partial^2}{\partial S^2}(G^2\Pi) \quad . \tag{39}$$

The Fokker-Planck equation of a marginal probability density function can be obtained by integrating (39) over a variable. Thus the probability density function of x and v,

$$\psi(x,v,t) = \int \Pi(x,v,S,t)dS$$

is given by

$$\frac{\partial\psi}{\partial t} = -v\,\frac{\partial\psi}{\partial x} - \frac{\partial}{\partial v}\,\{(-kv+K)\psi\} + B\,\frac{\partial^2\psi}{\partial v^2} \quad . \tag{40}$$

Some special cases of interest are found in (40).

(i) If ψ is an even function of x and K(x) is an odd function of x, integration of (40) over x leads to

$$\frac{\partial\Omega}{\partial t} = k\,\frac{\partial}{\partial v}(v\Omega) + B\,\frac{\partial^2\Omega}{\partial v^2} \tag{41}$$

where

$$\Omega = \Omega(v,t) = \int \psi(x,v,t)dx$$

which is the probability density function of velocity. For steady state

$$\Omega^*(v) = (k/2\pi B)^{\frac{1}{2}} e^{-\frac{k}{2B}v^2} \tag{42}$$

which is the Maxwellian velocity distribution.

(ii) If $k = K = 0$,

$$\frac{\partial \psi}{\partial t} = -v \frac{\partial \psi}{\partial x} + B \frac{\partial^2 \psi}{\partial v^2} \tag{43}$$

which was proposed by Imahori and Hori (1951) and Obukhov (1959) as a Lagrangian model for turbulent diffusion. Richardson's law (28) can be derived from (43).

Also, integration of (39) over x and v leads to

$$\frac{\partial \theta}{\partial t} = - \frac{\partial}{\partial S} (F_a(S)\theta) + C \frac{\partial^2}{\partial S^2} (G_a^2 \theta) \tag{44}$$

where

$$\theta(S,t) = \iint \Pi(x,v,S,t)dxdv$$

$$F_a(S) = \iint F(x,v,S)\Pi dxdv/\theta$$

$$G_a^2(S) = \iint G^2(x,v,S)\Pi dxdv/\theta \quad .$$

Multiplying S by (44) and integrating over S, we obtain the dynamical equation for $\overline{S} = \int S\theta dS$

$$\frac{d\overline{S}}{dt} = \overline{F_a(S)} \tag{45}$$

Finally, multiplying S by (39) and integrating over S, we obtain the advection-diffusion-reaction equation for $\overline{S}$ in the phase space

$$\frac{\partial \overline{S}}{\partial t} + v \frac{\partial \overline{S}}{\partial x} + \frac{\partial}{\partial v} \{(-kv + K(x))\overline{S}\} = B \frac{\partial^2 \overline{S}}{\partial v^2} + \overline{F(S)} \quad . \tag{46}$$

The equality (5) for the last term on the right-hand side of (46) may be assumed in practical applications.

LITERATURE CITED

Arnold, L. (1974). Stochastic Differential Equations: Theory and Applications. John Wiley & Sons, New York, 228 pp.

Batchelor, G.K. (1952). Diffusion in a field of homogeneous turbulence. II. The relative motion of particles. Proc. Cambridge Philos. Soc. 48:345-362.

Batchelor, G.K. (1967). An Introduction to Fluid Dynamics. Cambridge University Press, London-New York, 615 pp.

Bowden, K.F. (1965). Horizontal mixing in the sea due to a shearing current. J. Fluid Mech. 21:83-95.

Bulsara, A.R., Linderberg, K. and Shuler, K.E. (1982). Spectral analysis of a nonlinear oscillation driven by random and periodic forces. I. Linearized theory. J. Statist. Phys. 27:787-808.

Caughey, T.K. (1963). Equivalent linearization techniques. J. Acoustical Soc. Am. 35:1706-1711.

Chevray, R. and Venkataramani, K.S. (1979). Total dispersion of a scalar quantity in turbulent flow. Phys. Fluids 22:2284-2288.

Corrsin, S. (1962). Theories of turbulent dispersion. In: Mécanique de la Turbulence, 27-52. Centre National de la Recherche Scientifique, Paris.

Corrsin, S. (1974). Limitations of gradient transport models in random walks and in turbulence. In: Advances in Geophysics, vol. 18A:25-60. Academic Press, New York.

Csanady, G.T. (1973). Turbulent Diffusion in the Environment. D. Reidel Publishing Co., Dordrecht-Boston, 248 pp.

Imahori, K. and Hori, J. (1951). On the diffusion by turbulent motion. J. Meteorol. Soc. Japan 29:327-335.

Joseph, J. and Sendner, H. (1958). Über die horizontale Diffusion im Meere. Deut. Hydrogr. Z. 13:13-23.

Krasnoff, E. and Peskin, R.L. (1971). The Langevin model for turbulent diffusion. Geophys. Fluid Dynamics 2:123-146.

Lin, C.C. (1960a). On a theory of turbulent dispersion by continuous movements. Proc. Nat. Acad. Sci. USA 46:566-570.

Lin, C.C. (1960b). On a theory of turbulent dispersion by continuous movements. II. Stationary anistropic processes. Proc. Nat. Acad. Sci. USA 46:1147-1150.

Monin, A.S. (1962). On the Lagrangian equations of the hydrodynamics of an incompressible viscous fluid. J. Appl. Math. and Mech. 26:458-468.

Obukhov, A.M. (1959). Description of turbulence in terms of Lagrangian variables. In: Advances in Geophysics, vol. 6:113-116. Academic Press, New York.

Okubo, A. (1966). A note on horizontal diffusion from an instantaneous source in a nonuniform flow. J. Oceanogr. Soc. Japan 22:35-40.

Okubo, A. (1967). Study of turbulent dispersion by use of Lagrangian diffusion equation. Phys. Fluids, Suppl. 1967:S72-75.

Okubo, A. (1971). Horizontal and vertical mixing in the sea. In: Impingement of Man on the Oceans, 89-168, Hood, D.W. (ed.). John Wiley & Sons, New York.

Okubo, A. (1974). Some speculations on oceanic diffusion diagrams. Rapports et Procès-Verbaux des Réunions, vol. 167:77-85. International Council for the Exploration of the Sea.

Okubo, A. (1980). Diffusion and Ecological Problems: Mathematical Models. Springer-Verlag, Berlin-Heidelberg-New York, 254 pp.

Ozmidov, R.V. (1958). On the calculation of horizontal turbulent diffusion of the pollutant patches in the sea. Doklady Akad. Nauk SSSR 120:761-763.

Richardson, L.F. (1926). Atmospheric diffusion shown on a distance-neighbour graph. Proc. Roy. Soc. London A 110:709-727.

Stommel, H. (1949). Horizontal diffusion due to oceanic turbulence. J. Marine Res. 8:199-225.

Taylor, G.I. (1921). Diffusion by continuous movements. Proc. London Math. Soc. Ser. 2, 20:196-211.

TAXES IN CELLULAR ECOLOGY

by

Lee A. Segel

Department of Applied Mathematics
The Weizmann Institute of Science
Rehovot, 76100, Israel*

"Nothing is certain but death and taxes." — B. Franklin

Ecological aspects of chemotaxis in microorganisms have been reviewed by Chet and Mitchell (1976). More recently Carlile (1980) surveyed the implications for microbial ecology of various taxes and tropisms. Considerable material on ecological implications can be found in Levandowsky and Hauser's (1978) review of chemosensory responses in algae and protozoa, while the review by Lapidus and Levandowsky (1981) covers mathematical developments connected with microbial chemotaxis on cellular and population levels. These articles provide an excellent background to the present discussion. Thus, on occasion detailed references will not be cited for various points, when these can be found in the reviews just mentioned.

Here we shall concentrate on the ecological implications of various kinds of taxes, with emphasis on the possible contribution of theoretical developments. We will include within our review not only true one-celled microorganisms (and perhaps other small creatures) but also the interaction of cell populations in developing and mature higher organisms.

The Ubiquity of Taxes

We will define a taxis loosely, in accord with much common usage, as the tendency of an organism to move toward or away from some stimulus. Since the classical work of Frankel and Gunn (1940) many authors have taken pains to give precise definitions to different types of directed motions, but we shall not pursue this matter here.

<u>Chemotaxis</u> is perhaps the most studied variety of taxis. See for example

* Also at the Department of Mathematical Sciences, Rensselaer Polytechnic Institute, Troy, N.Y. 12181, U.S.A.

Gerisch's (1982) survey of chemotaxis in the cellular slime mold Dictystelium discoideum. Other common types are phototaxis, thermotaxis, and aerotaxis. Carlile (1980), for example, mentions more exotic varieties such as galvanotaxis (response to electric currents), negative geotaxis (tendency to swim upward), electrotaxis (response to electric fields), rheotaxis (upstream swimming), viscotaxis (tendency to move into regions of high viscosity) and attraction by vibrations (vibrataxis?).

We conjecture that any motile organism is tactic to at least one stimulus. (This may be called Franklin's law, in view of the citation that heads this article.) Once motility has evolved, it seems reasonable to suppose that many mutants will arise in which the motility is somewhat altered by certain agents, and that in some cases such alterations will benefit the organism and so will become fixed in the population.

Some Ecological Functions of Taxes

Although it may be difficult to provide conclusive demonstrations, many taxes seem to have fairly clear ecological functions. For example, positive and negative aero- and phototaxes are among the influences that presumably direct organisms to favorable physical environments. Sensory cues appear to guide sessile organisms to settling sites, or may lead parasites to particular hosts [as in the chemotactic migration of trematode miracidia to its intermediate snail host mentioned by Mansour (1979)] or even particular locations within particular hosts. A medically important illustration is the demonstration that Vibrio cholerae is strongly chemotactic toward mucosal surfaces, and that this enhances colonization success (Freter, O'Brien, and Macsai, 1979).

The use of taxes to lead organisms to their food can be exemplified by the chemotactic responses of the common marine bacterium Vibrio alginolyticus to material released by algae (Sjoblad and Mitchell, 1979), for extracellular algal material serves as nutrient for bacteria. Dictyostelium amoebae are probably attracted to their bacterial prey by means of chemotaxis to a bacterial secretion, folic acid (Gerisch, 1982).

Dictyostelium can also serve as an example of the role of taxes in sexual development, for they are attracted by immature macrocysts. A more classical example is "the chemotaxis of the male gametes of Allomyces, which occurs in response to sirenin emitted by the female gametangia and gametes but to no other compound, (and) can only have a role of fertilisation" (Carlile, 1980). In discussing chemotaxis of the spermatazoa of Muggiaea kochi (a type of siphonophore) Boon (1983) makes an order of magnitude estimate which indicates that the accumulating sperm have an appreciable stirring effect, thereby enhancing the spread of the putative attractant and facilitating further aggregation.

The functions of some of the more unusual taxes remain speculative. For example Maugh (1982) reports the hypothesis of R.P. Blakemore and R.B. Frankel that magnetotaxis along field lines helps marine bacteria locate the oxygen-depleted sediments where they flourish. In support of this notion they found that in the northern (southern) hemisphere the bacteria are predominately north (south) seeking, which indeed guides both sets of microorganisms toward the bottom. In the region of the magnetic equator the field lines are roughly parallel to the bottom, and here there are roughly equal numbers of north and south-seeking magnetotactic bacteria. It is conjectured that in this case magnetotaxis might help bacteria remain in the sediment once they have arrived there.

Particularly exotically named is necrotaxis, the tendency of certain white cells to move toward dead blood cells as a prelude to eating the corpses. Hu and Barnes (1970) provide a simple theory for necrotaxis, enabling certain parameter estimates to be made. The theory is subsumed in the works of Lauffenberger and his associates, to be mentioned below. What is relevant at this point is a first illustration of another reason for taxis by isolated cells. In one sense, the taxis provides preferential motion in the direction of food, but there is also a larger consideration, the survival of the organism of which the cell in question is a part.

Equations for Population Taxis

Phenomenological equations for population chemotaxis were put forward by Keller and Segel (1970). A more fully explained phenomenological derivation is given in the book edited by Segel (1980), Section 6.5. The equations were also derived from various models for individual cell behavior by investigators such as Patlak (1953), Segel (1977) and Alt (1980). For certain models corrections to the original phenomenological equations are required. In some cases measurements show the corrections to be negligible (Rothman and Lauffenburger, 1983).

The phenomenological derivation is based on the fundamental conservation equation for the population density b of the microorganism:

$$\partial b/\partial t = -\nabla \cdot J + Q \, . \tag{1}$$

The two terms on the right side provide contributions to the change in b that arise respectively from net flux J into a volume element, and net birth rate Q. If the motion is completely random then one expects a flux of the form

$$J = J_{random} = -\mu \nabla b \, . \tag{2}$$

The combination of (1) and (2) yields the usual diffusion equation, with the addition of the net creation term Q. The "motility" μ is a diffusion coefficient that can be estimated for microorganisms by simple experiments (Segel, Chet and Hennis, 1977). Effects of (1) and (2) with non-constant μ (when $Q = 0$) have been examined by Lapidus (1980a).

If there is an attractant of density c, one expects an additional chemotactic flux that is proportional to the gradient of c (for sufficiently small fluxes):

$$J = J_{random} + J_{chemotactic} \, , \quad J_{chemotactic} = \chi b \nabla c \, . \tag{3a,b}$$

For a given gradient of c, the flux should be proportional to the local density of microorganisms, when the density is sufficiently small. This explains the presence of the factor b in (3b). The chemotactic sensitivity χ measures the intrinsic response of a bacterium to a given gradient. Sometimes the chemotactic flux is written

$$J_{chemotactic} = Vb \tag{4}$$

where V is a drift velocity, analogous to convection. It should be borne in mind that $V = \chi \nabla c$ and so combines the effects of the environment (∇c) and of cellular behavior (χ).

A fairly general equation for the attractant concentration includes the possibilities (i) that attractant is consumed at a rate $k(c)$ per cell, (ii) that there is a net production of c by chemical reaction at rate $q(c)$, and (iii) that the attractant diffuses with diffusivity D (assumed constant for simplicity). These considerations lead to the following phenomenological equations for gross changes due to chemotaxis:

$$\partial b/\partial t = \nabla\cdot[\mu(c)\nabla b - \chi(c)b\nabla c] + Q \ , \qquad (5a)$$

$$\partial c/\partial t = -bk(c) + q(c) + D\nabla^2 c \quad .$$

There is a mild controversy as to the correct form of equation to use when the motility μ is not constant. Our remarks on this matter will provide views that may perhaps be regarded as in opposition to those of Lapidus and Levandowsky (1981) but it is better to think of them as an attempt to clarify the issues.

In discussing the random flux Lapidus and Levandowsky (1981) state that "to the extent that the motion of each cell is a random walk, with no interactions with other cells and no response to external sensory cues, ... the appropriate continuous expression" (in our notation) for the flux J is

$$J = -\nabla\cdot(\mu b) \ . \qquad (6)$$

This assertion is backed up by a (completely correct) derivation where a diffusion limit is taken of a random walk with a probability of moving from a given point that is a given function of a spatial variable x. The work "from" has been emphasized (and it was not by Lapidus and Levandowsky, 1981) for the results would be quite different if it were replaced by "to". Indeed, Skellam (1951) pointed out some years ago that (6) results from an unbiased walk where the probability of taking a step depends on conditions at the beginning of a step while the commonly used diffusive flux (2) is appropriate where conditions at the termination of a step determine the probability of moving.

There is no dispute as to the correct diffusion limit of a given biased random walk model. What remains is a difference in taste as to the most revealing way to arrange the terms. Lapidus and Levandowsky (1981) prefer to call (6) a random flux because it arises wholly "from differences in the isotropic motility at neighboring points of space". They reserve the notion of a drift for situations where at a given point there are different probabilities for motions in different directions. My view is that more understanding is gained if (6) is expanded to

$$J = -\mu\nabla b + bV \ , \quad V = -\nabla\mu \ . \tag{7}$$

The term $-\mu\nabla b$ is an isotropic flux, which is the same whether or not steps are influenced by conditions at their onset or their termination. The remaining term bV is an effective drift, indeed not brought about by intrinsic directional preferences at point x, but rather due to spatial differences in the vigor of random motion. For example, there will be a net drift in the positive x direction if organisms move randomly but with decreasing vigor as x increases $(d\mu/dx < 0)$ — because of smaller "steps" or more time between "steps".

A contribution to the drift velocity V can of course arise from fluid convection -- see the experimental study by Walsh and Mitchell (1978) and the related theoretical discussion, using (5), by Lapidus (1980b).

The phenomenological chemotaxis equation (5a) is expected to be valid only when (i) cell and nutrient densities are not high enough to promote interference effects, (ii) the gradients of these densities are small enough to permit linearization, and (iii) longer range effects are negligible so that higher derivative terms can be neglected. With respect to (i), Fu et al. (1982) conclude that cell interactions are in fact important in their experiments on combined thermo- and chemotaxis in leukocytes. Relevant to (ii) are experiments by Dahlquist, Elwell and Lovely (1976) on the chemotaxis of Salmonella typhimurium to L-serine. These authors found that the drift velocity V could be related to the attractant gradient ∇c by an expression of the Michaelis-Menten form

$$V = V_{max}\nabla c/(K + \nabla c) \ ,$$

where $V_{max} = 7\mu/\text{sec}$ and $K = 0.25 \ \text{mm}^{-1}$. According to these measurements, our assumption of proportionality between the drift velocity and the gradient $(V = \chi\nabla c)$ is suitable for situations in which the attractant concentration takes somewhat more than one centimeter to double.

Two other effects limit the accuracy of the chemotaxis equation (5a). One is that "memory" must be sufficiently short; otherwise the flux cannot be explicitly given as in (3), but instead requires an integration over past history (Segel, 1977). A second restriction is that members of the population must behave sufficiently similarly (Segel and Jackson, 1973). In this connection it has been found (Koshland, 1978) that "a bacterial culture that is genetically homogeneous and grown in homogeneous nutrient conditions nevertheless produces characteristically different individuals" in that they retain individual adaptations to pulses of chemoattractant. No estimates have been made as to whether the error in (5a) induced by the resulting dispersion of motilities and chemotactic response coefficients significantly affects population behavior.

Of interest in itself is speculation concerning the function of the built-in nongenetic variability. Koshland (1978) suggests that the advantage lies in the possibility that in rare circumstances, bacteria on the tails of the probability distribution would survive in some unusual distribution of toxins or nutrients.

We shall now survey some applications of system (5b), and suitable generalizations thereof, to band formation, to aggregation, and to the general understanding of the ecological roles of motility and taxes.

Bands

Research concerning travelling bands of tactic bacteria has become a major area of interaction between theory and experiment. Adler (1966) began the modern study of this subject, in the course of developing assays for chemotaxis. His work initiated what has proved to be an enormous effort in the molecular biological study of chemotaxis as an example of sensory transduction. For a recent review see Koshland (1980).

Repeating more quantitatively some 19th century experiments, Adler observed that bacteria inoculated into one end of a capillary tube would consume nutrient in their vicinity, creating a gradient. Pursuit of higher nutrient concentrations and continued consumption resulted in the formation of a band that moved steadily down the tube. In conditions of substrate excess, aerotaxis would bring about band formation.

Keller and Segel (1971) analyzed this phenomenon by looking for travelling wave solutions

$$b = b(x - \xi t) \ , \quad c = c(x - \xi t) \ , \quad \xi \text{ a constant "wave speed"} \ ,$$

to selected special cases of (5). Several other authors also examined conditions that would permit exact travelling wave solutions, and still others provided numerical analyses of representative problems. Novick-Cohen and Segel (1982) cite the earlier work in an analytic study which shows that deviations from conditions for exact travelling waves can be permitted at low nutrient concentrations, thereby admitting certain biologically realistic hypotheses that were previously excluded. Such deviations have the consequence that the band no longer exactly retains its shape and constant speed. Behavior of this nature is observed, for example in Adler's (1966) original experiments.

Qualitative descriptions of banding phenomena have been given by Smith and Doetsch (1969) in their study of chemotaxis away from acidic domains in Pseudomonas fluorescens, and by Allweis et al. (1977) in a demonstration that chemotaxis is important in the interaction of bacterial pathogens with mucosal surfaces. Reasonable agreement with the theory has been reported by Chen and his coworkers (Wang and Chen, 1981 is the latest paper in the series) who used laser scattering techniques to obtain accurate density profiles of the moving bacterial band. In an interesting recent investigation that combines absorption photometry measurements with theory, Boon and Herpigny (1982) found that the basic theoretical framework (5) had to be extended in order to explain events when a uniform distribution of E. coli are placed in a step gradient of glucose. The developing spatial inhomogeneity in the bacterial distribution brings about an inhomogeneity in oxygen concentration. This in turn affects glucose consumption -- and all this must be taken into account to explain the observed formation first of a broad band and later of a second sharp band. That chemotaxis is not necessary for band formation has been shown by Kennedy and Aris (1980). Just substrate-dependent growth and random motility suffice. Lauffenburger, Kennedy and Aris (1982) consider the combined effects of net growth, random motility, and chemotaxis with emphasis on band formation as a mechanism for effective utilization of nutrient. Puzzling aspects remain after comparison with the experimental work of Chapman (1973).

In unpublished work Odell (1982) has made several contributions to the theoretical study of bands. In particular he examined a situation where exploiters and victims assume the roles of bacteria and substrate in (5), with $Q = 0$, $D = 0$, $\chi(c) = \delta/c$, $q(c) = \alpha c(\beta - c)$; $\delta, \alpha, \beta, \mu, k$ constants. Thus the "victaxis" function χ was assumed to obey the Weber-Fechner law and the victim reproduction was taken as logistic. Odell found conditions under which there are exact solutions that take the form of travelling wave trains. Here a group of exploiters nearly wipes out the victim population — which then grows up from its low level only to be nearly consumed again by the next band of exploiters, etc. It would be interesting to know whether there are concrete examples of this multiple-herd behavior. Levandowsky (private communication) has pointed out that there are suggestive similarities to the Odell theory in Huffaker's (1958) study of predator-prey orange mite systems. Ziegler's (1977) discrete simulations provide another approach.

Aggregation

After their food supply runs out, free-living cellular slime mold amoebae (such as Dictyostelia) aggregate as the first step in the eventual formation of a multicellular organism. In its multicellular state, a typical cellular slime mold consists of a spherical collection of spores perched atop a slender, tapering stalk composed of dead cellulose-filled cells. The aggregation is mediated by a chemoattractant (identified for several species as cyclic-AMP) that the amoebae secrete in the hours following starvation. Keller and Segel (1970) showed how a version of equations (5) could produce aggregation as an instability of a uniform layer of cells and attractant. Further studies along this line are cited by Lapidus and Levandowsky (1981). Noteworthy recent contributions include Hagan and Cohen's (1981) treatment of spiral and target aggregation patterns, and Childress and Percus's (1981) examination of whether chemotactic collapse according to the Keller-Segel (1971) version of (5) can result in total aggregation to a delta-function distribution of cells.

The theoretical framework (5) has been used by Edelstein (1971) to provide an explanation for aggregation-like cell sorting in morphogenesis. This framework has also been employed by Lauffenburger and Keller (1982) in a model for the mutual approach of swarms of *Chondromyces apiculatus*, and for the coherence of individual swarms.

We will soon describe another study of aggregation, in a medical context.

Models for Ecological Functions of Motility

Earlier we mentioned some ecological functions for taxes such as the guidance of organisms toward food, favorable habitats, and sexual partners. Leaving aside the question of the experimental verification of these ideas, their very reasonableness seems to leave little scope for the theoretician. Thus we now turn to situations of more subtlety, where modelling has contributed or promises to contribute something to our understanding.

Purely random motility can be included within our discussion of taxes as a limiting case of completely undirected motion. It seems intuitively clear that motility in itself would be advantageous to an organism, but theoretical investigations have shown that this is not necessarily so. For example, one might think that the diffusive flux of nutrient into an organism would be enhanced by motion, for then the organism would continually leave behind areas depleted by its own feeding. Berg and Purcell (1977) pointed out, however, that for organisms as small as bacteria this effect has a negligible influence on uptake. The reason is that in the associated low Reynolds number flow the microorganism continually drags along an appreciable portion of its fluid environment.

When nutrients are inhomogeneously distributed, motility can indeed affect the feeding rate of microorganisms, by bringing them into regions of differing nutrient concentrations. Here one's intuition might be that motility is certainly beneficial for it will scatter concentrations of organisms and thereby provide a better average environment for each. That this is not necessarily correct is shown by Lauffenburger, Aris and Keller (1981) in their examination of bacterial growth in a confined region $0 \leq x \leq L$ with a fixed nutrient concentration c_0 at $x = L$ and a wall impermeable to nutrient c at $x = 0$. Assuming an interaction between equations for uptake, growth, and death (rate d) that are familiar from chemostat theory, and postulating boundaries that are impermeable to bacteria, Lauffenburger et al. (1981) study the mathematical problem

$$\partial b/\partial t = [r(c) - d]b + \mu \partial^2 b/\partial x^2 \ ; \tag{8a}$$

$$\partial c/\partial t = -Y^{-1} r(c)b + D\partial^2 c/\partial x^2 \quad ; \tag{8b}$$

$$\text{at} \quad x = 0 \;, \; \partial b/\partial x = \partial c/\partial x = 0 \quad ; \tag{8c,d}$$

$$\text{at} \quad x = L \;, \; \partial b/\partial x = 0 \;, \; c = c_0 \;. \tag{8e,f}$$

The birth rate $r(c)$, expected to be saturating function with $r(0) = 0$, is approximated by the step function

$$r(c) = r_m \;, \quad c \geqslant c^* \;; \quad r(c) = 0 \;, \quad c \leqslant c^* \;. \tag{9}$$

The governing problem is thus piecewise linear and can be explicitly solved. The major result is that at steady state the total bacterial population is a <u>decreasing</u> function of the bacterial motility parameter μ. This is because random motility disperses bacteria from the nutrient-rich fast growth zone near the substrate source at $x = L$. The effect is found to be significant when $(\mu k)^{\frac{1}{2}} < L$, i.e. when in a time of order one generation bacteria can move into a significantly different nutrient environment.

Possible effects of motility on competition between two species has been examined by Lauffenburger and Calcagno (1982) for the same consumption and growth type of assumptions, under the same geometric conditions, that were postulated in (8). The authors remark that the model could "represent situations as diverse as bacteria growing in water films around soil particles or hydrocarbon droplets, or in the mammalian gastrointestinal tract". As was to have been anticipated in view of the earlier results it is shown that steady coexistence is possible even though one species has a smaller growth rate, providing that its motility is sufficiently weak. An investigation of the stability of coexistence is promised in a future paper.

Models for Ecological Functions of Chemotaxis

It is to be expected that chemotaxis can override the deleterious effects of motility in moving bacteria away from peak nutrient concentrations. This is demonstrated by Lauffenburger, Aris and Keller (1982) in an analysis wherein the bacterial equation (8a) is supplemented by a chemotactic term of the form (3b). The analysis shows how strong the chemotaxis must be before its effect is significant. This paper also demonstrates the point (that seems consistent with observations) that populations are larger if the motility μ decreases at higher nutrient concentrations.

There is growing medical interest in cellular chemotaxis. For example, detailed studies have been made of chemotactic response in leukocytes (Zigmond and Sullivan, 1979) and granulocytes (Gerisch and Keller, 1981). Evidence has been presented that metastasis of malignant cells to bone is mediated by a chemotactic factor (Orr et al., 1979).

Particularly elegant is the role of leukocyte chemotaxis in the body's defense against disease (Snyderman, 1981). An obvious function can be ascribed to the finding that leukocytes are chemotactic to degradation products of bacteria. The leukocytes are also chemotactic to a fragment C5a of complement, the assemblage of molecules that leukocytes deploy to breach the cell walls of invaders and thereby to kill them. Apparently the "smoke" of an initial skirmish attracts defenders to the scene of an impending battle between them and the invading cells.

The same complement fragment has two further effects. It brings about a contraction of smooth muscle in post capillary venules and promotes the appearance of endothelial gaps. The first effect apparently decreases the dissipative effects of blood flow on the attractant gradient while the second opens the gates for leukocytes to leave the blood stream and move up the gradient.

Lauffenburger and Aris (1978), Lauffenburger (1983) and Rothman and Lauffenburger (1983) discuss how to estimate the motility and chemotactic parameters μ and χ of the phenomenological equations (5) from the "under-agarose assay" used for leukocyte migration. These parameters play a major role in a study of inflammation by Lauffenburger and Keller (1979) that quantifies the host's ability to overcome bacterial invaders. In an extension of this work, Lauffenburger and Kennedy (1983) postulated the following equations for the concentrations of bacteria $b(x,t)$ and leukocytes $w(x,t)$, where the k_i are constants:

$$\frac{\partial b}{\partial t} = \mu_b \frac{\partial^2 b}{\partial x^2} + \frac{k_1 b}{1+b/k_2} - \frac{k_3 bw}{k_4+b} , \tag{10a}$$

$$\frac{\partial w}{\partial t} = \mu_w \frac{\partial^2 w}{\partial x^2} - \chi \frac{\partial}{\partial x} \left(w \frac{\partial b}{\partial x}\right) - k_6 w + k_7(1+k_8 b) . \tag{10b}$$

The bacteria are randomly motile, grow with a self-damped birth rate, and are consumed by the leukocytes. Leukocytes possess both random and chemotactic (toward the bacteria) aspects to their motion and they die off at rate k_6. In the macroscopic perspective of the analysis, the last term in (10b) models the entry of leukocytes into the arena of conflict by passage through venular walls, at a rate that increases with bacterial population.

Lauffenburger and Kennedy (1982) show that the uniform solution of (10) can become unstable for realistic parameter values. This can result in a rather strongly nonuniform steady state with pockets of bacteria that are reminiscent of clinical observations. As the authors point out, the aggregation here differs from that found in slime mold amoebae in that the chemotactic cells are agents for stability in the present case while they promote instability in the case of the slime mold.

Need for deeper study of the ecological effects of taxes is suggested by striking results of I. Chet and R. Mitchell (private communication). These investigators checked about 20 different species of motile marine bacteria for chemotaxis to approximately 10 different chemicals. They found that every species responded chemotactically to at least one chemical, but each species seemed to have a different mixture of strong or weak positive responses, no response, and strong or weak negative responses.

These results suggest that there is more to the ecological function of taxes than avoidance of harmful agents and preferential motion in the direction of favorable locations. There appears to be a complex dynamic situation in which a spectrum of taxes enables a community of organisms to exploit different temporal-spatial niches. As a first step in illustrating such a situation I have begun an investigation into the possible coexistence of generalists and specialists. [Heller (1980) found conditions for such coexistence in an interaction between predators and sometimes harmful prey.]

Consider the competition between specialists A and B and generalist C. A and B respectively consume nutrients α and β, and move chemotactically toward relatively high concentrations of these nutrients. Generalist C can handle both nutrients, but in each case less efficiently than the specialists. A model for this interaction could take the following form, a generalization of (8a) and

(8b):

$$\partial A/\partial t = [r_A(\alpha)-d]A + \mu\nabla^2 A - \chi_A\nabla\cdot(A\nabla\alpha) ,$$

$$\partial B/\partial t = [r_B(\beta)-d]B + \mu\nabla^2 B - \chi_B\nabla\cdot(B\nabla\beta) ,$$

$$\partial C/\partial t = [r_{C1}(\alpha) + r_{C2}(\beta) - d]C + \mu\nabla^2 C - \chi_{CA}\nabla\cdot(C\nabla\alpha) - \chi_{CB}\nabla\cdot(C\nabla\beta) ,$$

$$\partial\alpha/\partial t = -Y_A^{-1}r_A(\alpha)A - Y_{C1}^{-1}r_{C1}(\alpha)C + D_\alpha\nabla^2\alpha + S_\alpha(x,t) ,$$

$$\partial\beta/\partial t = -Y_B^{-1}r_B(\beta)B - Y_{C2}^{-1}r_{C2}(\beta)C + D_\beta\nabla^2\beta + S_\beta(x,t) .$$

$$r_A(\alpha) > r_{C1}(\alpha) , \quad r_B(\beta) > r_{C2}(\beta) ,$$

$$\chi_A > \chi_{CA} , \chi_B > \chi_{CB} , Y_A > Y_{C1} , Y_B > Y_{C2} . \tag{11}$$

I suspect that coexistence will be possible, at least for nutrient source terms $S_\alpha(x,t)$ and $S_\beta(x,t)$ which provide localized pulses of nutrient at points that are randomly distributed in space and time. I have started an investigation of a compartmental version of (11) but have as yet no results to report.

If indeed taxes play a subtle role in community organization, then one can anticipate corresponding subtle effects of pollutants that clog the sensory apparatus. A start on examining sublethal effects of pollutants has been made for example by Chet and Mitchell (1976) in an experimental study of chemotaxis, and by Segel and Ducklow (1982) in a theoretical examination of some aspects of coral ecology.

Overview

In pulling together the strands of this discourse, I would like to stress reasons for paying more attention to cellular ecology, by which I mean not only "traditional" microbial ecology but also the ecology of cells in higher organisms. The importance of this field is enormous, dealing as it does with myriads of living organisms that have profound influences on the biosphere in general, and on human development and disease in particular. With the advent of genetic engineering futuristic applications such as "biomining" (Brierly, 1982) become possible, and require knowledge of population interactions. There are common threads, such as the phenomenological chemotaxis equation (5), that link the community behavior of bacteria, slime mold amoebae and blood cells. Perhaps most important to professional ecologists is the fact that microorganisms - especially bacteria - offer relatively simple opportunities to formulate general principles in ways that are amenable to experimental test.

The study of bacterial chemotaxis seems to offer an unmatched opportunity for an integrated approach to understanding the interaction of genetics, physiology, and ecology. Molecular biologists are intensively studying the detailed mechanisms of bacterial chemotaxis in an attempt to unravel in full detail at least one process of sensory transduction. As anticipated by Adler (1966) when he started the study, existing detailed knowledge of bacterial genetics and the ease of inducing mutations in microorganisms have been indispensable tools. Impressive progress has been made in the understanding of the flagellar motor and of the receptor, but much remains to be done, especially on the connections between the receptor and the motor. Still, for the ecologist and the student of evolutionary theory, concentration on the effects of bacterial chemotaxis offers the enormous advantage of dealing with an environmental response for which details such as "cost", number of genes involved, and mutation frequency are becoming better and better understood.

Acknowledgements. The author has benefitted from conversations with a number of colleagues, notably I. Chet, R. Mitchell and S. Levin. D. Lauffenburger made valuable comments on an earlier version of the manuscript. This work was partially supported by NSF grant MCS-8203246. The kindness of the Conference hosts prompts the dedication of this paper to a symbol of those who aid people struggling to be creative — the Triestine benefactress of Rilke, Princess Marie von Thurn und Taxis-Hohenlohe.

References

Adler, J. (1966). Chemotaxis in bacteria. Science 153: 708-716.

Allweis, B., Dostal, J., Carey, K.E., Edwards, T.F. and Freter, R. (1977). The role of chemotaxis in the ecology of bacterial pathogens of mucosal surfaces. Nature 266: 448-450.

Alt, W. (1980). Biased random walk models for chemotaxis and related diffusion approximations. J. Math. Biol. 9: 147-177.

Berg, H. and Purcell, E.M. (1977). Physics of chemoreception. Biophys. J. 20: 193-219.

Boon, J.P. (1983). Motility of living cells and micro-organisms. The Application of Laser Light Scattering to the Study of Biological Motion (J.C. Earnshaw and M.W. Steer, eds.), New York: Plenum Press.

Boon, J.P. and Herpigny, B. (1982). Bacterial chemotaxis and band formation: response to the simultaneous effect of two attractants. Faculté des Sciences, CP. 231, Université Libre de Bruxelles, Preprint.

Brierly, C.L. (1982). Microbiological mining. Scientific American 247 (2): 42-51.

Carlile, J.J. (1980). Positioning mechanisms -- the role of motility, taxis and tropism in the life of microorganisms. Contemporary Microbial Ecology (D.C. Ellwood et al., eds.), London: Academic Press, 55-74.

Chapman, P. (1973). Chemotaxis in bacteria. Ph.D. Thesis, Univ. of Minnesota.

Chet, I. and Mitchell, R. (1976). Ecological aspects of microbial chemotactic behavior. Ann. Rev. Microbiol. 30: 221-239.

Childress, S. and Percus, J.K. (1981). Nonlinear aspects of chemotaxis. Math. Biosci. 56: 217-237.

Dahlquist, F.W., Elwell, R.A. and Lovely, P.S. (1976). Studies of bacterial chemotaxis in defined concentration gradients. A model for chemotaxis toward L-serine. J. Supramol. Struct. 4: 329(289)-342(302).

Edelstein, B.B. (1971). Cell specific diffusion model of morphogenesis. J. Theoret. Biol. 30: 515-532.

Fraenkel, G. and Gunn, D. (1940). The Orientation of Animals. London: Oxford Univ. Press. (Reprinted by Dover Publications, 1961).

Freter, R., O'Brien, P.C.M. and Macsai, M.S. (1979). Effect of chemotaxis on the interaction of cholera vibrios with intestinal mucosa. Am. J. Clinical Nutrition 32: 128-132.

Fu, T.K., Kessler, J.O., Jarvik, L.F. and Matsuyama, S.S. (1982). Philothermal and chemotactic locomotion of leukocytes. Cell Biophysics 4: 77-95.

Gerisch, G. (1982). Chemotaxis in Dictyostelium. Ann. Rev. Physiol. 44: 535-552.

Gerisch, G. and Keller, H.U. (1981). Chemotactic reorientation of granulocytes stimulated with micropipettes containing fMet-Leu-Phe. J. Cell Sci. 52: 1-10.

Hagan, P.S. and Cohen, M.S. (1981). Diffusion-induced morphogenesis in the development of Dictyostelium. J. Theoret. Biol. 93: 881-908.

Heller, R. (1980). Foraging on potentially harmful prey. J. Theoret. Biol. 85: 807-813.

Hu, C-L. and Barnes, F.S. (1970). A theory of necrotaxis. Biophys. J. 10: 958-969.

Huffaker, C.B. (1958). Experimental studies on predation: dispersion factors and predator-prey oscillations. Hilgardia 27: 343-383. Reprinted in W.E. Hazen, Readings in Population and Community Ecology, Saunders.

Keller, E.F. and Segel, L.A. (1970). Initiation of slime mold aggregation viewed as an instability. J. Theoret. Biol. 26: 399-415.

Keller, E.F. and Segel, L.A. (1971). Travelling bands of chemotactic bacteria: a theoretical analysis. J. Theoret. Biol. 30: 235-248.

Kennedy, C.R. and Aris, R. (1980). Travelling waves in a simple population model involving growth and death. Bull. Math. Biol. 42: 397-429.

Koshland, D.E. (1978). Heredity, environment, and chance in the responses of an individual cell. Birth Defects: Original Article Series 14: 401-415.

Koshland, D.E., Jr. (1980). Bacterial Chemotaxis as a Model Behavioral System. N.Y.: Raven Press.

Lapidus, I.R. (1980a). "Pseudochemotaxis" by micro-organisms in an attractant gradient. J. Theoret. Biol. 86: 91-103.

Lapidus, I.R. (1980b). Microbial chemotaxis in flowing water in the vicinity of a source of attractant or repellent. J. Theoret. Biol. 85: 543-547.

Lapidus, I.R. and Levandowsky, M. (1981). Mathematical models of behavioral responses to sensory stimuli by protozoa. Biochemistry and Physiology of Protozoa 4 (2nd ed.): 253-260.

Lauffenburger, D. (1983). Measurement of phenomenological parameters for leukocyte motility and chemotaxis. Proc. 1st Int. Conf. on Leukocyte Locomotion and Chemotaxis, AAS 12, Agents and Actions Supplement 12 (H.U. Keller and G.O. Till, eds.), Basel: Birkhäuser Verlag: 34-53.

Lauffenburger, D. and Aris, R. (1979). Measurement of leukocyte motility and chemotaxis parameters using a quantitative analysis of the under-agarose migration. Math. Biosci. 44: 121-138.

Lauffenburger, D., Aris, R. and Keller, K.H. (1981). Effects of random motility on growth of bacterial populations. Microb. Ecol. 7: 207-227.

Lauffenburger, D., Aris, R. and Keller, K.H. (1982). Effects of cell motility and chemotaxis on microbial population growth. Biophys. J. 40: 209-219.

Lauffenburger, D. and Calcagno, P.B. (1982). Competition between two microbial populations in a non-mixed environment: effect of cell random motility. Biotechnology and Bioengineering, submitted.

Lauffenburger, D. and Keller, K.H. (1979). Effects of leukocyte random motility and chemotaxis in tissue inflammatory response. J. Theoret. Biol. 81: 475-503.

Lauffenburger, D. and Keller, K.H. (1982). An hypothesis for approaching swarms of myxobacteria. Unpublished preprint.

Lauffenburger, D. and Kennedy, C.R. (1983). Localized bacterial infection in a distributed model for tissue inflammation. J. Math. Biol. 16: 141-163.

Lauffenburger, D., Kennedy, C.R. and Aris, R. (1982). Travelling bands of motile bacteria in the context of population survival. Bull. Math. Biol., in press.

Levandowsky, M. and Hauser, D.C.R. (1978). Chemosensory responses of swimming algae and protozoa. Int. Rev. Cytology 53: 145-210.

Mansour, T.E. (1979). Chemotherapy of parasitic worms: new biochemical strategies. Science 205: 462-469.

Maugh, T.H. (1982). Magnetic navigation an attractive possibility. Science 215: 1492-1493.

Novick-Cohen, A. and Segel, L.A. (1982). A gradually slowing travelling band of chemotactic bacteria. J. Math. Biol., in press.

Odell, G. Periodic waves of predators chasing logistically reproducing immobile prey. Preprint. RPI Dept. Math. Sci., Troy, N.Y. 12181, U.S.A.

Orr, W., Varani, J., Gondek, M.D., Ward, P.A. and Munday, G.R. (1979). Chemotactic responses of tumor cells to products of resorbing bone. Science 203: 176-178.

Patlak, C.S. (1953). Random walk with persistence and external bias. Bull. Math. Biophys. 15: 311-338.

Rothman, C. and Lauffenburger, D. (1983). Analysis of the linear under-agarose chemotaxis assay. Ann. Biomed. Eng'g., in press.

Segel, L.A. (1977). A theoretical study of receptor mechanisms in bacterial chemotaxis. SIAM J. Appl. Math. 32: 653-665.

Segel. L.A. (1980). Mathematical models in molecular and cellular biology. Cambridge: Cambridge University Press.

Segel, L.A., Chet, I. and Hennis, Y. (1977). A simple quantitative assay for bacterial motility. J. Gen. Microbiol. 98: 329-337.

Segel, L.A. and Ducklow, H.W. (1982). A theoretical investigation into the influence of sublethal stresses on coral-bacterial ecosystem dynamics. Bull. Marine Sci. 32: 919-935.

Segel, L.A. and Jackson, J.L. (1973). Theoretical analysis of chemotactic movement in bacteria. J. Mechanochem. and Cell Motility 2: 25-34.

Sjoblad, R.D. and Mitchell, R. (1979). Chemotactic responses of Vibrio alginolyticus to algal extracellular products. Canadian J. Microbiol. 25: 964-967.

Skellam, J.G. (1951). Random dispersal in theoretical populations. Biometrika 38: 196-218.

Smith, J.L. and Doetsch, R.N. (1969). Studies on negative chemotaxis and the survival value of motility in Pseudomonas fluorescens. J. Gen. Microbiol. 55: 379-391.

Snyderman, R. (1981). Molecular and cellular mechanisms of leukocyte chemotaxis. Science 213: 830-837.

Walsh, F. and Mitchell, R. (1978). Bacterial chemotactic responses in flowing water. Microbiol. Ecol. 4: 165-168.

Wang, P.D. and Chen, S.H. (1978). Quasi-elastic light scattering from migrating chemotactic bands of Escherichia coli II. Analysis of anisotropic bacterial motions. Biophys. J. 36: 203-219.

Ziegler, B.P. (1977). Persistence and patchiness of predator-prey systems induced by discrete event population exchange mechanisms. J. Theoret. Biol. 67: 687-713.

NONLINEAR DIFFUSION PROBLEMS IN AGE-STRUCTURED POPULATION DYNAMICS

Stavros Busenberg* and Mimmo Iannelli†

Harvey Mudd College
Claremont, California

Abstract. We describe a method for treating a broad class of nonlinear age-dependent population problems which may involve spatial diffusion. This method effectively decouples the age-dependent part of the problem from the population interaction terms, allowing the treatment of a number of age-structured problems that had not been otherwise accessible. We illustrate our method by giving a general asymptotic behavior result as well as by applying it to some specific models.

1. Introduction. Age plays an important role in the dynamics of populations of many biological organisms. The fertility and mortality of individuals are two basic parameters that depend on their age. Some other effects that are often age-dependent include the propensity of individuals to succumb to certain diseases, and their ability to avoid predators or to compete for limited resources. Yet many population dynamics models either avoid to incorporate the age-structure of the population, or else include it via the indirect paths of introducing time delays or of separating the population into discrete age groups. This is often the case because the continuous age-structured versions of such models seem to present mathematical difficulties. Our aim here is to show that, for a rather broad class of possibly nonlinear population dynamics models, the inclusion of the age structure does not add major mathematical difficulties to those already present because of the nonlinear dynamic interactions or the spatial diffusion that may be part of the model without age structure. This is due to the use of

*This author was partially supported by NSF grant MCS-8301905.

†Permanent address: University of Trento, Italy. This author was partially supported by CNR grant #81.01942.01.

a basic transformation that effectively decouples the age-dependent dynamics from the population interaction part of the model. In this decoupling, the part of the problem that involves age-dependence can be completely analyzed, and in fact, remains the same for a wide class of models. The remainder of the problem which includes the dynamics of the interactions of the population requires a separate analysis, as would be the case even when age-structure is not included in the model. This basic decoupling method was developed by Busenberg and Iannelli (1982a) and was then used by them to analyze some specific nonlinear diffusion problems in a population with age-structure (Busenberg and Iannelli 1982b, c). Here we describe how this basic transformation applies to a class of models that is sufficiently broad to include age-dependent generalizations of many classical models of ecology, epidemics and demography. We do not aim for the maximal generality, and it will be clear to the reader that the method readily extends to cases we do not discuss here. We try to include equations which are sufficiently general to tempt the readers to study age-dependent versions of some of their favorite population models.

The paper is organized as follows. In the next section we present the equations that we will treat, we derive the decoupled problem, and we discuss the approach for analyzing it. We also give two results that describe the solutions of the age-dependent part of the decoupled problem, and in particular, give detailed information on the asymptotic large time behavior of these solutions. The third section is devoted to some examples of the use of our method in the analysis of specific equations of population dynamics.

<u>2. The Basic Decoupling of Age-Dependent Models</u>. We consider N interacting populations each with its own intrinsic age-dependent birth and death dynamics. For a general discussion of the classical equations of age-dependent population dynamics we refer the reader to Hoppensteadt (1975) and Cushing (1979). The basic variables that we use are the chronological age a, the time t, the spatial position x, and the densities $u_i(a,t,x)$, $i = 1,2,\dots,N$. Here, $u_i(a,t,x)$

denotes the number of individuals, per unit age and unit spatial volume (i.e., length, if space is one-dimensional), who are of age a at time t and at position x. By $\mu_i(a)$ and $\beta_i(a)$ we denote the intrinsic death rate per unit age, and the intrinsic birth rate per unit age, of individuals in the ith population who are of age a. The total ith population density is given by

$$P_i(t,x) = \int_0^\infty u_i(a,t,x)da.$$

Then the basic equations governing the dynamics of the ith population are

$$(P)\qquad \begin{aligned} &\frac{\partial u_i}{\partial t} + \frac{\partial u_i}{\partial a} + \mu_i u_i = F_i \\ &u_i(0,t,x) = \int_0^\infty \beta_i(a)u_i(a,t,x)da, \quad u_i(a,0,x) = u_i^0(a,x) \\ &u_i(a,t,x) \to 0 \quad \text{as} \quad a \to \infty, \quad u_i(a,t,x) \geq 0 \end{aligned}$$

+ Appropriate boundary conditions on the x variable

The terms F_i contain the description of the dynamic interaction of the various populations as well as the spatial diffusion operators. We shall specify the form that these terms may take a little later. For now let us note that the traditional approach to problem (P) involves the use of the variables

$$(P_i,B_i), \qquad B_i(t,x) = \int_0^\infty \beta_i(a)u_i(a,t,x)da$$

and under rather strong restrictions on the F_i leads to systems of Volterra integral equations. If instead of the initial data $u_i(a,0,x) = u_i^0(a,x)$ (census type data) one were to specify the data $u_i(0,t,x) = u_i^0(t,x)$ for $t \leq 0$ (birth record type data), then one obtains a delay differential equation formulation of this problem. These reductions to Volterra or delay equations work for only special forms of the μ_i, β_i and F_i. In our approach we use the pairs of variables

$$(P_i(t,x),w_i(a,t,x)), \quad \text{where} \quad w_i(a,t,x) = u_i(a,t,\phi)/P_i(t,\phi),$$

and ϕ is an appropriately defined space-time characteristic. The w_i are a natural variable describing the proportion of individuals of age a in the total population at time t and at position ϕ, where ϕ measures the drift of these individuals in space. So, in the first of our two variables $P_i(t,x)$, we use an "Eulerian" coordinate with the position of observation x fixed for all t, while in the second $w(a,t,x)$, we use a "Lagrangian" coordinate with the position ϕ of observation of $u(a,t,\phi)/P(t,\phi)$ following the space-time drift of the individuals.

In order to make matters more precise we need to specify the types of nonlinearities F_i in (P) that we will treat. For simplicity, we assume that the spatial domain is one-dimensional, and consider spatial diffusion where the flux of individuals depends on total population pressures. We also allow each of the F_i to depend on M quantities that have the form

$$S_j(t,x) = \int_0^\infty g_j(a,t,x,\vec{u}(a,t,x))da, \quad j = 1,\dots,M, \tag{2.1}$$

where $\vec{u}(a,t,x) = (u_1(a,t,x),\dots,u_N(a,t,x))$. For example, $S_j(t,x)$ may have the form $\int_0^\infty \gamma(a)u_j(a,t,x)da$, and, in a predator-prey situation, may measure the effective trophic value of the prey taken by the predators. So, we let

$$\begin{aligned} F_i &= b_i(t,x,\vec{S}(t,x),\vec{P}(t,x), \frac{\partial\vec{P}}{\partial x}(t,x))\frac{\partial u_i}{\partial x}(a,t,x) \\ &+ c_i(t,x,\vec{S}(t,x),\vec{P}(t,x), \frac{\partial\vec{P}}{\partial x}(t,x), \frac{\partial^2\vec{P}}{\partial x^2}(t,x))u_i(a,t,x) \end{aligned} \tag{2.2}$$

where $\vec{S} = (S_1,\dots,S_M)$, is the M-vector of the functions defined by (2.1), and $\vec{P} = (P_1,\dots,P_N)$ is the N-vector of total populations.

Now, let

$$w_i(a,t,x) = \frac{u_i(a,t,\phi_i(t,0,x))}{P_i(t,\phi_i(t,0,x))}, \tag{2.3}$$

where $\phi_i(t,t_0,x)$ solves the ordinary differential equation initial value problem

$$(2.4)\quad \begin{aligned} &\frac{d}{dt}\phi_i = -b_i(t,\phi_i,\vec{S}(t,\phi_i),\vec{P}(t,\phi_i),\vec{P}_x(t,\phi_i)) \\ &\phi_i(t_0,t_0,x) = x \quad . \end{aligned}$$

Then a direct computation following the lines laid out in Busenberg and Iannelli (1982a, b) leads to the following equation describing the behavior of w(a,t,x)

$$(P_1^i)\quad \begin{aligned} &\frac{\partial w_i}{\partial t} + \frac{\partial w_i}{\partial a} + \mu_i w_i = w_i \int_0^\infty [\mu_i(a)-\beta_i(a)]w_i(a,t,x)da \\ &w_i(0,t,x) = \int_0^\infty \beta_i(a)w_i(a,t,x)da \\ &w_i(a,0,x) = w_i^0(a,x) = u_i^0(a,x)/P_i^0(x), P_i^0(x) = \int_0^\infty u_i^0(a,x)da \\ &\int_0^\infty w_i(a,t,x)da = 1,\quad w_i(a,t,x) \geq 0,\quad w_i(a,t,x) \to 0,\quad \text{as}\quad a \to \infty, \end{aligned}$$

which are N completely decoupled problems which can be studied independently (see Theorem 2.1 and Theorem 2.2, below). Once the w_i are determined from (P_1^i) the following, now known, quantities are defined

$$\lambda_i(t,x) = \int_0^\infty [\beta_i(a)-\mu_i(a)]w_i(a,t,x)da,$$

which, as we shall see later, can be interpreted as "transient Malthusian coefficients", and we have the following problems for the P_i and the ϕ_i:

$$(P_2^i)\quad \begin{aligned} \frac{\partial P_i(t,x)}{\partial t} &= b_i(t,x,\vec{\sigma}(t,x,\vec{\phi}(0,t,x),\vec{P}(t,x)),\vec{P}(t,x),\frac{\partial \vec{P}(t,x)}{\partial x})\frac{\partial P_i}{\partial x}(t,x) \\ &+ c_i(t,x,\vec{\sigma}(t,x,\vec{\phi}(0,t,x),\vec{P}(t,x)),\vec{P}(t,x),\frac{\partial \vec{P}}{\partial x}(t,x),\frac{\partial^2 P}{\partial x^2}(t,x))P_i(t,x) \\ &+ \lambda_i(t,\phi_i(0,t,x))P_i(t,x) \\ P_i(0,x) &= P_i^0(x) \end{aligned}$$

where $\vec{\sigma}$ is a vector whose components are given by

$$\sigma_j(t,x,\vec{y},\vec{z}) = \sigma_j(t,x,y_1,\ldots,y_N,z_1,\ldots,z_N)$$

$$= \int_0^\infty g_j(a,t,x,w_1(a,t,y_1)z_1,\ldots,w_N(a,t,y_N)z_N)da.$$

Moreover, we have the following problems for $\vec{\phi}$

$$(P_3^i)\qquad \begin{aligned} &\frac{d\phi_i}{dt}(t,t_0,x) = -b_i(t,\phi_i,\sigma(t,\phi_i,\vec{\phi}(0,t,\phi_i),\vec{P}(t,\phi_i)),\\ &\qquad\qquad \vec{P}(t,\phi_i),\vec{P}_x(t,\phi_i))\frac{\partial P_i}{\partial x}(t,\phi_i)\\ &\phi(t_0,t_0,x) = x\ . \end{aligned}$$

The basic problems that remain to be analyzed are (P_2^i)-(P_3^i). These may look formidable, but their complication is totally due to the non-linear interaction terms and the diffusion terms and is not related to the age-dependence which has been resolved in the separated problems (P_1^i). As we shall see in the examples of the next section, for a number of the more traditional interaction terms (say predator-prey interactions), one obtains problems of the form (P_2^i)-(P_3^i) which have already been extensively studied, and hence, the basic imbedding of (P) into the decoupled problems (P_1^i) and (P_2^i)-(P_3^i) totally settles the original problem.

We conclude this section by presenting the two basic results that describe the solutions of (P_1^i). The first is the result that guarantees existence and smoothness of the solutions of (P_1^i) and the other gives a description of the asymptotic behavior of these solutions as $t \to \infty$. The proofs of these results are found in Busenberg and Iannelli (1982b). Because of the inclusion of a spatial diffusion term in (P^i), there are some rather technical but natural conditions that must be imposed on the initial data w_0 in order that sufficiently smooth solutions may exist. We refer the reader to the original source (Theorem 3.1 of Busenberg and Iannelli 1982b) for these conditions, and here refer to them simply as hypotheses H.

Theorem 2.1. Under hypothesis H on the initial data w_0^i, the problem (P_1^i) has a unique solution which exists for all $(a,t) \in [0,\infty) \times [0,\infty)$ and for all x in the spatial domain under consideration.

Before stating the next result we need to introduce some basic terminology. Let

(2.5) $$\Pi^i(a) = \exp[-\int_0^a \mu_i(s)sa],$$

so $\Pi^i(a)$ is the probability of an individual surviving up to age a. Also, let p^{*i} be the unique real number satisfying

(2.6) $$\int_0^\infty \beta_i(a)\Pi^i(a)e^{-p^{*i}a}da = 1.$$

If $p^{*i} = 0$, then the net reproductive rate of the ith population is one, that is, the ith population would, if left alone $(F_i \equiv 0)$ settle to a steady state. If $p^{*i} < 0$ this net reproductive rate is less than one and if $p^{*i} > 0$ it is greater than one.

Theorem 2.2. Let p^{*i} be given by (2.6), then if there exists $A > 0$ with $w_1^0(a,x) = 0$ for all $a > A$, we have

(2.7) $$\lim_{t\to\infty} w(a,t,x) = \Pi^i(a)e^{-p^{*i}a}/\int_0^\infty e^{-p^{*i}a}\Pi^i(a)da,$$

and this limit is uniform in x.

This result is essentially Theorem 6.2 in Busenberg and Iannelli (1982b) where its proof can be found. In (2.7) the right hand side is taken to be zero whenever $\int_0^\infty e^{-p^*a}\Pi(a)da = \infty$. We now proceed to the examples which illustrate the usefulness of this method of treating the problem (P^i).

3. Applications to Some Specific Equations of Population Dynamics. We present some examples of specific models that can be treated by the method that we have

been describing. The degree of difficulty in analyzing these models varies with the complexity of the nonlinear dynamics and the spatial diffusion process that they incorporate, but it is not affected by the fact that age dependence is included.

Example 1. *A single population with a logistic effect on the mortality.* In this model we consider an age-structured population with a death rate having a term that increases with the total population, say the specific mortality takes the form

$$\mu(a) + kP^{\alpha}(t) \tag{3.1}$$

$\alpha > 0$ and $k > 0$ constants. We have dropped the variable x from (3.1) because there is no spatial dependence in this model. This model corresponds to equation (P^i) with $i = 1$ (we hence drop the subscript i from our variables: $u_1 = u$, $P_1 = P$), and $F_1 = -kP^{\alpha}(t)u(a,t)$, i.e., in (2.2) we take $b_1 \equiv 0$ and $c_1 = -kP^{\alpha}u$. Because $b_1 \equiv 0$, the problem (P_3^i) is trivial with solution $\phi_1 \equiv x$, and plays no role in the solution. The problem (P_2^1) is also simple and is

$$\frac{dP(t)}{dt} = -kP^{\alpha+1}(t) + P(t)\int_0^{\infty} [\beta(a)-\mu(a)]w(a,t)da \quad . \tag{3.2}$$

Writing $A(t) = \int_0^{\infty} [\beta(a)-\mu(a)]w(a,t)da$, we note that (3.2) can be integrated by elementary techniques to give

$$P(t) = \exp[-\int_0^t A(s)ds][1+\alpha kP_0^{\alpha}\int_0^t \exp(-\alpha\int_0^u A(s)ds)du]^{-1/\alpha}P_0 \quad . \tag{3.3}$$

Now, if $\int_0^{\infty} \Pi(a)e^{-p^*a}da < \infty$, Theorem 2.2 yields $A(t) \to p^*$ as $t \to \infty$, provided that the initial population has no individuals of infinite age. Consequently, from (3.3) we obtain directly the following asymptotic behavior:

$$\lim_{t\to\infty} P(t) = \begin{cases} (\frac{p^*}{k})^{\frac{1}{\alpha}} & \text{if } p^* > 0 \\ 0 & \text{if } p^* < 0. \end{cases} \tag{3.4}$$

Thus, from Theorems 2.1 and 2.2

$$\lim_{t\to\infty} u(a,t,x) = \begin{cases} (\frac{p^*}{k})^{\frac{1}{\alpha}} \Pi(a)e^{-p^*a} / \int_0^\infty e^{-p^*a}\Pi(a)da \\ 0 \quad \text{if} \quad p^* < 0 . \end{cases} \tag{3.5}$$

So the population tends to zero if its net reproductive rate is less than one and tends to a non-trivial steady state age distribution explicitly given in (3.5) if the net reproductive rate is greater than one.

A special case of this result $(\alpha = 1)$ is treated by Marcati (1982), and by Busenberg and Iannelli (1982b), under different conditions on the $w^0(a)$. The more general form of interaction

$$\mu(a)u(a,t) + f(P(t))u(a,t) \tag{3.6}$$

can be studied in a manner analogous to the above, provided that the resulting ordinary differential equation for P:

$$\begin{aligned} &\frac{dP}{dt}(t) = -P(t)f(P(t)) + A(t)P(t) \\ &P(0) = P^0, \quad A(t) = \int_0^\infty [\beta(a)-\mu(a)]w(a,t)da, \end{aligned} \tag{3.7}$$

can be analyzed.

Example 2. *An age dependent predator-prey model.* In this model we consider the classical predator-prey equations and make the assumption that the predators' ability to catch the prey is directly proportional to a weighted average over the age distribution of the predator population u_2. So, the rate of capture of the prey u_1 is given by $u_1(a,t) \int_0^\infty \gamma_1(a)u_2(a,t)da$. The value of the captured prey for the predator can be assumed to have any one of several forms that have interesting interpretations. In the first, we assume that death rate of the predators is reduced by a term proportional to a weighted average over age of the available prey, i.e., by a term of the form $u_2(a,t) \int_0^\infty \gamma_2(a)u_1(a,t)da$. The resulting equations for u_1 and u_2 are

$$\frac{\partial u_1}{\partial a} + \frac{\partial u_1}{\partial t} + \mu_1 u_1 = -u_1 \int_0^\infty \gamma_1(a) u_2(a) da$$

(3.8)

$$\frac{\partial u_2}{\partial a} + \frac{\partial u_2}{\partial t} + \mu_2 u_2 = u_2 \int_0^\infty \gamma_2(a) u_1(a) da$$

with same birth and initial data as in the problem (P^i). Since there is no spatial diffusion, the problems (P_3^i) are again trivial, while the problems (P_2^i) are

$$\frac{dP_1}{dt} = P_1[\alpha_{11}(t) - P_2\alpha_{12}(t)]$$

(3.9)

$$\frac{dP_2}{dt} = P_2[\alpha_{22}(t) + P_1\alpha_{21}(t)]$$

where

$$\alpha_{ii}(t) = \int_0^\infty (\beta_i(a) - \mu_i(a)) w_i(a,t) da, \quad \alpha_{ij}(t) = \int_0^\infty \gamma_i(a) w_j(a,t) da.$$

Thus the equations for (P_1, P_2) are the standard predator-prey relations with time varying coefficients. From Theorem 2.2 we can conclude that $\alpha_{ii}(t) \to p^{*i}$ and $\alpha_{ij}(t) \to q^i$ where $q^i = \int_0^\infty e^{-p^{*j}a} \gamma_i(a) \pi^j(a) da / \int_0^\infty e^{-p^{*j}a} \pi^j(a) da$, so the limiting form of (3.9) is indeed the classical predator-prey model.

A slightly different model is obtained if we assume that the value of the prey is proportional to a weighted average of the captured prey and reduces the mortality of the predator by the factor

$$u_2(a,t)\left(\int_0^\infty \gamma_2(a) u_1(a,t) da\right)\left(\int_0^\infty \gamma_1(a) u_2(a,t) da\right).$$

The resulting equations for u_1 and u_2 are

$$(3.10)\quad \begin{aligned} \frac{\partial u_1}{\partial t} + \frac{\partial u_1}{\partial a} + \mu_1(a)u_1 &= -u_1 \int_0^\infty \gamma_1(a)u_2(a,t)da \\ \frac{\partial u_2}{\partial t} + \frac{\partial u_2}{\partial a} + \mu_2(a)u_2 &= u_2\left(\int_0^\infty \gamma_2(a)u_1(a)da\right)\left(\int_0^\infty \gamma_1(a)u_2(a)da\right) . \end{aligned}$$

The resulting equations for P_1 and P_2 are

$$(3.11)\quad \begin{aligned} \frac{dP_1}{dt} &= P_1[\alpha_{11}(t) - P_2\alpha_{12}(t)] \\ \frac{dP_2}{dt} &= P_2[\alpha_{22}(t) + P_1P_2\bar{\alpha}_{21}(t)] \end{aligned}$$

where α_{11}, α_{12} and α_{22} are given above, and

$$\bar{\alpha}_{21}(t) = \left(\int_0^\infty \gamma_2(a)w_1(a,t)\right)\left(\int_0^\infty \gamma_1(a)w_2(a,t)da\right).$$

Again, the coefficients $\alpha_{ij}(t)$ and $\bar{\alpha}_{21}(t)$ tend to limits that can be determined by using Theorem 2.2, and the large time behavior of (3.11) can be analyzed. Once P_1 and P_2 are known, then the u_i are obtained from the relation $u_i(a,t) = P_i(t)w_i(a,t)$. We will forgo a more detailed discussion of these models at this point.

Example 3. *A single population with group migration and an Allee affect.* We consider a population with an Allee effect in the death rate, that is, there is intraspecific mutualism for low population densities while at higher densities the usual logistic competition comes in. For a discussion of the Allee effect see Watt (1968). Moreover, the population is assumed to be diffusing in a one-dimensional spatial domain with a flux proportional to the gradient of the total population. The Allee effect is incorporated in a term of the form $k(1-P)u$, k a constant, while the diffusion mechanism takes the form of the term

$$\frac{\partial}{\partial x}\left(\frac{u}{P}\frac{\partial P}{\partial x}\right) = \frac{\partial}{\partial x}\left(\frac{u}{P}\right)\left(\frac{\partial P}{\partial x}\right) + \frac{u}{P}\frac{\partial^2 P}{\partial x^2} .$$

This type of diffusion mechanism has been analyzed by Busenberg and Iannelli (1982b), and it turns out that, in discussing the asymptotic behavior of u, one can consider only the term $\frac{u}{P}\frac{\partial^2 P}{\partial x^2}$ thus neglecting the terms that are quadratic in the gradient of the population. For simplicity, we make this approximation and also assume that $\mu(a) = \beta(a)$, that is the birth and death rates are equal, and $\int_0^\infty \Pi(a)da < \infty$. In this case $p^* = 0$ (see equation (20)) and the equations for u and P are

$$(3.12) \qquad \frac{\partial u}{\partial a} + \frac{\partial u}{\partial t} + \mu u = \frac{u}{P}\frac{\partial^2 P}{\partial x^2} + k(1-P)u$$

and

$$(3.13) \qquad \frac{\partial P}{\partial t} = \frac{\partial^2 P}{\partial x^2} + k(1-P)P, \quad P(0,x) = P_0(x) .$$

Here we consider the population in a one-dimensional infinite spatial domain. The problem (P_3) is trivial in this case and yields the constant characteristic $\phi(0,t,x) = x$, for all $t \geq 0$. Now, (3.13) when considered as an initial value problem with $0 \leq P_0 \leq 1$, has solutions obeying $0 \leq P \leq 1$, and in fact, admits travelling wave solutions, that is, solutions of the form

$$(3.14) \qquad P(t,x) = \psi(x-\alpha t)$$

with ψ satisfying

$$(3.15) \qquad 0 \leq \psi \leq 1, \quad \lim_{x\to-\infty} \psi(x) = 1, \quad \lim_{x\to\infty} \psi(x) = 0.$$

A discussion of these results can be found in Aronson and Weinberger (1975) together with other basic properties of such solutions. So, if P_0 is the initial datum leading to the solution (3.14), the problem (P) with (3.12) as dynamic equation, admits a travelling wave solution of the form

(3.16) $$u(a,t,x) = \Pi(a)\psi(x-\alpha t)/\int_0^\infty \Pi(a)da$$

which has the limiting behavior $\lim_{t\to\infty} u(a,t,x) = \Pi(a)/\int_0^\infty \Pi(a)da.$

Example 4. *A single population with a degenerate diffusion mechanism.* We consider a population undergoing spatial diffusion in the domain $x \in [0,1]$ according to a mechanism where the flux of individuals is proportional to the total population gradient. The boundary is considered to be completely inhospitable, and we assume $u(a,t,0) = u(a,t,1) = 0$ as boundary conditions. This model is discussed by MacCamy (1981) who also studies two special cases, the first with β and μ constants (hence, there is no age dependence) and the second where $\beta = \bar{\beta}e^{-\alpha a}$, μ = constant and $\bar{\beta}$ and α constants. These restrictions imply that the birth rate is maximum at age zero and the death rate is constant. The case with β and μ arbitrary continuous and bounded non-negative functions is analyzed in Busenberg and Iannelli (1982c). The resulting equation for u is

(3.17) $$\frac{\partial u}{\partial t} + \frac{\partial u}{\partial a} + \mu u = \frac{\partial}{\partial x}(u\frac{\partial P}{\partial x})$$

and the problems (P_2) and (P_3) now are

(3.18)
$$\frac{\partial P}{\partial t} = \frac{\partial}{\partial x}(P\frac{\partial P}{\partial x}) + P\int_0^\infty [\beta(a)-\mu(a)]w(a,t,\phi(0,t,x))da$$
$$P(0,x) = P^0(x), \qquad P(t,0) = P(t,1) = 0$$

and

(3.19) $$\frac{d\phi(t,t_0,x)}{dt} = \frac{\partial P}{\partial x}(t,\phi(t,t_0,x)), \quad \phi(t_0,t_0,x) = x.$$

Because of the degeneracy of the diffusion mechanism (i.e., the term $\frac{\partial}{\partial x}(P\frac{\partial P}{\partial x})$ stops being elliptic when $P = 0$), this problem involves considerable mathematical difficulties and we refer the reader to the original papers cited above for its analysis. Here, we look at the simple case where $\beta(a)-\mu(a) = k$, with $k > 0$, a constant. Then, the problem (3.18) becomes (recall that $\int_0^\infty w(a,t,x)da = 1$)

$$(3.20)\qquad \begin{aligned} &\frac{\partial P}{\partial t} = \frac{\partial}{\partial x}\left(P\frac{\partial P}{\partial x}\right) + kP \\ &P(0,x) = P^o(x), \quad P(t,0) = P(t,1) = 0, \end{aligned}$$

which is now decoupled from (3.19). Now, (3.20) is a problem treated by MacCamy (1981) when he studies the case μ = constant, β = constant, and he shows that it has a nontrivial stationary solution $P^s(x)$ such that any solution with $P^o(x) \geq 0$, $P^o(x) \not\equiv 0$ for $x \in [0,1]$, obeys $\lim_{t\to\infty} P(t,x) = P^s(x)$. From this and Theorem 2.2 we can conclude that

$$(3.21)\qquad \begin{aligned} \lim_{t\to\infty} u(a,t,x) &= \lim_{t\to\infty} P(t,x)w(a,t,\phi(0,t,x)) \\ &= P^s(x)\pi(a)e^{-p^*a}/\int_0^\infty \pi(a)e^{-p^*a}da. \end{aligned}$$

Moreover, it is easy to see that $p^* = k$ in this case, and the asymptotic behavior of $u(a,t,x)$ is completely determined when $\beta(a) = \mu(a) + k$, with β and μ arbitrary continuous nonnegative bounded functions of a.

4. Summary. Age dependence is an important factor in population dynamics models but is often not included in such models because of the mathematical difficulties that its inclusion entails. We present a class of age dependent population models whose structure allows the effective separation of the age dependence from the dynamics of the interacting population. These models also include the possibility of spatial diffusion of the populations. We give two basic results that constitute the first steps in the analysis of these models, and illustrate our method by applying it to some specific examples of age dependent population models. Specifically, we show that this method gives a straightforward way of studying the

asymptotic behavior of the solutions, and also brings out a natural set of variables that ease the interpretation of the mathematical results obtained for such models.

The two key points of the method that we have presented are: (i) The proportion of individuals of age a in the total population is shown to converge to a stable stationary age profile which is independent of the interaction and diffusion terms. (ii) The total population part of the problem is, as a consequence of (i), asymptotically autonomous and has as limiting equation the corresponding age-independent model. Hence, in effect, the age-dependent models of the special class (P) - (2.2) that we have treated have an asymptotic behavior that can often be described by assuming a stable age distribution and studying the resulting age-independent problem. This rule cannot be applied indiscriminately, and care should be taken to insure that the limiting behavior of the non-autonomous age-dependent problem for the total population is the same as that of the autonomous problem to which it converges as time increases.

We conclude by noting that an interesting recent paper by M. Gurtin and R. MacCamy ((1982) Product solutions and asymptotic behavior for age-dependent, dispersing populations, Math. Biosci. 62: 157-167) considers the limiting behavior of a problem included in the class (P) - (2.2) that we treat, using techniques that are similar in spirit to some of those that we have described.

References

[1] Aronson, D.G. and H.F. Weinberger (1975). Nonlinear diffusion in population genetics, combustion and nerve propagation, in: Proceedings of the Tulane Program in Partial Differential Equations and Related Topics, Lecture Notes in Biomathematics 446, Springer.

[2] Busenberg, S. and M. Iannelli (1982a). A method for treating a class of nonlinear diffusion problems, Rendiconti Acad. dei Lincei. To appear.

[3] Busenberg, S. and M. Iannelli (1982b). A class of nonlinear diffusion problems in age-dependent population dynamics, J. Nonlinear Analysis T.M.A. To appear.

[4] Busenberg, S. and M. Iannelli (1982c). A degenerate nonlinear diffusion problem in age-structured population dynamics. (Preprint).

[5] Cushing, J. (1979). Volterra integrodifferential equations in population dynamics. In M. Iannelli (ed.), Mathematics in Biology, C.I.M.E. conference proceedings, Liguori, Napoli.

[6] Hoppensteadt, F. (1975). Mathematical Theory of Population Demographics, Genetics and Epidemics, CBMS-NSF Regional Conference Series in Applied Mathematics 20, Society for Industrial and Applied Mathematics, Philadelphia.

[7] MacCamy, R.C. (1981). A population model with nonlinear diffusion, J. Diff. Eq. 39: 52-72.

[8] Marcati, P. (1982). On the global stability of the logistic age-dependent population growth. (Preprint).

[9] Watt, K.E.F. (1968). Ecology and Resource Management, McGraw-Hill, New York.

A MATHEMATICAL MODEL OF POPULATION DYNAMICS INVOLVING DIFFUSION AND RESOURCES

by Adolf Haimovici
Iaşi-Romania

The dynamics of an age structured population involving diffusion have been considered by many authors (see e.g. di Blasio, G. (1979), di Blasio, G. and Lamberti, L. (1979), Levin, S. (1976). Our aim is to approach the same problem, under the added assumption that the population's dynamics depend on the dynamics of a vector of "resources". This is to be interpreted in a general manner, so that if a component of this vector is negative, it will be viewed as a deterrent to growth. Thus we will not impose sign restrictions on this vector.

A new equation is introduced in the usual system, which accounts for the dynamics of the resources. Specifically we will suppose that the growth rate of the resources is a linear functional of the age-specific population density.

§1. Statement of the problem. In what follows, we will denote by

i) $u(t,a,x)$, the density of individuals of age a, at the moment t, and the point x of the space;

ii) $Du=\lim_{h\to o} \frac{u(t+h,\ a+h) - u(t,x)}{h}$ (see f.i. Gurtin, M. and Mc Camy, R.C. (1974), Hoppensteadt, A. (1975))

iii) A, the diffusion coefficient, which we will suppose constant;

iv) $r(t,x)$, the n-dimensional vector, depending on time t and point x of the space, representing the resources, and $r_o(x)$ the initial distribution of these resources;

v) λ a demographic coefficient, depending on time, age and resources, i.e. $\lambda = \lambda(t,a,r(t,x))$;

vi) μ, the birth coefficient, depending on time, age and on the point x, $\mu = \mu(t,a,x)$;

vii) φ, the initial distribution of the population $\varphi = \varphi(a,x)$;

viii) $B=B(t,x)$ the total birth rate at time t and point x;

ix) $S=S(t,a,x)$ an n vector $R_+ \times R_+ \times R \to R^n$;

x) $g=g(t,x)$: $R_+ \times R \to R$ the inherent growth rate of resources in the absence of the population.

The space in which diffusion takes place will be $\mathbb{R}$. The equations of our model are:

(1.1) $$Du - A^2 \frac{\partial^2 u}{\partial x^2} = \lambda(t,a,r(t,x))u,$$

(1.2) $$B(t,x) = \int_0^\infty \mu(t,\alpha,x)u(t,\alpha,x)d\alpha\,,$$

(1.3) $$\frac{\partial r}{\partial t} = \int_0^\infty S(t,\alpha,x)u(t,\alpha,x)d\alpha + g(t,x),$$

(1.4) $$u(o,a,x)=\varphi(a,x), \qquad r(o,x)=r_o(x),$$

(1.5) $$\varphi(o,x) = \int_0^\infty \mu(o,\alpha,x)\varphi(\alpha,x)d\alpha$$ - an obvious compatibility condition.

Remarks. 1. If $A=o$, i.e. if there does not exist diffusion, then λ is the death rate of the age $\underline{a}$ population and obviously $\lambda < o$. In our case, this condition is no longer valid, since the contribution of the diffusion can lead to the fact that at some points x, $\lambda \geqslant o$.

2. The equation (1.2) is the same as that to be found in Hoppensteadt (1975).

3. The third equation expresses the fact that the speed of the growth rate of the resources is a linear functional of age-specific population density depending on time and on space. If a certain component i of S is positive, and $r_i > o$, then this means that the resource r_i is produced by the population; if $S_i < o$, then this means that the component i of r is a pollutant, which grows with the

population density.

§2. Hypotheses. We suppose

A_1. λ is continuous on $R_+ \times R_+ \times R^n$ and bounded

$$|\lambda(t,a,r)| \leq \lambda(t,a) \leq \Lambda, \text{ and } \int_0^\infty |\lambda(t+s,a+s)|\, ds \leq \Lambda, \quad o \leq \Lambda < 1/2$$

for every $r \in R^n$, $\|r\| \leq M$.

A_2. λ is lipschitz with respect to the last variables, i.e.

$$|\lambda(t,a,r_1) - \lambda(t,a,r_2)| \leq Le^{-ka} \|r_1 - r_2\| ,$$

where L, k are positive constants, and $\|.\|$ is the norm in R^n.

B_1. μ is continuous in $R_+ \times R_+ \times R$, and satisfies (see also Hoppensteadt (1975))

$$\mu_1 a e^{-ha} \leq \mu \leq \mu_0 a e^{-ha}, \quad o \leq \mu_1 < \mu_0, \text{ h - positive constants.}$$

C. The vector S is continuous and bounded on $R_+ \times R_+ \times R$ and g is continuous on R x R, and

$$\|S\| \leq S_0(t)^{-sa}, \quad \text{s=const.} \qquad \int_0^\infty S_0(t)dt = S_1 = \text{const.}$$

$$\|g(t,x)\| \leq g_1 \quad \text{and} \quad \int_0^\infty \|g(t,x)\|\, dt \leq G = \text{const.}$$

D. The function φ is continuous on $R_+ \times R$ and bounded

$$o \leq \varphi \leq \varphi_0.$$

E. The function r_0 is continuous on R and bounded

$$\|r_0\| \leq R.$$

F. The above constants satisfy

$$R + \left(\frac{1}{s} + 1 + G\right) \frac{\varphi_0 S_1}{s} \leq M, \quad S_1 Q/h \leq 1.$$

§3. Transformation of the system (1.1) - (1.5). As usual, we consider first a pair of arbitrary positive, but fixed numbers $(t_0, s_0) \in R_+ \times R_+$, and denote

$$u(t_0+\tau, a_0+\tau, x) = \bar{u}(\tau, x),$$

$$r(t_0+\tau, x) = \bar{r}(\tau, x),$$

(3.1) $$\lambda(t_0+\tau, a_0+\tau, \bar{r}(\tau, x)) = \bar{\lambda}(\tau, \bar{r}),$$

$$\mu(t_0+\tau, a_0+\tau, x) = \bar{\mu}(\tau, x),$$

$$\varphi(a_0+\tau, x) = \bar{\varphi}(\tau, x).$$

Then, one sees easily that

(3.2) $$Du(t_0+\tau,\ a_0+\tau, x) = \frac{\partial \bar{u}}{\partial \tau}(\tau, x),$$

and (1.1) becomes

(3.3) $$\frac{\partial \bar{u}}{\partial t} - A^2 \frac{\partial^2 \bar{u}}{\partial x^2} = \bar{\lambda}(\tau, \bar{r})\bar{u} .$$

Taking into account the thermic potentials, we can rewrite (3.3) in the form

(3.4) $$\bar{u}(\tau, x) = \int_0^{\tau} d\sigma \int_{-\infty}^{\infty} E(\tau, x; \sigma, \xi)\, \bar{\lambda}(\sigma, \bar{r}(\sigma, \xi))\bar{u}(\sigma, \xi) d\xi + \int_{-\infty}^{\infty} E(\tau, x; o, \xi)\bar{u}(o, \xi) d\xi ,$$

where

$$E(t, x; \sigma, \xi) = \frac{1}{2A\sqrt{\pi(t-\sigma)}} \exp\left(-\frac{(x-\xi)^2}{4A^2(t-\sigma)}\right).$$

Suppose now that $\bar{u}(o, \xi)$ is a known function and denote

(3.5) $$v(\tau, x) = \int_{-\infty}^{\infty} E(\tau, x; o, \xi)\ \bar{u}(o, \xi) d\xi .$$

The equation (3.4) is a linear integral equation with kernel $E(\tau, x; \sigma, \xi)\bar{\lambda}(\sigma, \bar{r}(\sigma, \xi))$ where $\bar{\lambda}$, as a consequence of A_1, satisfies

(3-6) $$\int_{-\infty}^{\infty} |\bar{\lambda}(\sigma, \bar{r}(\sigma, x))|\, d\sigma \le \Lambda .$$

We take now into account the identity

(3.7) $$\frac{(x-x_1)^2}{\tau-\tau_1} + \frac{(x_1-x_2)^2}{\tau_1-\tau_2} = \frac{(x-x_2)^2}{\tau-\tau_2} + \frac{(x_1-\tilde{x})^2}{\frac{(\tau-\tau_1)(\tau-\tau_2)}{\tau-\tau_2}} ,$$

with

$$\tilde{x} = \frac{(\tau_1 - \tau_2)x_1 + (\tau_1 - \tau_2)x_2}{\tau - \tau_2},$$

and, suppose that r is a known function with $\|r\| \leq M$=const. Then we denote:

$$\Lambda_0 = 1$$

$$\Lambda_n(\tau, x; \sigma, \xi; r) = \int_\sigma^\tau d\tau_n \int_{-\infty}^{\infty} \bar{\lambda}(\tau_n, \bar{r}(\tau_n, \eta_n)) \cdot$$

$$\frac{\Lambda_{n-1}(\tau, x; \tau_n, \eta_n; r)}{2A\sqrt{\frac{(\tau-\tau_n)(\tau_n-\sigma)}{\tau-\sigma}}} \exp\left(-\frac{(\eta_n - \tilde{\eta}_n)^2}{4A^2 \frac{(\tau-\tau_n)(\tau_n-\sigma)}{\tau-\sigma}}\right) d\eta_n,$$

$$\tilde{\eta} = \frac{(\tau-\tau_n)\xi + (\tau_n - \sigma)x}{\tau-\sigma}.$$

With these notations, the solution of the equation (3.4) is:

(3.8) $$\bar{u}(\tau, x) = \int_{-\infty}^{\infty} \sum_0^\infty \Lambda_k(\tau, x; o, \eta; \bar{r}) E(\tau, x; o, \eta) \bar{u}(o, \eta) d\eta.$$

The sum $\sum_1^\infty \Lambda_k(\tau, x; o, \eta; \bar{r})$, is positive, as a consequence of the hypothesis A_1. Indeed, we have

$$|\Lambda_n| \leq \Lambda^n, \quad \left|\sum_1^\infty \Lambda_k\right| \leq \frac{\Lambda}{1-\Lambda},$$

so that:

(3.9) $$0 \leq \frac{1-2\Lambda}{1-\Lambda} \leq \sum_0^\infty \Lambda_k \leq \frac{1}{1-\Lambda}.$$

For $a \leq t$ and

$$a_0 = o, \quad t_0 = t-a, \quad \tau = a,$$

(3.8) leads to

(3.10) $$u(t, a, x) = \int_{-\infty}^{\infty} \sum_0^\infty \Lambda_k(a, x; o, \eta; \bar{r}) E(a, x; o, \eta) B(t-a, \eta) d\eta,$$

and for $a > t$, and

$$a_0 = a-t, \quad t_0 = o, \quad \tau = t,$$

we obtain from (3.8):

$$(3.11)\quad u(t,a,x) = \int_{-\infty}^{\infty}\sum_{0}^{\infty} \Lambda_k(t,x;o,\eta;\bar{r})E(t,x,o,\eta)\,\varphi(a-t,\eta)\,d\eta .$$

§4. Write now (1.2) in the form:

$$(4.1)\quad B(t,x)= \int_0^t \mu(t,\alpha,x)u(t,\alpha,x)d\alpha + \int_t^{\infty}\mu(t,\alpha,x)u(t,\alpha,x)d\alpha .$$

Using (3.10) and (3.11) this equation leads to

$$(4.2)\quad B(t,x)= \int_0^t \mu(t,t-\alpha,x)d\alpha\int_{-\infty}^{\infty}\sum_{0}^{\infty}\Lambda_k(t-\alpha,x;o,\eta,\bar{r})$$

$$E(t-\alpha,x;o,\eta)B(\alpha,\eta)d\eta \,+$$

$$+\int_0^{\infty}\mu(t,t+\alpha,x)d\alpha\int_{-\infty}^{\infty}\sum_{0}^{\infty}\Lambda_k(t+\alpha,x;o,\eta,\bar{r})E(t+\alpha,x;o,\eta)\,\varphi(\alpha,\eta)d\eta ,$$

which is a new linear integral equation with B as unknown function; the kernel of this equation - as a consequence of (3.9) - is positive, and so is also the last member of (4.2); it follows that the solution B(t,x) of (4.2) is positive. It results that also $\bar{u}(\tau,x)$ from (3.8) - and, as a consequence - u(t,a,x) from (3.10) and (3.11) are positive.

We can now state the

<u>Theorem A. If r is a vector function satisfying $\|r\| \le M$ = const and the hypotheses of §2 are satisfied, then the equation (1.1), (1.2), (1.4), (1.5) have positive solutions u and B.</u>

The kernel of the equation (4.2)

$$K(t,\alpha,x,\eta)= \mu(t,t-\alpha,x)\ \sum_{0}^{\infty}\Lambda_k(t-\alpha,x;o,\eta;\bar{r})E(t-\alpha,x;o,\eta)$$

satisfies the inequalities:

$$(4.3)\quad \nu_1(t-\alpha)e^{-h(t-\alpha)}E(t-\alpha,x;o,\eta)\le K(t,\alpha,x;\eta)\le$$
$$\le \nu_0(t-\alpha)e^{-h(t-\alpha)}E(t-\alpha,x;o,\eta),$$

where

$$\nu_0^2=\frac{\mu_0}{1-\Lambda},\qquad \nu_1^2=\frac{(1-2\Lambda)\mu_1}{1-\Lambda},$$

and the last term of (4.2) satisfies

(4.4) $$F(t,x) \leq \nu_0^2 (ht+1)e^{-ht} \varphi_0,$$

It follows that the resolvent kernel K satisfies

(4.5) $$\sum_0^\infty \nu_1^n \frac{(t-\alpha)^{2n-1}}{(2n-1)!} e^{-h(t-\alpha)} E(t,x;\alpha,\xi) \leq$$

$$\leq \mathcal{K}(t,\alpha;x,\xi) \leq \sum_0^\infty \nu_0^n \frac{(t-\alpha)^{2n-1}}{(2n-1)!} e^{-h(t-\alpha)} E(t,x;\alpha,\xi).$$

We obtain finally

(4.6) $$B(t,x) \leq \nu_0^n \rho \frac{\varphi_0}{h^2} \exp((\nu_0-h)t), \qquad \rho = \max(h, \nu_0).$$

Analogously, if we denote:

$$\varphi_1(t,x) = \int_0^\infty d\beta \int_\beta^\infty e^{-h\beta} E(t,x;o,\xi) \varphi(\beta,\xi) d\xi,$$

$$\varphi_2(t,x) = \int_0^\infty d\beta \int_{-\infty}^{-\infty} e^{-h\beta} E(t,x;o,\xi) \varphi(\beta,\xi) d\xi,$$

and

$$\tilde{\varphi}(t,x) = \min(\nu_1 \varphi_1(t,x), \varphi_2(t,x)),$$

it follows:

(4.7) $$B(t,x) \geq \nu_1 \tilde{\varphi}(t,x) \exp[(-h+\nu_1)t].$$

Remark 1. Taking into account that, by hypotheses, $\tilde{\varphi}$ is bounded, it follows from (4.6) and (4.7), if $\nu_0 < h$, that the birth rate tends to zero as t tends to infinity; and if $\nu_1 > h$, then the birth rate tends to infinity as t tends to infinity, in each compact of $\mathbb{R}$.

2. If $\nu_0 < h$, then $\rho = h$, and

$$B(t,x) \leq \frac{\nu_0 \varphi_0}{h} \leq \varphi_0,$$

and from (4.3) we obtain, when a <t:

(4.8) $$u(t,a,x) \leq \frac{1}{1-\Lambda} \varphi_0,$$

so that, this inequality remains valid for all a,t.

§5. To study the equation (1.3) we will assume in the following that $\nu_o < h$; then let r_1, r_2 be two vectors in R^n, depending on t and x; for $\bar{u}_1$, $\bar{u}_2$ from (3.4), we get

$$|\bar{u}_1(\tau,x)-\bar{u}_2(\tau,x)| \le \int_o^{\tau} d\sigma \int_{-\infty}^{\infty} E(\tau,x;\sigma,\xi) \Big\{ |\bar{\lambda}(\sigma,\bar{r}_1(\sigma,\xi)) - \bar{\lambda}(\sigma,\bar{r}_2(\sigma,\xi))| \quad \bar{u}_1(\sigma,\xi) + |\bar{\lambda}(\sigma,\bar{r}_2(\sigma,\xi))| \; |\bar{u}_1(\sigma,\xi)-\bar{u}_2(\sigma,\xi)| \Big\} d\xi + \int_{-\infty}^{\infty} E(\tau,x;o,\xi) \; |\bar{u}_1(o,\xi)-\bar{u}_2(o,\xi)| \; d\xi$$

and, after hypothesis A_2:

$$(1-\Lambda)\max|\bar{u}_1(\tau,x)-\bar{u}_2(\tau,x)| \le \frac{\Lambda L}{(1-\Lambda)k} |||r_1-r_2||| \int_{-\infty}^{\infty} E(\tau,x;o,\xi)\bar{u}(o,\xi)d\xi + \max_{\xi\in R} |\bar{u}_1(o,\xi) - \bar{u}_2(o,\xi)| ,$$

where

$$|||r_1 - r_2||| = \max_{\substack{x\in R \\ t\in R^+}} \|r_1-r_2\| ,$$

from which, <u>for a <t</u>, we deduce, with regard to (3.9) and (3.10):

(5.1)
$$\max_{\substack{a\le t \\ x\in R}} |u_1(t,a,x)-u_2(t,a,x)| \le$$
$$\frac{\Lambda L}{(1-\Lambda)^2 k} |||r_1-r_2||| \cdot \int_{-\infty}^{\infty} E(a,x;o,\xi)B(t-a,\xi)d\xi + \frac{1}{1-\Lambda} \max_{\substack{a\le t \\ x\in R}} |B_1-B_2|$$

and, with (4.5)

$$\max_{\substack{a\le t \\ x\in R}} |u_1(t,a,x)-u_2(t,a,x)| \le$$
$$\frac{\Lambda L}{(1-\Lambda)k} \varphi_o |||r_1-r_2||| \exp[(\nu_o-h)(t-a)] + \frac{1}{1-\Lambda} \max_{\substack{a\le t \\ \xi\in R}} |B_1(t-a,\xi)-B_2(t-a,\xi)|$$

<u>and for</u> $a > t$:

(5.2) $$\max_{\substack{a>t\\ x\in R}} |u_1(t,a,x)-u_2(t,a,x)| \le \frac{\Lambda L}{(1-\Lambda)^2 k} \; ||| r_1-r_2 ||| \, .$$

Further, from (4.1) we have

$$|B_1(t,x)-B_2(t,x)| \le \int_0^t \mu(t,\alpha,x) \; |u_1(t,\alpha,x)-u_2(t,\alpha,x)| \; d\alpha \; +$$
$$+ \int_t^\infty \mu(t,\alpha,x) \; |u_1(t,\alpha,x)-u_2(t,\alpha,x)| \; d\alpha \; ;$$

and taking into account (5.1), (5.2), and the hypothesis B_1, we have

$$|B_1(t,x)-B_2(t,x)| \le \frac{L}{(1-\Lambda)^2 k} \varphi_0 \mu_0 \; ||| r_1-r_2 ||| \; \frac{1}{\nu_0^2} e^{\nu_0 t} \; +$$

$$+ \frac{1}{\nu_0^{\,2}}(th+1) \; e^{-ht} + \frac{1}{(1-\Lambda)h^2} \max_{\substack{t\in R_+\\ x\in R}} |B_1-B_2| \, .$$

Supposing in addition $(1-\Lambda)h^2 < 1$, we deduce:

(5.3) $$\max |B_1-B_2| \le P \; ||| r_1-r_2 ||| \quad , \text{ with } P = \frac{3\Lambda L \varphi_0 h^2}{[(1-\Lambda)h^2-1] \; k} \, .$$

Comming back to (5.1) and (5.2), we obtain

(5.4) $$\max |u_1(t,a,x) - u_2(t,a,x)| \le Q \; ||| r_1-r_2 ||| \, ,$$

where

$$Q = \frac{\Lambda L}{(1-\Lambda)^2 k} \varphi_0 \; + \frac{1}{1-\Lambda} P \le 4 \frac{\Lambda L \varphi_0}{(1-\Lambda)^2 \, [\, (1-\Lambda)h^2-1]} \, .$$

§6. We come now to equation (1.3), which we write in the form

(6.1) $$r(t,x) = r_0(x) + \int_0^t d\tau \int_0^\infty S(\tau,\alpha,x) u(\tau,\alpha,x) d\alpha + \int_0^t g(\tau,x) d\tau \, .$$

For this equation we will prove that there exists a <u>continuous bounded solution</u>.

To this end, we consider the operator $\mathcal{A}$, defined by

$$\tilde{r}(t,x) = (\mathcal{A} r)(t,x) = r_0(x) + \int_0^t d\tau \int_0^\infty S(\tau,\alpha,x) u(\tau,\alpha,x) d\alpha + \int_0^t g(\tau,x) d\tau$$

on the sphere $\|r\| \le M$, of the n-dimensional space with metric induced by the euclidian norm; this space is obviously complete.

First we see that, taking into account (3.10), (3.11) and (4.6) with $\nu_0 < h$, and the hypothesis E, we have:

$$\|\tilde{r}(t,x)\| \leq R+S_0 \frac{\rho\nu_0\varphi_0}{h^2}\int_0^t \left\{ \frac{1}{(h-\rho+s)}(1-\exp(-\rho\tau))\exp(-s\tau)+\right.$$

$$\left.+\varphi_0 S_0 \frac{1}{S_0 s}\exp(-s\tau)\right\} d\tau + C \leq R+\frac{\varphi_0 S_0}{s}\left(\frac{\rho\nu}{h-\rho+s}+1\right)+G,$$

which after hypothesis is smaller than M, i.e. the sphere $\|r\| \leq M$ is transformed by the operator defined in the left-hand side of (6.2), into itself.

Concerning the contraction property of this operator, we see that if r_1 and r_2 are two vector-functions of the above sphere and $\tilde{r}_1$ and $\tilde{r}_2$ their transforms through $\mathcal{A}$, then, taking into account (5.4), we have:

$$\|\tilde{r}_1-\tilde{r}_2\| \leq \int_0^t d\tau \int_{-\infty}^{\infty} S(\tau,\alpha,x)Q\ |||r_1-r_2|||\ d\alpha \leq$$

$$+\int_0^t d\tau \int_0^{\infty} S_0 e^{-s\alpha} Q\ |||r_1-r_2|||\ d\alpha \leq \frac{S_1}{s} Q\ |||r_1-r_2|||\ ;$$

since hypothesis, $S_1 Q/h \,/\, 1$, the existence follows from Banach's fixed point theorem.

§7. <u>The influence of the diffusion</u>. To study the influence of the diffusion, we must see the dependence of the solution on the coefficient A. To this end, we add to the hypotheses in §2, the following ones:

A_3. λ has bounded derivatives with respect to the components of the third variable

$$\left|\frac{\partial\lambda}{\partial r}\right| \leq \bar{\Lambda}_1,$$

B_2. μ has a bounded derivative with respect to the third variable

$$\left|\frac{\partial\mu}{\partial x}\right| \leq \mu_0 e^{-pa} \qquad \mu_0,\ p\text{-constants},\ \mu_0, p > 0.$$

C_2. The components of the vectors S and g are differentiable and satisfy

$$\int_0^{\infty} \left\| \frac{\partial S}{\partial x} \right\| d\alpha \leq \bar{S}_1 \bar{e}^{mt}, \qquad \left\| \frac{\partial g}{\partial x} \right\| \leq \gamma .$$

E_2. r_o is differentiable and

$$\left\| \frac{\partial r_o}{\partial x} \right\| \leq R_1$$

F_2. $$1 - \Lambda - \frac{L \varphi_o}{(1-\Lambda)k} - \frac{\mu_o}{h^2} > o .$$

We obtain (Remark 2, §4)

$$\| u \| \leq \frac{1}{1-\Lambda} \varphi_o ,$$

$$\left| \frac{\partial u}{\partial x} \right| \leq \begin{cases} k \dfrac{2\varphi_o}{A} \left(\sqrt{a} + \dfrac{1}{\sqrt{a}} \right), & \text{if } a \leq t, \\ k \dfrac{2\varphi_o}{A} \left(\sqrt{t} + \dfrac{1}{\sqrt{t}} \right), & \text{if } a > t, \end{cases} \tag{7.1}$$

$$\left| \frac{\partial B}{\partial x} \right| \quad M_1 + \frac{1}{A} M_2 \Phi_1(t), \tag{7.2}$$

$$\left\| \frac{\partial r}{\partial x} \right\| \quad Q_1 + \frac{1}{A} Q_2 \Phi_2(t), \tag{7.3}$$

Φ_1, Φ_2 being two increasing on t functions, and M_1, M_2, Q_1, Q_2 constants.

Then, taking into account the differentiability with respect to x of the functions λ, μ, S, we have, from (3.4):

$$\begin{aligned} \bar{u}(\tau, x) = {} & \frac{1}{\sqrt{\pi}} \int_0^{\tau} d\sigma \int_{-\infty}^{\infty} e^{-\xi^2} \bar{\lambda}(\sigma, \bar{r}(\sigma, x)) \bar{u}(\sigma, x) d\xi + \\ & + \frac{1}{\sqrt{\pi}} \int_{-\infty}^{\infty} e^{-\xi^2} \bar{u}(o, x) d\xi + \\ & \frac{1}{\sqrt{\pi}} \int_0^{\tau} d\sigma \int_{-\infty}^{\infty} e^{-\xi^2} \left\{ \frac{\partial \bar{\lambda}}{\partial r}(\sigma, \bar{r}(\sigma, x)) \frac{\partial \bar{r}}{\partial x}(\sigma, x^*) \bar{u}(\sigma, x^*) + \right. \end{aligned} \tag{7.4}$$

$$+ \bar{\lambda}(\sigma, \bar{r}(\sigma, x^*)) \frac{\partial \bar{u}}{\partial x}(\sigma, x^*) \Big\} 2A \sqrt{\tau - \sigma}\, d\xi$$

$$+ \frac{1}{\sqrt{\pi}} \max_{x^*} \frac{\partial \bar{u}}{\partial x}(o, x^*) \quad 2A\sqrt{\tau} \,,$$

where x^* is a value between x and $x+2A \sqrt{\tau - \sigma}\,\eta$; after considering the above estimates we find

$$\left| \bar{u}(\tau,x) - \int_o^{\tau} \bar{\lambda}(\sigma, r(\sigma,x)) \bar{u}(\sigma,x) d\sigma - \bar{u}(o,x) \right| \leq C_1 A \sqrt{\tau^3} + C_2 F(\tau),$$

where C_1, C_2 are constants depending on the constants involved in the hypotheses, but independent of A, and F is an increasing on t function, also independent of A.

Consider now the system (1.1) - (1.5) with A=o, and denote by (v(t,x), b(t,x), R(t,x)) its solution. By analogous notations and operations, we obtain

$$\text{(7.5)} \quad \begin{aligned} \bar{v}(\tau,x) &= \int_o^{\tau} \bar{\lambda}(\sigma, R(\sigma,x)) \bar{v}(\sigma,x) + \bar{v}(o,x), \\ b(t,x) &= \int_o^{\infty} \mu(t,\alpha,x) v(t,\alpha,x) d\alpha, \\ \frac{\partial R}{\partial t} &= \int_o^{t} S(t,\alpha,x) v(t,\alpha,x) d\alpha + G(t,x), \\ v(o,a,x) &= \varphi(a,x), \quad R(o,x) = r_o(x). \end{aligned}$$

We can now state the

<u>Theorem</u>. <u>The difference between the solution u of the system (1.1) - (1.5) and the solution v of (7.5) is</u>

$$|u - v| \leq C_1 A \sqrt{t^3} + C_2 F(t),$$

C_1 <u>and</u> C_2 <u>being constants and</u> F <u>an increasing function of t, with</u> F(o)=o.

<u>Proof</u>. Comparing (7.4), with the first equation (7.5), we obtain:

$$\left| \bar{u}(\tau,x) - \bar{v}(\tau,x) \right| \leq \int_o^{\tau} L e^{-k(a_o+\sigma)} \, \| r-R \| \, \frac{\varphi_o}{1-\Lambda} d\sigma +$$

$$+ \max_{\sigma} \Lambda \left| \bar{u}(\sigma,x) - \bar{v}(\sigma,x) \right| + \left| \bar{u}(o,x) - \bar{v}(o,x) \right| + C_1 A \sqrt{\tau^3} + C_2 F(t),$$

which means:

<u>for a < t</u>

$$(7.6) \quad (1-\Lambda)\max_{a\le t} |u(t,a,x)-v(t,a,x)| \le \frac{L\varphi_o}{(1-\Lambda)k} |||r-R||| + $$

$$+ \max_{a\le t} |B(t-a,x)-b(t-a,x)| + C_1 A \sqrt{t^3} + C_2 F(t);$$

<u>for a > t</u>

$$(7.7) \quad (1-\Lambda)\max_{a>t} |u(t,a,x)-v(t,a,x)| \le \frac{L\varphi_o}{1-\Lambda)k} |||r-R||| + C_1 A \sqrt{t^3} + C_2 F(t).$$

From (1.3) and the third equation (7.5), we obtain

$$|||r-R||| \le \frac{\bar{S}_1}{s} \max |u-v|, \quad |B-b| \quad \frac{\mu_o}{h^2} \max |u-v|$$

So that, finally we have

$$\max |u-v| \le c_1 \quad A \sqrt{t^3} + c_2 F(t),$$

where

$$c_i = C_i / [1-\Lambda - \frac{L\varphi_o}{(1-\Lambda)k} - \frac{\mu_o}{h^2}]$$

as stated.

References

di Blasio, G. (1979). Non linear Age-dependent Population Diffusion. <u>J. of Math.Biology</u> <u>8</u>: 265-284

di Blasio, G. and Lamberti, L. (1979). An initial Boundary Value Problem for Age-dependent Population Diffusion, <u>SIAM J. Appl.Math.</u> <u>35</u>: 593-615.

Gurtin, M. and Mc Camy, R.C. (1974). Non-linear age dependent population dynamics. <u>Arch.Rational Mech.Anal.</u> <u>54</u>: 281-300

Haimovici, A. (1978). On the Growth of a Population Dependent on Ages and involving Resources and Pollution. <u>Math. Biosci.</u> <u>6</u>

Hoppensteadt, F., (1975). Mathematical Theories of Populations. Demographics, Genetics and Epidemics. Soc. Ind. Appl. Math. Philadelphia, 1-35.

Levin, S. (1976). Spatial Patterning and the Structure of Ecological Communities, Some Math. Questions in Biology: VII.

PART VII

SPATIAL PATTERN AND DIFFUSION MODELS

Critical Patch Size for Plankton and Patchiness*

Akira Okubo

Marine Sciences Research Center
State University of New York
Stony Brook, New York 11794 USA

1. Introduction

Patchy distributions of plankton in the sea and lakes have been well documented (Tonolli and Tonolli, 1960; Cassie, 1963; Platt et al., 1970; Steele, 1976, 1977; Harris, 1980). However, the precise mechanism from which plankton patchiness arises is still a subject of considerable controversy. To name some proposed mechanisms they are: (i) mechanical retention in wind-generated convective cells or frontal zones (Langmuir, 1938; Stommel, 1949; Bowman and Esaias, 1978; Floodgate et al., 1981), (ii) behavioral reaction to distributions of environmental parameters such as temperature, salinity, light, and nutrients (Cassie, 1959; Forward, 1976; Heaney and Eppley, 1981), (iii) exclusion of certain zooplankton by phytoplankton (Bainbridge, 1953), (iv) food-chain association in predator-prey relations (Tonolli, 1958), in particular the phenomenon of spontaneous pattern generation through diffusive instability (Levin and Segel, 1976), and (v) aggregative behaviors (swarming, schooling) for breeding and feeding (Clutter, 1969; Hamner and Carleton, 1979).

In fact, patchiness most likely arises from a variety of mechanisms and processes under various conditions. However, many cases appear to share a single common process which acts as an "anti-patchiness" agent. This process is diffusion due to turbulence in surrounding media or random movements of organisms. Generally speaking, diffusion tends to counteract the formation of organism aggregation, and to give rise to a more uniform distribution (note an exception of diffusion-induced instability in predator-prey interactions). Thus an endless interplay occurs between the aggregative process of organism growth or reproduc-

*Contribution No. 350 from the Marine Sciences Research Center, State University of New York at Stony Brook.

tion and the antiaggregative process of diffusion, and a dynamical balance may be established in such a manner that the rate of growth of organisms within a patch is equal to the rate of loss of organisms due to diffusion into the surroundings where organisms cannot survive or where the net growth rate is negative. Since the rate of growth is proportional to the volume of the patch and the diffusion rate is proportional to the surface area of the patch, the patch size at the dynamical balance must be a minimum critical size or simply critical size, below which the population of phytoplankton cannot be maintained and the patch disappears.

2. Model by Kierstead, Slobodkin, and Skellam ("KISS model")

A simple, nonetheless useful mathematical model for the critical patch size was given by Kierstead and Slobodkin (1953) and Skellam (1951) independently. This model is based on a simple diffusion and exponential growth equation for plankton concentration, which is expressed, in one-dimensional space, by

$$\frac{\partial S}{\partial t} = D\frac{\partial^2 S}{\partial x^2} + \alpha S\ , \qquad 0 < t,\ 0 \leq x \leq L \tag{1}$$

subject to

$$\begin{aligned} &\text{at } t = 0;\ S = S_o(x) \\ &\text{at } x = 0 \text{ and } x = L;\ S \equiv 0\ . \end{aligned} \tag{2}$$

The solution is given by

$$S(x,t) = \sum_{n=1}^{\infty} a_n \sin\frac{n\pi}{L}x \exp\{\alpha - \frac{n^2\pi^2}{L^2}D\}t \tag{3}$$

with

$$a_n = \frac{2}{L}\int_o^L S_o(x)\sin\frac{n\pi}{L}x dx\ .$$

If $L < \pi(D/\alpha)^{1/2}$, $S(x,t)$ will approach zero as time progresses, while if $L > \pi(D/\alpha)^{1/2}$, $S(x,t)$ will increase indefinitely with time, thus a plankton bloom may occur. Therefore the critical (minimum) patch size L_{cr} is determined from the equality condition, i.e.,

$$L_{cr} = \pi(D/\alpha)^{1/2} . \tag{4}$$

The above is applied for a linear habitat. For two-dimensional space, e.g., a circular habitat, the critical diameter is obtained by

$$L_{cr} = 4.81(\pi/\alpha)^{1/2} . \tag{5}$$

This model has been a cornerstone in the later development of the subject. Hereafter it is referred to as the KISS model (KIerstead-Slobodkin-Skellam), and the critical size (4) as the KISS scale denoted by L_o.

The further development has been made in various directions. It includes (i) generalization of the growth rate function and relaxation of the boundary condition, (ii) scale and density dependencies in diffusivity, (iii) addition of advection, (iv) nonlinear population dynamics, and (v) multispecies interactions. Interestingly enough these generalizations and improvements on the original model have shown that with a few exceptions, the KISS model is robust as a mathematical theory for the critical size problem.

In this article I will briefly review the main result of those later developments and present a new model which attempts to eliminate the arbitrariness of the boundary condition involved in the original KISS model.

3. Generalization of the KISS model

Platt and Denman (1975) and Wroblewski *et al*. (1975) included the effect of herbivore grazing on phytoplankton in the KISS model. Using an Ivlev type grazing function, they calculated the critical size to be

$$L_{cr} = \pi\{\frac{D}{(\alpha - R_m\lambda)}\}^{1/2} , \qquad \alpha - R_m\lambda > 0 \tag{6}$$

where R_m is the maximum grazing ration of the herbivore and λ is the Ivlev constant. As is expected, the effect of herbivore grazing reduces the net growth rate of the population, so that the critical size becomes larger than the KISS scale. If $\alpha \leq R_m \lambda$, all patches must vanish as time progresses.

The boundary condition of the KISS model that the plankton patch is surrounded by completely unsuitable water is rather arbitrary. In reality the transition from favorable to unfavorable conditions is gradual, and organisms move more or less freely both into and out of the favorable region. Ludwig et al. (1979) and Evans (1978, unpublished manuscript) calculated a critical size assuming that organisms can survive but not grow outside the favorable habitat, so that the diffusion-reaction equation outside is expressed by

$$\frac{\partial S}{\partial t} = D \frac{\partial^2 S}{\partial x^2} - \beta S \quad \text{for} \quad |x| > \frac{L}{2} \tag{7}$$

where $\beta \geq 0$ is the death rate of organisms. Equation (7) is solved by coupling with (1) subject to the continuity conditions for the population density and its flux at the boundaries, $|x| = L/2$. As a result the following criterion for critical size is obtained:

$$L_{cr} = L_o \{ \frac{2}{\pi} \tan^{-1} (\frac{\beta}{\alpha})^{\frac{1}{2}} \} \tag{8}$$

where $\tan^{-1}(\beta/\alpha)^{\frac{1}{2}}$ takes the value of the principal branch. For infinite β the critical scale is reduced to the KISS scale, L_o. For finite β/α, L_{cr} is less than L_o, decreasing to zero as β/α approaches zero; this limiting situation corresponds to a Malthusian population surrounded by reflecting barriers.

In reality for phytoplankton, a reasonable value for β/α might be 0.1 when population decay is simply due to plankton respiration outside the favorable area. For this value L_{cr} is 20% of L_o. Thus the critical size does not differ much from the KISS scale when the conditions in the surrounding water are relaxed.

Gurney and Nisbet (1975) consider a model in which the growth rate varies continuously in space, in particular varying with distance from a habitat center in a parabolic fashion, i.e.,

$$\alpha(x) = \alpha_o \{ 1 - (\frac{x}{x_o})^2 \} \tag{9}$$

where α_o is the maximum growth rate at the center. The favorable region extends from $-x_o$ to x_o, and beyond this region the death rate progressively dominates. Unlike the KISS model the boundary conditions are applied only at infinity, i.e., $S \to 0$ at $|x| \to \infty$. Gurney and Nisbet solved the problem to obtain the minimum size of the favorable region for the population to survive. It is given by $2x_{cr} = 2(D/\alpha_o)^{\frac{1}{2}}$, which is not significantly different from the KISS scale.

In the following I will show that the critical size problem with spatially variable growth rates can be treated in a general way. To this end consider a largely hostile environment containing a single central region of viable habitat; the growth rate r(x) is given by

$$\begin{aligned} &r(x) > 0, \ -x_o \leq x \leq x_o \\ &r(x) < 0, \ x_o < |x| \end{aligned} \tag{10}$$

with the assumption that all the properties of the space are symmetrical about the origin x = 0 so that the appropriate boundary conditions for S read

$$\begin{aligned} &\text{at } x = 0, \ \partial S/\partial x = 0 \\ &\text{at } x \to \infty, \ S \to 0 \ . \end{aligned} \tag{11}$$

The basic equation for S can be written as

$$\frac{\partial S}{\partial t} = D \frac{\partial^2 S}{\partial x^2} + r(x)S \ , \quad 0 \leq x \ . \tag{12}$$

For simplicity we further assume r(x) to be a non-increasing function of x. Let $\alpha > 0$ be the (maximum) value of r at x = 0, and express r(x) in terms of a non-decreasing function V(x) as

$$r(x) \equiv \alpha - \alpha V(x) \tag{13}$$

with

$$\begin{aligned} &V(0) = 0, \ V(x) \leq 1 \text{ for } 0 \leq x \leq x_o, \ V(x) > 1 \text{ for } x_o < x \ , \\ &V(\infty) = \beta/\alpha \equiv \nu^{-1} \ (0 < \alpha < \beta) \ . \end{aligned} \tag{14}$$

Equation (12) is rewritten as

$$\frac{\partial S}{\partial t} = D \frac{\partial^2 S}{\partial x^2} + \{\alpha - \alpha V(x)\} S . \tag{15}$$

Defining

$$\begin{aligned} t &\equiv \alpha^{-1}\tau \\ x &\equiv (D/\alpha)^{\frac{1}{2}}\xi \equiv \ell\xi \\ x_o &\equiv \ell\xi_o \end{aligned} \tag{16}$$

we nondimensionalize (15) with respect to the independent variables

$$\frac{\partial S}{\partial \tau} = \frac{\partial^2 S}{\partial \xi^2} + \{1 - V(\xi;\ \xi_o,\ \nu)\} S . \tag{17}$$

Assume a product solution:

$$S = e^{\lambda\tau}\phi(\xi) . \tag{18}$$

Substituting (18) into (17) we obtain

$$\frac{d^2\phi}{d\xi^2} + \{E - V(\xi)\}\phi = 0 \tag{19}$$

with

$$E = 1 - \lambda . \tag{20}$$

Equation (19) is subject to

$$\begin{aligned} &\text{at } \xi = 0,\ d\phi/d\xi = 0 \\ &\text{at } \xi \to \infty,\ \phi \to 0 . \end{aligned} \tag{21}$$

Equation (19) is exactly analogous to the Schrödinger wave equation with potential $V(\xi)$ and energy E (Schiff, 1955). Thus an eigenfunction ϕ that satisfies the boundary conditions (21) can exist for a particular value of E. There may be a finite or infinite number of discrete values of E, but our concern is only with the minimum value of E ($= E_m$), which depends only on ℓ and ξ_o. The minimum of E corresponds to the maximum of λ, i.e., the largest eigenvalue λ_o which determines whether the population ultimately grows or decays in an exponential fashion. Therefore $\lambda = \lambda_o = 0$, i.e., $E_m = 1$ determines the critical size x_{cr} for x_o

$$E_m(x_{cr}/\ell, \nu) = 1$$

or

$$\begin{aligned} x_{cr} &= \ell \; fn(\nu) \\ &= (D/\alpha)^{1/2} \; fn(\nu) \; . \end{aligned} \tag{22}$$

I now give some examples.

a. KISS model

$$V(\xi) = \begin{cases} 0 \; , & 0 \leq \xi \leq \xi_o \\ \infty \; (\text{or } \nu = 0) \; , & \xi_o < \xi \end{cases} \tag{23}$$

This corresponds to the case of one-dimensional square well potential with a perfectly impenetrable wall at ξ_o. The minimum eigenvalue E_m is obtained by

$$E_m^{1/2} = \pi/2\xi_o$$

so that the critical size for x_o is formed by setting $E_m = 1$.

$$x_{cr} = \frac{\pi}{2} \left(\frac{D}{\alpha}\right)^{1/2}$$

or

$$L_{cr} = 2x_{cr} = \pi(D/\alpha)^{1/2} = L_o$$

which recovers the KISS scale.

b. Evans and Ludwig <u>et</u> <u>al</u>. model

$$V(\xi) = \begin{cases} 0 \; , & 0 \leq \xi \leq \xi_o \\ \nu^{-1} = \beta/\alpha \; , & \xi_o < \xi \end{cases} \tag{24}$$

This is the case of one-dimensional square well potential with finite potential step, and the minimum eigenvalue E_m is obtained by

$$\tan(E_m^{1/2}\xi_o) = (1 + \nu^{-1} - E_m)^{1/2} .$$

Setting $E_m = 1$, we get

$$x_{cr} = (D/\alpha)^{1/2} \tan^{-1}(\beta/\alpha)^{1/2}$$

or

$$L_{cr} = 2(D/\alpha)^{1/2} \tan^{-1}(\beta/\alpha)^{1/2} = L_o \{2/\pi \tan^{-1}(\beta/\alpha)^{1/2}\}$$

which agrees with Equation (8).

c. Gurney-Nisbet model

$$V(\xi) = (\xi/\xi_o)^2 . \tag{25}$$

This corresponds to the familiar potential for a harmonic oscillator, and the minimum eigenvalue E_m is given by

$$E_m = \xi_o$$

so that

$$x_{cr} = \ell = (D/\alpha)^{1/2}$$
$$L_{cr} = 2(D/\alpha)^{1/2} .$$

d. Sverdrup's model (Sverdrup 1953)

In this model r(x) is taken as $r(x) = \alpha e^{-kx} - \beta (0 < \beta < \alpha)$, so that

$$V(\xi) = 1 - e^{-k'\xi} + \nu^{-1} , \quad k' \equiv k\ell = (D/\alpha)^{1/2} \ln \nu^{-1}/x_o . \tag{26}$$

It seems that the Schrödinger equation with this potential has not been studied.

$$\frac{d^2\phi}{d\xi^2} + \{E - (1 + \nu^{-1} - e^{-k'\xi})\}\phi = 0 .$$

The solution satisfying the boundary condition at infinity is obtained by

$$\phi = AJ_{2\gamma/k'}(2/k' \, e^{-k'\xi/2})$$

with A as constant and $\gamma \equiv 1 + \nu^{-1} - E$.

The condition at $\xi = 0$ leads to

$$J'_{2\gamma/k'}(2/k') = 0 \; . \tag{27}$$

The smallest positive root of (27) along with $\gamma(E = 1) = 1 + \nu^{-1}$ determines critical size for x_o:

$$x_{cr} = (D/\alpha)^{\frac{1}{2}} fn(\nu)$$

or

$$L_{cr} = 2(D/\alpha)^{\frac{1}{2}} fn(\nu) \; .$$

Since the oceanic motion consists of a wide range of eddies, a mathematical model appropriate for oceanic diffusion cannot be described by a constant diffusivity; a more appropriate model should account for the scale dependence of diffusion. A crude approach is to introduce an appropriate scale dependence of diffusivity to the critical size formula and to solve for the size. For instance take the KISS scale formula for two dimensional space

$$L_{cr} = 4.81 \; (D/\alpha)^{\frac{1}{2}} \tag{28}$$

and use a scale dependent diffusivity

$$D = P \; L_{cr}/2 \tag{29}$$

where the linear dependence of D on the scale is after Joseph and Sendner (1958) whose oceanic diffusion model is characterized by the diffusion velocity P of order 1 cm/sec. Combining (28) with (29), we obtain

$$L_{cr} = 11.5 \; P/\alpha \; . \tag{30}$$

A more rigorous derivation relies on the following diffusion equation

$$\frac{\partial S}{\partial t} = \frac{1}{r} \frac{\partial}{\partial r} (Pr^2 \frac{\partial S}{\partial r}) + \alpha S \tag{31}$$

where Joseph and Sendner's theory is applied to the two-dimensional horizontal diffusion with $D = Pr$. Equation (31) is subject to the boundary condition that at $r = L_{cr}/2$, $S \equiv 0$ and to an appropriate initial condition. Solving (31) under these conditions, we obtain

$$L_{cr} = 7.34\ P/\alpha \tag{32}$$

(Okubo, 1978). For more general treatments with $D = cr^m$ (c: constant, $m \geq 0$), the reader is invited to consult Okubo (1978).

Estimated values of the critical size of phytoplankton for various oceanic diffusion models range from 1 km to 2 km for α = 1 div/day and 20 to 50 km for α = 1 div/10 days. These theoretical models of phytoplankton support the general observations that plankton patches appear to occur at scales of the order of 10-100 km in the open sea (Steele, 1976) and at 1-10 km in a semi-enclosed bay (Platt _et al._, 1970). It is important to note that no matter what the model may be, the essential feature of the KISS model is preserved in the formulation of the critical size, i.e., the size is determined by the balance of diffusion rate and net growth rate (McMurtrie, 1978).

When advection occurs in addition to diffusion, the critical patch size may be quite different from the KISS scale. For example, when a patch of plankton is placed in a zone of convergence, a flow pattern (advection) which tends to act against diffusion is present. It then becomes obvious to expect that the size of plankton bloom may be significantly smaller than the KISS scale.

A mathematical model for this case is given by

$$\frac{\partial S}{\partial t} = D\frac{\partial^2 S}{\partial x^2} + \frac{\partial}{\partial x}(US) + \alpha S \tag{33}$$

where U(x) is the amplitude of the current converging to the center, x = 0. Equation (33) is subject to the same boundary conditions as before. For a uniformly converging flow, U(x) = v (constant), the critical size is obtained by

$$L_{cr} = \pi(D/\alpha)^{\frac{1}{2}}\mu \tag{34}$$

where μ is the smallest positive root of the equation

$$m^{\frac{1}{2}}\tan\{(1-m)^{\frac{1}{2}}\pi\mu/2\} = (1-m)^{\frac{1}{2}} \tag{35}$$

with $m = v^2/4\alpha D$ (see Okubo, 1978 for detail). As $m \to 0$, $\mu \to 1$ and as $m \to \infty$, $\mu \to 0$. This model may explain the existence of very narrow bands of phytoplankton bloom in frontal zones (Bainbridge, 1957; Simpson and Pingree, 1978).

On the other hand, the critical size becomes larger than the KISS scale when a patch is placed in a diverging flow; as a matter of fact the critical size may become infinite, and a patch can never exist (Okubo, 1978). For a uniformly diverging flow, μ in (34) is the smallest positive root of the equation

$$m^{\frac{1}{2}} \tan\{(1 - m)^{\frac{1}{2}}\pi\mu/2\} = -(1 - m)^{\frac{1}{2}} . \qquad (36)$$

Thus, for $m \geq 1$, i.e., $v \geq 2(D\alpha)^{\frac{1}{2}}$, μ becomes infinite and so is L_{cr}. For example, if $D = 10^5$ cm^2/sec, $\alpha = 1$ div/day, and $v = 2$ cm/sec, then $L_{cr} \to \infty$, whereas the KISS scale is about 1 km. Even a relatively weak convergence of a few centimeters per second in the sea is capable of destroying phytoplankton blooms.

In passing, density-dependent diffusion may create another dramatic change in the critical size problem. Mathematical models for density-dependent dispersal of population have been developed by Gurney and Nisbet (1975), Gurtin and MacCamy (1977), Shigesada and Teramoto (1978), Shigesada et al. (1979), and Shigesada (1980) among others. In particular Gurney and Nisbet (1975) investigated the effect of density-dependent diffusion on the spatial distribution of a population undergoing spatially varying growth and death. For purely density-dependent diffusion their model equation reads

$$\frac{\partial S}{\partial t} = \frac{\partial}{\partial x}(\gamma S \frac{\partial S}{\partial x}) + \alpha_o\{1 - (x/x_o)^2\}S \qquad (37)$$

where diffusivity is assumed to depend linearly on the population density S, $D = \gamma S$. Gurney and Nisbet (1975) then show that regardless of the size of the favorable region there always exists a non-trivial, non-negative, stable steady-state solution to (29), i.e., there is no minimum size of habitat for population survival.

The density-dependent diffusion plays an important role in animal dispersal (Morisita, 1971). Shigesada (1980) compared her mathematical model of density-dependent dispersal favorably with experimental data on ant lion dispersion by Morisita. Zooplankton might as well exhibit density-dependent dispersal. Even though phytoplankton are considered to be passive in their movement, many dinoflagellates are able to migrate vertically by phototactic responses (Forward, 1976; Blasco, 1978; Heaney and Eppley, 1981). Whether or not any density dependence operates in the dinoflagellate movement is unclear. However, an analysis of the self-shading effect on algal vertical distribution by Shigesada and Okubo (1981) suggests that the self shading may give rise to some density dependence in the advective flow of dinoflagellates, so that the dispersal of dinoflagellates can be modelled by a Burgers' type equation (Levin and Okubo, 1983).

The critical size problem has also been extended to more general population growth processes such as logistic growth (Skellam, 1951; Levandowsky and White, 1977; Ludwig et al., 1979) and asocial population growth (Bradford and Philip, 1970a, b).

For a population undergoing logistic growth and diffusion the KISS scale is still a minimum size of habitat required for survival. To accommodate more individuals, the larger habitat size is required. To maintain the level of population at the carrying capacity of the resources, the habitat size must be infinitely large. An asocial population is characterized by negative growth at small population densities, positive growth at intermediate densities, and negative growth at large densities. The critical habitat size is obtained as $c(D/\alpha_1)^{\frac{1}{2}}$, where α_1 is a growth rate at an intermediate population density and c is a numerical constant of order unity. Again a formula similar to the KISS scale is applied to the critical size for asocial populations.

4. Vertical-horizontal coupling in phytoplankton patchiness: a new model

Though fairly robust, the KISS model is by no means immune to flaws. A particular weakness of the model is found in its arbitrary boundary conditions, i.e., two distinct water masses, one favorable and the other unfavorable for organisms. Even the improved models which can relax this boundary condition are still unable to interpret the occurrence of plankton patches in the sea where the oceanographical conditions are seemingly uniform in the horizontal direction.

The previously mentioned models, including the KISS, are concerned primarily with the horizontal diffusion of plankton patches by horizontal turbulence. This is the main reason why the model needs the arbitrary boundary conditions. In the sea the vertical and horizontal processes are strongly coupled in such a manner that the combined action of vertical shear in horizontal currents and transverse vertical mixing can produce an effective dispersion in the horizontal direction (Bowden, 1965). This "shear effect" claims that the vertical and horizontal processes of dispersion cannot be treated independently, but rather the advection-diffusion in the vertical direction is responsible for the dispersion of plankton in the horizontal direction.

Analyzing chlorophyll-a data obtained in central Long Island Sound, Wilson and Okubo (1980) demonstrated that the horizontal distribution of phytoplankton was generated by the interaction of the vertical structure with a vertical shear in horizontal currents at semi-diurnal and lower frequencies and possibly by short period internal waves. In short, *horizontal patchiness is a manifestation of vertical patchiness*.

Steele (1976, 1978), Evans *et al*. (1977), Evans (1978), and Kullenberg (1978) attempted to couple the vertical and horizontal processes in interpretation of plankton patchiness. Yet a unified mathematical model has not been developed to evaluate the critical size of phytoplankton in the sea. In this article I will outline a new model of vertical-horizontal coupling; a full exploration of the model will be presented elsewhere.

The concept of a *critical depth* for phytoplankton blooming is relatively old (Gran and Braarud, 1935). Sverdrup (1953) first presented a mathematical model dealing with conditions for the vernal blooming of phytoplankton, coincidentally in the same year as Kierstead and Slobodkin's work. Sverdrup's approach is based on a comparison between the depth of upper mixed layer and the "critical depth" at which the total production of plant beneath a unit surface is equal to the total respiration.

In the upper layer of the ocean the photosynthesis exceeds the destruction by respiration, while in the lower layer the loss exceeds the production. The two layers are separated by the depth of compensation at which the rate of production exactly balances the rate of loss. In this sense the modified KISS model with relaxed boundary conditions can be applied to determine the critical depth of compensation for phytoplankton blooming. (Note that this critical depth of compensation is not the same as Sverdrup's critical depth.) To this end let a and b be respectively the net growth rate in the upper layer and net loss rate in the lower layer, and K_z be vertical diffusivity. Then the same argument as before leads to the following result for the critical depth of compensation

$$H_{cr} = (K_z/a)^{1/2} \tan^{-1}(b/a)^{1/2} . \qquad (38)$$

If the observed depth of compensation z_c is shallower than H_{cr}, phytoplankton population tends to decay. On the other hand, if z_c is deeper than H_{cr}, phytoplankton blooming may occur.

To make only an order of magnitude estimate for H_{cr}, we take $K_z = 10$ cm^2/sec, $a = 10^{-5} \sim 10^{-6}/sec$, $b/a \sim 1$, as typical for the open ocean. These values yield

$$H_{cr} = \begin{cases} 10 \text{ m for 1 cell division per day} \\ 30 \text{ m for 1 cell division per 10 days} \end{cases} \qquad (39)$$

As a result of the shear effect this critical depth can manifest itself into the horizontal patch scale. A simple shear effect model will explain the process (Okubo, 1980). Consider a small patch of phytoplankton starting to diffuse in a layer where exists a uniform vertical shear given by the velocity profile $u = \Omega z$. In time t the patch will diffuse vertically a distance of the order of $(K_z t)^{1/2}$. Accordingly the effective shear, i.e., vertical velocity difference, acting on the patch amounts to $\Omega(K_z t)^{1/2}$, and the patch will disperse horizontally, in time t, a distance equal to $\Omega(K_z t)^{1/2} t$. In other words the horizontal scale of the patch L increases with time as $L \sim \Omega K_z^{1/2} t^{3/2}$, while the vertical scale grows with time as $H \sim K_z^{\frac{1}{2}} t^{\frac{1}{2}}$, so that we find the following relationship between L and H by eliminating t from both expressions

$$L = \Omega K_z^{-1} H^3 \tag{40}$$

so far as the order of magnitude is concerned.

The critical horizontal scale L_{cr} is obtained when $H = H_{cr}$, i.e.,

$$L_{cr} = \Omega K^{-1} H_{cr}^3 \ . \tag{41}$$

Using (38) for H_{cr}, we can rewrite (41) as

$$L_{cr}/H_{cr} = \Omega/a\ (\tan^{-1}(b/a)^{\frac{1}{2}}) \sim \Omega/a \tag{42}$$

since usually $\tan^{-1}(b/a)^{\frac{1}{2}}$ is order unity.

The above expression indicates that the horizontal critical scale is proportional to the (vertical) critical depth of compensation; the proportionality constant is simply the ratio of the vertical current shear (physical parameter) to the net growth rate (biological parameter) in the upper layer. Typically $\Omega = 10^{-3}$/sec. For $a = 10^{-5}$/sec (1 div/day), $L_{cr}/H_{cr} \sim 100$, and for $a = 10^{-6}$/sec (1 div/10 days), $L_{cr}/H_{cr} \sim 1000$. Thus the magnification factor ranges from 100 to 1000. Using (39) for H_{cr} we estimate

$$\begin{aligned} L_{cr} &\sim 1 \text{ km for } a = 1 \text{ div/day} \\ L_{cr} &\sim 30 \text{ km for } a = 1 \text{ div/10 days} \ . \end{aligned} \tag{43}$$

Hence this coupling model predicts minimum scales of horizontal patchiness of roughly the right order of magnitude.

A more rigorous mathematical analysis of the vertical-horizontal coupling in patchiness depends on the following shear diffusion equation with population growth:

$$\frac{\partial S}{\partial t} = -u(z)\frac{\partial S}{\partial x} + K_x\frac{\partial^2 S}{\partial x^2} + K_z\frac{\partial^2 S}{\partial z^2} + r(z)S \ , \quad t > 0,\ -\infty < x < \infty,\ 0 \le z \tag{44}$$

where u(z) is the horizontal velocity varying with depth, K_x and K_z are horizontal and vertical diffusivities, respectively, and r(z) is the depth variable net growth rate. Equation (44) is subject to

$$\begin{aligned} &\text{at } t = 0, \quad S = \delta(x)\delta(z) \\ &\text{at } z = 0, \quad \partial S/\partial z = 0 \\ &\text{at } |x| \to \infty, \text{ or } z \to \infty, \quad S \to 0 . \end{aligned} \tag{45}$$

Since our concern is primarily with the relationship between vertical and horizontal scales associated with a patch of plankton, we transform (44) into a set of moment equations. Define the jth moments ($j \geq 0$) by

$$m_j(z,t) \equiv \int_{-\infty}^{\infty} x^j S(x,z,t)dx . \tag{46}$$

Multiplying (44) by x^j and integrating over x, we obtain

$$\frac{\partial m_j}{\partial t} = ju(z)m_{j-1} + j(j-1)K_x m_{j-2} + K_z \frac{\partial^2 m_j}{\partial z^2} + r(z)m_j \tag{47}$$

subject to

$$\begin{aligned} &\text{at } t = 0, \; m_j = \delta(z)\delta_{jo} \quad (\delta_{jo}\text{: Kronecker delta}) \\ &\text{at } z = 0, \; \partial m_j/\partial z = 0 \\ &\text{at } z \to \infty, \; m_j \to 0 . \end{aligned} \tag{48}$$

The set of equations (47) subject to (48) can be solved successively starting with the zero-order moment equation

$$\frac{\partial m_o}{\partial t} = K_z \frac{\partial^2 m_o}{\partial z^2} + r(z)m_o \tag{49}$$

$$\begin{aligned} &\text{at } t = 0, \quad m_o = \delta(z) \\ &\text{at } z = 0, \quad \partial m_o/\partial z = 0 \\ &\text{at } z \to \infty, \; m_o \to 0 . \end{aligned} \tag{50}$$

The vertical variance σ_z^2 associated with m_o is obtained from

$$\sigma_z^2(t) = \int_o^{\infty} z^2 m_o(z,t)dz \tag{51}$$

and the horizontal variance σ_x^2 from

$$\sigma_x^2(t) = \int_0^\infty m_2(z,t)dz - \{\int_0^\infty m_1(z,t)dz\}^2 . \tag{52}$$

The square root of the variance gives us the scale of the patch in the corresponding direction.

5. Further extensions

For proper modelling of an ecosystem it is often necessary to deal with interacting populations of two or more species. Thus Dubois (1975) attempted to explain the horizontal structure of zooplankton-phytoplankton populations in environments containing a patch of physiologically suitable water surrounded by hostile conditions, although he did not pay attention to the critical size problem. Robert Armstrong (personal communication) has made an attempt to remove the arbitrary boundary conditions of the KISS model by coupling phytoplankton and nutrient dynamics. Armstrong's model allows both patch expansion by diffusion and limitation of growth by nutrient exhaustion.

Levin and Segel (1976) and Okubo (1974, 1978) suggest that plankton patchiness may arise from diffusive instability, in which initially stable uniform distributions of predator and prey populations, e.g., zooplankton and phytoplankton, are destabilized by the differential dispersal rates of the species. The basic idea derives from the celebrated work by Turing (1952) on morphogenesis and was first advanced in an ecological context by Segel and Jackson (1972) and Levin (1974).

Levin and Segel's model (1976) is based on the system

$$\frac{\partial S}{\partial t} = S(a_1 + c_1 S - b_1 Z) + D_1 \frac{\partial^2 S}{\partial x^2}$$

$$\frac{\partial Z}{\partial t} = Z(-c_2 Z + b_2 S) + D_2 \frac{\partial^2 Z}{\partial x^2} \tag{53}$$

where S and Z are respectively phytoplankton and zooplankton population densities, D_1 and D_2 are respective diffusivities, and a_1, b_1, b_2, c_1, c_2 are all positive constants. This predator-prey system admits a spatially uniform stable equilibrium (S^*,Z^*) in the absence of diffusion:

$$S^* = a_1c_2/(b_1b_2 - c_1c_2)$$
$$Z^* = a_1b_2/(b_1b_2 - c_1c_2)$$

provided $1 > \frac{c_1}{b_2} < \frac{b_1}{c_2}$.

We now impose to the stable state small perturbations of the initial form

$$S' = p \cos kx, \quad Z' = q \cos kx \ .$$

The diffusive instability of Turing's sense arises when

$$D_2/D_1 > \{(b_1/c_2)^{\frac{1}{2}} - (b_1/c_2 - c_1/b_2)^{\frac{1}{2}}\}^{-2} \equiv R_{cr} \ . \tag{54}$$

For R slightly greater than R_{cr}, a perturbation will destabilize the system and start to grow if its initial wavelength is approximately equal to L_{cr}:

$$L_{cr} = 2\pi(D_1/a_1)^{\frac{1}{2}}\{(b_1/c_2)^{\frac{1}{2}} - (b_1/c_2 - c_1/b_2)^{\frac{1}{2}} - 1\}^{-1} \ . \tag{55}$$

It is important to remark that the above expression for the critical wavelength is very much analogous to the KISS scale, although in Levin and Segel's model no such assumption is made that a viable water is surrounded by totally hostile environments. As a matter of fact the original state is spatially uniform and the initial perturbation is assumed to be periodic in space.

Segel and Levin (1976) also developed a nonlinear analysis for the system (52) and have shown that the destabilized equilibrium is replaced by a spatially non-uniform steady state. Thus a pattern of patchiness can arise from a uniform pattern of phytoplankton-zooplankton system due to the effects of diffusivity-driven instability.

Mimura and Murray (1979) and Mimura et al. (1979) considered a generalized predator-prey model with dispersal to study in detail the development of patchiness.

LITERATURE CITED

Bainbridge, R. (1953). Studies on the interrelationship of zooplankton and phytoplankton. J. mar. biol. Assoc. U.K. 32:385-447.

Bainbridge, R. (1957). The size, shape and density of marine phytoplankton concentrations. Biol. Rev. Camb. Philos. Soc. 32:91-115.

Blasco, D. (1978). Observations on the diel migration of marine dinoflagellates off the Baja California coast. Mar. Biol. 46:41-47.

Bowden, K.F. (1965). Horizontal mixing in the sea due to a shearing current. J. Fluid Mechanics 21:83-95.

Bowman, M.J. and Esaias, W.E. (eds.) (1978). Oceanic Fronts in Coastal Processes. Springer-Verlag, Berlin-Heidelberg-New York, 114 pp.

Bradford, E. and Philip, J.P. (1970a). Stability of steady distributions of asocial populations in one dimension. J. theor. Biol. 29:13-26.

Bradford, E. and Philip, J.P. (1970b). Note on asocial populations dispersing in two dimensions. J. theor. Biol. 29:27-33.

Cassie, R.M. (1959). An experimental study of factors inducing aggregation in marine plankton. N.Z. J. Sci. Technol. 2:339-365.

Cassie, R.M. (1963). Microdistribution of plankton. Oceanogr. Mar. Biol. Annu. Rev. 1:223-252.

Clutter, R.I. (1969). The microdistribution and social behavior of some pelagic mysid shrimp. J. Exp. Mar. Biol. Ecol. 3:125-155.

Dubois, D.M. (1975). A model of patchiness for prey-predator plankton populations. Ecol. Model. 1:67-80.

Evans, G. T. (1978). Biological effects of vertical-horizontal interaction. In: Spatial Pattern in Plankton Communities, 157-179, Steele, J.H. (ed.). Plenum, New York.

Evans, G.T., Steele, J.H. and Kullenberg, G.E.B. (1977). A preliminary model of shear diffusion and plankton populations. Scottish Fisheries Research Rep. No. 9, Dept. Agriculture and Fisheries for Scotland, Aberdeen.

Floodgate, G.D., Fogg, G.E., Jones, D.A., Lochte, K. and Turley, C.M. (1981). Microbiological and zooplankton activity at a front in Liverpool Bay. Nature 290:133-136.

Forward, R.B. (1976). Light and diurnal vertical migration: photobehavior and photophysiology of plankton. Photochemical and Photobiological Reviews, vol. 1:157-209, Smith, K.C. (ed.). Plenum, New York.

Gran, H.H. and Braarud, T. (1935). A quantitative study of the phytoplankton in the Bay of Fundy and the Gulf of Maine. J. Biol. Bd. Can. 1:210-227.

Gurney, W.S.C. and Nisbet, R.M. (1975). The regulation of inhomogeneous populations. J. theor. Biol. 52:441-457.

Gurtin, M.E. and MacCamy, R.C. (1977). On the diffusion of biological populations. Math. Biosci. 33:35-49.

Hamner, W.J. and Carleton, J.H. (1979). Copepod swarms: attributes and role in coral reef ecosystems. Limnol. Oceanogr. 24:1-14.

Harris, G.P. (1980). Temporal and spatial scales in phytoplankton ecology. Mechanisms, methods, models, and management. Can. J. Fish. Aquat. Sci. 37:877-900.

Heaney, S.I. and Eppley, R.W. (1981). Light, temperature and nitrogen as interacting factors affecting diel vertical migration of dinoflagellates in culture. J. Plankton Res. 3:331-344.

Joseph, J. and Sendner, H. (1958). Über die horizontale Diffusion im Meere. Dtsch. Hydrogr. Z. 11:49-77.

Kierstead, H. and Slobodkin, L.B. (1953). The size of water masses containing plankton bloom. J. Mar. Res. 12:141-147.

Kullenberg, G.E.B. (1978). Vertical processes and the vertical-horizontal coupling. In: Spatial Pattern in Plankton Communities, 43-71, Steele, J.H. (ed.). Plenum, New York.

Langmuir, I. (1938). Surface motion of water induced by wind. Science 87:119-123.

Levandowsky, M. and White, B.S. (1977). Randomness: time scales, and the evolution of biological communities. In: Evolutionary Biology, vol. 10:69-161, Hecht, M.K. et al. (eds.). Plenum, New York.

Levin, S.A. (1974). Dispersion and population interactions. Am. Nat. 108:207-228.

Levin, S.A. and Okubo, A. (1983). Mathematical models for phototactic aggregation of dispersing organisms (in preparation).

Levin, S.A. and Segel, L.A. (1976). Hypothesis for the origin of planktonic patchiness. Nature 259:659.

Ludwig, D., Aronson, D.G. and Weinberger, H.F. (1979). Spatial patterning of the spruce budworm. J. Math. Biol. 8:217-258.

McMurtrie, R. (1978). Persistence and stability of single-species and prey-predator systems in spatially heterogeneous environments. Math. Biosci. 39:11-51.

Mimura, M. and Murray, J.D. (1979). On a planktonic prey-predator model which exhibits patchiness. J. theor. Biol. 75:249-262.

Mimura, M., Nishiura, Y. and Yamaguti, M. (1979). Some diffusive prey and predator systems and their bifurcation problem. Ann. N.Y. Acad. Sci. 316:490-510.

Morisita, M. (1971). Measuring of habitat value by the "environmental density" method. In: Statistical Ecology, vol. 1:379-401, Patil, G.P. et al. (eds.). Penn. State Univ. Press, University Park, Pennsylvania.

Okubo, A. (1974). Diffusion-induced instability in model ecosystems. Chesapeake Bay Institute, The Johns Hopkins University, Tech. Rep. No. 86. Baltimore, Maryland.

Okubo, A. (1978). Horizontal dispersion and critical scales for phytoplankton patches. In: Spatial Pattern in Plankton Communities, 21-42, Steele, J.H. (ed.). Plenum, New York.

Okubo, A. (1980). Diffusion and Ecological Problems: Mathematical Models. Springer-Verlag, Berlin-Heidelberg-New York, 254 pp.

Platt, T. and Denman, K.L. (1975). A general equation for the mesoscale distribution of phytoplankton in the sea. Mem. Soc. R. Sci. Liege, 6th Ser., vol. 7:31-42.

Platt, T., Dickie, L.M. and Trites, R.W. (1970). Spatial heterogeneity of phytoplankton in a near-shore environment. J. Fish. Res. Bd. Can. 27:1453-1473.

Schiff, L.I. (1955). Quantum Mechanics. McGraw-Hill, New York, 544 pp.

Segel, L.A. and Jackson, J.L. (1972). Dissipative structure: an explanation and an ecological example. J. theor. Biol. 37:545-559.

Segel, L.A. and Levin, S.A. (1976). Application of nonlinear stability theory to the study of the effects of diffusion on predator-prey interactions. In: Topics in Statistical Mechanics and Biophysics: A memorial to Julius L. Jackson, AIP Conf. Proc. No. 27, 123-152, Piccirelli, R.A. (ed.). Am. Inst. Physics, New York.

Shigesada, N. (1980). Spatial distribution of dispersing animals. J. Math. Biol. 9:85-96.

Shigesada, N., Kawasaki, K. and Teramoto, E. (1979). Spatial segregation of interacting species. J. theor. Biol. 79:83-99.

Shigesada, N. and Okubo, A. (1981). Analysis of the self-shading effect on algal vertical distribution in natural waters. J. Math. Biol. 12:311-326.

Shigesada, N. and Teramoto, E. (1978). A consideration on the theory of environmental density. Jap. J. Ecol. 28:1-8.

Simpson, J.H. and Pingree, R.D. (1978). Shallow sea fronts produced by tidal stirring. In: Oceanic Fronts in Coastal Processes, 29-42, Bowman, M.J. and Esaias, W.E. (eds.). Springer-Verlag, Berlin-Heidelberg-New York.

Skellam, J.G. (1951). Random dispersal in theoretical populations. Biometrika 38:196-218.

Steele, J.H. (1976). Patchiness. In: The Ecology of the Sea, 98-115, Cushing, D.H. and Walsh, J.J. (eds.). W.B. Saunders Co., Philadelphia.

Steele, J.H. (1977). Plankton patches in the northern North Sea. In: Fisheries Mathematics, 1-19, Steele, J.H. (ed.). Academic Press, New York-London.

Steele, J.H. (1978). Some comments on plankton patches. In: Spatial Pattern in Plankton Communities, 1-20, Steele, J.H. (ed.). Plenum, New York.

Stommel, H. (1949). Trajectories of small bodies sinking slowly through convection cells. J. Mar. Res. 8:24-29.

Sverdrup, H.U. (1953). On conditions for the vernal blooming of phytoplankton. J. Cons. Cons. Int. Explor. Mer 18:287-295.

Tonolli, V. (1958). Ricerche sulla microstructura di distribuzione della zooplankton nel Lago Maggiore. Mem. Ist. Ital. Idrobiol. 10:125-152.

Tonolli, V. and Tonolli, L. (1960). Irregularities of distribution of plankton communities: considerations and models. In: Perspectives in Marine Biology, 137-143, Buzzati-Traverso, A.A. (ed.). Univ. California Press, Berkeley, California.

Turing, A.M. (1952). The chemical basis of morphogenesis. Philos. Trans. R. Soc. Lond. B237:37-72.

Wilson, R.E. and Okubo, A. (1980). Effects of vertical-horizontal coupling on the horizontal distribution of chlorophyll a. J. Plankton Res. 2:33-42.

Wroblewski, J.S., O'Brien, J.J. and Platt, T. (1975). On the physical and biological scales of phytoplankton patchiness in the ocean. Mem. Soc. R. Sci. Liege, 6th Ser., vol. 7:43-57.

Spatial Distribution of Rapidly Dispersing Animals in Heterogeneous Environments

Nanako Shigesada

Department of Biophysics, Kyoto University, Kyoto, Japan

1. Introduction

Ecological models incorporating spatial heterogeneity of habitats are of profound importance in understanding the movements of organisms and their effects on the stability of spatial distributions of populations under natural circumstances. Equations describing the time development of the spatial distribution of a population in a heterogeneous environment fundamentally involve two terms, dispersal and growth, which are both functions of space. There have been several distinct approaches to the analysis of such models depending on the system under investigation and the type of method being applied (See reviews by Okubo (1980) and Levin (1981)). Among them, models for a single species in one dimensional space have been extensively studied for various types of ecological systems. Okubo (1980) analyzed effects of various kinds of spatially varying dispersal on the spatial structure of populations. Gurney and Nisbet (1974) and Namba (1980) included a spatially varying growth term in their model. In population genetics, Fleming (1975) , Nagylaki (1975) and May et al. (1975) studied the effect of environmental heterogeneity on the viability of individuals of a single species and presented the condition for the existence of clines in a one-dimensional space. As for two-species systems, the effect of dispersal with directed movements was taken into consideration by Comins and Blatt (1974) and Shigesada et al. (1979), and the effect of spatially varying growth was considered by Pacala and Roughgarden (1982) and Kawasaki and Teramoto (1979). However, these models incorporated the effect of heterogeneity either in dispersal or growth, but not in both processes.

Recently, Fife and Peletier (1981) studied a single species model in population genetics, in which effects of environmental heterogeneity were incorporated in both dispersal and growth processes. Shigesada and Roughgarden (1982) also considered these effects in a two-competing species model. The latter authors analyzed the time development of spatial distribution and its stability for the special case in which the dispersal process occurs very rapidly compared with the growth process of the species. In this paper we examine Shigesada and Roughgarden's model on a more mathematical basis by using a multiple scale method, and present a general formula for the time development of spatial distributions of populations. From the assumption of rapid dispersal, the original partial differential equation is reduced to an ordinary differential equation so that the analysis becomes much easier. We apply this method to a few systems with a single and with two competing species, and compare the results from this method with those from computer calculations.

2. Application of the multiple scale method to a single species population dynamics with dispersal

Consider a single-species population in a bounded heterogeneous habitat. Individuals of the species undergo dispersal both by random motion and by directed movement toward favorable places in the habitat. The population density also changes due to birth and death, which are usually dependent on both the position and population density. If $n(t,x)$ denotes the population density at time t and position x, the dynamical equation for the spatial distribution of the population in a one-dimensional bounded region $\sigma \equiv [0,L]$ is given by

$$\frac{\partial}{\partial t}n(t,x) = -\frac{\partial}{\partial x}J(x,n) + \varepsilon G(x,n), \qquad x \in \sigma, \tag{1}$$

where

$$J(x,n) = -\alpha\frac{\partial}{\partial x}n - \frac{d}{dx}U(x)\cdot n \;. \tag{2}$$

The term $J(x,n)$ is the flux of population due to the dispersal process. Namely, $-\alpha\frac{\partial}{\partial x}n$ (α is a positive constant) represents the flow associated with random movements of individuals, and the term $-\frac{d}{dx}U(x)\cdot n$ (where we assume that $U(x)$ has a continuous derivative) represents the flow due to directed movements of individuals toward favorable environments. Here we designate $U(x)$ as the "environmental potential", which induces the advection velocity, $-\frac{d}{dx}U(x)$, toward favorable regions. The second term of (1), $\varepsilon G(x,n)$, represents the net growth rate due to birth and death. For convenience in later discussions, we express the net growth rate by the product of the factor ε and G so that the dispersal term, $-\frac{\partial}{\partial x}J$, and G are of the same order of magnitude. We assume here that $G(x,n)$ is a bounded piecewise continuous function of x and has a continuous derivative with respect to n. We also incorporate an intraspecific competition in the function G such that G becomes negative as n exceeds a certain positive number k.

We assume that some animals are located initially in the closed region σ, where there is no population flow through the boundaries, so that the model is subject to the initial and boundary conditions:

$$\begin{aligned} n(0,x) &= s(x) \geq 0 \quad (\not\equiv 0)\ , \\ J(x,n) &= 0 \quad \text{at } x = 0 \text{ and } L, \end{aligned} \tag{3}$$

where $s(x)$ is continuous in σ.

In some ecological situations, dispersal and growth processes take place in different time scales. It is frequently seen in nature that the change in population density as a result of the dispersal process occurs more rapidly than the change due to the growth process. For example, some animals undergo daily migration, seeking resources and settling places, while they reproduce once or twice a year. Here we focus our attention on the cases where ε is small enough so that a rapid change in the spatial distribution of population due to the dispersal process occurs initially, followed by a slow long-term change in the population size due to the growth process. We can choose the scales of independent variables so as to set the order of magnitude of the dispersal rate to be $O(1)$. In this case we can analyze our model by the multiple scale (two-timing) method (Nayfeh,1973), and obtain a truncated expansion valid for all t up to $O(1/\varepsilon)$.

Let us introduce the following two different time scales, T_0, T_1 defined as

$$T_0 = t, \qquad T_1 = \varepsilon t \ .$$

We consider the solution of (1) as a function of these two time scales, $n(t,x) = n(T_0,T_1,x;\varepsilon)$ and we attempt to find the solution in the following form, which is valid for times as large as $O(1/\varepsilon)$:

$$n(t,x) = n(T_0,T_1,x;\varepsilon) = n^0(T_0,T_1,x) + \varepsilon n^1(T_0,T_1,x) + \cdots, \tag{4}$$

where the remainder is $O(\varepsilon^2)$ and n^1 is bounded for all T_0. We now carry out a perturbation procedure by noting that the time derivative is transformed according to

$$\frac{\partial}{\partial t} = \frac{\partial}{\partial T_0} + \varepsilon\frac{\partial}{\partial T_1} \quad . \tag{5}$$

Upon inserting (4) and (5) into (1) and equating coefficients of like powers of ε, we obtain

$$\frac{\partial}{\partial T_0}n^0 = -\frac{\partial}{\partial x}J(x,n^0), \tag{6}$$

$$n^0(0,0,x) = s(x),$$

$$J(x,n^0) = 0 \quad \text{at } x=0 \text{ and } L;$$

$$\frac{\partial}{\partial T_0}n^1 + \frac{\partial}{\partial x}J(x,n^1) = -\frac{\partial}{\partial T_1}n^0 + G(x,n^0), \tag{7}$$

$$n^1(0,0,x) = 0,$$

$$J(x,n^1) = 0 \quad \text{at } x=0 \text{ and } L.$$

The general solution of (6) is written in the form

$$n^0(T_0,T_1,x) = N^0(T_1)f(T_0,x) \ , \tag{8}$$

where $f(T_0,x)$ is the solution of the following equation,

$$\frac{\partial}{\partial T_0}f = -\frac{\partial}{\partial T}J(x,f), \tag{9}$$

$$f(0,x) = \frac{s(x)}{\int_\sigma s(x)dx} \ ,$$

$$J(x,f) = 0 \quad \text{at } x=0 \text{ and } L.$$

By integrating (9) over σ, we find $\int_\sigma f(T_0,x)dx = 1$. Thus, $f(T_0,x)$ may be regarded as the probability density of the spatial distribution of population in σ, since $f(T_0,x) \geq 0$. Equation (9) is a so-called regular Sturm-Liouville problem, and its solution is written as

$$f(T_0,x) = \sum_{i=1}^{\infty} c_i \exp\{-\lambda_i T_0\} \varphi_i(x) , \tag{10}$$

where φ_i are the eigenfunctions of (9) and λ_i are the eigenvalues, which are nonnegative and can be arranged in the following increasing sequence (Berg and McGregor,1966):

$$0 = \lambda_1 < \lambda_2 < \lambda_3 < \cdots$$

Thus $f(T_0,x)$ is bounded for all x and T_0 and asymptotically approaches an equilibrium $f^*(x)$, which is the solution of $J(x,f^*) = 0$:

$$f(T_0,x) \xrightarrow{T_0 \to \infty} f^*(x) = \frac{\exp\{-\frac{U(x)}{\alpha}\}}{\int_\sigma \exp\{-\frac{U(x)}{\alpha}\}\,dx} . \tag{11}$$

The function $N^0(T_1)$ remains arbitrary, but we can determine it at the next stage of the perturbation.

To this end, let us integrate (7) over σ and put $N^1(T_0,T_1) = \int_\sigma n^1(T_0,T_1,x)dx$. We then have

$$\frac{\partial}{\partial T_0} N^1(T_0,T_1) = -\frac{\partial}{\partial T_1} N^0(T_1) + \int_\sigma G(x, N^0(T_1) f(T_0,x))dx . \tag{12}$$

Since $n^1(T_0,T_1,x)$ is required to be bounded for all T_0, $N^1(T_0,T_1)$ should also be bounded for all T_0. However, the solution of (12), $N^1(T_0,T_1)$ will become unbounded, because of the occurrence of secular terms, unless we require the right hand side of (12) to tend to zero as $T_0 \to \infty$. So let us try to set the unknown function $N^0(T_1)$ equal to the solution of the following equation,

$$\begin{aligned} \frac{\partial}{\partial T_1} N^0(T_1) &= \int_\sigma G(x, N^0(T_1) f^*(x))\, dx , \\ N^0(0) &= \int_\sigma s(x)dx, \end{aligned} \tag{13}$$

which is obtained if we substitute $f^*(x)$ into $f(T_0,x)$ in (12) and set the right hand side of (12) equal to zero. The solution of (13), $N^0(T_1)$, is bounded for all T_1 because we imposed the condition that $G(x,n)$ becomes negative for large n.

Now let us examine whether the solution of (12), $N^1(T_0,T_1)$, is actually bounded for all T_0. By substituting (13) into (12), and integrating over T_0, we obtain the equation,

$$N^1(T_0,T_1) = \int_0^{T_0}\int_\sigma \{-G(x, N^0 f^*(x)) + G(x, N^0 f(T_0',x))\}dx dT_0' \quad . \tag{14}$$

The right hand side is verified to be bounded for all T_0 if we apply the mean value theorem and take Eq.(10) into consideration. Thus the assumption of N^0 as the solution of (13) proves to be appropriate. However, it should be noted here that $N^1(\infty,T_1)$ becomes divergent as the length of the habitat L becomes infinite, since the eigenvalue $\lambda_2 \to 0$ as $L \to \infty$.

To summarize the above analysis, we can conclude that:

The solution of (1), which is valid for times up to $O(1/\varepsilon)$, is given by

$$n(t,x) = N^0(\varepsilon t)f(t,x) + O(\varepsilon) \quad , \tag{15}$$

where $f(t,x)$ is the probability density of the spatial distribution given by (10), and $N^0(\varepsilon t)$ is the solution of the ordinary differential equation (13), which usually can be solved easily, as will be shown in the following sections.

Note here that $\int_\sigma n(t,x)dx = N^0(\varepsilon t) + O(\varepsilon)$, since $\int_\sigma f(t,x)dx = 1$, so that we can regard $N^0(\varepsilon t)$ as the total size of the population in σ. Thus we can see from (15) that when we focus our attention on the behavior of the rapid dispersal process (in the time range of $O(1)$), the distributional pattern changes so as to satisfy (10), approaching an equilibrium $f^*(x)$ without change in the total population size ; on the other hand, when we turn our attention to the long-term behavior (in the time range of $O(1/\varepsilon)$), the total population size $N^0(\varepsilon t)$ changes so as to satisfy Eq.(13), while the probability density of the spatial distribution always remains in the stationary state, $f^*(x)$. In the following section, we will apply the above method to a typical system of a single species.

3. Population with general logistic growth

Here we will consider the case in which the growth term is of the general logistic type,

$$G(x,n) = \{ a(x) - b(x)n \} n \tag{16}$$

where $a(x)$ and $b(x)$ (>0) are assumed to be piecewise continuous functions of x in σ. The intrinsic growth rate $a(x)$ may have both positive and negative values in the habitat and if regions satisfying $a(x) < 0$ predominate in σ, the population may fail to grow in this habitat as a whole. Thus we are interested in the conditions under which the population can grow in the habitat and how the total population size changes to approach an equilibrium state.

Substituting (16) into (13), we have the equation for the total population size $N^0(\varepsilon t)$:

$$\frac{d}{dT_1} N^0(T_1) = (A - BN^0)N^0 \quad , \tag{17}$$

$$N^0(0) = \int_\sigma s(x)dx \quad ,$$

where

$$A = \int_\sigma a(x)f^*(x)dx, \qquad B = \int_\sigma b(x)f^{*2}(x)dx,$$

$$f^*(x) = \exp\{-\frac{U}{\alpha}\} / \int_\sigma \exp\{-\frac{U}{\alpha}\} dx \quad .$$

The solution of (17) is given by

$$N^0(\varepsilon t) = \frac{AN^0(0)}{BN^0(0) + (A - BN^0(0)) \exp\{-\varepsilon tA\}} \tag{18}$$

which becomes, as $\varepsilon t \to \infty$,

$$\begin{aligned} N^0(\varepsilon t) &\to \frac{A}{B} \qquad \text{when} \quad A \geq 0 \\ &\to 0 \qquad \text{when} \quad A < 0. \end{aligned} \tag{19}$$

Thus if the average growth rate with respect to $f^*(x)$ is positive i.e.

$$A = \int_\sigma a(x)f^*(x)dx > 0, \tag{20}$$

then the population can grow in this habitat, and otherwise, the population becomes extinct. In other words, the condition $A > 0$ represents the invasion condition for the population. The value of A depends on the function U(x), and hence even if the average of the intrinsic growth rate a(x) over σ, $\int_\sigma a(x)dx$, is negative, A may be positive when U(x) has such an appropriate form that directed movements are induced toward the region where a(x) is positive.

As we noted previously, the two-timing expansion (15) is applicable as long as we are concerned with the time scale up to $O(1/\varepsilon)$, so that it is not necessarily uniformly valid for all time. However, in the case of the logistic growth of (16), it turns out to be a fairly good approximation even for a longer time range when ε is sufficiently small. Fig.1 shows that the truncated solutions (18) agree well with numerical results derived from (1).

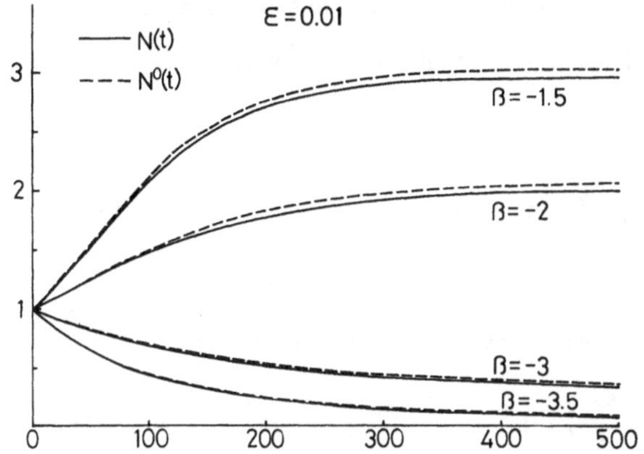

Fig. 1. Time variation of the total population sizes of single species. The solid curves are $N(t)=\int_\sigma n(t,x)dx$ derived from (1) by computer calculation. The broken curves are the truncated solutions of (18), N^0. Parameter values are $\alpha=1$, $b(x)=1$, $\varepsilon=0.01$, L=4 and advection velocity, $-dU/dx=1$, so that animals are attracted in the positive direction of the x axis. The intrinsic growth rate is $a(x)=\beta+x$. Results for $\beta=-1.5$, -2, -3 and -3.5 are plotted. The critical value of β for invasion in the multiple scale method is -3.074 at which A=0.

To further examine the validity of our expansion, we will compare the invasion condition (20) with the exact one which is analytically derived from Eq.(1) combined with (16). When the population is rare throughout the whole habitat, we obtain the invasion condition for the population satisfying Eq.(1) with (16) by analysing the following linearized equation about the solution n=0:

$$\frac{\partial}{\partial t}n = -\frac{\partial}{\partial x}J(x,n) + \varepsilon a(x)n \quad , \quad x \in \sigma \tag{21}$$

$$J(x,n) = 0 \quad \text{at } x=0 \text{ and } L.$$

If the equilibrium state n=0 is dynamically unstable, the population can grow in the habitat; namely, the population, even when rare, can invade the habitat. If n=0 is stable, on the other hand, the population finally becomes extinct in the habitat. Previously, Fleming (1975) performed a stability analysis of (21) for the special case of U(x)=0 and has presented a useful theorem on the stability condition. By applying his theorem with slight modifications to our model, we have invasion conditions for our system as follows:

i) $A > 0$,

or (22)

ii) $A < 0$ and $\varepsilon > \inf\left\{ \dfrac{\int_\sigma n_x^2 \exp\{-\frac{U}{\alpha}\}\,dx}{\int_\sigma a(x) n^2 \exp\{-\frac{U}{\alpha}\}\,dx} : \int_\sigma a(x) n^2 \exp\{-\frac{U}{\alpha}\}\,dx > 0 \right\}$.

Now if ε tends to zero, the above condition is reduced to $A > 0$, which exactly coincides with our conclusion (20).

To comfirm the above result for a specific example, let us consider the special case that has been analytically solved by Pacala and Roughgarden (1982):

$$\begin{aligned} a(x) &= s_1 > 0 \text{ (const.)} \qquad \text{for } 0 \le x < L_1 , \\ &= s_2 < 0 \text{ (const.)} \qquad \text{for } L_1 \le x \le L_1 + L_2 , \\ U(x) &= 0 . \end{aligned} \tag{23}$$

The invasion condition derived from (21) was given by the above authors as

$$\sqrt{s_1}\,\mathrm{Tan}\left(L_1\sqrt{\frac{\varepsilon s_1}{\alpha}}\right) - \sqrt{-s_2}\,\mathrm{Tanh}\left(L_2\sqrt{\frac{-\varepsilon s_2}{\alpha}}\right) > 0 . \tag{24}$$

When we are concerned with the case,

$$L_1\sqrt{\frac{\varepsilon s_1}{\alpha}} \ll 1, \qquad L_2\sqrt{\frac{-\varepsilon s_2}{\alpha}} \ll 1,$$

(24) is expanded as

$$A + \frac{\varepsilon}{3\alpha}\left(L_1^3 s_1^2 + L_2^3 s_2^2\right) + \cdots > 0 ,$$

where $A = s_1 L_1 + s_2 L_2$. As expected, we have again $A > 0$ as the invasion condition with a correction of order $O(\varepsilon)$.

4. Multi-species system

We now extend the previous study to multi-species systems. Consider an M-species system which satisfies the following equation,

$$\frac{\partial}{\partial t}n_i = -\frac{\partial}{\partial x}J_i(x,n_i) + \varepsilon G_i(x,n_1,n_2,\ldots,n_M), \quad x\in[0,L]\equiv\sigma \tag{25}$$
$$\text{for } i = 1,2,\ldots,M,$$

where we put

$$J_i(x,n_i) = -\alpha_i\frac{\partial}{\partial x}n_i - \frac{d}{dx}U_i(x)\cdot n_i . \tag{26}$$

$n_i(t,x)$ is the population density of the i-th species, and α_i, $U_i(x)$ and εG_i are respectively the diffusion constant, the environmental potential and the growth rate of the i-th species. These parameters are defined in the same way as in the single species system. The model is subject to the following initial and boundary conditions:

$$n_i(0,x) = s_i(x) \geq 0 \ (\not\equiv 0),$$

$$J_i(x,n_i) = 0 \quad \text{at } x=0 \text{ and } L.$$

The multiple scale method can be applied to the above equation in a similar way as in the case of the single species system, and the solution, which is valid for all t up to $O(1/\varepsilon)$, is given by

$$n_i(t,x) = N_i^0(t)f_i(t,x) + O(\varepsilon), \quad i=1,2,\ldots,M, \tag{27}$$

where $f_i(t,x)$ is the probability density for the spatial distribution of the i-th species and satisfies the following equation:

$$\frac{\partial}{\partial t}f_i = -\frac{\partial}{\partial x}J_i(x,f_i), \qquad i=1,2,\ldots,M \tag{28}$$
$$f_i(0,x) = \frac{s_i(x)}{\int_\sigma s_i(x)dx},$$
$$J_i(x,f_i) = 0 \quad \text{at } x = 0 \text{ and } L.$$

$N_i^0(t)$ is the total population size of the i-th species in σ and satisfies the dynamical systems,

$$\frac{d}{dt}N_i^0 = \varepsilon\int_\sigma G_i(x,\ N_1^0 f_1^*,\ N_2^0 f_2^*,\ \ldots,\ N_M^0 f_M^*)dx, \qquad i=1,2,\ldots,M \tag{29}$$

$$N_i^0(0) = \int_\sigma s_i(x)dx,$$

where

$$f_i^*(x) = \exp\{-\frac{U_i}{\alpha_i}\} \,/\, \int_\sigma \exp\{-\frac{U_i}{\alpha_i}\}\,dx \quad .$$

Now, let us consider a special case of a two competing species system, which has the following generalized Lotka-Volterra type growth functions,

$$G_i(x,n_1,n_2) = \{\,a_i(x) - \sum_j b_{ij}(x)n_j\,\}\,n_i, \qquad i=1,2, \tag{30}$$

where we assume that $a_i(x)$ and $b_{ij}(x)(>0)$ depend on position x.

As is well known, if neither of these two species undergoes dispersal (namely when $J_1=J_2=0$), they can coexist at a position x if and only if

$$\frac{b_{11}(x)}{b_{21}(x)} > \frac{a_1(x)}{a_2(x)} > \frac{b_{12}(x)}{b_{22}(x)} \tag{31}$$

and otherwise, one of the species always becomes extinct at x. Since the environment is heterogeneous, (31) may be satisfied at some places in the habitat, but not at other places. In such a case, we are interested in how the total population sizes of the two species in the habitat change with time, if both species undergo dispersal according to equation (26).

The equations for the population sizes in the multiple scale method are obtained by substituting (30) into (29):

$$\frac{d}{dt}N_i^0 = \varepsilon\{A_i - \sum_j B_{ij}N_j^0\}N_i^0\ , \qquad i=1,2, \tag{32}$$

where

$$A_i = \int_\sigma a_i(x)f_i^*(x)dx$$

$$B_{ij} = \int_\sigma b_{ij}(x)f_i^*(x)f_j^*(x)dx \quad .$$

Thus if

$$\frac{B_{11}}{B_{21}} > \frac{A_1}{A_2} > \frac{B_{12}}{B_{22}}\ , \tag{33}$$

the two species can coexist at least for times as large as $O(1/\varepsilon)$, and otherwise, one of the species tends to extinction.

Here it should be noted that Eq.(32) is analogous to the niche-partitioning theory of MacArthur and Levins (1967), if we take the real habitat space as the niche space. Namely, we can see that $f_i^*(x)$ and $\varepsilon a_i(x)$ correspond to their utilization function and resource function. Thus (32) may be interpreted as a behavioral version of the MacArthur-Levins formula for habitat partitioning by competing species.

Now we carry out a numerical calculation of (25) for a two competing species system with the following parameters:

$$J_1 = -0.5\frac{\partial}{\partial x}n_1 - 0.2n_1 , \quad J_2 = -0.5\frac{\partial}{\partial x}n_2 + n_2 , \quad L = 2,$$
$$G_1 = (1 - 0.1x - n_1 - n_2)n_1 , \quad G_2 = (1 + 0.5x - n_1 - n_2)n_2 . \tag{34}$$

We also calculate the total population sizes $N_1(t) = \int_\sigma n_1(t,x)dx$, $N_2(t) = \int_\sigma n_2(t,x)dx$, and compare them with the truncated solution by the multiple scale method given by (32), $N_1^0(t)$ and $N_2^0(t)$. With the parameters chosen in (34), $a_1(x)/a_2(x) < b_{11}/b_{21}$, b_{12}/b_{22} for any $x \in \sigma$, and hence only the 2nd species can survive everywhere in σ in the absence of dispersal. However, if both species undergo dispersal, (32) has a stable positive equilibrium state, because the condition (33) is satisfied (A_1=0.91, A_2=1.77, B_{11}=0.53, B_{12}=B_{21}=0.40, B_{22}=1.04), so that both species tend to coexist as a whole in the multiple scale method. In Fig.2, we show the time developments of the population sizes for three cases, $\varepsilon = 0.01$, 1 and 10. The solid and broken lines represent the numerical solutions of $N_1(t)$ and $N_2(t)$ derived from (25) for the case of (34) (hereafter called the exact solution), and the truncated solutions of (32), respectively. From this figure, we can see that for ε=0.01, the dynamical behavior of the truncated solution closely coincides with that of the exact solution, and the coincidence persists even for times longer than $O(1/\varepsilon)$. However, as ε increases so that the growth term becomes dominant in Eq.(25), the truncated solution deviates from the exact one with the lapse of time, and finally the first species becomes extinct in the exact solution, whereas in the truncated solution it remains positive for all time (the case of ε=10). In the above example, we chose parameters such that the two species undergo dispersal with directed movement toward different favorable places. Thus they segregate their habitats from each other, occupying those places in which they can grow at higher rates. This segregation facilitates the coexistence of species by relaxing the competition between the species.

Further computer calculations with various parameter values have shown that as long as ε is small, the time developments of population sizes according to (32) approximate exact ones fairly well for various kinds of potential functions $U_i(x)$ and growth function G_i.

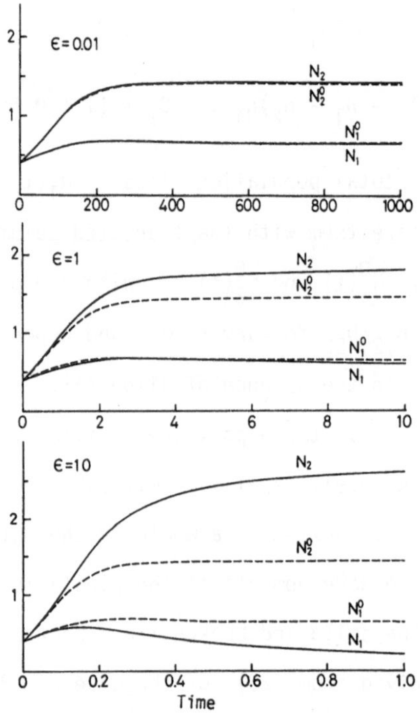

Fig. 2. Time variations of the total population sizes of the two competing species. The solid curves are $N_1(t) = \int_\sigma n_1(t,x)dx$ and $N_2(t) = \int_\sigma n_2(t,x)dx$ derived from Eq.(25) for the case of M=2 by computer calculation. The broken curves are the truncated solutions derived from (32), $N_1^0(t)$, $N_2^0(t)$. Parameter values are given by (34). Results for ε= 0.01, 1 and 10 are plotted. The agreement between the two curves becomes better as ε decreases.

Acknowledgments: The author would like to thank Dr. K. Kawasaki for his help in numerical computation and valuable comments. She also wishes to thank Drs. E. Teramoto and A. Okubo for their encouragement and interest in this work. Support from a Grant-in-Aid for Special Project Research on Biologicl Aspects of Optimal Strategy and Social Structure from the Japan Ministry of Education, Science and Culture is gratefully acknowledged.

References.

Berg, P.W. and McGregor, J.L. (1966). "Elementary Partial Differential Equations", Holden-Day, San Francisco.

Comins, H.N. and Blatt, D.W.E. (1974). Prey-predator models in spatially heterogeneous environments. J. Theor. Biol. 48, 75-83.

Fife, P.C., and Peletier, L.A. (1981), Clines induced by variable selection and migration. Proc. Roy. Soc. London B. 214, 99-123.

Fleming, W.H. (1975). A selection-migration model in population genetics, J. Math. Biol. 2, 219-233.

Gurney, W.S.C. and Nisbet, R.M. (1975). The regulation of imhomogeneous populations, J. theor. Biol. 52, 441-457.

Kawasaki, K. and Teramoto, E. (1979). Spatial pattern formation of prey-predator populations. J. Math. Biol. 8, 33-46.

Levin, S.A. (1981). In "Differential Equations and Applications in Ecology, Epidemics, and Population Problems". (ed. Busenberg, S.N. and Cooke, K.) Academic Press.

MacArthur, R.H. and Levins, R. (1967). The limiting similarity, convergence and divergence of coexisting species. Amer. Natur. 101, 377-385.

May, R.M. (1975). Gene frequency clines in the presence of selection opposed by gene flow. Amer. Natur. 109, 159-676.

Nagylaki, T. (1975). Conditions for the existence of clines, Genetics 80, 595-615.

Nayfeh, A.H. (1973). "Perturbation Methods", John Wiley & Sons, New York.

Namba, T. (1980). Density-dependent dispersal and spatial distribution of a population. J. Theor. Biol. 86, 351-363.

Okubo, A. (1980). "Diffusion and Ecological Problems: Mathematical Models". Berlin-Heidelberg-New York: Springer Verlag.

Pacala, S. and Roughgarden, J. (1982). Spatial heterogeneity and interspecific competition. Theor. Pop. Biol. 21, 92-113.

Shigesada, N., Kawasaki, K. and Teramoto, E. (1979). Spatial segregation of interacting species. J. theor. Biol., 79, 83-99.

Shigesada, N. and Roughgarden, J. (1982). The role of rapid dispersal in the population dynamics of competition. Theor. Pop. Biol., 21, 353-373.

SPATIAL DISTRIBUTION OF COMPETING SPECIES

Masayasu Mimura

Department of Mathematics, Hiroshima University, Hiroshima, 730, Japan

1. Introduction

To understand the mechanism of spatial patterning of ecological communities has been a central problem in population biology. Most often, spatial diversity in connected habitats has been assumed to be related to some heterogeneity in the environment (for instance, Fleming(1975), Shigesada and Roughgarden(1982)). On the other hand, it has also been found that stable patterns can exist in homogeneous environments (Levin(1979)).

In this paper, we will be concerned with the occurrence of stable patterns in a homogeneous environment. Suppose that n species are competing with each other and moving about by diffusion and advection. Then the population density of the i th species at time t and a position x, $u_i(t,x)$ may be described by the so-called competition-diffusion-advection equations

$$\frac{\partial}{\partial t}u_i = (r_i - \Sigma_{j=1}^{n} a_{ij}u_j)u_i + \nabla(d_i\nabla u_i - c_iu_i) \quad (i=1,2,\dots,n), \tag{1}$$

where r_i is the intrinsic growth rate of the i th species, a_{ii} and a_{ij} $(i\neq j)$ are coefficients of intra- and inter-specific competition, d_i and c_i are the diffusion coefficient and the advection velocity of the i th species, respectively. The term $d_i\nabla u_i - c_iu_i$ represents the flow due both to dispersal and direct movements of individuals. We envisage a habitat Ω bounded in $\mathbb{R}^s$.

First of all, we define a basically *homogeneous environment* by the following assumptions:

i) r_i, a_{ij}, d_i and c_i are explicitly independent of t and x, but may depend on $u = (u_1,u_2,\dots,u_n)$ and

ii) the boundary condition at the boundary is of no flux.

In the above situation, we will be discussing *stable non-uniform nonnegative steady state* (stable NUSS) solutions of the equation (1), (therefore disregarding spatial patterns possibly caused by environmental heterogeneity).

The simplest equation of type (1) takes the form

$$\begin{cases} \frac{\partial}{\partial t}u_1 = (r_1 - a_{11}u_1 - a_{12}u_2)u_1 + d_1\nabla^2 u_1 \\ & t > 0,\ x \in \Omega \\ \frac{\partial}{\partial t}u_2 = (r_2 - a_{21}u_1 - a_{22}u_2)u_2 + d_2\nabla^2 u_2 \end{cases} \tag{2}$$

subject to the initial and the no flux boundary conditions, where r_i, a_{ij} and d_i $(i,j=1,2)$ are all positive constants. There are a number of results on (2): For arbitrarily fixed d_i, r_i and a_{ij} except $a_{11}/a_{21} < r_1/r_2 < a_{12}/a_{22}$, there exists no NUSS solution (de Mottoni(1979), Brown(1982)). That is, a solution of (2) tends to be flat for large time. On the other hand, for r_i and a_{ij} satisfying $a_{11}/a_{21} < r_1/r_2 < a_{12}/a_{22}$, it is possible for NUSS solutions to exist for suitable d_i $(i=1,2)$. However, these are *unstable* when Ω is a convex domain (Kishimoto and Weinberger(1982)). These results indicate that spatial patterns of two competing species are not physically realizable if the domain is convex and the coefficients are constant.

This suggests several possibilities for the occurrence of stable NUSS solutions in the framework of the equation (1):

1) *Non-convex region*: Matano and Mimura(1983) have constructed a stable NUSS solution of (2) in a dumbbell-shaped domain with a suitably narrow middle domain when a_{ij} and r_i $(i,j=1,2)$ satisfy $a_{11}/a_{21} < r_1/r_2 < a_{12}/a_{22}$.

2) *Cross-diffusion*: Shigesada et al.(1979) incorporated the movement of non-linear dispersive forces given by

$$d_1 = 1 + \alpha_1 u_2 \quad \text{and} \quad c_1 = -\alpha_1 \nabla u_2 \quad (\alpha_1 \text{ is constant})$$

into (2). Then the equation takes the form

$$\begin{cases} \frac{\partial}{\partial t}u_1 = (r_1 - a_{11}u_1 - a_{12}u_2)u_1 + \nabla^2((1 + \alpha_1 u_2)u_1), \\ \frac{\partial}{\partial t}u_2 = (r_2 - a_{21}u_1 - a_{22}u_2)u_2 + d_2\nabla^2 u_2. \end{cases} \tag{3}$$

Then it is shown that (3) can exhibit non-uniform spatial patterns through the cross-diffusion-induced instability (Mimura and Kawasaki(1980)).

3) n (≥ 3) *competing species models* Evans(1980) and Kishimoto(1982) showed that the simplest equation of type (1) for n = 3

$$\frac{\partial}{\partial t}u_i = (r_i - \Sigma_{j=1}^{3} a_{ij}u_j)u_i + d_i\nabla^2 u_i \quad (i=1,2,3) \tag{4}$$

has stable NUSS solutions through diffusion-induced instability. This fact is somewhat surprising, because competitive type interaction is not, generally speaking, of activator-inhibitor type, which is the typical ingredient for two components diffusion-induced instability, as originally discovered by Turing(1952).

In this paper, we deal with spatial or spatial and temporal patterns which arise in 3 or 4 competing species models, exploiting differences in the diffusion rates.

The work reported in this paper has been done jointly with P. Fife of University of Arizona (Sec.2) and with K. Kishimoto of Hiroshima University and K. Yoshida of Kumamoto University (Sec. 3).

2. 3 competing species models and singular perturbation analysis

Consider the stationary problem of (4) and take the domain to be the one dimensional interval I = (0,1).

$$\begin{cases} 0 = (r_i - \Sigma_{j=1}^{3} a_{ij}u_j)u_i + d_i\dfrac{d^2}{dx^2}u_i, & x \in I \\ & (i=1,2,3) \\ \dfrac{d}{dx}u_i = 0, & x \in \partial I. \end{cases} \tag{5}$$

We first impose the following three conditions on r_i, a_{ij} and $d_i (i,j=1,2,3)$.

(A1) $$\frac{a_{22}}{a_{32}} < \frac{a_{23}}{a_{33}} < \frac{r_2}{r_3},$$

which implies that in the absence of u_1, there are no NUSS solutions, and if $u_2(0,x) \not\equiv 0$, then $\lim_{t\to\infty} (u_2(t,x), u_3(t,x)) = (r_2/a_{22}, 0)$.

(A2) $$\frac{a_{12}}{a_{22}} < \frac{r_1}{r_2} < \frac{a_{11}}{a_{21}} \quad \text{and} \quad \frac{a_{13}}{a_{33}} < \frac{r_1}{r_3} < \frac{a_{11}}{a_{31}},$$

which means that u_1 and u_2 (resp. u_3) can coexist in the absence of u_3 (resp. u_2) (see Fig. 1).

(A3) $d_1 = 1$, $d_2 = \varepsilon^2$, $d_3 = d\varepsilon^2$

for sufficiently small ε (>0) and some positive constant $d = O(1)$.

In addition to (A1)-(A3), we will further impose some conditions on r_i a_{ij} (i,j=1,2,3) later on.

We use an asymptotic analysis based on the smallness of ε in order to find the possible stable non-uniform solutions of (5). We first consider the reduced problem of (5) by setting $\varepsilon = 0$,

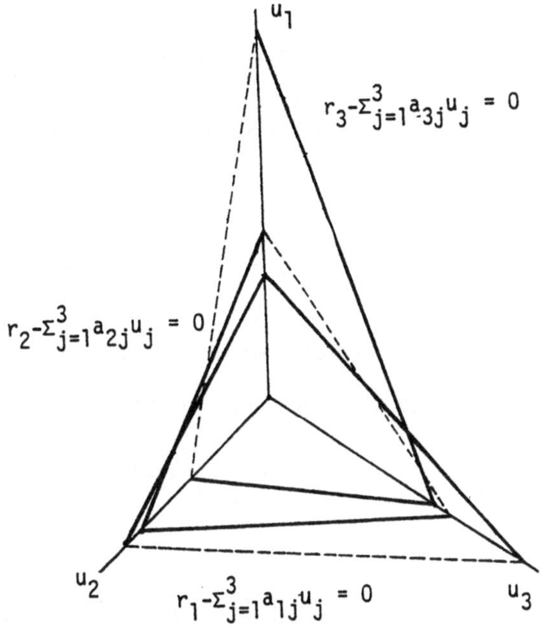

Fig. 1

$$
\begin{cases}
0 = (r_1 - \Sigma_{j=1}^{3}a_{1j}v_j)v_1 + \dfrac{d^2}{dx^2}v_1, \quad x \in I, \\
0 = (r_2 - \Sigma_{j=1}^{3}a_{2j}v_j)v_2, \\
0 = (r_3 - \Sigma_{j=1}^{3}a_{3j}v_j)v_3, \\
\dfrac{d}{dx}v_1 = 0, \quad x \in \partial I.
\end{cases}
\tag{6}
$$

From the second and third of (6), it follows that

(i) $v_2 = v_3 = 0$, or

(ii) $v_2 = (r_2 - a_{21}v_1)/a_{22}$, $v_3 = 0$, or

(iii) $v_2 = 0$, $v_3 = (r_3 - a_{31}v_1)/a_{33}$, or

(iv) $v_2 = v_2^*$, $v_3 = v_3^*$, which are solutions of $r_i - a_{i1}v_1 = \Sigma_{j=2}^{3}a_{ij}v_j$ (i=2,3).

Substituting one of the cases (i)-(iv) into the first equation of (6), we obtain a scalar equation with respect to v_1. For instance, the case (i) leads to the equation

$$0 = \frac{d^2}{dx^2}v_1 + (r_1 - a_{11}v_1)v_1,$$

which can be easily analyzed. On the other hand, a scalar equation for v_1 are also obtained by "piecing together" two conditions taken from (i)-(iv). Suppose

$$\text{(A4)} \qquad \xi_* = \frac{r_1a_{22} - a_{12}r_2}{a_{11}a_{22} - a_{12}a_{21}} < \frac{r_1a_{33} - a_{13}r_3}{a_{11}a_{33} - a_{13}a_{31}} = \xi^*$$

Fixing $\xi \in (\xi_*,\xi^*)$, we may fulfill (6) by using (ii) for $v_1 \in (\xi_*,\xi)$ and (iii) for $v_1 \in (\xi,\xi^*)$: then v_1 has to satisfy

$$0 = \frac{d^2}{dx^2}v_1 + f(v_1;\xi)v_1, \quad x \in I, \tag{7}$$

where

$$f(v_1;\xi) = \begin{cases} r_1 - a_{11}v_1 - (r_2 - a_{21}v_1)a_{12}/a_{22} & (\xi_* < v_1 < \xi), \\ r_1 - a_{11}v_1 - (r_3 - a_{31}v_1)a_{13}/a_{33} & (\xi < v_1 < \xi^*) \end{cases}$$

(Fig. 2). It is noted that $f(v_1;\xi)$ has a discontinuity of the first kind at $v_1 = \xi$. It is proved in Mimura et al.(1979) that the equation (7) with the no flux boundary conditions has multiple non-uniform, nonnegative (weak) solutions $v_1(x;\xi)$ satisfying

$$\xi_* < v_1(x;\xi) < \xi^*, \quad x \in \bar{I}.$$

Starting from $v_1(x;\xi)$, we obtain multiple non-uniform solutions of the reduced problem (6), $(v_1(x;\xi),v_2(x;\xi),v_3(x;\xi))$ whose second and third components take the form

$$\begin{cases} v_2(x;\xi) = (r_2 - a_{21}v_1(x;\xi))/a_{22} & \\ v_3(x;\xi) \equiv 0 & (v_1(x;\xi) < \xi), \end{cases}$$

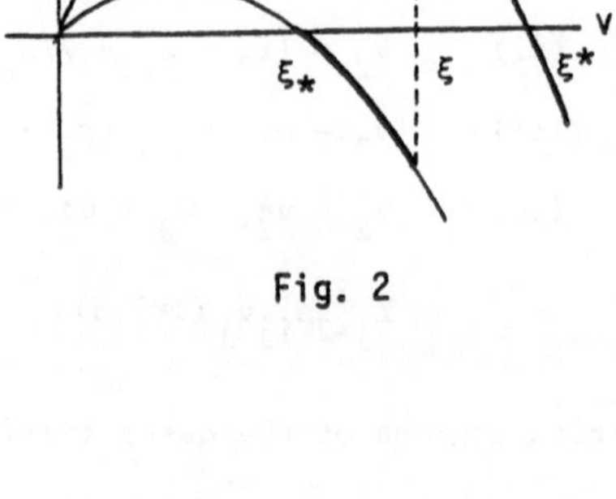

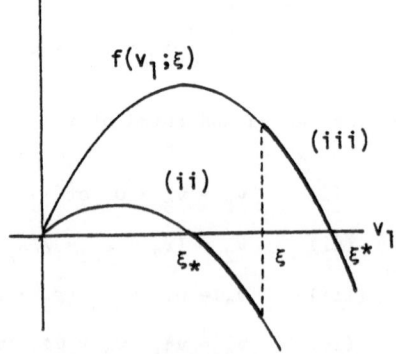

Fig. 2

and

$$\begin{cases} v_2(x;\xi) \equiv 0 & \\ v_3(x;\xi) = (r_3 - a_{31}v_1(x;\xi))/a_{33} & (\xi < v_1(x;\xi)). \end{cases}$$

Thus, it turns out that $v_2(x;\xi)$ and $v_3(x;\xi)$ are discontinuous at some point, say $x = x^*$ where $v_1(\bar{x}^*,\xi) = \xi$(Fig. 3). The solutions show that v_2 and v_3 are co-existing with spatial structure of segregation.

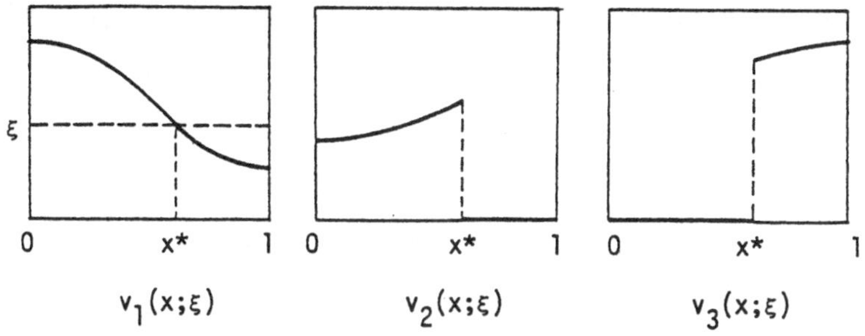

Fig. 3

This result is interesting, because in the absence of v_1, there exists no non-uniform nonnegative solution (5).

When ε is not zero but sufficiently small, we expect the appearance of internal transition layers (where large gradients occur in v_2 and v_3) in a small neighborhood of $x = x^*$. The method of singular perturbation is probably applicable for this problem although it has not yet been completely carried out (Fife(1976)). Mimura and Fife(1983) noted that for some $\bar{\xi} \in (\xi_*,\xi^*)$, $(v_1(x;\bar{\xi}),v_2(x;\bar{\xi}),v_3(x;\bar{\xi}))$ can serve as valid lowest-order approximations to solutions of the original problem (5) when ε is not zero but sufficiently small. Here we only show numerical experiments to confirm that the NUSS solutions constructed here exist *stably* when two

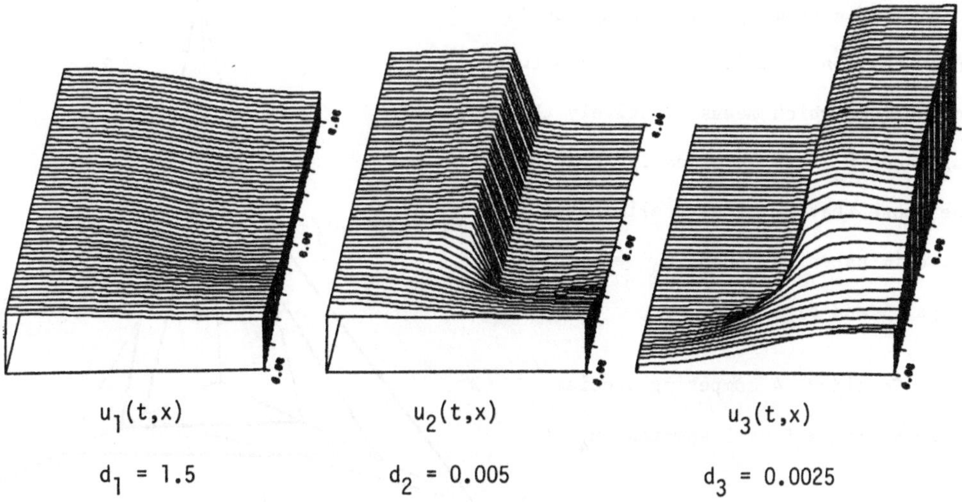

Fig. 4

diffusion coefficients are small (Fig. 4).

The result discussed here is interpreted in ecological terms as follows: Two competing species, u_2 and u_3, which cannot coexist in the absence of u_1, may coexist with spatially segregated distributions, in the presence of u_1, provided the diffusion rate of u_1 is much larger than those of u_2 and u_3.

3. 4 competing species models

In this section, we will show that spatial and temporal patterns are possible in a 4 competing species model

$$\frac{\partial}{\partial t}u_i = (r_i - \Sigma_{j=1}^{4}a_{ij}u_j)u_i + d_i\frac{\partial^2}{\partial x^2}u_i, \quad t > 0, \quad x \in I \quad (i=1,2,3,4)$$

subject to no flux boundary conditions at $x \in \partial I$.

We impose crucial assumptions on r_i and a_{ij} $(i,j=1,2,3,4)$.

(A5) Consider first 3 competing species, say u_1, u_2 and u_3 in the absence of u_4, governed, in the absence of diffusion, by

$$\frac{d}{dt}u_i = (r_i - \Sigma_{j=1}^{3}a_{ij}u_j)u_i, \quad t > 0 \quad (i=1,2,3). \tag{8}$$

Assume the coefficients are such that there is a (stable or unstable) limit cycle in $\mathbb{R}^3$ (May and Leonard(1975), Nakajima(1977), for instance).

An example showing this assumption is given by the *non-transitive* relation between three species, which means intutively that u_1, u_2 and u_3 are "cyclically" related in the sense that u_1 always eliminates u_2, u_2 eliminates u_3, and u_3 eliminates u_1 (Figs. 5 and 6)

(A6) Consider 4 competing species model by adding a fourth species u_4 to (8)

$$\frac{d}{dt}u_i = (r_i - \Sigma_{j=1}^{4}a_{ij}u_j)u_i \tag{9}$$

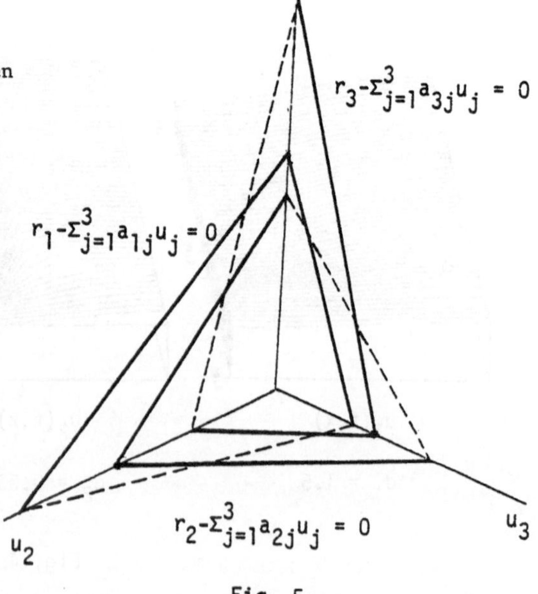

Fig. 5

(i=1,2,3,4) and suppose the coefficients are such that there is only one stable critical point of coexistence.

Let me show, for instance, the values of the parameters a_{ij}, r_i (i,j=1,2,3,4) so that (A5) and (A6) hold:

$$(a_{ij}) = \begin{bmatrix} 49 & 23 & 41 & 18 \\ 24 & 13 & 52 & 29 \\ 40 & 12 & 35 & 17 \\ 26 & 17 & 41 & 22 \end{bmatrix} \quad \text{and} \quad (r_i) = (131, 118, 104, 106)^t.$$

(Figs. 6 and 7)

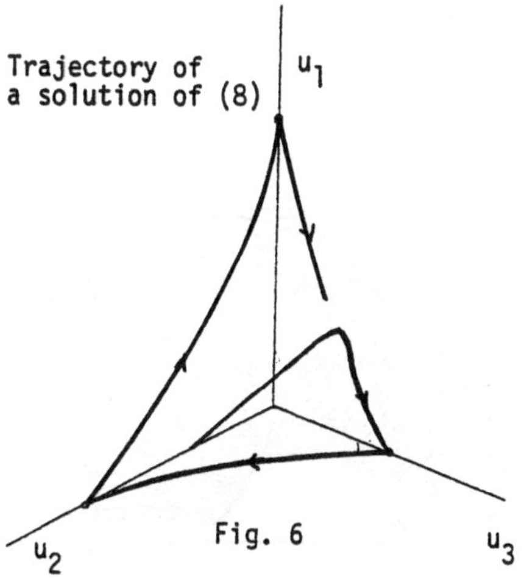

Fig. 6

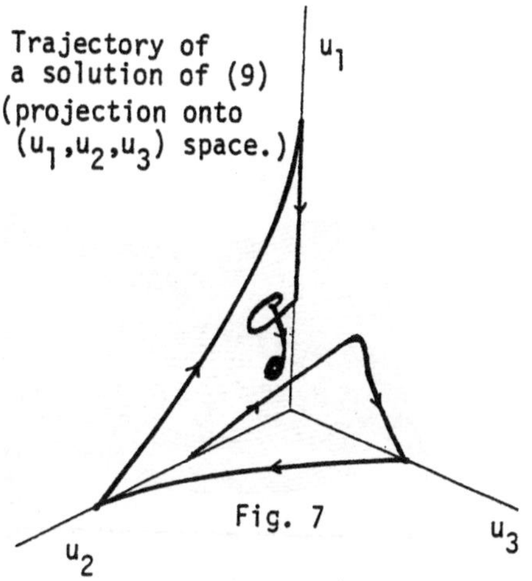

Fig. 7

Under these assumptions and appropriate "smallness" assumptions of some of the diffusion coefficients, much in the spirit of (A3), we can prove that a (Hopf) bifurcation phenomenon occurs where non-uniform periodic solutions branch off the uniform solution of coexistence referred to in (A6). Here we show the results of numerical simulations, providing evidence for large amplitude structures which exhibit spatial and temporal partitioning of the habitat (Fig. 8). The precise discussion will be reported in Kishimoto, Mimura and Yoshida(1983).

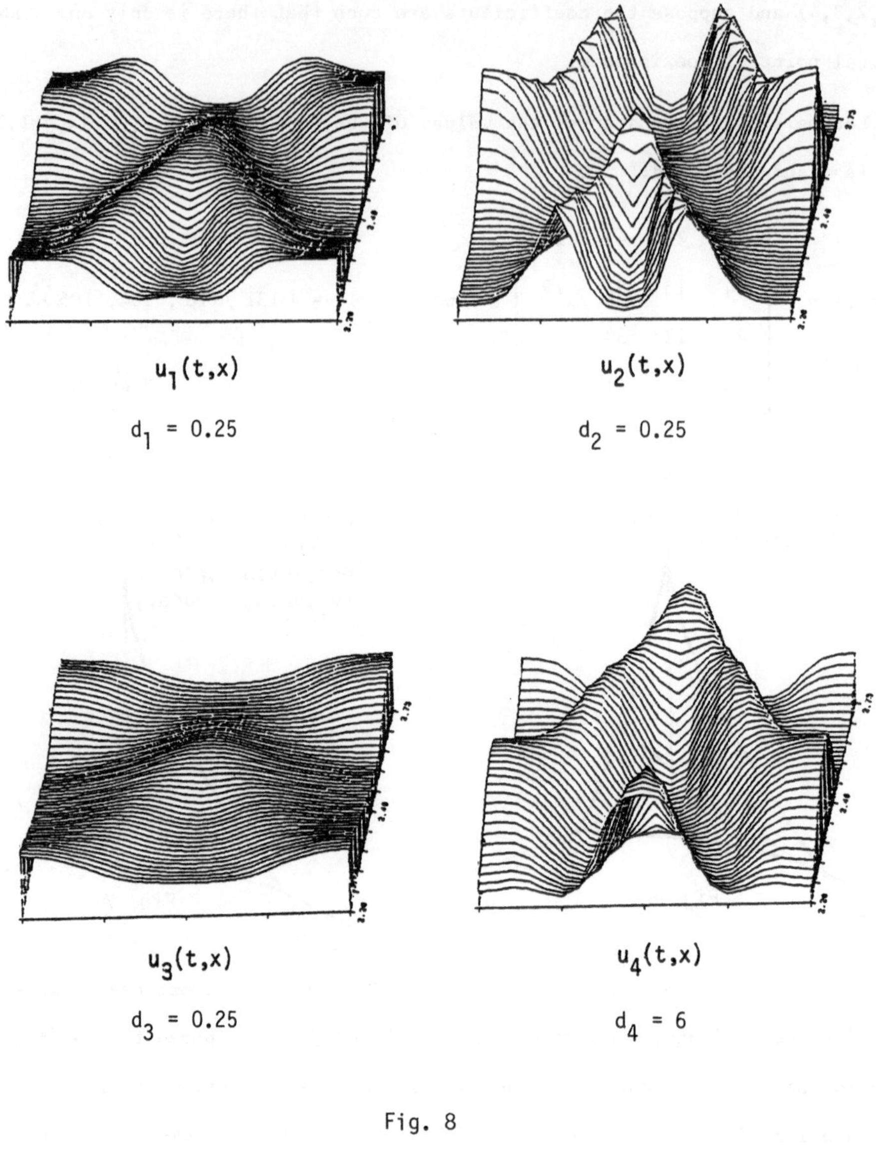

$u_1(t,x)$ $d_1 = 0.25$

$u_2(t,x)$ $d_2 = 0.25$

$u_3(t,x)$ $d_3 = 0.25$

$u_4(t,x)$ $d_4 = 6$

Fig. 8

References.

Brown, P. N. (1982). Decay to uniform states in competitive systems, preprint.

de Mottoni, P. (1979). Qualitative analysis for some quasilinear parabolic systems, Institute of Math. Polish Academy Sci. zam. 11/79, 190.

Evans, G. T. (1980), Diffusive structures: Counterexamples to any explanation ?, J. Theor. Biol., 82, 313-315.

Fife, P. C. (1976). Pattern formation in reacting and diffusing systems, J. Chem. Phys., 64, 554-564.

Fleming, M. H. (1975). A selection-migration model in population genetics, J. Math. Biol., 2, 219-233.

Kishimoto, K. (1982). The diffusive Lotka-Volterra system with three species can have a stable non-constant equilibrium solution, J. Math. Biol., 16, 103-112.

Kishimoto, K., Mimura, M. and Yoshida, K. (1982). Stable spatio-temporal Oscillations of diffusive Lotka-Volterra systems with three or more species, preprint.

Kishimoto, K. and Weinberger, H. (1982). The spatial homogeneity of stable equilibria of some reaction-diffusion systems on convex domains, preprint.

Levin, S. (1979). Non-uniform stable solutions to reaction-diffusion equations: Applications to ecological pattern formation, in "Pattern formation by dynamic systems and pattern recognition" (H. Haken ed.), 201-222, Springer, Berlin.

Matano, H. and Mimura, M. (1983). Pattern formation in competition-diffusion systems in non-convex domains, in preparation.

May, R. and Leonard, W. (1975). Nonlinear aspects of competition between three species, SIAM J. Appl. Math., 29, 243-253.

Mimura, M. and Kawasaki, K. (1980). Spatial segregation in competitive interaction-diffusion equations, J. Math. Biol., 9, 49-64.

Mimura, M. and Fife, P. C. (1983). A threee component system of competition and diffusion, preprint.

Mimura, M., Tabata, M. and Hosono, Y. (1980). Multiple solutions of two-point boundary value problems of Neumann type with a small parameter, SIAM J. Math. Anal., 11, 613-631.

Nakajima, H. (1977). Stability and periodicity in model ecosystem, Bussei Kenkyu, 28, 245-387.

Shigesada, N. and Roughgarden, J. (1982). The role of rapid dispersal in the population dynamics of competition, Theor. Pop., Biol., 21, 353-372.

Turing, A. (1952). The chemical basis of morphogenesis, Philos, Trans. Soc. London Ser. B237, 37-72.

SPACE STRUCTURES OF SOME MIGRATING POPULATIONS

Piero de Mottoni

Instituto di Matematica Applicata, Università dell'Aquila

1. Introduction

The most primitive mathematical models of ecology describe populations as space-homogeneous entities; or, equivalently, space-averages rather than the spatial structure of the populations are taken into account. In recent years, several efforts have been made to overcome such a simplified representation: in particular by trying to provide a mathematical description of segregation effects, by which, in the long run, different populations occupy different sub-areas. Among the resulting mathematical models, we shall focus here on those involving quasilinear (or semilinear) parabolic systems, i.e., systems of partial differential equations describing the time course of communities subject both to kinetic interaction and to linear (respectively, nonlinear) diffusion.

Generally speaking, one may think of several possible sources of space inhomogeneities; these play a central role especially when leading to stable stationary solutions, in view of the "observable" character of the latter. Among them, we shall list three main items: 1) "spontaneous" formation of stable, space-dependent stationary solutions ("patterns") in systems whose coefficients are space-independent, and which are supplemented by no-flux (i.e., Neumann homogeneous) boundary conditions. In such cases, space-homogeneous solutions, in particular stationary space-homogeneous solutions, do exist, but they are destabilized in the presence of diffusion. This comes about, typically, under one of the following circumstances: 1a) two (or more) components interact via <u>kinetics of special form</u> (activator-inhibitor), and linear diffusion occurs with markedly different coefficients; 1b) <u>three</u> (or more) <u>components</u> interact via more general kinetics, with diffusion as above; 1c) two (or more) components interact via a more general kinetics, but are subject to <u>crossed diffusion</u>; 1d) at least two components interact with general kinetics, but the space domain has a special form .

2) Next, we may have appearance of patterns in systems with space-dependent kinetics and/or diffusion: this seems to be a quite natural setting for obtaining space structures induced by biotic factors: As a matter of fact, in such cases space-homogeneous stationary solutions do not, in general, exist; however, proving the existence and stability of spatially dependent stationary solutions is in such case a technically difficult problem. Finally, let us mention the occurrence of patterns in 3) systems with space-independent coefficients, but subject to zero Dirichlet boundary conditions. Since in this case any non-trivial solution must exhibit a space-dependence, the most interesting questions here regard the existence, uniqueness (or multiplicity) of non-negative stationary solutions, and, moreover, the detailed analysis of their space structure, focusing in particular on segregation effects. Amont the increasing literature on the subject, let us mention, for a general discussion Levin (1979), and more specifically, for 1a) Fife (1979), for 1b), 1c) Mimura (1981) and (1983), for 1d) Matano and Mumura (1983). For 2) we may refer to Kawasaki and Teramoto (1979), Rothe (1982). For 3) we may quote Hadeler, an der Heiden, and Rothe, (1974), Leung, (1980), deMottoni, Schiaffino, and Tesei (1983).

In the following, in the spirit of item 3) above, and pursuing the analysis carried out in deMottoni, Schiaffino, Tesei (1983), we shall deal with existence, multiplicity and stability of regular non-negative stationary solutions associated with the quasilinear parabolic problem

$$
\begin{aligned}
&u_t = a\,\Delta u + u(b_1 - c_1 u - d_1 v) && \text{in } (0,\infty)\times G\\
&v_t = v(b_2 - c_2 v - d_2 u) && \\
&u = 0 \text{ in } (0,\infty)\times \partial G && \\
&u = u_0,\ v = v_0 \quad \text{in} \quad \{0\}\times G &&
\end{aligned}
\tag{1.1}
$$

where G is an open bounded domain in $\mathbb{R}^n$, with smooth boundary ∂G; a, b_i, c_i, d_i (i = 1,2) are positive constants, and u_0, v_0 are non-negative continuous functions. The case of small diffusion on the second component has been treated, in the one- dimensional case, by Mimura and Ito (Mimura, 1983), and in the bidimensional case (but Neumann homogeneous boundary conditions) by Matano and Mimura (1983). In the general case, very little seems to be known, although a convergence result in the limit of vanishing diffusion on the second component has been proved in deMottoni, Schiaffino and Tesei (1983).

The above system is suggested by the familiar Volterra-Lotka competition model, modified to account for migration of one species only, in a habitat surrounded by a hostile environment which prevents such species from surviving at the boundary.

We shall distinguish between two main cases, namely

I) $d_2/c_1 \leq c_2/d_1$, and

II) $d_2/c_1 > c_2/d_1$.

In terms of the underlying kinetic systems, the first case implies that a coexistence stationary solution (i.e., a stationary solution having both components positive), if it exists, is asymptotically stable, while the second case excludes the occurrence of stable coexistence stationary solutions.

Case I, which is discussed in detail in deMottoni, Schiaffino, Tesei (1938), gives rise to two significant subcases:

I-a) $b_2/b_1 < d_2/c_1 \leq c_2/d_1$ and

I-b) $d_2/c_1 < b_2/b_1 \leq c_2/d_1$.

Always in terms of the underlying space-clamp system, I-a) means that the only nontrivial stationary solutions are $(b_1/c_1,0)$ - a stable node - and $(0,b_2/c_2)$ - a saddle point - while I-b) entails the instability of both the above-mentioned stationary solutions and the existence and stability of the (unique) coexistence stationary solution $(\hat{u}, \hat{v})$, $\hat{u}, \hat{v} > 0$. (Note that the remaining case, $d_2/c_1 \leq c_2/d_1 < b_2/b_1$ implies that the stationary solutions are $(b_1/c_1,0)$ - a saddle - and $(0, b_2/c_2)$ - a stable node - but, as essentially unaffected by the diffusion on the first component, this case is not very interesting in the present context).

Case II, in turn, gives rise to the following two relevant subcases:

II-a) $d_2/c_1 > c_2/d_1 > b_2/b_1$, and

II-b) $d_2/c_1 > b_2/b_1 > c_2/d_1$.

In case II-a) the stationary solutions of the kinetic system are the same as in I-a) above (stability included); while in case II-b) the solutions $(b_1/c_1, 0)$ and $(0, b_2/c_2)$ are both stable nodes, but a new (unique) coexistence solution appears, which is a saddle point (cf. Fig. 5 below).

The purpose of the present note is to study how the above sketched situation for the kinetic system is modified by "switching on" diffusion on the u-component, i.e., to turn our attention to the stationary solutions associated with (1.1), for various values of the diffusion coefficient a. It turns out that the modifications introduced by diffusion are particularly sharp in cases I-a) and II-a), since diffusion not only gives rise to solutions having a marked segregation structure, but also produces coexistence stationary solutions where, in its absence, they are forbidden. The full information we obtain in such cases on the regular nonnegative stationary solutions (RNSS) is synthetized below:

Case I-a) i) for moderate diffusion ($0 < a < \hat{a}$) a coexistence RNSS exists (u^*, v^*), $u^* > 0$ in G, $v^* \geq 0$ in G, $v^* \neq 0$, exhibiting a segregation structure which, for small a, is particularly marked, since the v-species is allowed to survive only near the boundary, where the other one is kept at a low level because of the boundary condition. Moreover, (u^*, v^*) is unique within a certain class Π of RNSS a class which in fact exhausts all of the RNSSs if the boundary ∂G is connected. In addition, (u^*, v^*) enjoys "global" attractivity properties. ii) If diffusion is increased ($a > \hat{a}$), then the only RNSS which is stable is $(0, b_2/c_2)$. The critical parameter value $\hat{a}$ has an explicit expression, involving the remaining parameters and the principal eigenvalue of the Laplacian. It should be stressed that RNSSs in the set Π play a distinguished role even if not every RNSS belongs to Π; for it can be proved that stationary solutions which do not belong to Π cannot be stable.

Case II-a) Here again the class Π plays a crucial role, much as above; yet uniqueness of stationary solutions within Π typically fails, and in general exact multiplicity results are not available; rather, minimal sets of stationary solutions can be characterized. To simplify our exposition, we confine our attention to the one-dimensional case, in which an exact description of the RNSS is possible: For moderate diffusion, a coexistence RNSS exists, enjoying properties much alike those valid in case I-a). If diffusion is increased, but does not exceed a further threshold ($\hat{a} < a < a'$), while the above-mentioned coexistence RNSS continues to exist and remains (locally) attractive, the solution $(0, b_2/c_2)$ becomes stable, and an unstable coexistence RNSS appears, thus giving rise to a typical bistable regime. Further increase in the diffusion coefficient (a > a') causes both coexistence solutions to disappear, and the only stable RNSS is $(0, b_2/c_2)$.

The above results are visualized in the bifurcation graphs shown in Figs. 1, 2. Fig 1 describes the RNSS in case I-a) (any space dimension), Fig. 2 those in case II-a) in one space dimension. In more space dimensions this could be regarded as a minimal bifurcation graph in the sense of (P.L. Lions, 1982).

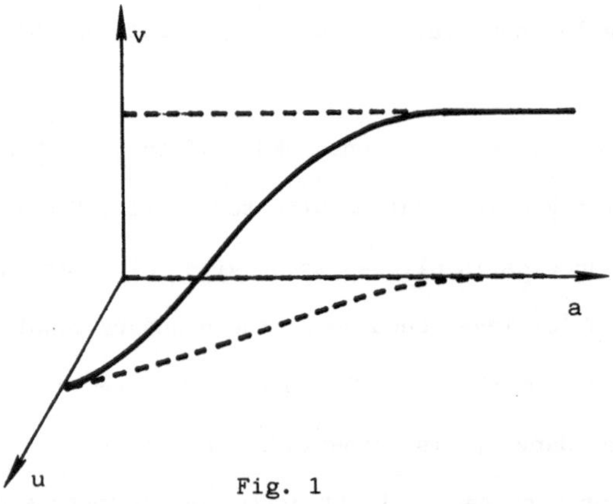

Fig. 1

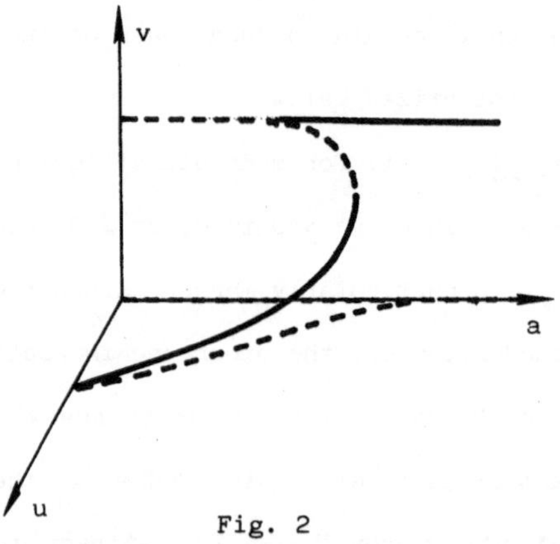

Fig. 2

For the sake of completeness, we also give the bifurcation graphs for cases I-b) and II-b), the latter with a caveat parallel to the above one.

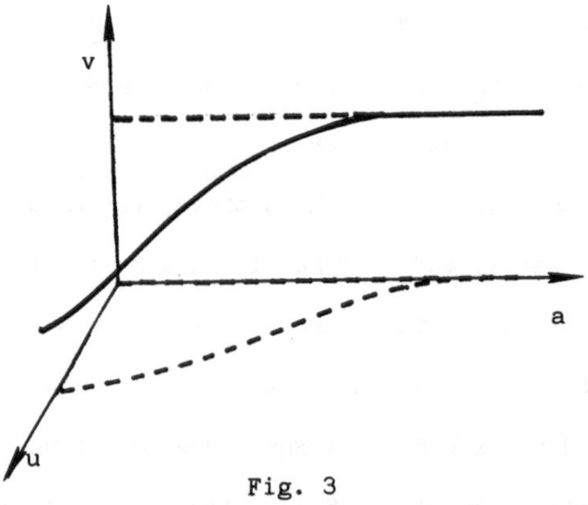

Fig. 3

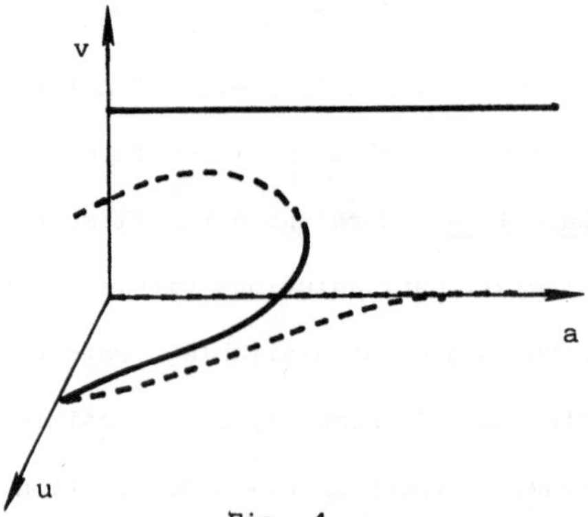

Fig. 4

2. General properties of the stationary solutions

Solutions to (1.1) will always be understood in the classical sense. Standard arguments ensure indeed that (1.1), under the assumptions introduced above, always admits a unique solution, and that the latter satisfies $u(t,x) \geq 0$, $v(t,x) \geq 0$ on $\mathbb{R}^+ \times G$;

moreover, $u(t,x) > 0$ on $\mathbb{R}^+ \times G$, whenever $u_o(x)$ is not identically zero.

A regular nonnegative stationary solution (RNSS) associated with (1.1) is a couple $(u, v) \in C^{2,\alpha}(\bar{G}) \cap C_o(\bar{G}) \oplus C(\bar{G})$ with $u \geq 0$, $v \geq 0$ in $\bar{G}$, satisfying

$$\begin{aligned} 0 &= a\Delta u + u(b_1 - c_1 u - d_1 v) \\ 0 &= v(b_2 - c_2 v - d_2 u) \end{aligned} \quad \text{in } G, \quad u = 0 \text{ on } \partial G. \tag{2.1}$$

Note that because of the maximum principle, every RNSS satisfies

$$0 \leq u \leq b_1/c_1 , \qquad 0 \leq v \leq b_2/c_2 \qquad \text{on } \bar{G}. \tag{2.2}$$

We shall denote by λ_o, ϕ_o the principal eigenvalue, respectively eigenfunction of the Laplacian Δ on G with Dirichlet homogeneous boundary conditions, and we shall normalize ϕ_o to one in the <u>sup</u> norm:

$$\Delta\phi_o + \lambda_o\phi_o = 0 \text{ in } G, \quad \phi_o = 0 \text{ on } \partial G, \; \phi_o > 0 \text{ in } G, \; \max \phi_o = 1. \tag{2.3}$$

Among the RNSSs, a special role will be played by the class Π, consisting of the RNSS of the form

(Π) $v = (1/c_2)(b_2 - d_2 u)^+$, where $(x)^+ = \max(0, x)$, and u is a nonnegative solution of $\quad a\Delta u + u(b_1 - c_1 u - (d_1/c_2)(b_2 - d_2 u)^+) = 0$ in G, $u = 0$ on ∂G. (2.4)

It is easy to see that the class Π can be characterized alternatively as follows: Define, for a given RNSS (u, v), the sets G_1, G_2 as follows:

$$G_1 := \{x \in G, \; v(x) = 0\}, \qquad G_2 := \{x \in G; \; b_2 - c_2 v(x) - d_2 u(x) = 0\}. \tag{2.5}$$

Then

(Π') $\quad$ (u, v) is in Π if and only if $\quad u(x) \geq b_2/d_2$ on G_1 .

The relevance of the class Π is motivated by the following

<u>Property 2.1</u> If the boundary ∂G is connected, and either I) or II) holds, then any RNSS associated with (1.1), having v not identically zero, belongs to Π.

The p r o o f makes essential use of the fact that, in all cases under consideration (except case I-b), (in which $G_1 = \emptyset$, so (Π') is trivially verified),

$$(\Delta + h)\, u \leq 0 \quad \text{in } G, \text{ h being a nonpositive function.} \tag{2.6}$$

To establish (2.6), it suffices to observe that $u(b_1 - c_1 u - d_1 v)$ equals either $u(b_1 - c_1 u) \geq 0$, or $u(b_1 - c_1 u - (d_1/c_2)(b_2 - d_2 u)^+)$. The latter quantity is nonnegative in cases I-a) and II-a) since, in case I-a), $b_1 - (d_1 b_2/c_2)$ is nonnegative and $(b_1 - (d_1 b_2/c_2))/(c_1 - (d_1 d_2/c_2)) > b_1/c_1$; in case II-a), the quantity $b_1 - (d_1 b_2/c_2)$ is again nonnegative, and $c_1 - (d_1 d_2/c_2) < 0$. Thus, in cases I-a) and II-a), (2.6) holds with h = 0. In case II-b), it is easily seen that (2.6) is valid with h equal either to zero or to $b_1-(d_1b_2/c_2)<0$. (That in case (I-b) $G_1=\emptyset$, as asserted above, follows from the fact that, on G_2, $v=(d_1/c_2)(b_2-d_2u)^+\geq(d_1/c_2)(b_2-(d_2b_1/c_1))\geq 0$, so that, by the continuity of v, the claim follows.)

In proceeding further, the following facts play a crucial role:

(i) $G=G_1\cup G_2$; $G_1\cap G_2\cap\partial G=\emptyset$ (obvious, recalling the continuity of v);

(ii) $\partial G\subseteq G_2$, that is, $v=b_2/c_2$ on the whole of ∂G. This is a consequence of the assumed connectedness of ∂G: for details, see de Mottoni, Schiaffino, Tesei, (1983).

Let us now complete the proof of the proposition: since $v=b_2/c_2$ on ∂G, denote by C the connected component of ∂G in $\{x\in G;\ v(x)>0\}$. Also, define $\Gamma=C\cap\partial G$ (note that $\partial C=\partial G\cup\Gamma$), so that, if $\Gamma=\emptyset$, then G_1 is void, too, and (Π') is trivially verified. On the other hand, if $\Gamma\neq\emptyset$, then, by the definition of C, $u=b_2/d_2$ on $\partial\Gamma$. On the strength of the maximum principle, because of (2.6), we then deduce $u\geq b_2/c_2$ on $G\sim C$, and the claim is proved.

Besides Property 2.1, another feature highlights the relevance of the class Π; in particular, we can prove:

Property 2.2 Assume either I) or II) to hold. Let (u', v') satisfy $u'\in W^{2,p}(G)\cap C_o(\bar{G})$ (for some p > 1), $u'\geq 0$; $v\in L^\infty(G)$, $v'\geq 0$ a.e. in G, and be a weak solution of (2.1). Assume that the set $G':=\{x\in G;\ v'(x)=0 \text{ and } u'(x)<b_2/d_2\}$ has positive measure. Then (u', v') is unstable relative to the $C_o(\bar{G})\oplus L^\infty(G)$ norm.

In other words, it follows from this property that solutions such that $v\neq(1/c_2)\cdot(b_2-d_2u)^+$ (which can occur either if ∂G is not connected, or if v is discontinuous) are necessarily unstable, so that they play a minor role in describing the asymptotics of the system under consideration.

The proof is based on a straightforward application of comparison results to the system which governs the evolution of $u(t,x)-u'(x)$, $v(t,x)-v'(x)$. Indeed, it turns out that if $u(0,x)=u'(x)$ and $v(0,x)=v'(x)+\sigma q$, with $\operatorname{supp} q\subseteq G'$, and $\sigma>0$, then v(t,x) cannot remain in an arbitrarily fixed L^∞-neighbourhood of v'(x). However, small σ is chosen; for more details, see de Mottoni, Schiaffino, Tesei (1983).

3. More about the stationary solutions

Due to Property 2.1, the problem of existence and multiplicity of stationary solutions to the full system (2.1) is reduced to the simpler corresponding problem for the equation governing the u-component (cf. (2.4)), namely

$$a\,\Delta u + u(b_1 - c_1 u - (d_1/c_2)(b_2 - d_2 u)^+ = 0 \quad \text{in } G, \quad u = 0 \text{ in } \partial G. \tag{3.1}$$

This problem, in turn, has a straightforward answer in case I), because the nonlinearity is, in such a case, concave (Krasnosel'skii, 1964). Thus, at most one nonnegative nontrivial solution to (3.1) exists: this solution, indeed, exists if and only if $a < \hat{a} = \lambda_0^{-1}(b_1 - (d_1 b_2/c_2))$. Moreover, in case I), the function ψ:

$$\psi(u) = b_1 - c_1 u - (d_1/c_2)(b_2 - d_2 u)^+ \tag{3.2}$$

is nonnegative whenever u is a nonnegative stationary solution, so that it is easy to investigate the behaviour of the nonzero stationary solution as $a \to 0^+$, for the solution depends on a in a (pointwise) monotonic way. This leads to the bifurcation graphs displayed above (note in particular tht $u \to b_1/c_1$ uniformly on compact subsets of G as $a \to 0^+$).

In case II), the situation is much more complicated, and no general results are available, although "minimal" bifurcation graphs for (3.1) can be derived, using e.g. the results in (P.L. Lions, 1982): In particular, in case II-a), there is some <u>a'</u>, $a' > \hat{a}$ such that no nonnegative nontrivial solution exists for $a > a'$; for $0 < a < \hat{a}$ there is at least one nonnegative nontrivial solution; while for $\hat{a} < a < a'$ at least two nonnegative nontrivial solutions can be found. A similar situation prevails in case II-b), the threshold value <u>$\hat{a}$</u> being now replaced by zero. Moreover, in case II-a), one can argue as in case I) to prove that the maximal solution existing for $0 < a < \hat{a}$ converges to b_1/c_1 (uniformly on compact subsets of G) as $a \to 0^+$.

The different shapes the function ψ (cf. (3.2), (3.1)) takes on in the four cases under consideration are shown, along with the corresponding location of the isoclines in the (u,v) phase-plane for the underlying kinetic system, in Fig. 5.

To obtain a more detailed insight into the structure of the solutions to (3.1), one can specialize to the one-dimensional case: G = (0, L). In fact, the solutions of (3.1) which are positive in (0, L) are in a one-to-one correspondence with the roots μ of the equation $L\,(a/2)^{1/2} = T(\mu)$, where T is the "time-map":

$$T(\mu) = \int_0^1 dt\,((\Psi(\mu) - \Psi(t\mu))/\mu)^{-1/2}, \quad \Psi \text{ denoting a primitive of } \psi(u)u.$$

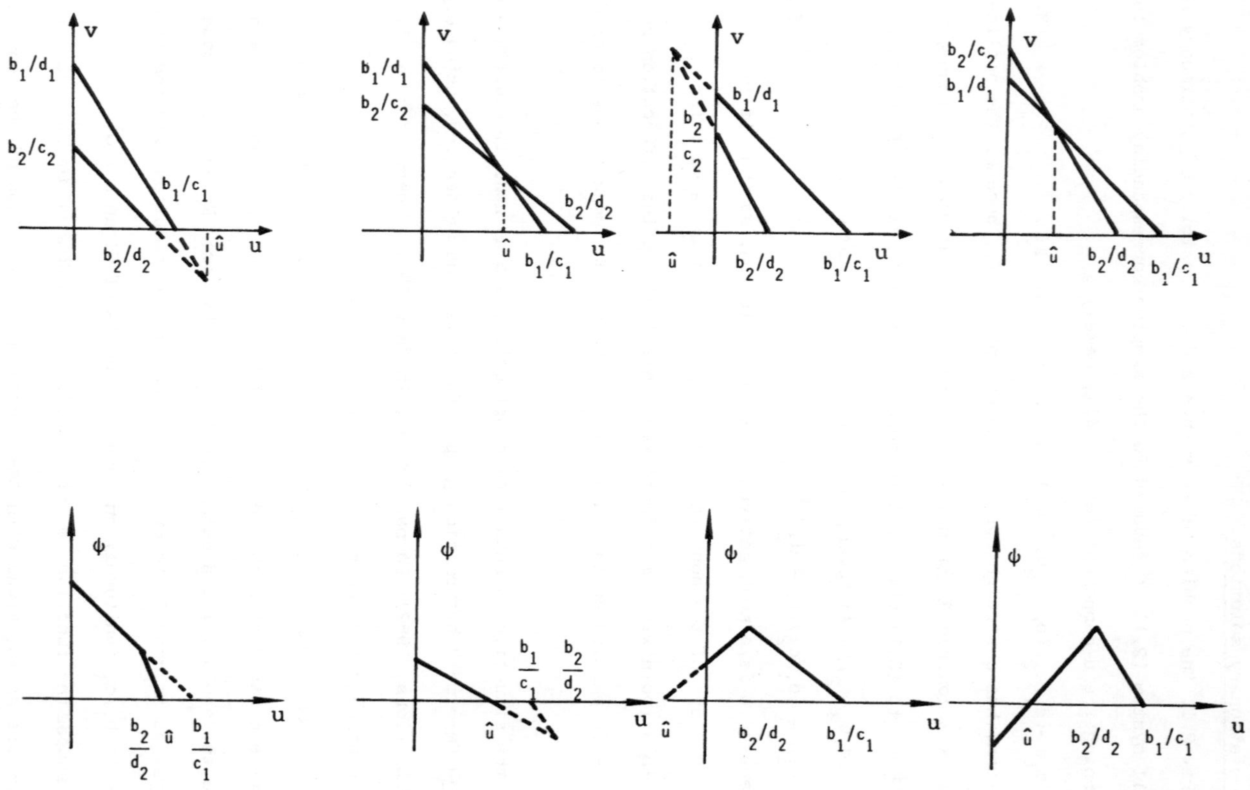

Fig. 5: Phase-plane for the underlying kinetic system (above), and corresponding behaviour of the function $\psi(u)$ (note that $\psi(0) = b_1 - d_1\, b_2/c_2$); notation: $\hat{u} := (b_1c_2 - d_1b_2)/(c_1c_2 - d_1d_2)$.

After a number of lengthy, but not difficult calculations, one can evaluate the limiting values of T at the boundary of the interval of definition, and show that, on that interval, T has exactly one minimum, which leads to the bifurcation diagrams displayed in Figs. 3, 4.

4. Asymptotics of the solution of the evolutionary problem

While, excepting case I-b, methods based on the construction of a Lyapunov functional are not particularly handy, monotonicity methods prove useful, since, given an upper (respectively, lower) solution $\bar{u} \geq 0$ of (3.1), the couple $(\bar{u}, (b_2-d_2\bar{u})^+/c_2)$ turns out to be an upper-lower (respectively, lower-upper) solution to (2.1) (cf. Sattinger, 1971-2). On the other hand, in view of the quasi-monotonic structure of (1.1), to prove attractivity of a C_o -isolated stationary solution (u^*,v^*) associated with (1.1) it suffices to find an upper-lower solution $(\bar{u}, \bar{v})$ and a lower-upper solution $(\underline{u}, \underline{v})$ of (2.1), satisfying $\bar{u} \geq u^* \geq \underline{u}$, $\bar{v} \leq v^* \leq \underline{v}$, and with $\|\bar{u}-\underline{u}\|_\infty+\|\bar{v}-\underline{v}\|_\infty$ small enough . This can be easily achieved in case I), obtaining, in addition, "global" attractivity results as well, as discussed at length in de Mottoni, Schiaffino, Tesei (1983). In case II, one should first answer the question of isolatedness of RNSS s of (1.1), or at least, of the maximal such solution, a question which, in general, remains open. However, in one space dimension, the direct investigation of the stationary solutions alluded to above shows that there are precisely zero, one or two solutions, according to the value of a. Focusing on the solution of largest norm, u^* (for a fixed value of $a \in (0, a'))$, it is easy to find a sequence of upper solutions $\{u_n\}$ satisfying $u_n \geq u^*$ and converging to u^* uniformly (they are in fact, defined recursively in a standard way starting from $u_1 = b_1/c_1$). As to lower solutions, they can be constructed by taking solutions (of largest norm) of the same problem, but on a proper subinterval of (0,L), continued to zero in the residual set. Having obtained in this way upper and lower solutions to (3.1), recalling the above described connection between (3.1) and (2.1), we can set up lower-upper and upper-lower solutions to (2.1) suitable to prove the attractivity of the stationary solution whose u-component has the largest norm, thus justifying the claims made in Sect. 1.

5. Spatial structure of stationary solutions

The "segregation" effect characterizing the RNSSs can be read off directly from the fact that, if (u,v) is a RNSS, then $v = (b_2 - d_2 u)^+/c_2$. The marked spatial structure which, as announced in the Introduction, prevails (except case I-b)) for small va-

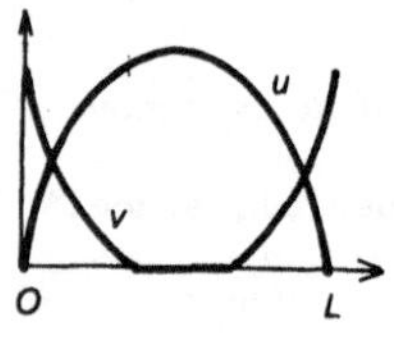

lues of a, can be proved as follows: On the one hand, we know that in a stationary solution the v-component has to be zero whenever the u-component exceeds b_2/d_2. But, as already observed, the (maximal) stationary solution has a u-component which approaches b_1/c_1 as a goes to zero. Since, except case I-b), $b_1/c_1 > b_2/d_2$, we see that for a small enough the subset of G where v is zero is actually non-empty; thus the solution (u,v) of (2.1) has the aspect represented, in the one-dimensional case, in the above figure.

References

Fife, P.C. (1979) The Mathematics of Reacting and Diffusing Systems. Lecture Notes in Biomathematics, Springer, Berlin - Heidelberg - New York.

Hadeler, K.-P., an der Heiden, U., and Rothe, F. (1974) Nonhomogeneous spatial distributions of populations, Journ. Math. Biol. 1: 165-176.

Kawasaki, K. and Teramoto, E. Spatial pattern formation of prey-predator populations, Journ. Math. Biol. 8: 33-46.

Krasnosel'skii, M.A. (1964) Positive Solutions of Operator Equations, Noordhoff, Groningen (transl. form the Russian edition, Moscow, 1962).

Leung, A. (1980) Equilibria and stabilities for competing species equations with Dirichlet boundary data, Journ. Math. Anal. Appl. 73: 2&4-218.

Levin, S.A. (1979) Non uniform steady state solutions to reaction-diffusion equations: applications to ecological pattern formation. In: Pattern Formation by Dynamical Systems and Pattern Recognition, H. Haken ed., Springer Berlin-Heidelberg-New York: 210-225.

Lions, P.L. (1982) On the existence of positive solutions of semilinear elliptic equations, SIAM Review 24 (4): 441-467.

Matano, H. and Mimura, M (1983). Preprint. University of Hiroshima.

Mimura, M. (1981) Stationary pattern of some density-dependent diffusion systems with competitive dynamics, Hiroshima Math. Journ.11: 621-635.

Mimura, M (1983). Private communication.

de Mottoni, P., Schiaffino, A., and Tesei, A. (1983). On stable space-dependent stationary solutions of a competition system with diffusion. Zeitschrift für Analysis und ihre Anwendungen (Leipzig), to appear.

Rothe, F. (1982). A priori estimates, global existence and asymptotic behaviour for weakly coupled systems of reaction-diffusion equations. Habilitationsschrift, Universität Tübingen.

Sattinger, D.H. (1971-2) Monotone methods in nonlinear elliptic and parabolic equations, Indiana University Math. Journ. 21: 979 -1000.

Piero de Mottoni
Istituto di Matematica Applicata
Università dell'Aquila
I - 67040 Poggio di Roio (L'Aquila)
Italy

Lecture Notes in Biomathematics